Transport in Plants

Wilhelm Pfeffer

* 1845 in Grebenstein
† 1920 in Leipzig

Discoverer of the osmotic laws; postulation of a semipermeable living *'Plasma-haut'* (plasma membrane) at the surface of the cytoplasm; experiments and theories on permeation and transport suggesting *"dass ein Körper in chemischer Verbindung mit den aufbauenden Theilchen der Plasmahaut ins Innere befördert und wieder abgespalten wird"* thus preconceiving the carrier theory of membrane transport.

Ulrich Lüttge
Noe Higinbotham

Transport in Plants

Springer-Verlag
New York   Heidelberg   Berlin

Professor Dr. Ulrich Lüttge
Institut für Botanik
Technische Hochschule Darmstadt
Schnittspahnstr. 3-5
D-6100 Darmstadt
Federal Republic of Germany

Professor Noe Higinbotham
Washington State University
Pullman, Washington 99163
USA
And:
University of Washington
Friday Harbor Laboratories
Friday Harbor, Washington 98250
USA

With 180 figures.

Cover drawing by Frau Doris Schäfer: The scheme of a higher plant showing short-, medium-, and long-distance transport, and interactions with the environment, symbionts, and parasites as mediated by transport.

The use of general descriptive names, trade names, trademarks, etc. in this publication, even if the former are not especially identified, is not to be taken as a sign that such names, as understood by the Trade Marks and Merchandise Marks Act, may accordingly be used freely by anyone.

Library of Congress Cataloging in Publication Data
Lüttge, Ulrich.

   Transport in plants.

   Bibliography: p.
   Includes index.
   1. Plant translocation. I. Higinbotham, Noe,
1913-   joint author. II. Title.
QK871.L84   581.1'1   79-992

Softcover reprint of the hardcover 1st edition 1979

9 8 7 6 5 4 3 2 1

ISBN-13: 978-1-4615-9649-3   e-ISBN-13: 978-1-4615-9647-9
DOI: 10.1007/978-1-4615-9647-9

Preface

This book is addressed to all biologists seeking a review of the various transport processes of minerals and organic substances in plants from the level of cell organelles to the longer-distance movements in the largest trees. It is directed toward students having had some elementary physiology, but the attempt has been made to provide information of interest on the frontiers of current research. Doing this comprehensively, we wished to consider all of the points of view that appeared to be important; on the other hand, space and time were limited. Therefore, the presentation had to strike an intermediate ground between the style of a textbook giving representative treatments of selected problems and a comprehensive reference book covering all ramifications. The reader will notice that the pendulum will swing more toward one and then to the other. We did not want to avoid, and we felt it was not appropriate to neglect completely our own special research interests, which led to some emphasis on certain subjects.

The immediate origin of the book is the Heidelberger Taschenbuch 125 (HTB 125) *Stofftransport der Pflanzen* by U.L. (1973), which in turn was preceded by an earlier work, *Aktiver Transport: Kurzstreckentransport bei Pflanzen* Protoplasmatologia vol. VIII/7 b by U.L. (1969). At the Liverpool Workshop on Ion Transport in 1972 organized by W. Peter Anderson, and while in a jovial and expansive mood, the authors agreed to produce an English version. In the summer of 1975, thanks to a Guest Professorship of the Technische Hochschule Darmstadt for N. H. at the Institut für Botanik in Darmstadt, a literal translation into English of HTB 125 was accomplished. By this time our sober reflections indicated that the many advances in the field would require extensive revisions and additions to consider the current status of transport research. Drafts were prepared during the subsequent two years. A US Senior Scientist Award of the Alexander von Humboldt Foundation, Bad Godesberg, allowed N.H. once again to stay for a year (1977–78) in Darmstadt, during which time the final manuscript was completed.

Volume 2 of the *Encyclopedia of Plant Physiology (New Series), Transport in Plants II*, Part A Cells, Part B Tissues and Organs, Springer-Verlag (1976), in whose edition one of us (U.L.) was involved together with Michael G. Pitman (Sydney), proved to be both a help and a hindrance: a help, as it made a wealth of information readily available; a hindrance, as the Encyclopedia, with two additional volumes on Transport in Plants edited by M.H. Zimmermann and J.A. Milburn (Phloem Transport), and C.R. Stocking and U. Heber (Intracellular Interactions and Transport Processes), increased the difficulty of making decisions limiting the selection of material treated here.

The book consists of 13 chapters, i.e., an Introduction and 12 Chapters organized in four chapter groups. Although each chapter stands alone, the progression is from simpler to more complex models, and in a general way from more primitive organisms to those having evolved into more complicated systems and having a greater division of labor among intricately structured organs.

The authors are grateful for the help of a number of people in Darmstadt: Frau Doris Schäfer for the figures; Frau Tuja Krieger, Frau Inge Hill, Frau Irene Schmidt and others who did the typing. Dr. Ernst Steudle and Dr. Ulrich Zimmermann, Jülich, helped with the biophysics of water transport and Dr. Steudle provided some of the problems at the end of Chapter 2. Dr. Anton Novacky, Columbia, Missouri, critically read several chapters in a partially finished state and Dr. Michael G. Pitman, Sydney, reviewed Chapter 2. We also very much appreciate the many reprints and preprints supplied us by numerous authors.

Noe Higinbotham
Ulrich Lüttge

July, 1979

Contents

Chapter 1

The Starting Point

The transport of matter is among the most important tasks which all living organisms have to fulfill; transport is a basic process necessary for life. Transport processes are so important that they are not thoroughly dependent on external conditions. Rather they are strongly determined by the living organism itself, either passively by the structural properties of the organism, or actively by the metabolic energy transfer of the organism.

These statements may soon appear self-evident to the reader; however, they should be elucidated using two examples which are very different in regard to their degree of complexity. With this, some important points of view about transport will become evident.

1.1 The Flow of Matter and Energy Through a Higher Plant

As a starting point for our discussion let us first choose a higher plant. We have then to consider a highly differentiated organism which is composed of a number of parts or organs of different function. For typical terrestrial higher plants the external medium is not the same for all their parts. The root system is in exchange with the soil or substrate in which the plant grows. The shoot system is in exchange with the atmosphere. Problems of transport of matter result from the different functions of shoot and root systems. Let us for the moment stick to the prototype of the higher terrestrial plant and forget about special adaptations. The typical functions of the roots are uptake of mineral nutrients and water. These substances must be absorbed by the root cells and distributed to the different tissues of the root and, in part, they have to be transported eventually to the shoot. The shoot system serves for water and salt uptake only in special ecological situations which differ from the prototype situation, e.g., in aquatic plants. With the pigments of its green cells, the shoot system absorbs light from solar radiation and transforms the light energy by photosynthetic CO_2 assimilation into chemical energy, which then has to be distributed to all parts of the plant, including the root. This distribution is determined by the requirements of the different parts of the plant and transport occurs from the place of production to the place of consumption, i.e., from source-to-sink. The shoot system not only absorbs energy-rich radiation, it also exchanges heat with the environment. Of particular importance is gas exchange with the atmosphere depending on the metabolic situation which influences the gradient of partial pressure between the plant tissue and the atmosphere; CO_2, O_2, and H_2O vapor are released or taken up.

This is relevant not only for the coordination of organs in the whole plant, but rather transport processes must also make the connections between compartments within organelles, organelles within cells (chloroplasts, mitochondria), cells within a tissue, and tissues within an organ (Fig. 1.1).

In the case of source–sink coordination, transport deals with distribution of substrates. The integrating role of transport processes is emphasized by mentioning the specific distribution of growth regulators. These growth regulators, or phytohormones, are effective in minimal concentrations, or amounts, and integrate development, differentiation, and physiological activity of different parts of the plant in even more subtle ways than substrate transport. One possible starting point in the consideration of transport physiology is to aim at understanding the multiplicity of separate transport processes and their cooperation in the maintenance of the functions of the complex organism as a whole.

Exchange of matter and energy also occurs between the atmosphere and the soil; in particular exchange of gases O_2, CO_2, H_2O, and N_2, which is driven by physical forces, and the exchange of heat radiation. Furthermore, the atmospheric and edaphic conditions in which the plant lives are also influenced by the neighbors of the plant, as indicated in Figure 1.1, i.e., other organisms such as other plants, animals, and microbes, which live in the same environment. The environment of our prototype plant is determined by the total biotic community. In view of such extremely complex units the transport physiologist is faced with an almost hopeless task when he tries to explain the flow of material throughout the different parts of the entire system. We will see later that often the transport physiologist tries to reduce this problem by using simplifications. This, however, often brings him in danger of losing contact with the entirety of the problems in the whole plant and the whole ecosystem.

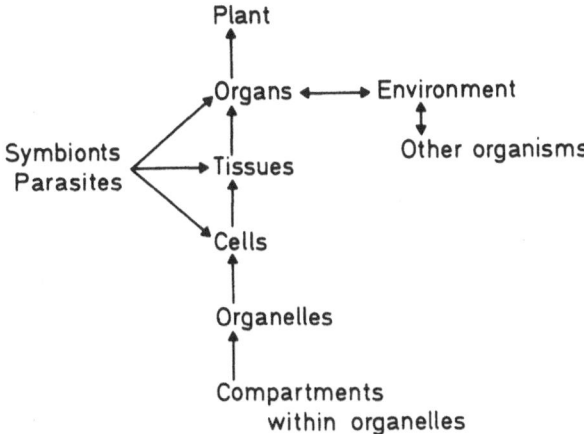

Fig. 1.1 The significance of transport processes: Correlating and integrating functions in relation to the plant interior, the environment, and to other organisms.

Alternatively, to start we can use simple models and imagine how the evolution of organisms may have commenced. Most of the researchers who speculate about the initiation of life agree that the most primitive cells must have originated in an aqueous solution, a suspension of various organic and inorganic substances, an original broth. The separation of small compartments from the bulk of the medium, i.e., the formation of coacervate droplets (Oparin, 1963a,b), must have been the decisive first step. In other words, a barrier appeared which one can imagine as a thin membrane or film formed possibly from lipid molecules. Such lipid molecules consist of both hydrophilic and hydrophobic parts, and on phase boundaries usually form regular layers (see Chap. 5). A total separation of the droplets from the environment, however, would not have led to the origin or to the differentiation of living organisms. Right from the beginning, an exchange of matter with the external medium must have been possible across the membrane barriers. One can exemplify this using a number of simple models. In the following we will designate the compartment which is surrounded by the membrane as "inside" (i) and the indefinitely large bulk of the medium "outside" (o). Let us consider a substance A which is distributed between inside and outside by the purely physical laws which are given in the next chapter. The movement of this substance A between the outside and the inside, or vice versa, may be hindered by the membrane; however, in the course of time the system approaches an equilibrium between inside and outside (Fig. 1.2a). In this way "a material emancipation from the environment" (Netter, 1959) cannot be achieved.

Let us now consider that A is subject to a chemical reaction in the interior of the coacervate droplet and in this reaction is changed to B, and let us also consider that this reaction is enhanced by a catalyzer, i.e., a pro-

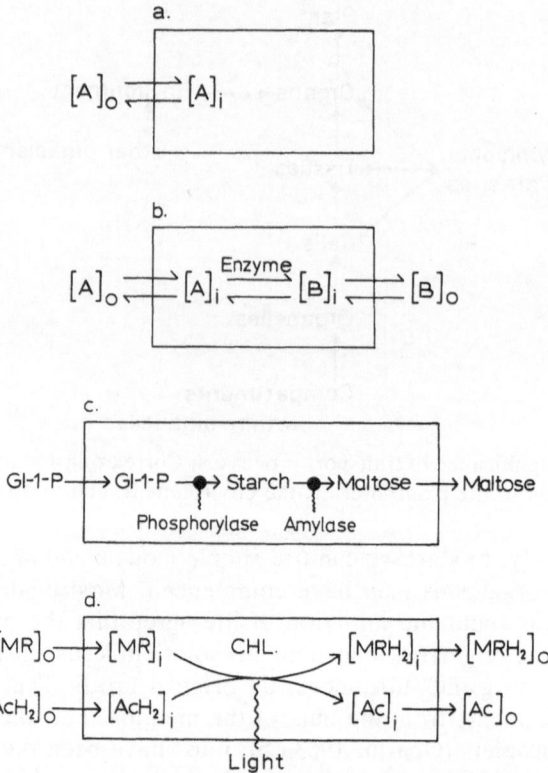

Fig. 1.2. Membrane transport and the "material emancipation from the environment" (Netter, 1959), here the achievement of a difference between inside and outside solutions. **a** simple flux equilibrium; **b** steady state fluxes in an open system; **c** and **d** models for steady state fluxes according to Oparin (1963a); (See text).

teinoid acting like an enzyme. If the catalyzer is located only in the interior of the compartment the reaction A → B will occur there much more rapidly than in the outer medium. If the external medium is infinitely large, compared to the internal space of the membrane-enclosed compartment, then A can be continuously taken into the interior, and B can be released to the exterior. Because of the accelerated consumption of A_i, and the associated formation of B_i, the concentration A_i will always be smaller than the concentration A_o, and the concentration of B_i will always be larger than the concentration B_o. This system will approach a steady state in which the concentrations A_i and B_i do not change any more unless the system itself is subject to drastic alterations, e.g., by modification of the properties of the membrane or by inactivation of the enzyme (Fig. 1.2b). This equilibrium, however, is very different from the system (Fig. 1.2a) considered above. In Figure 1.2a at equilibrium the uptake of A is equal to the release of A, thus the concentration $A_o = A_i$, and influx = efflux.

With the enzymatic reaction (Fig. 1.2b), in which A is transformed to B in the interior, the following holds: the concentration of A_o is larger than the concentration A_i. A influx is larger than A efflux, the concentration of B_i is larger than the concentration of B_o, and B efflux is larger than B influx. In the stationary state all these terms are independent of time (dynamic equilibrium).

Such processes are not mere hypothetical ideas. Examples of experimental models are shown in Figure 1.2c and d. Figure 1.2c shows a synthesis of starch from glucose $-1-$ P and the degradation of starch by amylase in the interior of the coacervate droplet so that the droplet takes up G–1–P and releases maltose. Figure 1.2d describes light-dependent redox processes in the coacervate droplet. Coacervate droplets containing chlorophyll are able to oxidize ascorbic acid, AcH_2, when illuminated and to transfer the hydrogen to methylene red, MR. The coacervate droplets take up AcH_2 and MR from the medium. They absorb light and they release MRH_2 and Ac to the exterior. The membrane-bound space is now differentiated in its contents from the medium. It is emancipated. One can speculate that in this way coacervate droplets which, by pure chance, have received certain catalytic substances inside, and have thereby developed into primal living cells.

It is not the intent in this discussion to describe the various hypotheses about the origin of life, which in detail necessarily remain very speculative. Rather it shows that even the most primitive organisms which one may imagine could neither live without transport processes nor without barriers. Without the controlled passage through their barriers, organisms could not originate and could not persist.

1.2 Summarizing Comparison

Two models have been dealt with in this chapter: the highly differentiated higher plant in its environment, which is modified by many factors, and the primal original cell in a more or less homogeneous aqueous medium. In different ways these two models initiate thinking about transport processes and investigations of transport. The following summarizing comparison of these two models illustrates a dilemma in research on transport in plants: there is a hierarchy in complexity of models applicable to the problems attacked. One should always remember this dilemma in weighing the significance of the individual results which are discussed in the following chapters.

The final aim should be an understanding of all transport processes in a system as complex as the intact higher plant. This should consider not only the mechanisms of all single transport processes, but also their mutual interdependence, their cooperation, and regulation. We will see that the research accomplished is very, very far from this objective but that there are first approximations explaining the coupling of a number of transport processes in complex systems. We will find the simple fact that it becomes increasingly difficult to describe single transport processes correctly when the system which one tries to understand becomes more complex. If transport studies start with an intact plant, or, at least with intact cells which are themselves extremely complex, being made up by numerous membranes around subcellular compartments or organelles, then one can call this a physiological point of view. In this case, the overall function of the organism is the center of interest.

In order to describe specific transport processes exactly, of course, the simple outside-inside model is much better suited. This is the starting point of most of the physical or physicochemical considerations because the simpler the system is, and the more exactly the compositions of the exterior and the interior phases and the chemical nature of the membrane are known, the more precisely one can describe the movement of a particular species of particle. Systems which are often used to investigate the thermodynamics of transport processes are built up of two solution phases separated by an artificial membrane.

The discrepancy between the starting points and aims of physiologists and physical chemists, respectively, is large. Accordingly the compromises which they are ready to make in order to attain the objectives of their research are different. The physical chemist is satisfied if it is possible to describe by precise mathematical laws the fluxes of solute particles which occur under exactly defined conditions. He then does not mind considering isolated fluxes only in a nonphysiological system in which the implications for physiologically complex processes cannot be elucidated clearly. On the other hand, the physiologist does not mind too much in some cases if exact mathematical formulation is impossible, as long as he obtains information on the localization of transport processes within a cell, or within an organism, and their coupling with metabolic reaction sequences.

In the following chapters we will try to discuss both points of view and to show what they might be able to achieve jointly. To unite both points of view is not at all easy; however, in a number of cases nature itself facilitates this task by providing among the multitude of its living species some cells, or organisms, which stand between the simple models of the physical chemist and the complex multicellular organisms.

Part I

Biophysical Background and the Substances Subject to Transport: Chapters 2 and 3

Chapter 2
Biophysical Relations

2.1 Chemical Potentials

2.1.1 Diffusion

Solute concentration differences in solutions otherwise homogeneous have a tendency to be lost during the course of time; a solute, S, tends to become uniformly distributed throughout the volume of the solvent. In other words, if the amount of S in a unit of volume in one part of a container is greater than in another part, then the ions or molecules of S will move kinetically, i.e., as a result of thermal energy, from the concentrated part to the more dilute part of the container until the whole process stops and the system does not change any more (Fig. 2.1). This type of transport of matter, the driving force of which is a gradient of concentration, or a chemical potential gradient, we call diffusion. The final state of the system, in which the concentration in all parts of the container is uniform, is called a diffusion equilibrium. When the diffusion equilibrium is established, new concentration differences in the system cannot occur unless external energy is applied. Thus diffusion is a typical irreversible process, a process which leads to an equilibrium eliminating all the concentration differences present in the system at the beginning. It is a natural process in the thermodynamic sense. Diffusion can also be used as an illustration of the second law of thermodynamics. Transport in the opposite direction across cell membranes, i.e., against the chemical potential gradient (uphill transport), in which the particles have to be moved by applying energy, has been called active transport; work is done in such a system. We will see later that this definition of active transport is often not sufficient, especially in many of the systems which are of interest in biological transport investigations.

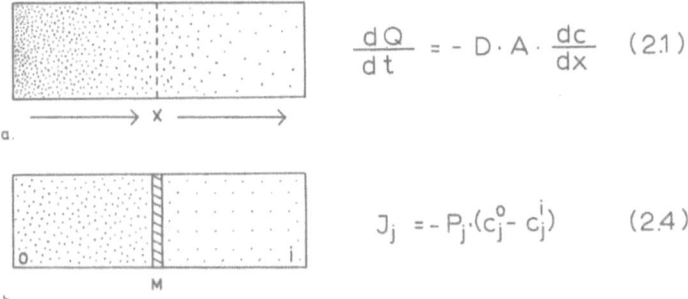

$$\frac{dQ}{dt} = -D \cdot A \cdot \frac{dc}{dx} \quad (2.1)$$

$$J_j = -P_j \cdot (c_j^0 - c_j^i) \quad (2.4)$$

Fig. 2.1. Diffusion resulting from a chemical potential gradient: **a** free diffusion; **b** diffusion through a membrane barrier.

Concentration gradients are important in biological transport in different ways. One important principle mentioned before is that of Figure 1.2 depicting the example of an enzymatic reaction which transforms a substance A to a substance B resulting in lowering the concentration of A and increasing the concentration of B and, in this way, regulating the net transport of A and B by diffusion. The diffusion results directly from concentration gradients, but indirectly from the enzymatic reaction. Concentration gradients can also be established by active transport. For instance, if a substance is transported actively from a compartment I into a compartment II, the concentration in I eventually will be smaller than the concentration in II. The substance can diffuse into another compartment, III, if its concentration is larger in II than in III, and so on (Fig. 2.2). By a combination of active and passive movements and enzymatic reactions, the plant, or the plant cell which is made up of many compartments, can solve many transport problems.

As a starting point for a quantitative description of diffusion, one can cite the statistical behavior of single diffusing molecules. A visible model for this is Brownian movement (Jacobs, 1967). This movement of small visible particles in suspension originates from the random impact of molecules moving kinetically in the medium. The thermodynamics of this process is too complex to discuss here (see Moore, 1957). The best-

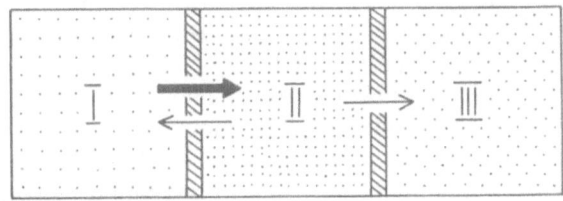

Fig. 2.2. A three-compartment model in series with two membrane barriers having active transport from *I* to *II* (*thick arrow*) and passive transport from *II* to *I* and from *II* to *III* (*thin arrows*).

known laws of diffusion were developed empirically by Fick in 1855 (see Jacobs, 1967) after he had found the analogy between the diffusion of particles in solution and the transfer of heat in solid matter. According to the first of Fick's laws:

$$\frac{dQ}{dt} = -D \cdot A \cdot \frac{dc}{dx} \qquad (2.1)$$

an amount of substance, dQ, diffuses during an amount of time, dt. A is the cross-sectional area through which the diffusion occurs, i.e., an area which is perpendicular to the direction of diffusion. The concentration gradient is $\frac{dc}{dx}$, in other words, the difference in concentration along a coordinate x which is perpendicular to the area, A. D is a constant which, under conditions of equal pressure and equal temperature depends on the solute and the solvent, and is called the diffusion constant. The minus sign means a positive downhill transport with a given negative concentration gradient, i.e., the direction of net diffusive transfer is from higher to lower concentration regions (see Fig. 2.1a).

In this form, Fick's law applies only to ideal solutions in which the interactions between (uncharged) solute and solvent molecules do not reduce activities, and uncorrected concentration values may be used. However, in real solutions the thermodynamically effective concentration, or activity, may be reduced. An example would be an electrolyte which fails to dissociate completely into separate ions. The chemical potential of a solute species, j, is determined by its activity, a_j, and the latter may be related to concentration, c_j, by use of an experimentally determined activity coefficient, γ_j:

$$a_j = \gamma_j c_j \qquad (2.2)$$

In dilute solutions γ_j is often close to 1, but at higher concentrations may be much less than 1; consequently for greater precision a_j rather than c_j is preferred.

2.1.2 Diffusion Across Membranes as a Special Case

As shown in Section 1.1, the diffusion across the membrane barrier is usually much more interesting for us than the transport of particles in a solution which is not compartmented.

We have considered diffusion across an imaginary area, A. If there is a membrane of finite area and negligible thickness, it is customary to replace the diffusion coefficient, D, by the permeability coefficient, P (Fig. 2.1b). If no further assumptions are made regarding the shape of the concentration gradient (see, however, Fig. 2.3) in analogy to Eq. (2.1) the following equation will hold for nonelectrolytes under isothermal and isobaric conditions:

$$\frac{dQ}{dt} = -P \cdot A \cdot (c_o - c_i) \qquad (2.3)$$

or, for a specific solute, j:

$$J_j = -P_j (c_j^o - c_j^i) \qquad (2.4)$$

where J_j is the net movement of j in mol cm^{-2} s^{-1}; c refers to concentration; o and i indicate outside or inside compartments. The permeability coefficient, P_j, in contrast with the diffusion coefficient, D_j, is dependent not only on the kind of particles in question (i.e., the particles of the solvent and the solute) but also the nature of the membrane, for example, on its thickness, Δx, and its molecular structure (Chap. 5). The permeability coefficient is a most important term in the investigation of biological transport; it is important to note that P_j may relate to a "Δx" of one, two, or more membranes. P_j may also be defined (Nobel, 1974) as:

$$P_j = \frac{D_j K_j}{\Delta x} \qquad (2.5)$$

where K_j is the partition coefficient [Sect. 5.1.2.1; Eq. (5.1)] of a solute between the phases such as oil and water and is a measure of relative solubility. Thus P_j incorporates the three terms to the right of Eq. (2.5). It should be noted that P_j values cited in the literature generally are not corrected for Δx.

2.1.3 Measurement of Permeability Coefficients

Usually the properties of the membrane are not known in detail. Nevertheless, P_j can be determined according to Eq. (2.4). First the flux of the particles must be measured, i.e., the amount of substance which is transported in a unit of time over a unit of area. Second, at the same time, the external and the internal concentrations have to be determined to provide ΔC_j, i.e., $(c_j^o - c_j^i)$. When this is done, however, then one must bear in mind that the surface phenomena described in Figure 2.3 complicate the measurement. Even if the liquid phases (solutions) on the two sides of the membrane are well stirred, the solute concentrations, in the immediate vicinity of the membrane (C_o' and C_i') will not be the same as in the bulk of both liquid phases (C_o and C_i). A concentration profile will build up (Dainty, 1963). In the case of Figure 2.3, in which $C_o > C_i$, $C_o > C_o'$ and $C_i < C_i'$, this phenomenon may lead to an underestimation of permeability coefficients determined according to Eq. (2.4) because $\Delta C > \Delta C'$ (taking $\Delta C = C_j^o - C_j^i$) and the actual membrane permeability coefficient P_j^M is not equal to $-J_j/\Delta C$ but:

$$P_j^M = -J_j/\Delta C' \qquad (2.6)$$

where

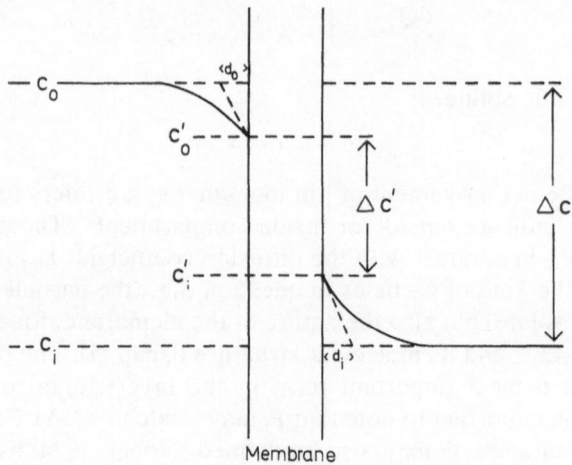

Fig. 2.3. A concentration profile at a membrane showing the effect of an unstirred layer. (Modified from Slatyer's, 1967, original from Dainty, 1963.)

$$\Delta C' = C_j'^o - C_j'^i.$$

The concentration profile of Figure 2.3 shows that there is always a layer of solution at a surface which is not affected by stirring. The effective thickness of this layer is d_o and d_i, respectively. This is the so-called "unstirred layer." In living cells the stirring effect inside may be provided by cyclosis (cytoplasmic streaming). If in addition one bears in mind the compartmentation of the cytoplasm, it becomes clear that for living cells d_i is not a useful parameter because it is not readily measurable.

Surface phenomena are further complicated by the fact that charged particles may interact with fixed electrical charges in the membrane or in the cell wall. This leads to greater or lesser effective concentrations locally and to electrical effects (Donnan systems, see Sect. 2.2.2.2).

2.1.4 Water Potential, Permeability, Reflection Coefficient, and Osmosis

Permeabilities are important also in osmosis. An ideal osmotic cell is surrounded by an ideally semipermeable membrane which allows water to penetrate readily but is entirely impermeable to solutes (Fig. 2.4).

The water potential, Ψ_W, is a function of the chemical potential of H_2O (Nobel, 1974). The highest water potential is that of chemically pure water and is defined as zero (atm., bar, Pa or dynes cm^{-2}). Thus water potentials of biological systems usually have more or less negative values. Ψ_W is composed of various components, a pressure term, Ψ_P, a term due to the presence of solutes, Ψ_S, and a matrical term, Ψ_τ due to water associated with the matrix, i.e., imbibition (by solids, colloids, wall, etc.):

$$\Psi_W = \Psi_P = \Psi_S + \Psi_\tau. \tag{2.7}$$

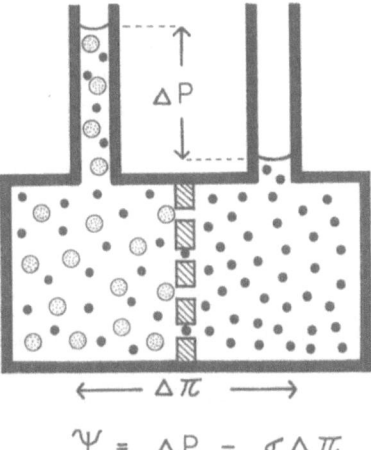

$$\Psi = \Delta P - \sigma \Delta \pi$$

Fig. 2.4. Osmotic system: two compartments separated by a semipermeable membrane. Semipermeability of the membrane is symbolized by pores and particle sizes, where *larger dotted circles* represent solute particles unable to pass through the pores and *smaller closed circles* represent permeating solvent particles (e.g., water). See text for symbols and Eqs. (2.8) and (2.10).

Ψ_P, the pressure, can be zero but, of course, with normally turgid cells has positive values; Ψ_S and Ψ_τ can be zero but usually have negative values. The water potential of a cell surrounded by a semipermeable membrane is often written as an equation, where Ψ_W is given by the gradient of hydrostatic pressure, ΔP, and of osmotic pressure $\Delta \pi$, and by a gravitational term

$$\Psi_W = \Delta P - \Delta \pi + \rho_W gh \qquad (2.8)$$

where ΔP and $\Delta \pi$ correspond to gradients of Ψ_P and Ψ_S, respectively. ρ_W is density, g is the gravitational acceleration constant, and h is height. The gravitational term may be neglected in osmotic relations of cells since water movement is over a short distance. Ψ_W, P, and π are generally given in atmospheres or bars. Ψ_W with respect to older terminology represents the negative value of suction pressure (Saugkraft) or the "diffusion pressure deficit"; P, hydrostatic pressure, is wall or turgor pressure; and π is the osmotic potential (or osmotic pressure). A system for demonstrating the hydraulic pressure generated by osmosis is shown in Figure 2.4.

Throughout the plant and its adjacent environment, water movement is in accord with its potential (or activity) gradient. Thus with larger plants, in the ascent of xylem sap (and perhaps the descent of solution in the phloem) the term $\rho_W gh$ becomes important. An ascent of 10 m represents a potential difference of about 1 bar.

Plasmolysis of plant cells is explained in terms of water potential gradients. If plant cells are transferred into a concentrated solution of a suitable substance (plasmolyticum) the cells lose water (exosmosis), the volume of the cells, especially the volume of the vacuole, is reduced. Consequently, the wall pressure, P, is also reduced; with the water loss the cytoplasm may withdraw from the cell wall; this phenomenon is plasmolysis. With transfer to a weak solution or H_2O the protoplast swells due to H_2O uptake (endosmosis).

According to the Van't Hoff relation as verified in experiments by W. Pfeffer (1877):

$$\pi_S = RT\Sigma_j \gamma_j c_j \tag{2.9}$$

where R is the gas constant (liter-atmosphere $mol^{-1} K^{-1}$) and T is absolute temperature (K). Since at 20°C RT is approximately 24-liter-atm mol^{-1} a 0.3 M solution of an undissociated solute gives a value for π_S of about 7.2 atm; corn and pea root cells have about this value, since incipient plasmolysis is reached at an external glucose concentration of 0.3 M.

The correlations between Ψ_S, π, and P and cell volume are shown in Figure 2.5. The ideal osmotic behavior described here is due to absolute impermeability of the membrane for the solute particles. If, however, the membrane allows some permeation of the solute particles, this means that it is not a fully effective barrier because the permeation of the particles reduces the gradient. This difference from the ideal behavior can be expressed by the so-called selectivity or reflection coefficient σ.

$$\Psi_W = \Delta P - \sigma \Delta \pi. \tag{2.10}$$

In the case of an absolute selectivity, solutes are "reflected" from the membrane, i.e., do not permeate at all, and $\sigma = 1$. With membranes which are permeable equally for water and solute, σ becomes 0. There-

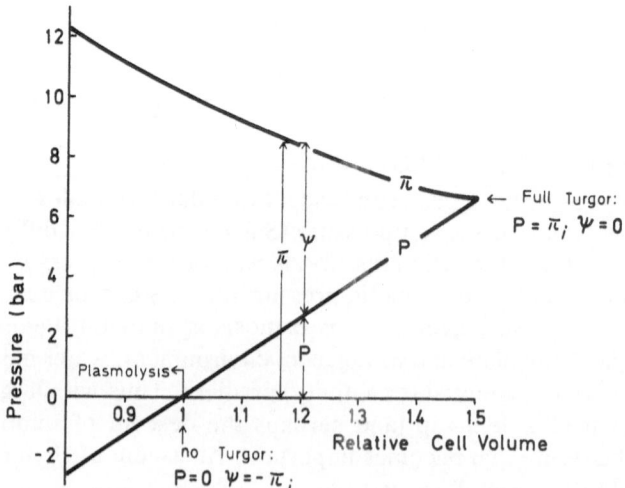

Fig. 2.5. Interdependence between water potential, Ψ, hydraulic (turgor, or wall) pressure, P, osmotic concentration, π, and cell volume. (Slatyer, 1967).

fore, if a cell is plasmolyzed in a solution in which $0 < \sigma < 1$ plasmolysis will be reversed in the course of the experiment. This so-called deplasmolysis can be partial or total. It is due to the permeation of the solute which thus increases π in the cell and reduces $\Delta\pi$. The volume changes which are associated with such processes, and which are dependent on the permeability of the solute particles and the cell wall elasticity, can be measured rather exactly with the microscope (plasmomentry: Stadelmann, 1966): or, in the case of cells in suspension, by measuring light dispersion. In this way plasmolysis research yields values of permeabilities. The approach has been criticized, however, because the membrane itself may change properties during plasmolysis due to dehydration. Thus membrane permeability of plasmolyzed cells may be different from permeability of nonplasmolyzed cells (Dainty, 1963). For other procedures of measuring the parameters of osmotic processes see Section 2.2.3.5.

2.2 Electrical and Electrochemical Potentials

2.2.1 Diffusion of Electrolytes

Thus far we have considered only the movement of electrically neutral solutes along a concentration gradient. If the movement of charged particles, ions, is caused solely by concentration differences, i.e., if no additional driving force is present (e.g., an electrical field), the same laws hold and Fick's Law can be applied; such cases are very rare. However, imagine a long container (see Fig. 2.1a) with a homogeneous solution of KCl. Consider that at one end of the container a small amount of HCl is added so that the HCl concentration is much smaller than the KCl concentration. With the addition of HCl the Cl^- concentration at this end of the container will be changed relatively little as compared with more distant parts of the container. The concentration of H^+, however, will be changed considerably, so that for H^+ alone a concentration gradient is established. The H^+ ions will then be the only ones showing a diffusion gradient and their transport will be largely determined by the concentration differences. This example becomes still clearer if instead of HCl a small amount of radioactive KCl (^{42}KCl or $K^{36}Cl$) is used. Then one can measure the diffusion of labeled K^+ or Cl^- ions in a solution which, in respect to the K^+ and Cl^- concentrations, is homogeneous. The K^+ or Cl^- diffusion in a homogeneous KCl solution is a measure of the so-called self-diffusion of the K^+ or Cl^- ions.

The maximum rate of diffusion is attained when the mobility of anions and cations is equal. Or, in other words, when the anion A^- and the cation C^+ of the electrolyte have the same diffusion coefficient, i.e., when $D_{A^-} = D_{C^+}$. To a certain extent the electrostatic attraction of the oppositely charged ions prevents the establishment of large differences in the mobilities of C^+ and A^-. Anions and cations are not able to migrate entirely independently of each other, and considerable force has to be applied to separate them (for example, see electrogenic pump below, Sects. 2.2.2.2 and 2.2.2.3). On the other hand the A^- and the C^+ mobilities may be quite different. This expresses itself, for instance, in the different abilities of electrolyte solutions of various salts to transmit an electric current. Here the ions migrating in an electric field are the carriers of charge, and electrical conductivity is a measure of the mobility of the ions. Solutions of salts with the same cation but different anions and salts with the same anion but different cations have different equivalent conductivities, e.g.,

NaCl	108.99 cm^2 Ω^{-1} eq^{-1}
KCl	130.10 cm^2 Ω^{-1} eq^{-1}
KNO$_3$	126.50 cm^2 Ω^{-1} eq^{-1}

(from Moore, 1957). This phenomenon is described by Kohlrausch's law which relates the equivalent conductivity, Λ, of an electrolyte solution with the equivalent anion and cation conductivity λ^- and λ^+ respectively:

$$\Lambda = \lambda^+ + \lambda^-. \tag{2.11}$$

The equivalent conductivity $\lambda^\pm$ and the ion mobility $u^\pm$ are related as follows:

$$u^\pm = \frac{\lambda^\pm}{F} \tag{2.12}$$

where F is the Faraday, which is the amount of electricity carried by one equivalent of ions. The migrations of ions in diffusion and in an electrical field, respectively, are restricted by the same resistances. Thus a close correlation is established between the diffusion coefficient, D, and the ion mobility, $u^\pm$; D is proportional to $u^\pm$. The diffusion coefficient for an electrolyte C^+A^- is given according to Nernst for the condition D_{C^+} unequal to D_{A^-} as follows:

$$D = \frac{2D_{C^+} \cdot D_{A^-}}{D_{C^+} + D_{A^-}}. \tag{2.13}$$

For a binary electrolyte composed of univalent ions, the relation between the diffusion coefficient and ion mobility can be described as:

$$D = \frac{2RT}{F} \cdot \frac{u^+ \cdot u^-}{u^+ + u^-}. \tag{2.14}$$

A more general form of the equation is obtained from Eqs. (2.12) and (2.14) as follows:

$$D = \frac{RT}{F^2} \cdot \frac{\lambda^+ \cdot \lambda^-}{\lambda^+ + \lambda^-} \cdot \left(\frac{1}{z^+} + \frac{1}{z^-}\right) \tag{2.15}$$

where $z^{\pm}$ is the valence of the ions. Different mobilities of anions and cations of an electrolyte ($u^+ \neq u^-$) may lead to an unequal distribution of electrical charge, and this results in a diffusion potential:

$$\Delta E = \frac{RT}{F} \cdot \frac{u^+ - u^-}{u^+ + u^-} \cdot \ln \frac{da}{dx}. \qquad (2.16)$$

From these considerations an important conclusion follows, i.e., for the transport of ions there are two components, namely, the activity, or the chemical potential, gradient, and the gradient of electrical potential. In addition, for plant cells, a term would have to be introduced which takes hydrostatic pressure into account. This term, however, can be neglected for most practical purposes when changes in pressure are small. A driving force affecting ion transport is then the electrochemical driving force (μ) for an ion:

$$\mu_j = \mu_j^s + RT \ln a_j + z_j FE. \qquad (2.17)$$

Simply put, Eqs. (2.16) and (2.17) mean that ions having greater mobility tend to leave their co-ions behind and some charge separation occurs leading to an electrical potential gradient; this tends to propel the lagging ion. The direction and magnitude of this force is given by the difference between μ_j values, e.g., $\mu_{j^+} - \mu_{j^-}$. The reference point for the electrical potential has to be agreed upon conventionally, referring the electrochemical potential to a standard state in which the chemical potential equals μ^s.

2.2.2 Transport of Ions Across Membranes

2.2.2.1 The Nernst Equation

As with diffusion of noncharged particles, the movement of ions across a membrane barrier appears as a special case of diffusion. The properties of the membrane become important in addition to the electrochemical driving force. With a binary electrolyte having a univalent cation and anion an electrical membrane potential will be built up as a diffusion potential when the permeability of the membrane for anions and cations is different, i.e., if $P_{A^-} \neq P_{C^+}$. From Eq. (2.16) it becomes clear immediately that in addition to the ion concentration in the two phases any electrical potential across the membrane would affect the distribution of ions between the two phases which are separated by the membrane. Or, in other words, the electrochemical potential gradient determines the equilibrium which the system approaches in the course of time. At equilibrium the driving forces for the ion movements from outside to inside and from inside to outside must be equal, i.e., $\mu_j^{io} = \mu_j^{oi}$. Making use of Eq. (2.17) this leads to the following:

$$\mu_j^s + RT \ln a_j^o + z_jFE_o = \mu_j^s + RT \ln a_j^i + z_jFE_i \qquad (2.18)$$

or

$$E_j = \frac{RT}{z_jF} \ln \frac{a_j^o}{a_j^i}. \qquad (2.19)$$

This is the Nernst equation in the form frequently used in studies of transport in plants. Often concentrations instead of activities are used because the activity coefficients are not known exactly (e.g., in the cytoplasm). However, when the ion concentrations in the cytoplasm exceed ~100 mM it becomes important to use activities, although it must be assumed that cytoplasm has little effect on the activity coefficient.

In most cases, the potential difference, PD (E_{vo}) between the interior (vacuole) of plant cells and the external solution bathing the cells is −50 to −250 mV, interior negative (see Sect. 2.2.3.2). The electropotential gradient is essentially a sum of E_j values modified by P_j values [see Eq. (2.28)].

When the ions are distributed passively, strictly according to the electrochemical potential gradient, the system at equilibrium must obey Eq. (2.19). An ion, j, diffusing independently (unaffected by other ions or by interactions with the membrane), may have a Nernst potential, E_j, equal to E_M, the membrane potential; K^+ often approaches this but in cells the E_j values of other ions are usually far removed from the value predicted by Eq. (2.19). This leads to the conclusion that in addition to electrochemical potential gradients, other driving forces are involved. In other words, when the measured E_M is different from E_j calculated from known ion concentrations outside and inside, this suggests that an active ion transport must occur from inside to outside, or in the opposite direction. This means that the demonstration of movement against a chemical potential gradient is not sufficient to indicate active transport; rather the movement against an electrochemical potential gradient has to be demonstrated. The equations show that ions can move passively against a chemical gradient if there is a large enough electrochemical potential gradient in the opposite direction. In ion transport studies with plants, we generally rely on the criterion of active transport given by Eq. (2.19), although there are a number of limitations as we will see below.

2.2.2.2 Membrane Potentials, Donnan Potentials

There are three main mechanisms which may cause the build-up of membrane potentials: (1) passive diffusion of ions; (2) Donnan systems; (3) active ion transport (or electrogenic pumps). As we have seen above, a membrane potential can be build up by *diffusion* if the permeability coefficients for anions and cations are unequal and there is a concentration gradient. When there is a concentration gradient and one ionic species of a system studied has an extremely low permeability coefficient so that for

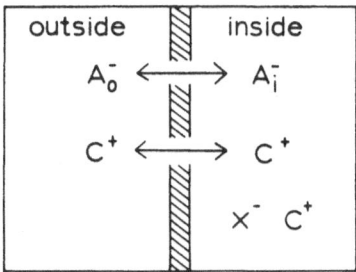

Fig. 2.6. Outside–inside model with the inner compartment as a Donnan phase. X_i^-, an anion inside, which cannot permeate the membrane, while C^+ and A^- do permeate equally. The text describes how such a system can approach equilibrium where

$$E_{Don} = \frac{RT}{F} \ln \frac{a_o^{C^+}}{a_i^{C^+}}$$

[See text and Eq. (2.20)]

practical purposes the flux of this species is zero, but the co-ion diffuses readily, a PD is generated. This may occur when a membrane is entirely impermeable for a certain ion, or when these ions are bound to cell structures as fixed charges and thus are immobile. In the latter case, a membrane is not even required, the phase boundary is given by the range of the influence of the charges of the fixed ions. In the first case, the phase boundary is marked by the membrane. Such a system is shown in Figure 2.6. In order to describe how such a system may approach equilibrium in an imaginary experiment, let us start by assuming that the concentration of the electrolyte $C^+ A^-$ is equal on both sides of the membrane. The cation C^+ shall be the only cation present in the system and, therefore, it is the counter ion for the indiffusible anion X^-; let us assume that initially our system is in a state of electrical neutrality, i.e., there is no electrical potential, then

$$\{[C^+][A^-]\}_i = \{[C^+][A^-]\}_o, \quad \text{but}$$
$$[C^+]_i > [C^+]_o, \quad \text{because}$$
$$[C^+]_i = [A^-]_i + [X^-]_i$$

Thus there is a chemical gradient for C^+ which will drive C^+ diffusion from inside to outside. Outward diffusion of C^+ will generate a potential difference (inside negative), which will tend to drive A^- outwards. Gradually an electrochemical potential equilibrium is established in which the sums of the driving forces (electropotential gradient plus concentration gradient) are equal for anion and cation influx and efflux, respectively. Using the Nernst equation [Eq. (2.19)] then for the *Donnan potential* one arrives at:

$$E_{Don} = \frac{RT}{zF} \ln \frac{[C^+]_o}{[C^+]_i} = \frac{RT}{zF} \ln \frac{[A^-]_o}{[A^-]_i}. \tag{2.20}$$

The direction of the Donnan potential, E_{Don}, is always given by the charge of the immobile ion, i.e., in our case, of an indiffusible anion, the inner phase is negative with respect to the outer phase. In the equilibrium the following applies:

$$\frac{[C^+]_i}{[C^+]_o} = \frac{[A^-]_o}{[A^-]_i} = r. \tag{2.21}$$

This is called the Donnan equilibrium. The letter r represents the Donnan quotient. With CA we have assumed a binary electrolyte of univalent ions. Briggs et al. (1961) give more detailed examples for various other Donnan systems, including some divalent ions.

As a result of the establishment of the Donnan equilibrium, the phase containing indiffusible ions has a higher total concentration than the exterior phase, and therefore a higher osmotic pressure. This is a fact which plays an important role in transport through a region of fixed charges as in cell walls. It can also be used to demonstrate that development of a higher concentration of an ion does not necessarily depend on active transport. A third possible cause of membrane potentials is *active transport of an ion*. One can conceive of neutral pumps which actively transport a salt as a whole, which move anions and cations in equal amounts in one direction, or which transport ions by exchange in opposite directions in an electrically neutral one-for-one process as shown in Figure 2.7. If there is equal exchange of an ion inside for another ion outside, such as Na_i^+ for K_o^+, then "coupling" is suggested; that is, the movements of the two ions are interdependent. For many years this was believed to be true for certain animal cells in which the amount of Na^+ released appeared to equal the amount of K^+ taken up; Na^+ efflux was dependent on the presence of K^+ outside, and thus the concept of coupling developed. If there is any deviation from stoichiometry, however, then some separation of charge occurs and a voltage gradient develops (Fig. 2.7); the voltage depends on the degree of the disparity (see Sects. 2.2.2.4 and 2.2.3.2). Neutral pumps do not lead to a change in the electrical gradient across the membrane, whereas with electrogenic ion pumps a voltage gradient is generated (see Sect. 2.2.2.4). Other ions not subject to the pump may diffuse passively along the gradient built up by the active transport.

2.2.2.3 Goldman's Voltage Equation

With Eq. (2.19) we have arrived at the criterion for passive distribution of ions between the interior and exterior phase attained at a state of equilibrium when the driving forces μ_{oi} and μ_{io} for the influx J_{oi} and efflux J_{io} are equal, i.e.,

$$\mu_{oi} = \mu_{io} \text{ and } J_{oi} = J_{io}. \tag{2.22}$$

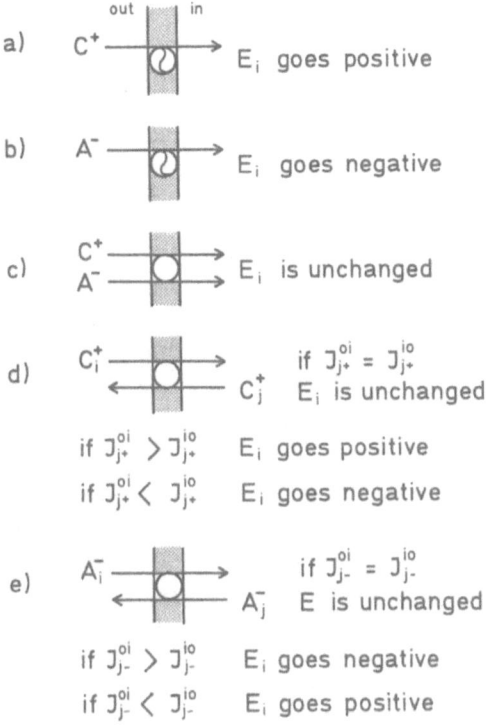

Fig. 2.7. Active transport or pumping of ions and electrogenesis across a membrane. In Figures **a** and **b** anions and cations, respectively, can diffuse passively along the electrical gradient set up by the pump; this mode of coupling is called electrogenic coupling. In Figure **c** the pump is electrically neutral since cation and anion uptake are equal. Figures **d** and **e** represent coupling of exchanges, e.g., by a carrier, but if the coupling is not stoichiometric electrogenesis occurs: nonstoichiometry could result from a difference in affinity between i and j for the sites along the transfer pathway.

For simple logic it follows that the net flux J_j is the difference between the influx and the efflux, whether the driving forces are electrochemical potentials or active transport mechanisms, or both, i.e.,

$$J_j = J_j^{oi} - J_j^{io}. \tag{2.23}$$

Under equilibrium conditions given above, J_j must be 0. If a system, however, is not in equilibrium, and has some net flux, and assuming that the electropotential gradient across the membrane is linear, the Goldman flux (or current) equation applies (Goldman, 1943; Hodgkin and Katz, 1949)

$$J_j = -\frac{z_j u_j E}{d} \cdot \frac{a_j^o - a_j^i e^{z_j FE/RT}}{1 - e^{z_j FE/RT}} \tag{2.24}$$

where d is the thickness of the membrane and the other symbols have the same meaning as before. Eq. (2.24) for the condition $J_j = 0$ reduces to the Nernst equation [Eq. (2.19)]. The ion mobility in the membrane, u, and the membrane thickness, d, are not known exactly for most biological membranes, so that it is more useful to take these parameters together and express them as a permeability coefficient, P_j:

$$P_j = \frac{RT\, u_j}{F\, d} \tag{2.25}$$

Thus Eq. (2.24) becomes:

$$J_j = -P \frac{z_j FE}{RT} \cdot \frac{a_j^o - a_j^i e^{z_j FE/RT}}{1 - e^{z_j FE/RT}} \tag{2.26}$$

This then means that electrical charges move through the membrane or that each ion flux is a current. The electrical current can be calculated by substituting F^2 for F in Eq. (2.25). If we assume that the total charge transported in such a system by anions and cations is equal, and that the ions participating are Cl^-, K^+, and Na^+, then electroneutrality is maintained and the net current, I, is equal to zero; thus:

$$I_{net} = J_{K^+} + J_{Na^+} + J_{Cl^-} = 0 \tag{2.27}$$

From Eqs. (2.26) and (2.27), after algebraic transformation, the equation based on the assumption of a constant field (Goldman's voltage equation) is obtained in the following form (Goldman, 1943; Hodgkin and Katz, 1949):

$$E = \frac{RT}{F} \ln \frac{P_K a_K^o + P_{Na} a_{Na}^o + P_{Cl} a_{Cl}^i}{P_K a_K^i + P_{Na} a_{Na}^i + P_{Cl} a_{Cl}^o} \tag{2.28}$$

where P_K, P_{Na}, and P_{Cl} are the permeability coefficients of their respective ions; a_K^i, a_{Na}^i, and a_{Cl}^i are the ion activities in the interior phase; and a_K^o, etc., the ion activities in the exterior phase.

The relative permeabilities are often more easily measured than the permeability coefficients proper. By measuring changes in E induced by changes in $[K^+]_o$, $[Na^+]_o$, and $[Cl^-]_o$ the following relationship can be used to obtain these estimates of relative permeabilities (Hope and Walker, 1961); with concentrations instead of activities, Eq. (2.28) becomes:

$$E = \frac{RT}{F} \ln \frac{[K^+]_o + \alpha[Na^+]_o + \beta[Cl^-]_i}{[K^+]_i + \alpha[Na^+]_i + \beta[Cl^-]_o} \tag{2.29}$$

where

$$\alpha = P_{Na}/P_K, \text{ and } \beta = P_{Cl}/P_K.$$

This equation can be extended for participation of further ions; however, it becomes more complex when polyvalent ions are introduced. If this equation is formulated for a single ion it becomes identical to Eq. (2.19). The constant field equation allows the determination of whether ions are distributed solely according to their electrochemical potential gradient, or there is active transport. When ion movement does not conform to this equation, active transport generally may be assumed; an equilibrium is not required.

2.2.2.4 Electrogenic Pumps

An ion transport which is driven by metabolism, and which generates an electropotential, is said to be electrogenic. Criteria for electrogenic pumps include the following (Higinbotham et al., 1970): (1) The measured membrane potential exceeds the maximum value possible from diffusion, e.g., predicted from Eq. (2.28). (2) Metabolic inhibitors rapidly and reversibly depolarize the cell. (3) Respiration rates and E_M values show some interdependence. (4) With the pump blocked, the Goldman voltage equation is met experimentally. Additionally: (5) Withholding an ion induces depolarization of the cell (Saddler, 1970a, b; Gradmann, 1970). (6) In green cells light may induce hyperpolarization (Saddler, 1970a, b; Gradmann, 1970; Spanswick, 1973). (7) Short-circuiting, i.e., making a zero electropotential gradient shows an electrical current accompanied by ion transport (Blount and Levedahl, 1960).

It is desirable to show that two or more of these criteria are met since any one, possibly excepting the first, may be fortuitous. The species for which there is good evidence of electrogenic pumps are given by Findlay and Hope (1976). A good example of an electrogenic pump is the inward transport of Cl^- in the marine alga *Acetabularia mediterranea* (Saddler, 1970a, b; Gradmann, 1970); lowering of Cl^- concentration, as well as application of metabolic inhibitors, causes depolarization from -170 to -90mV. Inhibitor experiments with a fungus *Neurospora* (Slayman, 1965, 1970; Slayman et al., 1970) and with higher plants (Higinbotham, 1970; Higinbotham et al., 1970; Fischer et al., 1976) also show a clear dependence of the membrane potential on energy from metabolism (Figs. 2.8, 2.9). One has to conclude from these investigations that metabolism-dependent electrogenic ion pumps can contribute considerably to the generation of membrane potentials.

In the case of an electrogenic pump, the Goldman voltage equation [Eq. (2.28)] does not hold true, and another term must be added for the active ion pump. In generating a potential, the electrogenic pump, of course, affects the passive diffusional relationships described by the Goldman equation. To describe the passive and electrogenic relationships some coupling between respiration and the pump must be assumed, yet the condition of no net current, Eq. (2.27), must be met. (The amount of current required to charge the membrane during hyperpolarization is negligible.

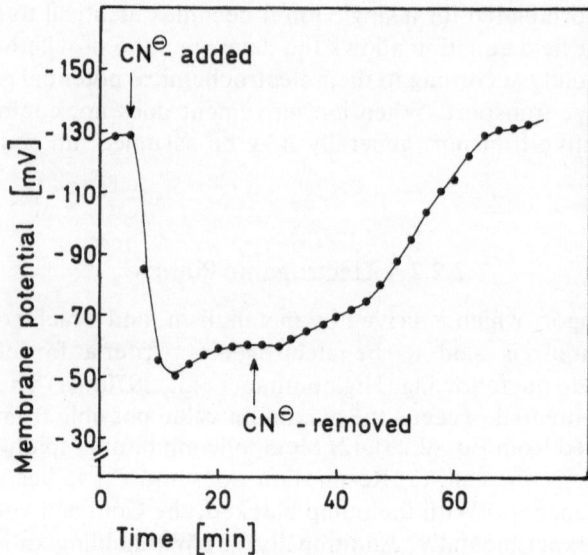

Fig. 2.8. Metabolism-dependence of the membrane potential in cells of pea epicotyl (*Pisum sativum*). Cyanide reversibly lowers the potential. Of the total PD of about -130 mV a portion (about -70 mV) is sensitive to CN^- and is considered to be electrogenic; the other portion (about -60 mV) is insensitive to CN^- and represents a diffusion PD. (Higinbotham et al., 1970).

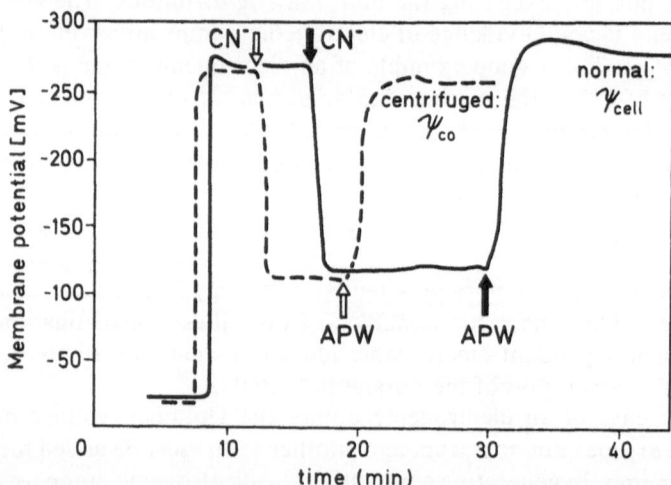

Fig. 2.9. Membrane potential recordings of *Mnium* leaf with the tip of the electrode in the cytoplasm of a centrifuged cell measuring E_{co} and in the vacuole of a noncentrifuged cell measuring E_{vo}. The effect of CN^- on depolarization and recovery on removal of CN^- is also, shown in each case. APW, artificial pond water (in mM: 1 NaCl, 0.1 KCl, 0.05 $CaSO_4$, 1 NaH_2PO_4 + Na_2HPO_4 at pH 7.0). (Fischer et al., 1976.)

See Nobel, 1974.) By use of irreversible thermodynamics, in order to account for the various interactions involved, the following has been derived (Higinbotham and Anderson, 1974; Spanswick, 1973):

$$E_M = E_G + \frac{I^a}{g_M} \qquad (2.30)$$

where E_M is the membrane potential; E_G is the potential calculated from Goldman's relation Eq. (2.28); I^a is the current of the active (electrogenic) pump; and g_M is the membrane conductance. Higinbotham and Anderson caution that g_M should be evaluated when the pump is blocked, e.g., by metabolic inhibitors. Spanswick (1973) considers that the pump sites are much more conductive than the domain of passive diffusion and that I^a can be assessed better by current-voltage relationships, with the pump operating and with E_K, the Nernst potential of K^+ [Eq. (2.19)], equal to E_G, in which condition the current from passive diffusion should be zero. Vredenberg (1973) found in *Nitella translucens* that electrical resistance increased with a rise in electrogenic pump activity, a result not consistent with the idea of a leaky pump. In view of the many complications involved with use of inhibitors and with possible effects of hyperpolarization on such properties as the permeability coefficients, etc., at present the best measure of E_P (E generated by pump) remains the measure of the degree of hyperpolarization over E_G or E_K. Experimentally this must be determined by blocking metabolism and measuring the depolarization. With the pump blocked, E_M should equal E_G and the polarization above this value found before inhibition should reflect the power of the electrogenic pump (Figs. 2.8, 2.9).

2.2.2.5 Action Potentials

External electrical, mechanical, or chemical stimuli can transiently depolarize the membrane potential. In analogy with animal systems we call these events action potentials. They are observed in algal coenocytes such as cells of Characeae and *Acetabularia,* and in excitable organs of higher plants like *Mimosa, Dionaea,* and *Drosera;* in the latter three cases movements of the organs are associated with the action potential (reviews: Bünning, 1939; Umrath, 1959; Sibaoka, 1969; Gradmann, 1970; Gradmann et al., 1973; Hope and Walker, 1975; Findlay and Hope, 1976; Haupt, 1977; Paszewski et al., 1977). Spontaneous electrical events have been recorded in membranes of various plant species (Pickard, 1973).

The basic molecular processes during action potentials in animals and plants do not seem to be different in principle, although in either case they are still poorly understood. The resting membrane potential usually is much more negative in plant than in animal cells. On the other hand responses and conduction of action potentials are much more sluggish in

Table 2.1. Comparison of action potentials in plant and animal systems. (Compiled from data in Bünning, 1939; Prosser, 1973; Findlay and Hope, 1976; it must be noted that all figures mark orders of magnitude only; the propagation in animal nerves highly depends on the cable characteristics such as structure and cross sectional area.)

	System	Duration of rising phase [s]	Velocity of propagation [mm · s⁻¹]	Duration of refractory state	
				absolute [s]	relative [s]
Plants	Charophyte intermodal cell	1	10–20	4–40	60–150
	Mimosa petiole of leaves	0.5	20–30		
	Dionaea trap lobes	0.1–0.2	60–170	0.6	<30
Animals	*Anodonta* (fresh water clam) nerve cord	0.1	45		
	Octopus mantle nerve	0.01	3,000		
	Eledone moschata mantle nerve	0.003	4,500		
	Mammalian nerve fibers	0.0004	100,000	0.0005	0.001–0.01

plant cells. Nevertheless, Table 2.1 shows that there are only differences of degree between plant and animal systems; the more primitive animal systems display responses on the same order as the fastest plant reactions. In animal nerves, the ionic basis of action potentials is given by fluxes, permeabilities, and electrochemical gradients of K^+ and Na^+, whereas in plant cells K^+ and Cl^- are involved. Ca^{2+} apparently is needed for establishment of membrane excitability, at least in some cases. Also in some animals Cl^- fluxes are involved, as in plants, for the action potential (Prosser, 1973).

Hope and Walker (1975) have described the course of events during an action potential in Charophyte cells after an above-threshold stimulation as follows (where small letters refer to points on the time scale in Fig. 2.10):

(a) stimulation and sharp increase of G_{Cl} (membrane conductivity for Cl^-), concomitantly with the onset of an inwardly directed electrical current J_{Cl} carried by a Cl^- efflux and a depolarization of the PD;

(a) to (b) the continuing depolarization increases G_{Cl} still further;

(b) to (c) G_{Cl} spontaneously begins to decrease again and J_{Cl} decreases;

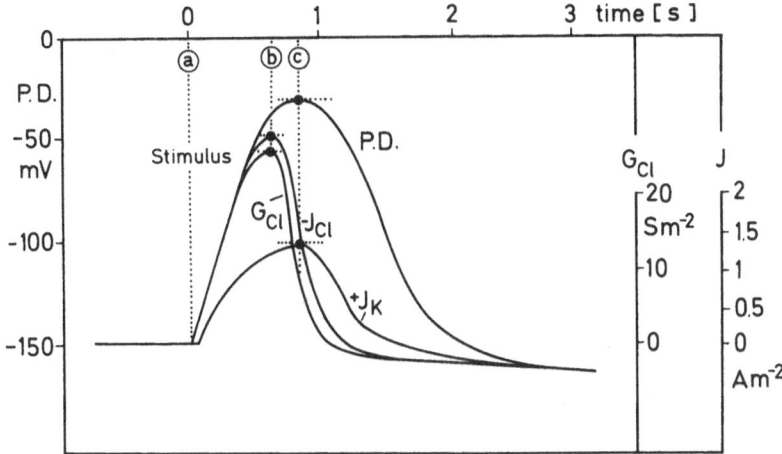

Fig. 2.10. Course of events during an action potential across the plasmalemma of a charophyte cell; see text. (Redrawn from Hope and Walker, 1975.)

(a) to (c) an outwardly directed current J_K, K^+ efflux, rises as the PD depolarizes;

(c) the peak depolarization is reached when $-J_{Cl} = +J_K$; after (c) G_{Cl} continues to decrease; the PD returns to its initial level; J_{Cl} and J_K decrease; for some time the membrane remains unexcitable or refractory, an absolute refractory state meaning no reaction at all can occur, a relative refractory state referring to a situation in which a stronger stimulus may cause a reaction, or a somewhat reduced reaction occurs (see also Table 2.1).

2.2.3 Measurement of Some Important Parameters

The usefulness of the quantitative interrelationships between chemical and electrical potentials and fluxes of ions in physiological transport investigations depends, of course, largely on the ability to measure in biological systems the characteristic parameters required by the equations. All parameters measured are not only subject to certain experimental errors, but are also subject to systematic errors because certain assumptions have to be made about the structure of the system under investigation. In this respect, biological systems are extremely different from physicochemical outside-inside models whose structure can be determined by the experimenter himself deliberately choosing certain concentrations and separating membranes whose properties are well known.

2.2.3.1 Measurement of Concentrations

Usually in experiments with plant cells or tissues, only the outer concentration in a certain range can be freely chosen by the experimenter. The large coenocytic cells of some algae (Characeae: *Chara, Nitella, Nitel-*

lopsis; Siphonales: *Valonia, Halicystis;* etc.) however, are characterized
by extremely large vacuoles often approaching a cubic centimeter in volume and having a multinucleate cytoplast lining the wall; in these, one can
use a perfusion technique by introducing microcapillaries and displacing
the cell sap with solutions of any chosen composition. Measurements can
also be made of the internal concentrations of the cells by directly obtaining the cell content and analyzing it with microanalytical methods
(Hope and Walker, 1975). Doing this, the vacuole, streaming cytoplasmic
layer, and the stationary wall-associated layer of cytoplasm can be analyzed separately. Some values of the ion concentrations are to be found in
Table 6.3.

Using tissues of higher plants we are limited to determinations of the
water content by comparing fresh and dry weight and to measurement of
ion contents by analysis of extracts.

Flame photometry is commonly used to assay Na^+ and K^+ concentrations and electrochemical titration to measure the Cl^-; these three ions
have been used frequently in transport studies. The results obtained are
subject to the error already discussed of working with concentrations instead of activities [see Eq. (2.2)]; there are some procedures for direct
measurements of ionic or molecular activities (see Sect. 2.2.3.6). Also
the plant cell or the plant tissue is treated as a uniform space with a continuously well-stirred aqueous solution of the solutes investigated. In reality plant cells consist of many compartments which are isolated from one
another; of course, the two large compartments are the vacuole and the
cytoplasm. The latter, however, is again compartmented into numerous
different spaces. The simple outside-inside model therefore represents a
very coarse simplification of the actual situation. In Chapter 6 we shall
consider attempts to use models involving compartments which increase
in complexity but which are still oversimplifications.

2.2.3.2 Measurement of Electropotentials
Across Cell Membranes

Electrical membrane potentials are measured by the aid of glass microelectrodes whose outer tip diameter is on the order of 1 μm or smaller for
small cells but 2–20 μm for large algal cells. The glass electrodes used for
insertion are filled usually with 3 M KCl solution; they are therefore
strictly micro-salt-bridges. In some cases weaker KCl solutions, e.g., 0.3
M KCl, are used to minimize KCl diffusion from the electrode tip into the
cell, which could lead to errors in measurement. The electrodes are introduced into the cell by the aid of micromanipulators so that the electropotential difference between the cell interior and the external solution can be
measured. It is, of course, important to know in which compartment the
tip of the measuring electrode is situated. Generally the electrode tip is in-

serted into the central vacuole giving the PD between vacuole and outside, E_{vo}; thus the PD is across the tonoplast and the plasmalemma as well
as the cell wall (Fig. 2.11). A reference electrode is immersed in the external solution; the reference electrode and the insertion electrode are
connected to a sensitive electrometer having high input resistance (or
impedance). Often a recorder is used; a customary circuit for plant cells is
shown in Figure 2.11. A more complete description for plant cells is provided by Nobel (1974), Hope and Walker (1975), and Findlay and Hope,
(1976).

Potential measurements with microelectrodes were originally done
predominantly with large coenocytes. In particularly favorable cases one
can introduce electrodes selectively into the cytoplasm or into the vacuole and then measure separately the PD's across the plasmalemma (E_{co})
and the tonoplast (E_{vc}); (see Table 2.2). Here it becomes clear that the
electropotential across the plasmalemma, E_{co}, always has a rather significant negative value. The PD across the tonoplast, E_{vc}, is usually zero or
has a low positive value. Therefore, most plant cells investigated show
the total electropotential between the vacuole and the external solution,

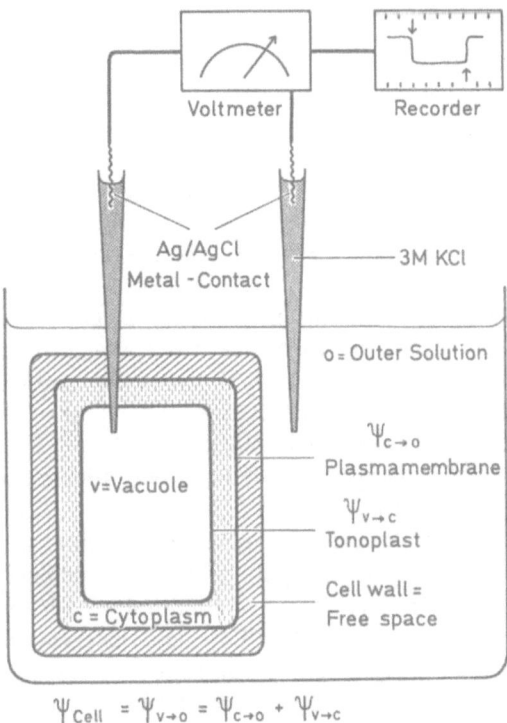

$$\Psi_{Cell} = \Psi_{v \to o} = \Psi_{c \to o} + \Psi_{v \to c}$$

Fig. 2.11. The apparatus for transmembrane electropotential measurements. The
microelectrodes have a resistance usually in the range of 5–50 megohms and the
millivoltmeter must have an input impedence of about 10^{10} ohms or more.

Table 2.2. Membrane potentials of algal cells (MacRobbie, 1970a). E_{vo} is the potential between vacuole and outer solution; E_{co} is the potential between cytoplasm and outer solution; and E_{vc} is the potential from vacuole to cytoplasm. $E_{vo} = E_{co} + E_{vc}$.

	E_{vo}	E_{co} [mV]	E_{vc}
Fresh- and brackish water algae:			
Nitella translucens	−122	−140	+18
Nitella flexilis	−155	−170	+15
Chara corallina	−152	−170	+18
Hydrodictyon africanum	− 90	−116	+26
Marine algae:			
Halicystis ovalis	− 80	− 80	± 0
Valonia ventricosa	+ 17	− 71	+88
Chaetomorpha darwinii	+ 10	− 70	+80
Griffithsia	− 55	− 80	+25
Acetubalaria mediterranea	−174	−174	± 0

E_{vo}, to be negative. This also applies to the cells of higher plants, with which measurements with microelectrodes are now used with increasing success. Although it seems clear in cells of higher plants that the plasmalemma is the major electrical barrier, measurements of PD across the tonoplast are relatively meager. E_{vo} is always negative in intact tissue and E_{vc} equals approximately zero (Etherton and Higinbotham, 1960; Denny and Weeks, 1968; Fischer et al. 1976; see Fig. 2.9) or is slightly positive by about 5 to 15 mV (Mertz, 1973). PD's (E_{vo}) of +17 and +10 mV, which have been measured in *Valonia* and *Chaetomorpha*, respectively, are exceptions (Findlay and Hope, 1976; see also Table 2.2), but in these cases there is a PD of about +88 to +80 mV across the tonoplast with the vacuole being positive relative to the cytoplasm; thus the cytoplasm relative to the outer solution is about −70 mV.

Positive PD's of isolated wall-less protoplasts have been observed by Heller and coworkers (Heller et al., 1974; Heller, 1977; Rona et al., 1977) and were also reported for isolated vacuoles (by Lin et al., see Racusen et al., 1977). The regeneration of a cell wall reduces the positive potential of isolated protoplasts (Heller, 1977) and eventually leads to restoration of a negative potential (Racusen et al., 1977). The observation of positive potentials of isolated protoplasts was confirmed by Racusen et al., (1977). However, Racusen et al., (1977) and Heller (1977) report different findings on plasmolyzed cells which still have a cell wall. In the measurements of Racusen et al., these also had positive potentials, whereas Heller (1977) recorded negative potentials as long as the cell wall was intact;

positive PD's were found only after rupturing the cell wall by aid of a micromanipulator. The reason for this discrepancy is not clear although different species are involved. The comparisons of PD's of intact cells, plasmolyzed cells with intact walls, protoplasts, and isolated vacuoles are important, since protoplasts and vacuoles can be stabilized only in plasmolyzing media of high osmotic strength. The positive PD's of isolated protoplasts and vacuoles suggest a severe change of membrane properties during isolation and limit the suitability of these preparations for assessment of membrane transport processes of intact cells and tissues.

In view of the fact that a number of plants accumulate acids, apparently into the vacuoles, more intensive studies of tonoplast properties are justified, in particular with respect to proton transport. The recently developed procedure for isolating intact vacuoles (Buser and Matile, 1977; Lin et al., 1977) may permit the necessary electrical and flux measurements (but see also the limitations discussed above).

2.2.3.3 Measurement of Fluxes and Permeabilities

The net flux, J_j, of a substance can be determined by measuring changes of the concentration of the substance in the interior or exterior phase or in both phases. When the net flux is negative there is a net efflux because the interior phase loses the solute to the exterior phase. If the net flux is positive there is a net influx. The influx J_j^{oi} and the efflux J_j^{io} are determined by the aid of radioactive isotopes. Such labeling in the exterior phase permits measurement of the rates of initial uptake of the isotope into the tissue and thus of the influx. Since the isotope concentration in the interior phase is, for practical purposes, equal to zero during the course of a short experiment, efflux can be neglected, $\Phi_{io} = \Phi$, and using Φ_j instead of J_j, Eq. (2.26) becomes:

$$\Phi_j^{oi} = -P_j \frac{z_j FE}{RT} \cdot \frac{a_j^o}{1 - e^{z_j FE/RT}}. \tag{2.31}$$

Efflux can be measured similarly if the cells or tissues are well loaded with isotope; then the external solution can be replaced by a nonlabeled solution and the rate of isotope loss from the tissue is measured. Of course, the concentration and the specific radioactivity (disintegration rate/mol) of the internal solution must be known. This has occasionally been overlooked, and it has been concluded superficially that efflux is very slow or equals zero if, in experiments, very little radioactivity is lost from the loaded tissues. For tracer ion efflux using Eqs. (2.24) and (2.26), the following equation may be obtained:

$$\Phi_j^{io} = P_j \frac{z_j FE}{RT} \cdot \frac{a_j^i e^{z_j FE/RT}}{1 - e^{z_j FE/RT}}. \tag{2.32}$$

Equations (2.31) and (2.32) are Goldman's (1943) flux, or current, equations. Net flux can also be obtained from influx and efflux rates determined using the isotope labeling techniques. Data for typical fluxes observed in plant cells are to be found in Chapter 6.

2.2.3.4 Measurement of Electrical Resistances

If the membrane potential, E_M, and the external concentration or activity, a_j^o, are known, influx or efflux is measured, and a_j^i has been determined, then P_j can be calculated [Eqs. (2.31), (2.32)]. Another method for determination of permeability is by measurement of the resistance, r, (or its reciprocal, conductivity, G) of a membrane. An electrode is introduced into a cell and a current is passed through the membrane. The following relationship, a form of Ohm's law, is obtained:

$$r = \Delta E_M \, A/I \qquad (2.33)$$

where r is resistance; ΔE_M is the change in PD induced by the current, I; and A is the surface area of the cell. Two circuits are usually required, one for passing current, the other for monitoring E_M; thus two microelectrodes must be inserted into the cell. Although, properly executed, this procedure works well with giant algal cells (Hope and Walker, 1975), it can be quite frustrating with higher plant cells and lead to rather widely ranging values (Higinbotham et al., 1964). We think that the difficulty in part lies in the fact that the insertion of a second electrode causes cell damage, e.g., perhaps a loss of turgor, and there may not be a flow of cytoplasm to permit a proper seal around the electrode. Another important factor is the occurrence of plasmodesmata connecting the cells so that the path of the current is not well defined; thus the current actually passing directly through the membrane of the cell in which the electrode is sited may not be accurately assessed; some current may pass through adjacent cells. This problem has been quite well evaluated by Spanswick (1972a).

More recently a technique has been developed to permit use of a single microelectrode for simultaneous measurement of membrane electrical potential and resistance (Anderson et al., 1974); the initial results show promise of better information on resistances.

Electrical conductance, $G \; (=1/r)$, should provide an independent measurement of permeability. The relations are (Hope and Walker, 1975):

$$g_M \, (\Delta E_M) = \frac{\Delta J_M}{\Delta E_M} = \Sigma G_j \, (\Delta E_M) \qquad (2.34)$$

where g_M is the chord conductance of the membrane measured by passing a current which produces the flow, ΔJ_M, giving a step change in voltage, ΔE_M. Of course g_M represents a sum of all the partial ionic conductances, G_j's. In the case of a univalent ion diffusing independently and in flux equilibrium (Hope and Walker, 1975):

$$G_j = \frac{\Delta J_j}{\Delta E_M} = P_j \frac{c_j^0 F^2 E_M (1 - e^{F\Delta E_M/RT})}{RT\Delta E_M (1 - e^{F\Delta E_M/RT})}. \qquad (2.35)$$

If the major partial ionic conductances are known, P_j can be calculated; conversely G_j can be determined if the P_j's are known.

Generally, ion permeabilities have been estimated from tracer fluxes and the electrochemical potential gradient using Goldman's flux equations [Eqs. (2.31) and (2.32)]. This led to data showing a considerable discrepancy between P_j values estimated from tracer fluxes and those from electrical conductance. Two possible major causes for this discrepancy advanced by Hope and Walker are: (1) that there are interactions of ions passing in file through narrow pores and thus there is not independent movement; and (2) that the H^+ ions contribute a considerable conductance, particularly if they are subject to electrogenic efflux pumping (see Kitasato, 1968; Spanswick, 1973).

Conductance may be directly related to tracer flux as follows (Hodgkin, 1951):

$$G_j = \frac{z_j^2 F^2 \Phi_j^{oi}}{RT}. \qquad (2.36)$$

In terms of tracer fluxes at 25°C this reduces to $\Phi_j = 958/r$ (Gottlieb and Sollner, 1968). This equation could be used more often to detect discrepancies between electrical conductance and tracer-measured fluxes. It may be, as Vredenberg (1973) has claimed, that under appropriate conditions the discrepancy is not as large as generally reported. If the discrepancy is due to interactions of solution particles with one another or with the cell membrane, as suggested by Hope and Walker, better evaluation is needed.

2.2.3.5 Measurement of Water Relation Parameters

It is a common laboratory procedure to measure osmotic relations by plasmolysis using a series of concentrations, thus determining the point at which incipient plasmolysis is observed. Alternatively, the cell or tissue sap can be isolated and its osmotic concentration determined, e.g., by measuring the freezing point, or using a vapor pressure method. This provides a measure of the osmotic concentration, π, as used in Eq. (2.8); at incipient plasmolysis $P = 0$ and Ψ_W, the cell water potential is equal to π. From these relationships P can be calculated assuming ideal conditions, i.e., the wall is completely rigid, that π is accurately measured, and that membrane permeability is not altered. Unfortunately, as will be seen later, these conditions are not met and the situation is much more complex (Zimmermann, 1977). The measurements of π are highly variable and, with higher plant cells in particular, are probably inaccurate.

With the lengthy internodal cells of Characean algae, the system of

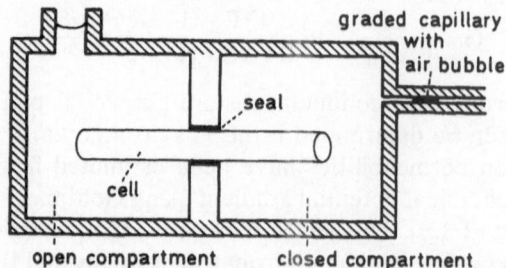

Fig. 2.12. Apparatus for measurement of transcellular osmosis. (Dainty, 1976.)

transcellular osmosis proved useful for determinations of water relation parameters (review Dainty, 1976). Cells are sealed in chambers between two compartments as shown in Figure 2.12. The left compartment is open, and the osmotic pressure of the solution can be suddenly changed in this compartment so that transcellular osmotic water flow through the cell is induced. The second compartment is closed; an air bubble in a capillary attached to this compartment allows the recording of volume flow. Using the appropriate equations, derived from osmotic laws, hydraulic conductivities and reflection coefficients can be calculated.

With higher plant tissues the water potential has been determined also, using the pressure-bomb developed by Scholander (Fig. 2.13). Ψ_W is measured as the external gas pressure required to press out fluid from the cut end of a plant organ. Again with the appropriate experimentation and mathematical formulation various water relation parameters can be obtained (review Dainty, 1976). The disadvantage of this approach is that only overall estimates for whole plant organs can be made.

A considerable improvement has been made recently with the advent of direct measurements of pressure, P, in large cells (Green and Stanton, 1968; Steudle and Zimmermann, 1974; Zimmermann and Steudle, 1975). The last two papers utilized an elegant technique in which a microcapillary (tip diameter about 60 μ) was inserted into *Nitella* or *Valonia* cells; the pressure, which can be increased or decreased by a plunger, was detected by a miniature pressure transducer yielding a voltage signal (Figs. 2.14, 2.15). By this technique the authors have been able to estimate reflection coefficients, σ, hydraulic conductivity, L_p, and the elastic modulus of cell walls, ε. Further, they have found a number of interactions which include (1) a direct influence of P on water permeability, (2) differences in rates between exosmosis and endosmosis, (3) the elastic modulus, ε, of the cell wall generally increases with pressure, and (4) other properties may be changed by pressure, e.g., electropotential and ion fluxes. When the changes of turgor induced are small with respect to the size of the absolute turgor, the artifacts suspected by Dainty in the plasmometric method are avoided. Some measurements of σ from this procedure are provided in Table 2.3.

Fig. 2.13. Diagram of a "pressure bomb" derived from the original version by Scholander et al. (1964). The vessel is made from thick metal to withstand high pressures. The screw-lid clamps the plate carrying the specimen onto the top of the vessel. The shoot is held into a hole made in a sheet of rubber which is then compressed by the bolts in the plate to seal the shoot. Compressed gas is fed into the vessel through the inlet tube and a measuring device used to record the pressure. A lens can be used to observe any solution flow from the cut end of the shoot. (Dainty, 1976.)

More recently the pressure probe sensing and pressure-regulating device has been refined to permit its use with higher plant tissue, the cells of which are much smaller (see Hüsken et al., 1978).

The hydraulic coefficient, L_p, and the reflection coefficient, σ, can be estimated from the following phenomenological equations describing the interactions in which coupled flows occur with one solute (Katchalsky and Curran, 1965):

$$J_v = L_p (\Delta P - \sigma_s \pi_s) \tag{2.37}$$
$$J_s = (1 - \sigma_s) \cdot \bar{c}_s \cdot J_v + \omega \Delta \pi_s \tag{2.38}$$

where J_v is volume flow; J_s is the solute flow driven by ΔP and $\Delta \pi$; $\bar{c}_s$ is the average concentration of s; ω is a coefficient of solute permeability; and other symbols are as given before. For details on the manner of arriving at accurate P and σ_s measurements the original papers should be consulted (see Zimmermann, 1977).

Further complications arise from the fact that cell walls are not rigid

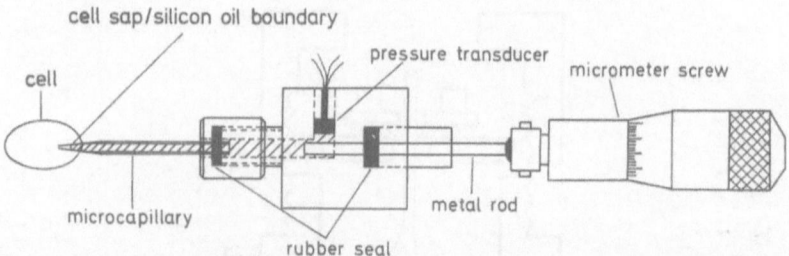

Fig. 2.14. Schematic diagram of the pressure probe for measuring turgor pressure, P, the volumetric elastic modulus of the cell wall, ε, and the hydraulic conductivity, L_p, in single cells. The pressure inside the cell is transmitted by the oil in the microcapillary and monitored by a pressure transducer mounted in the Plexiglas chamber. Interference effects arising from compression of the oil and temperature fluctuations in the system are excluded by means of an accurate adjustment of the cell sap/oil boundary in the tip of the microcapillary. This is achieved with an electronic feedback system (not shown) based on the measurement of the resistance between two electrodes (silver wires). A resistance electrode is fed into the tip of the microcapillary which changes its resistance as the boundary cell sap/oil moves. (Zimmermann and Steudle, 1974; for the model modified for cells of higher plants see Hüsken et al., 1978.)

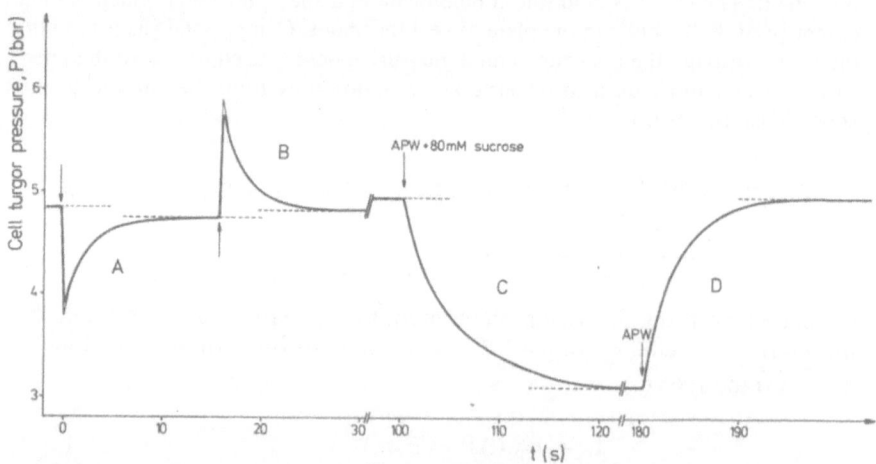

Fig. 2.15. Turgor pressure relaxation process in a cell of *Nitella flexilis* in response to changes in the turgor pressure (curves *A* and *B*) and in the osmotic pressure of the artificial pond water (APW), respectively (curves *C* and *D*). Turgor pressure was changed by means of the pressure probe, that is, by a displacement of the oil/cell sap boundary in the microcapillary (see Fig. 2.14). The osmolarity of the external solution was changed by adding 80 mM sucrose. (Zimmermann and Steudle, 1974.)

Table 2.3. Reflection coefficients, σ_s, of some nonelectrolytes for the internode of *N. flexilis* (with standard deviations, the number of cells is given in brackets). The nonelectrolyte concentration in the testing solution was $c_s = 160$ mM. For comparison the data for *N. translucens, C. australis* and *V. utricularis* are given. (Steudle and Zimmermann, 1974.)

Solute	Molecular radius (Å)	Reflection coefficients of			
		N. flexilis	*V. utricularis*[a]	*N. translucens*[b]	*C. australis*[b]
Sucrose	5.3[c]	0.97 ± 0.01 (5)	1	—	—
Glucose	4.4[c]	0.96 ± 0.07 (2)	0.95	—	—
Glycerol	2.74[d]	0.80 ± 0.04 (2)	0.81	—	—
Acetamide	2.27[d]	0.91 ± 0.02 (2)	0.79	—	—
Urea	2.03[d]	0.91 ± 0.01 (2)	0.76	1	1
Formamide	—	0.79 ± 0.04 (3)	—	1	1
Ethylene glycol	—	0.94 ± 0.02 (3)	—	1	1
Isopropanol	—	0.35 ± 0.05 (2)	—	0.27 (0.40)[f]	—
n-Propanol	—	0.17 ± 0.06 (2)	—	0.16	0.22
Ethanol	2.13[e]	0.34 ± 0.02 (2)	—	0.29 (0.44)[f]	0.27
Methanol	1.83[e]	0.31 ± 0.04 (2)	—	0.25 (0.50)[f]	0.30

[a] Data taken from Zimmermann and Steudle (1970).
[b] Data taken from Dainty and Ginzburg (1964b).
[c] Data taken from Durbin (1960).
[d] Data taken from Goldstein and Solomon (1960).
[e] Data taken from Villegas and Barnola (1961).
[f] Data corrected for unstirred layers (Dainty and Ginzburg. 1964a).

but rather with pressure they stretch; this property can be measured as the elastic modulus. ε:

$$\varepsilon = V \frac{dP}{dV} \simeq V \frac{\Delta P}{\Delta V} \qquad (2.39)$$

where P is hydraulic pressure against the wall, and V is cell volume. ε affects transport of both water and solutes. Unfortunately, the change with increasing P is not linear and to further complicate the problem it is affected by cell shape. In cylindrical cells, for example, ε is a function of both longitudinal and transverse elasticities (see Steudle and Zimmermann, 1974). Fortunately, with the pressure probe system it may be possible to estimate ε in cells, including those in higher plant tissues, without resolving the modulus into its component parts.

The phenomenon of dielectric breakdown of membranes is induced by a jolt of 0.5 to 1 V delivered in microseconds. During the breakdown, which occurs at a critical voltage, the membrane is highly permeable to small and large ions or molecules (Zimmermann et al., 1976); however, it restores itself within about 10 min. For both electrical and pressure studies on cells, Coster and Zimmermann (1975) have developed the concept of an electromechanical model to explain the membrane modifications in active and passive transports induced by voltage, pressure, or both. The membrane presumably is the site of turgor-sensing.

Osmoregulation may occur in response to osmotic stress by either ion transport processes or by biochemical processes. In the latter case, for example, a polymeric substance may be converted to substances of lower molecular weight, thus increasing π; the reverse reaction would decrease π. When salt accumulation occurs as an osmotic adaptation to an environment of low water potential, an additional problem arises due to the high sensitivity of cytoplasmic enzyme and organelle systems to salt. This problem among the higher plants is particularly well understood in halophytes accumulating NaCl. With the exception of some halophilic bacteria (Halobacteriales; Lanyi, 1974) high Na^+ concentrations are not compatible with enzyme proteins in any known cases of halophytes and nonhalophytes. Thus NaCl must be largely compartmented in the vacuoles. This may require solutes other than minerals in the cytoplasm which are compatible with enzyme proteins, and which balance the NaCl-dependent π in the vacuole. Such compatible solutes are synthesized during salinity adaptation; they include polyalcohols (such as glycerol, mannitol) and various nitrogen compounds (such as amino acids, particularly proline, and betaine, and related N-compounds; Cram, 1976; Wyn Jones et al., 1977; Storey and Wyn Jones, 1977.)

2.2.3.6 Measurement of Ion Activities Within Cells

Some workers have maintained that protoplasm has a highly ordered structure in which ions and water are in a bound state, thus not being free to diffuse readily and having a drastically lowered electrochemical activity (Ling, 1962; Troshin, 1961). If this were true, the theories based on membrane function in transport would have to be abandoned. In any case, it is important to be able to assign some value to the activities of the solvent, water, and solutes in living protoplasts.

There are several methods available for estimating the chemical and electrochemical activities of solution particles within the cell. These include measurements of osmotic values (P and π are based on chemical activities), electrical conductivity, and ion activity by use of fine-tipped specific ion electrodes, or by nuclear magnetic resonance (NMR). It may be concluded from various measurements that activities within the protoplast are quite high and largely as expected in a viscous solution.

In many plant tissues the vacuoles compose about 80% of the volume and have relatively little protein, although ions, mineral or organic, are often accumulated in appreciable amounts. By use of an ion-specific microelectrode Vorobiev (1968) made direct measurements of K^+ activities in the cytoplasm and vacuole of *Chara australis* ($= C.$ *corallina*); these activities corresponded closely with those calculated (using activity coefficients) from measured concentrations. Ion-specific microelectrodes for H^+, Na^+, and Cl^- are also available and others may be developed (see Andrianov et al., 1971; Bowling, 1976).

In etiolated pea stem tissue the activities of K^+ and Na^+ have been estimated by use of nuclear magnetic resonance (NMR). There was a close correspondence between the amounts of activity detected by NMR during ion accumulation and those determined by chemical analysis (Magnuson et al., 1973). It was concluded that in these tissues K^+ and Na^+ cannot be bound in a highly ordered structure; the results do not preclude a small amount of binding, but also they give no support for this idea.

The activity of water in protoplasm also can be measured by NMR. The results indicate that to a large extent water is free and not bound by a structured state.

2.2.4 The Ussing-Teorell Equation

From Eqs. (2.31) and (2.32) another important relationship was developed independently by Ussing (1949) and Teorell (1949):

$$\frac{\Phi_j^{oi}}{\Phi_j^{io}} = \frac{a_j^o}{a_j^i e^{z_j FE_M/RT}}. \qquad (2.40)$$

At flux equilibrium, when $\Phi_j^{oi}/\Phi_j^{io} = 1$, this equation reduces to the Nernst Eq. (2.19). If tracer fluxes do not conform to the Ussing-Teorell criterion it is generally concluded that active transport is occurring. The discrepancy for some ions can be very large (Pierce and Higinbotham, 1970). Conclusions based on small deviations may be doubtful and there is reason to think that for K^+ in cells the condition of independent ion movement may not be met. In squid nerve cells Hodgkin and Keynes (1955) found that K^+ diffusion fitted a modified empirical version of the flux ratio relation:

$$\ln \frac{\Phi_j^{oi}}{\Phi_j^{io}} = n \left(\ln \frac{a_j^o}{a_j^i} - z_j FE_M/RT \right) \qquad (2.41)$$

where n is a coefficient which in this case was 2.5. This formulation is one which is in accord with the hypothesis that ions, j (K^+ in this case), move in a file through a pore, n being the number of interacting events during passage. The net effect observed is that the electrochemical potential term is multiplied by n, which suggests a much stronger effect of the electropotential gradient than expected from tracer fluxes. A similar effect was observed by Walker and Hope (1969) when they examined the relation of the electropotential gradient by voltage clamping on K^+ and Na^+ fluxes across the membranes of *Chara corallina*; again n was about 2.5. In this study also the membrane electrical conductance exceeded by several-fold the conductance estimated from tracer fluxes.

The flux-ratio equation is still regarded as an excellent criterion for passive transport; deviations from it in themselves cannot be regarded as proof for active transport, but rather strongly suggest it. In such cases we must look carefully for a way to evaluate the fluxes in terms of possible interactions including those involving some coupling with metabolism.

2.3 Irreversible Thermodynamics and Onsager Coefficients

Many of the equations [e.g., Eq. (2.19)] given before are based on classical thermodynamics and hold true for processes in, or very near, equilibrium; they are not well suited to the processes, characteristic of living cells, which are often far from equilibrium. In addition they are based on ideal conditions in which each molecule or ion of the system behaves independently whereas in reality, as seen above, there may be frictional interactions between these particles. For this situation the formulations of irreversible thermodynamics were derived. A system in nonequilibrium moves spontaneously, and irreversibly, toward equilibrium as in the example of diffusion discussed before (Sect. 2.1.1).

In the view of irreversible thermodynamics the flux of j_1 is not dependent only on the chemical potential gradient but rather may be affected by forces exerted in the fluxes of j_2, j_3, etc., as well as being subject to other forces, e.g., the electrical gradient or enzymatic reactions of metabolism. Ion j moves in accord with the net force exerted on it. Interdependence of the fluxes of j_1, j_2, j_3, etc., may lead to coupled movements; and coupling of fluxes with metabolic events seems likely. An example of interdependence is the interaction between water and solute (Katchalsky and Curran, 1965; Slatyer, 1967):

$$J_W = L_{WW} \, \Delta\mu_W + L_{WS} \, \Delta\mu_S \qquad (2.42)$$
$$J_S = L_{SW} \, \Delta\mu_W + L_{SS} \, \Delta\mu_S \qquad (2.43)$$

where L represents a phenomenological coefficient for WS or SS interactions, and $\Delta\mu$ the chemical potential gradient of W or S. According to Onsager's reciprocity law $L_{WS} = L_{SW}$. The right hand side of the equation may be replaced by $\Delta P + \Delta\pi$ [see Eqs. (2.37) and (2.38)]. When J_W or J_S is zero the reflection coefficient, σ, is determined by:

$$\sigma = -\frac{L_{WS}}{L_{WW}} = \frac{\Delta P}{\Delta\pi}. \qquad (2.44)$$

If additional interactions are involved, e.g., if respiratory energy is required for solute movement, another equation and more coefficients are required (see Dainty, 1976; Katchalsky and Curran, 1965; Nobel, 1974).

The hydraulic coefficient, L_p, and the reflection coefficient, σ, mentioned previously are not simple Onsager coefficients, but rather represent the net sum of several interactions.

2.4 Osmoregulation

For many years our conception of osmotic phenomena in plant cells was essentially that of a physical process to which a plant was subjected, perhaps like transpiration. In recent years the concept of osmoregulation in plants has emerged. Such a process has been known for years in wall-less unicellular plants and animals in which the operation of contractile vacuoles and transport pumps prevent collapse or bursting; these cells are restricted to reasonably narrow limits with respect to pressure gradients compared to plant cells surrounded by a wall. It is now known, however, that plant cells with walls possess turgor-pressure-sensing mechanisms and respond by changes in hydraulic conductivity, ion transport, enzyme activity, and growth (Zimmermann, 1977). The membranes, particularly the plasmalemma, are sensitive to physical pressure, either hydraulic or electrical.

2.5 Criteria for Active Transport

2.5.1 Transport Against an Electrochemical Potential Gradient, a Criterion for Active Transport

In considering Fick's law [Eq. (2.1)], the Nernst equation [Eq. (2.19)], the constant field equation [Eq. (2.28)], and the Ussing-Teorell equation [Eq. (2.40)] we have already defined active transport in a negative way. All of these equations apply only to systems with passive transport processes. We have stated that active transport is very likely when a system investigated does not obey these laws; that means it departs from the passive diffusional processes described by these equations. The definition of active transport obtained in this way is the following: noncharged particles are actively transported if their net movement is against a concentration gradient, and electrically charged particles are transported actively if their net movement is against the gradient of their electrochemical potential. The term active transport implies that particles in these cases no longer move due to their own kinetic energy, but rather that they are driven by energy from metabolic processes. In this sense, the transport of species, j_1, is passive if it is distributed according to a gradient which has been built up by active transport of another particle species, j_2. For instance, suppose a substance, j_1, is dissolved in water, j_2, and j_1 is moved actively from one side of the membrane to the other; then if water molecules diffuse according to the water potential gradient following this active movement of the solute, j_1, only the transport of j_1 and not the transport of water, j_2, is active.

2.5.2 Passive Transport Against Chemical and Electrochemical Gradients

2.5.2.1 Congruent and Incongruent Transport

We have reached a definition of active transport which comprises net uphill transport as the most important criterion. It should be made clear, however, that under certain conditions purely passive movements against such a gradient may occur. Schlögl (1964; see this reference for further literature) has shown that in a closed outside-inside system at constant pressure and at constant temperature three main types of transport are possible. The diagram shown in Figure 2.16a is the situation usually expected with solely passive transport with both substances moving in the direction of their concentration gradient. This type is called congruent transport. The types drawn in Figure 2.16b and c are composed of congruent transport of one particle species and incongruent transport of the other. Incongruent transport can be made possible by coupling of the fluxes of different particles, i.e., when one particle species carries the other particle species along during its transport. The coupling can occur as a result of frictional interactions between the particles; they have generally been known as electrokinetic phenomena: electroosmosis or solute drag effects (Sect. 2.5.2.2). These interactions are expressed in Onsager coefficients (Sect. 2.3).

2.5.2.2 Electroosmosis and Streaming Potentials: Electrokinetic Phenomena

Two major forces driving transport of material are hydraulic pressure gradients and electropotential gradients. The flow of an electrolyte solution

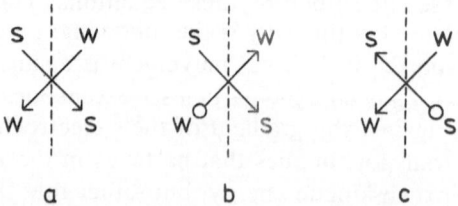

Fig. 2.16. Transport types in a system with two compartments separated by a membrane and two species of diffusing particles: W is water, the solvent, and S is the solute. **a** congruent transport of both components; **b** and **c** incongruent transport of W and S respectively.

through pores when an electropotential gradient exists is termed electroosmosis; the movement of ions may "drag" water with them. The converse of electroosmosis is a system in which electrolyte solutions are forced to flow by pressure gradients and thereby generate a streaming electropotential; water moves ahead of the ions and "solvent drag" occurs. In both cases a separation of charge occurs as a result of the imbalance of positive and negative sites along the pore walls. Both solvent and ions must move in the same pathway and for maximum effect there are restrictions on pore size. By strong forces of either pressure, or electropotential, water may appear to be driven against the osmotic gradient to yield what has been termed "negative osmosis"; however, this is simply a case of altering the forces stated in the equations (see House, 1974).

These processes are indeed interesting and it is not surprising that a number of attempts have been made to invoke electroosmotic phenomena as explanations of long-distance and short-distance transport in plants. Thus far, in general, it cannot be said that electroosmosis is a major factor in transport processes. Barry (1970) has found small but significant contributions of electroosmosis in transport between adjacent cells of *Chara corallina;* perhaps more important is his suggestion that the process may play a major role in rapid plant movements. Spanner (1975) has developed a hypothesis of phloem translocation based on an electrokinetic cycling of K^+ in sieve tube elements assuming electroosmosis across sieve plates. The major difficulties with such a system have been reviewed by MacRobbie (1971a; see also Chap. 11).

The water permeability for most biological membranes is much larger than the permeability for solutes. In cells of higher plants the permeabilities of urea and glycerine are 1/100 to 1/300 the permeability of water; sucrose permeability is 1/1000 that of water, as are also ions generally. Water permeability of artificial lipid films is up to 10^9 to 10^{10} times as large as the ion permeability.

Active water transport due to the direct effect of energy on the water molecule is therefore very unlikely in living membranes. Because of the high permeability it would make little sense for water movement in living cells and tissues to be regulated by active transport of water molecules. It is much more effective for this regulation to be mediated by the transport of solute particles.

2.5.2.3 Transport by Carriers

Another special case of the interaction of particles with the membrane, through which transport occurs, is considered by the carrier hypothesis. As in an enzymatic catalysis process the carrier, R, localized in a membrane, binds with the transported substance, S, on one side of the membrane; the carrier substrate compound SR diffuses across the membrane phase and is separated at the other side. In this way the carrier catalysis is

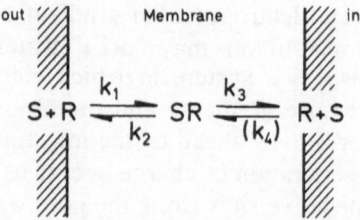

Fig. 2.17. Transport of a substrate, S, from outside to inside across a membrane catalyzed by a carrier, R, located in the membrane. Following formation of an associated complex, SR, at the outside and movement of SR across the membrane, dissociation of SR delivers S at the inside. According to the kinetic equation of Michaelis and Menten, reaction k_4 is nearly 0, thus the net process is unidirectional.

a transport of the substrate from one side of the membrane to the other (Fig. 2.17). This is catalyzed diffusion or facilitated diffusion. According to the kinetics formulated by Michaelis and Menten (1913) and Briggs and Haldane (1925) for certain enzymatic reactions the following equation applies for carrier transport:

$$\frac{v}{V_{max}} = \frac{[S]}{[S] + K_M} \qquad (2.45)$$

where v is the actual transport velocity measured at the concentration of substrate $[S]$ and V_{max} is the maximal rate of transport at saturation of the carrier sites by S. K_M is the Michaelis constant or the steady state constant of the dynamic equilibrium (see Fig. 2.17; Netter, 1959):

$$K_M = \frac{k_2 + k_3}{k_1} \qquad (2.46)$$

where the k's (Fig. 2.17) represent reaction constants. Note that in the Michaelis-Menten formulation, which describes only one of the several types of kinetics found in enzymatic reactions, the reaction rate k_4 is considered to be negligible. We will see later that in plant cells and tissues there are many transport processes observed which obey the Michaelis-Menten equation [Eq. (2.45)] and which appear to represent active transport (see Chap. 6).

Using the carrier hypothesis, the following two examples can be explained in which the coupling of different catalyzed transport processes mediates the movement of a substance against a gradient of concentration (Fig. 2.18).

Again let us consider the two-phase outside-inside model. The carrier, R, in the membrane catalyzes the transport of the substance S from one side of the membrane to the other. When the system is in equilibrium the concentration of the substance S is the same inside and outside, i.e., $[S_i] = [S_o]$. The same applies for the fluxes in both directions $\Phi_{Soi} = \Phi_{Sio}$.

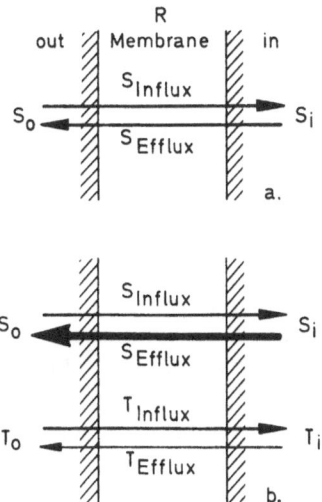

Fig. 2.18. Countertransport after Stein (1964). **a** equilibrium as a starting point; **b** fluxes after the addition of T on outside. The thicknesses of the *arrows* represent relative sizes of the fluxes under the hypothetical assumption that the active membrane component R has the same affinity for both substrates S and T and the external concentrations S and T are equal. (Lüttge, 1969.)

If one now adds to the system on one side (outside) a substance T which like S can serve as a substrate for the carrier R, then S_o and T_o will compete for the active carrier sites. Both substances will be transported to the inside according to the ratio of their concentrations, $[S_o]/[T_o]$, and according to the ratio of their affinities to the carrier R. If the affinity of the carrier is the same for both substrates each side of the membrane, and if the carrier is allowed to diffuse freely in the membrane so that it can form carrier substrate compounds equally on both phase boundaries, then Φ_{oi} will be reduced but Φ_{io} will be unchanged for a certain amount of time. This results from the fact that the concentration of T_i is low immediately after addition of T to the external solution and, depending on the size of the interior compartment, for a certain amount of time afterward so low that S_i is much larger than T_i. Therefore, for most of the T molecules transported from outside to inside an S molecule will be transported by exchange from the interior to exterior. In this way, S will be moved against its concentration gradient. Stein (1964) calls this coupling of catalyzed fluxes countertransport. Exchange diffusion is caused in a similar way. Again S_o and T_o compete for the active carrier sites. The addition of T in this case, however, occurs at a state far from equilibrium. The conditions are that S_o and S_i are large as compared to the Michaelis constant K_M and that the carrier R is nearly saturated with substrate. Because of this the net flux of S before the addition of T is almost zero. After the addition of competing substance T the influx of S will be inhibited immediately while the efflux of S will be reduced only after some time. The net flux of S becomes negative, i.e., it is directed from interior to exterior, and S is again transported against its concentration gradient. After some time the systems both with countertransport and exchange diffusion will approach a new equilibrium in which all concentrations and all fluxes in both directions are equal and the net fluxes are zero. Examples of countertransport and exchange diffusion have been investigated especially in erythrocytes. Carrier transport, as such, accordingly is not necessarily active transport. It is not justified to conclude from a transport process obeying the Michaelis-Menten formulation [Eq. (2.45)] that active transport is occurring; indeed, diffusion may simulate such kinetics (see Higinbotham, 1973a). On the other hand the concept of a carrier with specific attachment sites would provide a mechanism permitting coupling between transport processes and the metabolic reactions supplying energy for active transport. We will come back to this in Chapter 5.

2.5.3 Comparison of Electrochemical Theory With Carrier Kinetics

The usual experiments done to demonstrate carrier saturation and to determine rate constants under enzymatic kinetics theory have involved quite wide ranges of external concentrations. For an electrochemical point of view this raises a number of questions which render the kinetic data difficult, or impossible, to interpret properly. For example, in a KCl concentration range of 0.01 to 10 mM, and assuming that P_{Cl}/P_K is very low (true for most cells), the Nernst equation [Eq. (2.19)] predicts that the cell electropotential, E_M, will be markedly altered. If E_M is -180 mV with cells in 0.01 mM KCl then with passive diffusion $K_i^+ = 12.7$ mM:

$$E_M = -180(mV) = 58 \log \frac{[K^+]_o}{[K^+]_i} = 58 \log \frac{0.01}{12.7}.$$

Now if $[KCl]_o$ is raised to 10 mM the following relation results:

$$E_M = 58 \log \frac{10}{12.7} = -5.8(mV).$$

Thus it is seen that the cell PD, (E_M), an electromotive force tending to drive K^+ inward, has been drastically lowered. This depolarizing effect of higher external concentrations has been verified experimentally in plant cells (Higinbotham et al., 1964). To what extent this contributes to results interpreted as saturation of carrier uptake has not been properly assessed (however, see Mertz and Higinbotham, 1974).

Exposing cells to a wide range of external concentration results in a large change in ionic strength; also higher concentrations may have important osmotic effects and the membrane may be partially dehydrated.

According to the Van't Hoff relationship, [Eq. (2.9)], 50 mM KCl would give an osmotic pressure, approximately:

$$= (0.05 \times 2) \cdot 24 = 2.4 \text{ atm.}$$

Hastings and Gutknecht (1974) have found in *Valonia macrophysa* that a decrease in turgor pressure of 1 atmosphere could cause a four fold increase in K^+ influx (active in this species). For this reason the carrier kinetics theory requires some reassessments of methodology in order to make proper quantitative evaluations. Further comment on these problems will be made in Chapter 6.

2.5.4 Dependence on Metabolism as a Criterion for Active Transport

At the end of Section 2.5.1 it was stated that in active transport metabolic energy directly affects the movement of particles. The transport of a species of particles, j_1, due to a gradient established by active transport of a species of particles, j_2, in itself is not active transport. Biological systems usually are much more complicated than this simple example, and in many cases it is not possible to determine unequivocally which species of particle is in effect the primary object of active transport. It is usually much simpler to demonstrate that certain transport processes are somehow coupled with the energy supplying reactions of metabolism, for instance, by showing that they can be retarded by inhibitors of metabolic processes or by altering physiological conditions. Experimentally, the dependence of transport processes on metabolism can be demonstrated using inhibitors, anaerobiosis, or lowering the temperature. If the sites of action of the inhibitors are known exactly, certain conclusions can be made about the role in transport of specific metabolic reactions. In light-dependent transport processes it is important to determine the action spectrum and thus obtain knowledge about the pigment and reaction systems taking part in transfer of the absorbed light energy into other forms which can be utilized for transport processes (Chaps. 8 and 10).

The temperature dependence of transport processes can be expressed using the temperature quotient, Q_{10}, which determines how the transport rate, v, changes when the temperature is raised or lowered by 10°C:

$$Q_{10} = \frac{v_{(t + 10)}}{v_t}. \tag{2.47}$$

If the rate of transport is determined at two temperatures t_1 and t_2 then:

$$\ln Q_{10} = \frac{10}{t_2 - t_1} \ln \frac{v_2}{v_1}. \tag{2.48}$$

To plot the results obtained for the rate of transport at different temperatures, the Arrhenius graph is most suitable where the ordinate shows log v

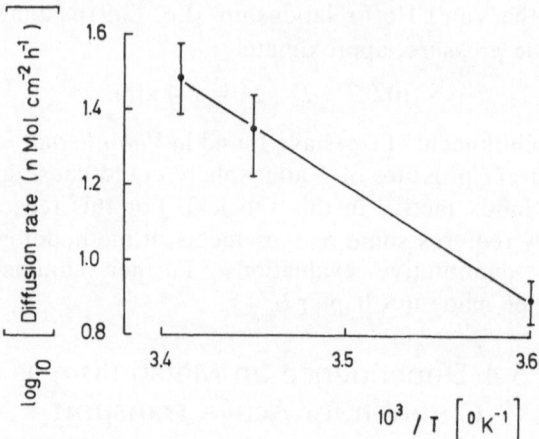

Fig. 2.19. An Arrhenius plot showing the dependence on temperature of passive glucose efflux from the epidermis of onion bulb scales between 5°–20°C. Between 10° and 20° C the Q_{10} is 2.4 and the energy of activation is 14.5 kcal mol^{-1}. (Steinbrecher and Lüttge, 1969.)

and the abscissa $1/T$ (T = absolute temperature in K; Fig. 2.19). Transport processes dependent on metabolism usually have high Q_{10} values similar to those obtained in enzymatic reactions. The Q_{10} of a diffusion process is usually lower. For diffusion in aqueous solution, which depends on the kinetic energy of the diffusing particles (thermal agitation) Q_{10} values between 1.2 and 1.5 are usually obtained. The Q_{10} values for diffusion across membranes are often higher and are found to be between 2 and 3. This will be understandable by considering Figure 2.20. The membrane barrier here is assumed to be a lipid phase. It becomes clear that the hydrophilic particle needs a considerable activation energy, μ_a, to move from the aqueous solution on the outside of the membrane into the lipid phase of the membrane. The activation energy for the exit from the lipid phase into the aqueous solution on the inside of the membrane is much lower, μ_b, (Fig. 2.20a). For hydrophobic or lipophilic particles the situation is reversed and μ_b is larger than μ_a (Fig. 2.20b). The activation energy required can be quite large under certain circumstances, so that even for purely passive downhill diffusion processes Q_{10} values larger than 3 have been measured. The Q_{10} values for the movement of butyramide and ethylene glycol through the cell membrane of eggs of the sea urchin *Arbacia*, for instance, are found to be between 3 and 4. The activation energy, μ_e, can be calculated from the rates of transport measured at various temperatures as:

$$\mu_e = \frac{R\, t_1\, t_2}{t_2 - t_1} \ln \frac{v_2}{v_1} \tag{2.49}$$

where R is the gas constant (in ergs degree^{-1}).

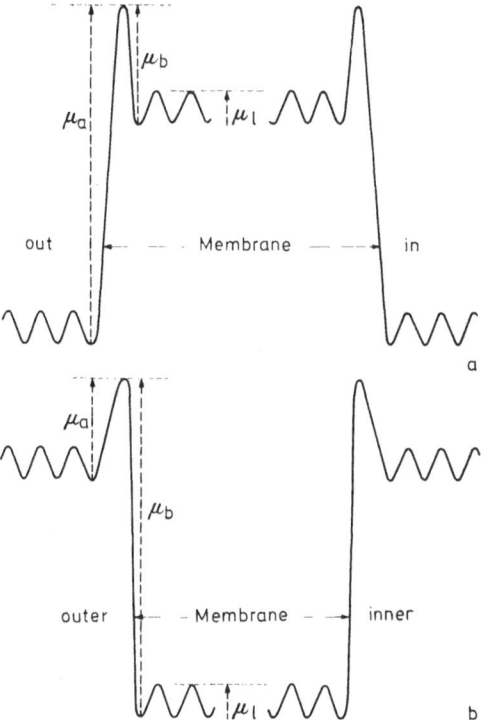

Fig. 2.20. Barriers for movement of a substance across a lipid membrane, after Danielli (from Jennings, 1963): **a** for a relatively hydrophilic and **b** for a less hydrophilic substance. $\mu=$ activation energy required for passing the various barriers, μ_a for entry of the lipid phase of the membrane, μ_b for exit of this phase, and μ_l for diffusion within the lipid phase.

Using Eqs. (2.48) and (2.49) the relation between activation energy and Q_{10} becomes clear:

$$\mu_c = \frac{R \, t_1 \, t_2}{10} \ln Q_{10}. \tag{2.50}$$

A high Q_{10} value therefore is not an unequivocal criterion for the participation of metabolic reactions in the transport process.

The question regarding the metabolic dependence of transport processes appears to be more general and far less rigorous than the specific question of thermodynamics for the immediate driving force of very distinct fluxes of particles. However, the extension of our knowledge to the complex interactions between metabolic reaction sequences involving energy transfer and transport processes leads to a detailed insight into the function of transport in living systems. Some questions we deal with involve thermodynamics, others are concerned with physiology or metabolism; both kinds of questions have to be asked. It is useful to distinguish the answers obtained using criteria of thermodynamics in defining active transport, from those obtained using physiological criteria such as metabolism dependence.

2.5.5 Definitions of Active Transport

Thus far we have demonstrated that a generally applicable criterion for active transport is difficult to find. From physiological investigations more or less extended lists of different criteria have been compiled. Such lists comprise, in addition to the inhibition by metabolic inhibitors and high Q_{10} values, the various properties of carrier transport, namely, obeying of enzyme kinetics [Eq. (2.45)], e.g., saturation, competitive inhibition, and effects of structurally similar substances (Hope and Walker, 1975; Jennings, 1963; Lüttge, 1969). It is very important to note that the systems described above (incongruent transport, negative osmosis, countertransport, and exchange diffusion), which appear to show passive uphill

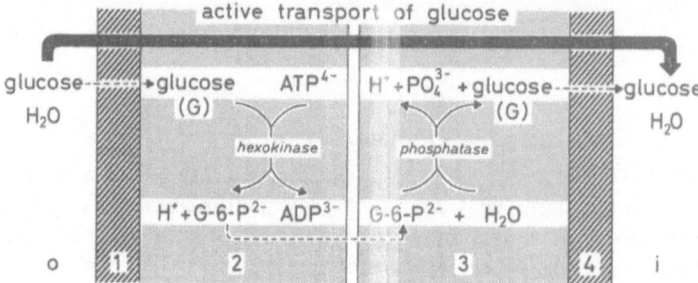

Fig. 2.21. Model of active transport in an asymmetric artificial membrane. The membrane consists of four layers. Layers 1 and 4 are cation exchange membranes which are impermeable to the anions involved in the system (G-6-P^{2-}, ATP^{4-}, ADP^{3-}, PO$_4{}^{3-}$). Layer 2 is a hydrophilic polymer gel phase containing hexokinase, layer 3 is a similar gel containing phosphatase. Both phase 2 and phase 3 contain ATP which cannot leave the sandwich membrane because of the ion exchange properties of layers 1 and 4. Glucose can penetrate these layers and can enter from the outer compartment (o) across layer 1 into layer 2. By phosphorylation of glucose to glucose-6-phosphate (G-6-P^{2-}), a concentration gradient of glucose between o and layer 2 is maintained. G-6-P^{2-} can diffuse into layer 3. By hydrolysis of the phosphate ester in layer 3 a concentration gradient of G-6-P^{2-} is maintained between layers 2 and 3; and glucose can diffuse into the inner compartment (i). As long as the ATP is not exhausted, glucose can move actively from o to i driven by concentration gradients established by the two enzyme reactions. It is important to note that the total amount of glucose present in the system is not affected by these reactions, i.e., the chemical reactions involving glucose in order to transport glucose do not utilize glucose to form other products from it. Glucose concentration is always smaller in 2 than in o, glucose concentration in i may be smaller or larger than in o. (Lüttge and Pitman, 1976; Redrawn from Néel, 1974.)

transport, proceed at the expense of stored energy, and approach a steady state as the concentrations become equal in the two compartments separated by the membrane. These systems are not capable of reaching a dynamic equilibrium or pseudo-steady state which is characteristic of open systems (Bertalanffy, 1953), i.e., the biological membrane systems with active transport in which the losses of energy are continuously replenished by metabolism (see Fig. 1.2b–d). A continuous flow of energy and matter occurs through all living systems. In the state of a dynamic equilibrium this does not alter the internal steady state levels, i.e., input equals output. The importance of metabolism is that its energy-providing role alone makes possible the steady states not in equilibrium with the environment. This aspect is taken into consideration by Kedem (1961) who approached the analysis of active transport by using the formalism of irreversible thermodynamic processes. Kedem assumes that active transport depends on metabolism-dependent chemical reactions which occur in the interior of the membrane, and that all fluxes of matter through the membrane may affect each other, for instance, by frictional interaction (Sect. 2.5.2.1). Such a coupling of chemical reactions and fluxes of matter is possible in asymmetric membranes. In a number of physicochemical laboratories experiments are performed using artificial membranes. The lipoprotein membranes of cells also quite certainly are asymmetric. Active transport occurs, according to Kedem, when a transport of particles is coupled to chemical reactions energetically and the particles themselves do not participate in the chemical reaction. This definition of active transport is well illustrated by the experiment depicted in Figure 2.21.

2.6 Symbols and Constants

A	area, e.g., cm²; or, electrical unit of current, amperes
A^-	univalent anion
a_j	chemical activity of ion j
atm	pressure: 76.0 cm Hg (at sea level); 1.013×10^6 dynes cm^{-2}; 0.1013 joules cm^{-3}; 1.013 bars; 98.7 kilo Pascals
bar	pressure; 0.987 atm; 10^6 dynes cm^{-2}
C	concentration, mol liter^{-1}; concentration also may be designated by brackets, e.g. [K$^+$]
C^+	univalent cation
c	concentration, mol liter^{-1}; or subscript or superscript designating cytoplasm
D	diffusion coefficient, cm² s^{-1}
d	dimension of length, e.g., cm
E	electrical potential, usually with reference to a point conventionally designated as zero; millivolts

e   base of natural logarithms, 2.71828

F   Faraday constant; amount of electricity required to discharge one ion equivalent; 96.487 coulombs mol^{-1}; 23.06 kilocal mol^{-1} $volt^{-1}$

Fa   the Falstaff; a unit of flow, liter s^{-1}

G   electrical conductance, mho cm^{-2}

g   gravitational acceleration constant, 980.6 cm s^{-2} (at sea level, at 45° latitude)

g   chord conductance of a membrane (see G)

h   height, meters

I   electrical current, amperes

i   index for interior or inner phase used as a subscript or superscript, e.g., c_i is inner concentration; Φ_j^{io} is the flux of j from inside to outside

J   net flux, mol cm^{-2} s^{-1}; J_j is the flux of species j

J_v volume flow, cm^3 cm^{-2} s^{-1} ($=cm$ s^{-1})

j   solute species, usually a subscript

K   equilibrium constant; K_M is the Michaelis-Menten constant, mol $liter^{-1}$; the concentration giving half-maximal velocity

K   temperature, Kelvin scale

k   reaction rate constants, k_1, k_2, etc.; s^{-1}

L_p coefficient of hydraulic conductivity, cm s^{-1} bar^{-1}

L_{WW}, L_{SS}, L_{WS}, etc. Onsager, or phenomenological, coefficients; flux per unit force

M   membrane, or molar

o   index for outer or exterior phase; subscript or superscript

Osmol   mol (of osmotically effective solute)

P   hydraulic pressure, turgor pressure; bars or atm

P, P_j permeability, permeability of solute j; cm s^{-1}

PD   electropotential difference, millivolts; ($=E_M$, or $E_o - E_i$)

Q   amount of substance, mol, equivalent, gram

Q_{10} temperature coefficient, dimensionless

R   gas constant; 8.3143 joules degrees Kelvin; 1.987 cal mol^{-1} K^{-1}; 0.083141 liter bar mol^{-1} K^{-1}

R   carrier for transported solutes

RT   2437 joules mol^{-1} at 20°C; 2437 liter bar mol^{-1} at 20°C

RT/F   25.3 mV at 20°C

r   Donnan ratio

r   electrical resistance

S, s   solute or substrate

[S]   concentration of S or other substance

T   degrees temperature on Kelvin scale, K

T   symbol for a transported solute

t   temperature, or time

u^+, u^- mobility of a cation or anion

V	volume, cm³
v	index for vacuole; subscript or superscript
v	velocity of uptake, usually in plant tissue micromols gram^{-1} hour^{-1}
w	water
X$^-$	indiffusible anion
x	vectorial coordinate; length, e.g., mm
z	valence of ion, + or −
α, β	coefficients, variable
γ	activity coefficient, dimensionless
Δ	difference value, gradient
δ	differential term
ε	elastic modulus or coefficient (of cell wall); bars
λ$^\pm$	electrical conductivity of C^+ (λ^+), of A^- (λ^-), or $C^+ A^-$ ($\lambda^\pm$)
Λ	equivalent conductivity
μ	chemical potential, joules mol^{-1}
μ_j^s	standard state of μ; joules mol^{-1}
π	osmotic pressure, or potential; bars
ρ_ω	density of water
$\rho_\omega g$	0.0979 bar m^{-1} (20°C, sea level, 45° latitude)
σ	reflection coefficient; dimensionless
τ	matrix
Φ	flux; used here for tracer-measured flux, e.g., Φ_j^{oi} is the flux of j from outside to inside
Ψ	water potential; bars
Ω	electrical resistance; ohm or ohm cm²
ω	(omega) coefficient of solute permeability

2.7 Problems and Answers

Problem 1. A plant tissue is allowed to reach flux equilibrium with an external solution containing 1mM each of K$^+$, Na$^+$, and Cl$^-$; analysis shows that the internal concentrations in mM are: 89 K$^+$, 10 Na$^+$, and 24 Cl$^-$. From the Nernst equation [Eq. (2.19)] what are the predicted Nernst electropotentials (E$_j$) of each ion?

Solution: Assuming that the concentrations are equal to the ion activities (at 20°C):

a. for K$^+$

$$E_K = \frac{RT}{z^+F} \ln \frac{[K^+]_o}{[K^+]_i} = 25.3 \ln \frac{1}{89} = 25.3 \, (-4.49)$$

$$E_K = -114 \text{ mV}$$

b. for Na$^+$

$$E_{Na} = 25.3 \ln \frac{1}{10} = 25.3 \, (-2.3) = -58 \text{ mV}$$

c. for Cl^-

$$E_{Cl} = \frac{RT}{z^-F} \ln \frac{1}{24} = (-25.3)(-3.18) = +80 \text{ mV}$$

Problem 2. With the same values as in Problem 1, above, the cell membrane electropotentials were measured and found to be -109 ± 5 mV. What is the driving force, E_j^p, inward or outward for each of the three ions?

Solution:

a. $E_K^D = E_M - E_K = (-109) - (-113) = +4$ mV
   (directed outward)
b. $E_{Na}^D = (-109) - (-58) = -51$ mV
   (directed inward)
c. $E_{Cl}^D = (-109) - (+80) = -29$ mV
   (directed outward)

Problem 3. The activity coefficients under the conditions as given in Problems 1 and 2 are (approximately; see Lattimer and Hildebrand, 1951): for K^+ at 1.0 mM, 0.98 and at 89 mM, 0.80; for Na^+ at 1.0 mM, 0.98 and at 10 mM, 0.92; for Cl^- at 1.0 mM, 0.98 and at 24 mM, 0.88. What are Nernst electropotential values when activities rather than concentrations are used? Compare with answers in 1a, b, and c.

Solution:

a. $E_K = 25.3 \ln \dfrac{0.98 \times 1}{0.80 \times 89} = -108$ mV

b. $E_{Na} = 25.3 \ln \dfrac{0.98 \times 1}{0.92 \times 10} = -57$ mV

c. $E_{Cl} = (-25.3) \ln \dfrac{0.98 \times 1}{0.88 \times 24} = +78$ mV

Problem 4. Suppose the permeability of K^+, P_K, determined by measuring radioactive tracer fluxes, is 2.0 picomoles cm s^{-1} (2.0×10^{-12} mol cm s^{-1}), and E_M is -104 mV under conditions otherwise as in Problems 1 and 3. To estimate P_{Na} it is possible to measure changes in E_M while keeping $[K^+]_o + [Na^+]_o$ constant but varying the K^+/Na^+ ratio. Such a measurement with 0.1 mM KCl plus 1.9 mM NaCl gave the $E_M = -137$ mV. What is P_{Na}? [See Eq. (2.29) Cl^- is constant and may be omitted for the solution.]

Solution: The α value is 0.1, i.e., P_{Na}/P_K. Since $P_K = 2$ pmol cm^{-2} s^{-1}, $P_{Na} = 0.1 \times 2$ pmol cm^{-2} s^{-1} or 0.2 pmol cm^{-2} s^{-1}.

Problem 5. The osmotic pressure of the cell sap of a plant cell is $\pi^i = 7$ bars (≈ 0.287) Osmol/l) and the osmotic pressure of the surrounding (medium) is $\pi^o = 0.5$ bars (≈ 20.5 mOsmol/l) at t = 20°C.

a. How large is the cell turgor pressure, P, at water flux equilibrium ($J_v = 0$), if the cell membrane is impermeable for all solutes? How large is turgor pressure at zero volume flow, if the reflection coefficients of the solutes in the cell with exosmosis, σ^i, and in the medium with endosmosis, σ^o, are $\sigma^i = 0.70$ and $\sigma^o = 0.90$, respectively?
b. Calculate the initial water (volume) flow, if the cell is suddenly transferred to a medium with an osmotic pressure of 0.3 Osmol/l ($\sigma^o = 0.90$; $L_p = 5 \cdot 10^{-7}$ cm s^{-1} bar^{-1}).
c. If σ^o in (b) were 0.50: which external osmotic concentration would be necessary to produce the same volume (water) flow as in (b)?
d. In another cell a change in cell turgor pressure, P, from 5 bar at equilibrium to 6.5 bar (e.g., by means of the pressure probe; Fig. 2.14) produced a water flow (J_v) of $3 \cdot 10^{-7}$ cm/s. Calculate the hydraulic conductivity, L_p, of the cell membrane.

$$0°C = 273.16 \text{ K}$$
$$R = 0.083143 \text{ 1 bar/mol degree}$$

Solution:

a. At flux equilibrium:

$$P = \sigma^i \pi^i - \sigma^o \pi^o$$

If σ^i, $\sigma^o = 1$:

$$P = \pi^i - \pi^o = 7 - 0.5 = 6.5 \text{ bar}$$

If $\sigma^i = 0.70$ and $\sigma^o = 0.90$:

$$P = 0.70 \cdot 7 - 0.5 \cdot 0.90 = 4.45 \text{ bar}$$

b. The change in the external osmolarity is $\Delta c^o = 300 - 20.5 = 279.5$ mOsmol/l, and the water flow (initially) produced is given by:

$$J_v = L_p \cdot \sigma^o \cdot \Delta c^o \cdot RT$$
$$J_v = 5 \cdot 10^{-7} \cdot 0.90 \cdot 0.2795 \cdot 0.083143 \cdot 293.16 = 3.07 \; 10^{-6} \text{ cm/s}$$

c. In this case the volume flows in both experiments (I and II) are equal. Thus:

$$L_p \cdot \sigma^o_I \cdot \Delta c^o_I \cdot RT = L_p \cdot \sigma^o_{II} \cdot \Delta c^o_{II} \cdot RT$$
$$\sigma^o_I \cdot \Delta c^o_I = \sigma^o_{II} \cdot \Delta c^o_{II}$$
$$\sigma^o_I = 0.90; \; \Delta c^o_I = 279.5 \text{ mOsmol/l}; \; \sigma^o_{II} = 0.5$$

Thus:

$$\Delta c_{ii}^o = \frac{0.90 \cdot 279.5}{0.5} = 503.1 \text{ mOsmol/l}$$

$$c_{ii}^o = 523.6 \text{ mOsmol/l}$$

d. The volume flow, J_v, is given by

$$J_v = L_p \cdot \Delta P,$$

since π^i and π^o remain constant.
P is the change in turgor pressure.
Thus:

$$L_p = \frac{J_v}{\Delta P}$$

$$L_p = \frac{3 \cdot 10^{-7}}{1.5}$$

$$= 2 \cdot 10^{-7} \text{ cm} \cdot \text{s}^{-1} \text{ bar}^{-1}$$

Problem 6. In a pressure relaxation experiment with a *Nitella* internode such as in Figure 2.15 the half-time, $t_{1/2}$, for water exchange between a plant cell and the medium was found to be 2.0 s. The osmotic pressure of the cell sap was $\pi^i = 6$ bars and the volumetric elastic modulus of the cell was $\varepsilon = 200$ bars.

a. Calculate the hydraulic conductivity, L_p, of the cell membrane from these data if the cylindrical internode has a diameter of $2r = 0.5$ mm.
b. During the experiment cell turgor pressure changes by 2 bar. Give the relative change in cell volume, $\Delta V/V$.
c. In an osmotic experiment such as in the last two experiments in Figure 2.15 the addition of 0.1 Osmol/l of an impermeable solute to the artificial pond water caused a water flow of $J_v^I(t = 0) = 7.5 \cdot 10^{-5}$ cm/s (initially), whereas 0.1 mOsmol/l of a permeable nonelectrolyte yielded $J_v^{II}(t = 0) = 5 \cdot 10^{-5}$ cm/s. Calculate the hydraulic conductivity of the cell membrane and the reflection coefficient, σ_s, of the permeable solute.

Solution:

a. L_p is given by [Zimmermann and Steudle, 1978]:

$$L_p = \frac{\ln 2 \cdot V}{A \cdot \pi_{1/2}(\varepsilon + \pi^i)},$$

where $V/A = \dfrac{\pi r^2 \cdot 1}{2\pi r \cdot 1} = r/2$ for a cylindrical cell (1 = cell length).

$$L_p = \frac{0.6931 \cdot 0.025}{2 \cdot 2 \cdot 206} \text{ cm} \cdot \text{s}^{-1} \cdot \text{bar}^{-1}$$

$$L_p = 2.10 \cdot 10^{-5} \text{ cm} \cdot \text{s}^{-1} \cdot \text{bar}^{-1}$$

b. The elastic coefficient, ε, is given by:

$$\varepsilon = V \frac{\Delta P}{\Delta V},$$

and thus

$$\Delta V / V = \Delta P / \varepsilon = \frac{2}{200} = 1\%$$

c. At $t = 0$ the driving forces for water flow are $\Delta \pi_i^0$ (impermeable solute) and $\sigma \cdot \Delta \pi_s^0$ (permeable solute). Thus, for the impermeable solute $J_v^I(t = 0)$ is given by:

$$J_v^I(t = 0) = L_p \cdot \Delta c_i^0 \cdot RT$$

$$L_p = \frac{J_v^I(t = 0)}{\Delta c_i^0 \cdot RT}$$

$$= \frac{7.5 \cdot 10^{-5}}{0.1 \cdot 0.083143 \cdot 293.16}$$

$$= 3.08 \cdot 10^{-5} \text{ cm s}^{-1} \text{ bar}^{-1}$$

For the permeable solute we may write:

$$J_v^{II}(t = 0) = L_p \cdot \sigma_s \, \Delta c_s^0 \cdot RT$$

$$\sigma_s = \frac{J_v^{II}(t = 0)}{L_p \cdot \sigma_s \cdot \Delta c_s^0 \, RT}$$

or

$$\sigma_s = \frac{J_v^{II}(t = 0)}{J_v^I(t = 0)} \frac{\Delta c_i^0}{\Delta c_s}$$

$$\Delta c_i^0 = \Delta c_s^0$$

$$\sigma_s = \frac{5 \cdot 10^{-5}}{7.5 \cdot 10^{-5}} = 0.667$$

Chapter 3

The Materials of Transport

3.1 Introduction

Many materials, inorganic and organic, may be absorbed and translocated by plants. Some time ago it was generally believed that photosynthetic autotrophs required only mineral elements for growth; however, it is now known that vitamin B_{12}, an organic compound containing cobalt, is essential for a number of algae (Wiessner, 1962). Heterotrophic plants also must be supplied with an external source of organic matter as well as minerals. Most woody plants and some other groups such as orchids have mycorrhizae, and it may be presumed that there is a complex exchange of materials between the two component species. Thus in the discussion to follow, minerals and organic substances are considered.

3.2 Minerals

3.2.1 The Environment as the Source

For land plants the chief source of water and minerals is the soil; equally important is the atmosphere, which supplies photosynthetic plants with CO_2 and receives O_2 and transpired water. Considerable amounts of cycling may occur, particularly of water and CO_2 in this plant–soil–atmosphere system. With aquatic plants, especially unicellular forms, absorption is more direct and cycling may be simpler.

Absorption and translocation is highly selective, so that the internal concentrations of mineral elements accumulated by the plant bear little or no relation to the concentrations available in the environment. For example, in plants the K^+/Na^+ ratio usually exceeds 1 and commonly is

5–20, whereas in the environment Na^+ generally exceeds K^+ (Epstein, 1972). This is also true for marine plants, halophytes, although in the ocean water the K^+/Na^+ ratio is about 0.022 (see Chap. 6, Table 6.3). Other elements low in concentration in the environment may be accumulated to quite high values in the plant. Elements may occur in excess in the environment, e.g., Na^+ and Cl^-; species or varieties may differ widely with respect to salt tolerance. The differences appear to be genetic and, of course, are very important ecologically as well as in crop physiology. Such adaptations extend to ions other than Na^+ and Cl^- (Epstein, 1972); some work has been done on the genetics of membrane transport and salt tolerance of higher plants (Sect. 5.3.2).

Cycling of elements is very important in several ways. Soluble minerals of soils, lakes, and oceans must come initially from the slow dissolution of parent rock material. Organisms absorb the elements necessary for their growth but on death and decay these elements are released again; however, they are generally retained somewhere in the relatively thin layer on earth which we term the biosphere. Without the heterotrophic organisms of decay we might imagine that most life would stop because of the accumulation of dead bodies on the earth's surface and in bodies of water; this would also lead to a depletion of available external nutrients (Fig. 3.1).

Each of the elements has a cycle but some appear to be more critical than others, e.g., the carbon and nitrogen cycles (Delwiche, 1965). Water turns over most rapidly and, in fact, J.B. van Helmont in the 17th century concluded that water alone was responsible for the growth of a willow tree to 164 lbs.; the 200 lbs. of soil in which the plant was grown for five

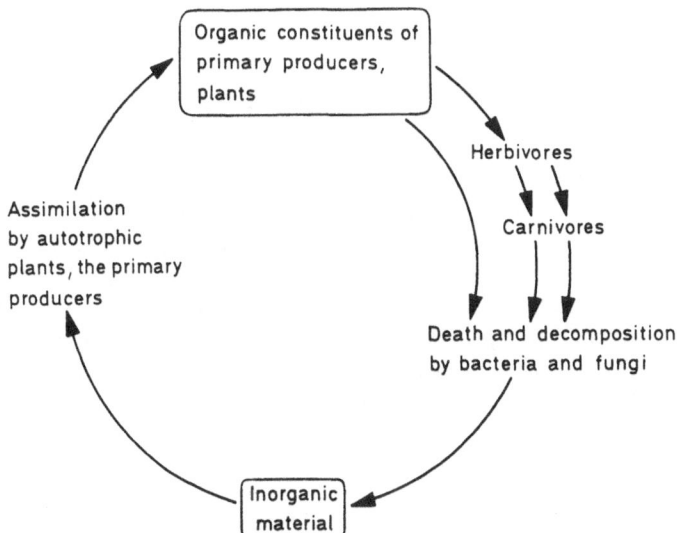

Fig. 3.1. The general cycle of elements in organisms and the environment

years lost only 2 ounces (Gabriel and Fogel, 1955). Of course, van Helmont did not know about carbon-fixation at the time but the experiment illustrates that most of the plant weight is due to water or to H and O compounds and relatively only small quantities of the soil minerals are required, essential as they are. Nonetheless, the amount of water in a plant is a small fraction of that passing through the plant at any given time during its development. Water is absorbed and is translocated throughout the plant carrying solutes up the xylem and down the phloem; in the leaves, because of the water potential gradient, most of the water is lost to the atmosphere by transpiration. A water molecule in rain reaching the plant root may have been many miles away the previous day; a water molecule evaporating from the leaf may be miles removed in a short time. Most of the other minerals, excepting CO_2, are not so mobile; however, wind currents from the ocean carry a very significant burden of NaCl, which tends to replenish these elements in the soil from which they and other ions are continually leached. Rain also may leach small amounts of minerals from the leaves of plants (Tukey, 1970) as well as from the soil.

The flow of water through the plant has been termed a necessary evil. Plants do vary with respect to their efficiency in water usage, that is the ratio between the mass of transpired H_2O per unit of dry matter produced. For example, this ratio may range from 74 in a C_4 plant to 840 in a C_3 plant (Lange et al., 1976) (for an explanation of C_4 and C_3 metabolism see Sect. 10.2.1).

Elements such as carbon, nitrogen, and sulfur occur in various states of oxidation and reduction. The cycles of nitrogen and sulfur are shown in Figures 3.2 and 3.3 respectively. It is an interesting fact that phosphorus is found only as PO_4 and although the charge may vary from 1–3, the PO_4 is not reduced.

3.2.2 Elements Required for Growth

Of all the mineral elements known, only a few are required for plant and animal life. In many cases it is difficult to establish the essentiality of certain elements and it now seems quite clear that there is not a proscribed list of elements essential for all species of living organisms, i.e., species may differ in their requirements. Generally speaking, however, the absence of an element essential to growth and reproduction leads to quite definite deficiency symptoms; in cultivated plants these symptoms have been carefully described (Baumeister, 1958; Stiles, 1958; Sprague, 1964; Baule and Fricker, 1967). Table 3.1 lists the usual nutrient elements and some related information.

3.2.2.1 Major Elements

With respect to plant nutrition it has been convenient to refer to those elements utilized in larger amounts from the substrate as major elements, or

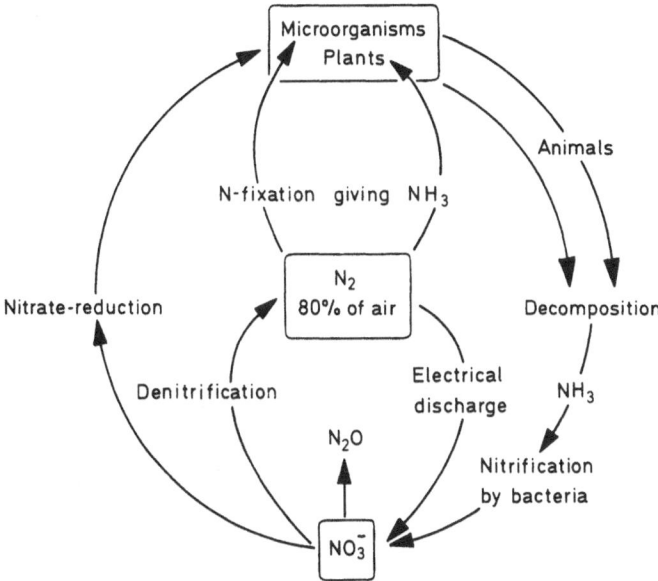

Fig. 3.2. The nitrogen cycle

macronutrients: minor elements, or micronutrients, are those required in much smaller quantities. Water and CO_2 are considered separately, but, of course, in this context would have to be called macronutrients. The major elements are K, Ca, Mg, N, P, and S; ordinarily N is absorbed as NO_3^- or NH_4^+, P as $H_2PO_4^-$, and S as SO_4^{2-}. Some lower forms, algae, fungi, and bacteria, appear not to need Ca (Epstein, 1972).

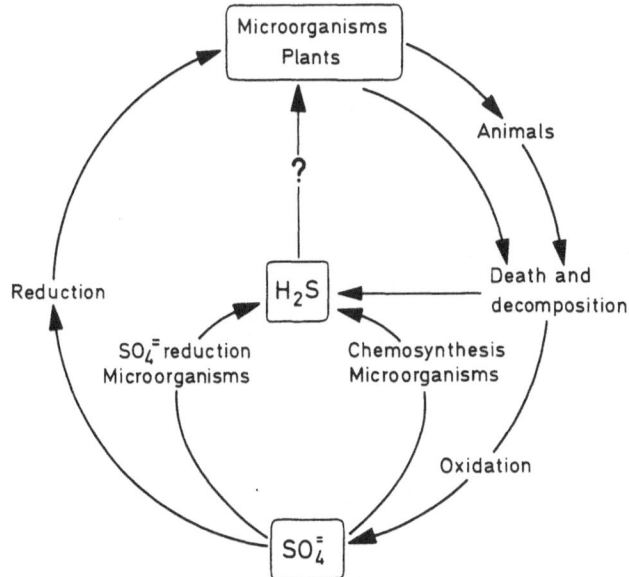

Fig. 3.3. The sulfur cycle

Table 3.1. Concentrations of nutrient elements in plant material at levels considered adequate.

Element	Chemical symbol	Atomic weight	Concentration in dry matter		Relative number of atoms with respect to molybdenum
			μmol/g	ppm or %	
				ppm	
Molybdenum	Mo	95.95	0.001	0.1	1
Copper	Cu	63.54	0.10	6	100
Zinc	Zn	65.38	0.30	20	300
Manganese	Mn	54.94	1.0	50	1,000
Iron	Fe	55.85	2.0	100	2,000
Boron	B	10.82	2.0	20	2,000
Chlorine	Cl	35.46	3.0	100	3,000
				%	
Sulfur	S	32.07	30	0.1	30,000
Phosphorus	P	30.98	60	0.2	60,000
Magnesium	Mg	24.32	80	0.2	80,000
Calcium	Ca	40.08	125	0.5	125,000
Potassium	K	39.10	250	1.0	250,000
Nitrogen	N	14.01	1,000	1.5	1,000,000
Oxygen	O	16.00	30,000	45	30,000,000
Carbon	C	12.01	40,000	45	40,000,000
Hydrogen	H	1.01	60,000	6	60,000,000

Source: Epstein (1972). Copyright 1965 by Academic Press, New York and London.

For halophytes Na and Cl also would have to be included as major elements since many species, a few terrestrial and many marine, absorb these elements in substantial amounts (Flowers et al., 1977). Most terrestrial and fresh water plants from the algae to flowering plants apparently can complete their life cycle without Na. Cl has been shown to be an essential minor element for tomatoes, presumably because it plays a role in photosynthesis (Warburg and Lüttgens, 1946). Because of the ubiquitous occurrence of elements such as Na and Cl, rigorous procedures were necessary to exclude Cl in order to establish its essentiality (Broyer et al., 1954). Brownell and Crossland (1972) have evidence that Na may be a micronutrient for plants having the C_4 photosynthetic pathway but not for C_3 plants. Later they (1974) found a need for Na in plants performing Crassulacean acid metabolism (CAM). Lunt (1966), in a review, also suggests that Na may function in photosynthesis. Although Na and Cl need not be supplied in substantial amounts these elements are generally tolerated well at macronutrient levels. Both halophytes and nonhalophytes (glycophytes) have developed Na^+ exclusion or sequestering processes (Lüttge, 1975).

3.2.2.2 Minor Elements

A number of elements are required in trace amounts; however, the occurrence of an element in an organism is not proof that it is essential. For plants generally it seems that the following elements must be present in small amounts: Fe, Mn, Zn, Cu, Cl, B, and Mo. Some algae must have one or more of the following: Co, Si, I, and V; some higher plants need Se and Si (Epstein, 1972). Certain flagellated algae require vitamin B_{12} which contains a Co molecule (Hutner et al., 1949), but species of blue-green algae may utilize inorganic Co (Holm-Hansen et al., 1954).

3.2.2.3 Nutrient Solutions

Table 3.1 provides some idea of the requirements qualitatively and quantitatively for growth of plants. Many years of research were spent, however, before it became possible to grow plants in artificial media apart from the soil, the fresh water, or the marine environment. The search for an artificial well-defined medium in a modern sense can be said to have begun with the work of Julius von Sachs and W. Knop; this search perhaps reached a certain peak of activity and sophistication in the laboratories of D.R. Hoagland and co-workers at Berkeley (Hoagland and Arnon, 1950; Johnson et al., 1957). Certainly in the latter case, nutrient solution requirements and procedures were so well defined that they became a useful tool for scientists generally wishing to culture plants.

A very widely used liquid culture medium for higher plants is the Hoagland's No. 2 solution (Hoagland and Arnon, 1950). The composition of this solution, modified by Johnson et al., (1957), plus some other information, are provided in Table 3.2 (from Epstein, 1972). The novice in making up nutrient solution should note that care in the mixing is required if precipitation does not result with additions of the aliquots of the stock solutions; maximum dilution is desirable when adding Ca^{2+}, SO_4^{2-}, $H_2PO_4^-$ and micronutrient stocks. Also the solution is best suited to plants beyond the seedling stage and it is not uncommon to use a $1/2$ or $1/10$ dilution of the final concentration for growth of seedlings. Sutcliffe (1976a, b) has shown that the seed reserves of various ions are important in early development and this may account for the differing requirements during early growth. Also seedlings generally require no micronutrients during the first 7–20 days, except possibly for Fe, as indicated by the growth response. Several other nutrient mixtures have been compiled by Hewitt (1963).

Many short-term experiments on transport with excised roots of tissue segments have been performed in solutions with only one or two salts (Epstein, 1966) or in the absence of micronutrients (Higinbotham et al., 1967). This perhaps is justified by the fact that the reserves in the tissue of the various nutrient salts, plus those of the solution containers, and the atmosphere often provide an adequate source. Special purification

Table 3.2. Nutrient solution for higher plants.

Macronutrients

Compound	Molecular weight	Concentration of stock solution, M	Concentration of stock solution, g/l	Volume of stock solution per liter of final solution, ml	Element	Final concentration of element, μM	Final concentration of element, ppm
KNO_3	101.10	1.00	101.10	6.0	N	16000	224
$Ca(NO_3)_2 \cdot 4H_2O$	236.16	1.00	236.16	4.0	K	6000	235
$NH_4H_2PO_4$	115.08	1.00	115.08	2.0	Ca	4000	160
$MgSO_4 \cdot 7H_2O$	246.49	1.00	246.49	1.0	P	2000	62
					S	1000	32
					Mg	1000	24

Micronutrients

Compound[a]	Molecular weight	Concentration of stock solution, mM	Concentration of stock solution, g/l	Volume of stock solution per liter of final solution, ml	Element	Final concentration of element, μM	Final concentration of element, ppm
KCl	74.55	50	3.728		Cl	50	1.77
H_3BO_3	61.84	25	1.546		B	25	0.27
$MnSO_4 \cdot H_2O$	169.01	2.0	0.338	1.0	Mn	2.0	0.11
$ZnSO_4 \cdot 7H_2O$	287.55	2.0	0.575		Zn	2.0	0.131
$CuSO_4 \cdot 5H_2O$	249.71	0.5	0.125		Cu	0.5	0.032
$H_2MoO_4(85\% \ MoO_3)$	161.97	0.5	0.081		Mo	0.5	0.05
Fe–EDTA[b]	346.08	20	6.922	1.0	Fe	20	1.12

Source: Epstein (1972) after Johnson et al. (1957), modified Hoagland solution No. 2.

[a] A combined stock solution is made up containing all micronutrients except iron.

[b] Ferrous dihydrogen ethylenediamine tetraacetic acid.

schemes are generally required to demonstrate the essentiality of a micronutrient. Nonetheless, any specific effect of most micronutrients, or the lack of such effects, on membrane integrity and transport processes remain to be established.

It appears that the various species of lower forms (algae, fungi, and bacteria) have a greater heterogeneity with respect to nutrient requirements; of course it is characteristic of the heterotrophes that either or both organic carbon and nitrogen sources must be supplied, often plus other compounds such as vitamins or unknown factors in addition to the inorganic elements.

O'Kelley (1974) has summarized current information on the requirement of algal species for inorganic nutrients. Epstein (1972) has provided a comparative overview of the mineral requirements of algae, fungi, and bacteria. For recent specific information on nutrient solutions for various algal species, one may write to the following: Indiana University Culture Collection (of algae), Bloomington, Indiana, U.S.A.; Sammlung der Algenkulturen, Pflanzenphysiologisches Institut der Universität Göttingen, 3400 Göttingen, W. Germany.

3.2.3 Other Factors Affecting Nutriculture

In addition to providing elements essential for growth, other factors must be within some limits: pH, temperature, O_2 supply, a suitable water potential gradient, and, for green plants, light of appropriate intensity and duration.

For pH control in solutions, the $H_2PO_4^-$ in the mixture may offer some buffering action. Solutions having an N supply of NO_3^- only may become quite alkaline because of rapid removal of this anion; the effect can be offset by providing some N in the form of NH_4^+, as is the case with Johnson's solution (Table 3.2). Until recent years the sharp changes in pH of the external solution were regarded as cases in which either cations or anions became depleted depending upon selectivity of absorption. Although this may still be true, essentially, the emphasis now is on the concept that H^+ and/or OH^- ions may be actively extruded; this may occur by an exchange transport process (porter system) in which, respectively, cations or anions are driven inward. Some possible relationships between NO_3^- uptake, nitrate reductase, and H^+ transport are discussed further in Chapter 13. If for a given plant and nutrient mixture there is a balance of cations and anions in absorption, then pH change is negligible. If, experimentally, it is desirable to use other mixtures, external pH may be controlled by frequent renewal, frequent titration, continual renewal, or buffering.

The other factors, temperature, light, etc., are dealt with reasonably well in the ensuing chapters.

3.2.4 Metabolic Functions of Elements

An ultimate objective of the biologist is to understand all processes at the molecular level. We are not close to achieving this objective with most elements at present, but have arrived at a framework permitting the formulation of more exact questions. Below we shall consider most of the elements believed to be essential and will try to provide some currently relevant suggestions regarding their metabolic significance; it may be pertinent to remark that without the least of these elements life might not be possible, but with too much of any, life may be equally impossible.

It might be said that a living organism is made of "constitutive" materials, a part of the basic framework, and of materials which are "nonconstitutive", i.e., materials necessary but which cycle through. Actually, no such clear distinction can be made because H_2O or H and O, clearly make up the bulk of each (Table 3.3). C, N, and S follow H_2O (or H and O) in the ranking of constitutive structural elements.

3.2.4.A Water, H_2O

Water is said to be the universal solvent and is the continuum of living protoplasm with the external environment. The only possible exceptions are those bacteria utilizing hydrocarbons such as petroleum and asphalt

Table 3.3. Elemental composition of some organisms and substances (Epstein, 1972).

| Element | Per cent of dry weight | | | | |
	Corn plant, *Zea mays*	Man, *Homo sapiens*	Carbohydrate	Fat	Protein
O	44.43	14.62	51.42	11.33	24
C	43.57	55.99	42.10	76.54	52
H	6.24	7.46	6.48	12.13	7
N	1.46	9.33	—	—	16
Si	1.17	0.005	—	—	—
K	0.92	1.09	—	—	—
Ca	0.23	4.67	—	—	—
P	0.20	3.11	—	—	—
Mg	0.18	0.16	—	—	—
S	0.17	0.78	—	—	1
Cl	0.14	0.47	—	—	—
Al	0.11	—	—	—	—
Fe	0.08	0.012	—	—	—
Mn	0.04	—	—	—	—
Na	—	0.47	—	—	—
Zn	—	0.010	—	—	—
Rb	—	0.005	—	—	—

(van der Linden and Thijsse, 1965) but even in these cases, evolution is still conceived to be continuous and all, or nearly all, of the enzymes are, at least partially, hydrophilic like those in protoplasts of other living species.

Water as H_2O makes up the bulk of most living organisms. In higher plants it is necessary for turgor pressure, without which the plant either cannot grow, or wilts and loses its typical form, and may lose the capacity for both short- and long-distance transport of materials. In this respect it is nonconstitutive, but, in other respects it is constitutive, particularly in the form of carbohydrates, and, as such, still composes the bulk of the "dry" matter of most plants.

H and O constitute an integral part of all organic carbon compounds; only in certain hydrocarbons, e.g., carotenoids, is O reduced to low levels or to none.

3.2.4.B Carbon, Nitrogen, Phosphorus, and Sulfur

The elements C, N, P, and S are essential constituents of organic compounds.

3.2.4.C Potassium

Generally speaking, K^+ is the most mobile element in plants and does not occur as part of any stable organic compound. The hypothesis that K^+ and Na^+ are firmly bound to protoplasmic constituents (Ling and Cope, 1969) has been shown to be contrary to the evidence, which indicates that they are present as free ions (Magnuson et al., 1973; Palmer and Gulati, 1976).

K^+ serves at least two roles: it is an important osmotic constituent of cell sap and is an activator or cofactor of many enzymes including ATPases. Osmotically, K^+ is important in growth since it is accumulated to quite high concentrations and, of course, turgor is required for cell enlargement. Similarly, K^+ transport appears to be a process essential for the motor activity in opening and closing of guard cells (Fujino, 1967; Humble and Raschke, 1971) and for leaf movements (Satter, Marinoff et al., 1970; Satter, Schrempf et al., 1977); in the latter case, about 0.6 of the K^+ movement is accompanied by Cl^- transport, apparently with other unidentified anions balancing the remaining 0.4 K^+. In both the guard cells and leaf pulvinus motor cells, the osmotic concentration of K^+ reaches values adequate to explain the mechanical movements by turgor changes. It is presumed that the K^+ transport system may be a general phenomenon in motor activity in plants. Gradmann and Mayer (1977) have recently challenged the view that K^+ is the osmotically important ion in the motor activity of guard cells and leaf pulvini. They argue that the permeability of K^+ is too high and that Cl^- and organic acids are more likely to serve as the important solutes osmotically.

As an enzyme activator—or, perhaps, as a cofactor—K^+ is effective for many different enzymes (Evans and Sorger, 1966). There seems to be some correlation between the rather high internal K^+ concentration in cells and the K^+ concentration in vitro giving maximal enzyme activity; higher concentrations of K^+, often up to 300 mM or more, fail to reduce enzyme activity and this is commensurate with K^+ values in cells. An interesting question is: did the primeval cell first develop a membrane and transport system resulting in high internal K^+ concentrations with a subsequent evolutionary adaptation of enzymes to such values? Or, conversely, did the evolution of the enzymes from the beginning require such concentrations of K^+ for maximum activity as well as for regulatory processes? The answer may well be that the evolution of the two were concurrent. Certainly, other univalent cations in the same series, e.g., Na^+, Rb^+, and Cs^+ generally fail to provide enzyme activation commensurate with that of K^+. Rb^+ is the best substitute for K^+ whereas Na^+ may actually be inhibitory. One might speculate that basically there are principles governing enzyme structure and activity which favor the less hydrated K^+ ion over that of the larger Na^+ ion (Evans and Sorger, 1966). In this case, evolutionary selection would favor the adaptation of cell membrane and special organs serving to extrude, exclude, or sequester Na^+ from the working machinery of the plant. Similar questions may be raised about the evolutionary significance of the transport of other ions, e.g., H^+ and Ca^{2+} (see Raven, 1977c; and later chapters).

3.2.4.D Calcium

Although Ca^{2+} must be considered a macronutrient, the concentration of free ions in the protoplast may be quite low (Jones and Lunt, 1967). A large proportion of the Ca^{2+} accumulated may be as pectate salts in the wall or as insoluble salts within cells, e.g., Ca-oxalate crystals. Since these crystals are membrane-bound (Schötz et al., 1970) they may represent a process regulating the amounts of free Ca^{2+}. Ca^{2+} is quite immobile, not being readily translocated in the phloem, and consequently can have a very uneven distribution in the plant (Loneragan and Snowball, 1969).

Perhaps the most important function of Ca^{2+} is in stabilizing cell membranes. Ca-deficient plants examined by electron microscopy show a loss of cellular membranes and impairment of their organization (Marinos, 1962; Marschner and Günther, 1964). With omission of Ca^{2+} from a nutrient solution the other ions appear to be toxic; this may be an effect on membrane integrity since under these conditions cells may be quite leaky. The exact mechanism by which Ca^{2+} externally sustains membrane stability is not known; it has been shown that Ca^{2+} may promote the absorption of other ions, e.g., K^+ (Viets, 1944) and, under some conditions, cause hyperpolarization of cell PD (Higinbotham et al., 1964) which tends to drive cations in.

It appears that with certain exceptions (see Table 3.4) Ca^{2+} may not be an important activator of enzymes but, on the contrary, may tend to inhibit some enzymatic processes which require Mg^{2+} nonspecifically (Evans and Sorger, 1966). In this respect, Ca^{2+} in the cytoplasm is subject to some control mechanism; the lack of mobility of Ca^{2+} in the plant suggests that there may be evolutionary adaptations of membranes tending to regulate the amount of free Ca^{2+} in the protoplast (Raven, 1977c). Macklon (1975b; Macklon and Sim, 1976) has evidence from fluxes and electrochemical potential gradients that both Ca^{2+} and Mg^{2+} enter cells by passive diffusion but are actively pumped out.

For other possible functions of Ca^{2+} the books by Epstein (1972) and Gauch (1972) should be consulted.

3.2.4.E Magnesium

The divalent Mg^{2+} ion behaves very differently from Ca^{2+} and appears not to have a stabilizing effect on the membrane; it is a nonspecific activator of many enzymes (Evans and Sorger, 1966) including most ATPases and other enzymes transferring PO_4 (Epstein, 1972). It is also a constituent of chlorophyll, and about half the Mg^{2+} in a leaf is in the chloroplasts (Stocking and Ongun, 1962).

3.2.4.F Sodium, Lithium, Rubidium, and Cesium

With the exception of Na^+, which is required for certain halophytes, these elements may occur in plants simply because they are in the environment but they do not appear to be essential. Certain ATPases of animal origin, which serve a selective transport function in pumping Na^+ out and K^+ in, are activated synergistically by Na^+ plus K^+ (with Mg^{2+}; Glynn et al., 1971). Similar ATPases in plants whose activities are enhanced by Na^+ and K^+ have been reported by Hansson and Kylin (1969) from microsomal isolations but not by Hodges (1976) from isolated plasmalemma vesicles. That plants have an Na^+ extrusion system seems clear, but an Na^+–K^+ transport ATPase like that of erythrocytes has not been found. With many enzymes the order of effectiveness in activation is: $K^+ >$ $Rb^+ > Ca^+$; NH_4^+, Na^+, and Li^+ often show some enhancement but not uncommonly Na^+ and Li^+ inhibit the reaction (Evans and Sorger, 1966). It has been suggested that this effect may be attributed to the hydrated ionic radius which for K^+, Rb^+, and Ca^{2+} is 0.532, 0.537, and 0.509 nm, respectively, whereas the radius for Na^+ is 0.79 nm and for Li^+, 1.00 nm. Such differences in size could well affect the conformation of enzymes and the accessibility of the substrate to reaction sites.

Like K^+ these univalent ions are quite mobile and are readily translocated throughout the plant.

3.2.4.G Other Elements

The functions of a number of elements which are essential for all or some plants are summarized in Table 3.4 (modified from Evans and Sorger, 1966).

There are a number of elements found in plants (including gold) which appear to be fortuitous since they have not been shown to be essential, according to the criteria developed by Hoagland and co-workers (Arnon, 1951). For details see Epstein (1972), Hewitt (1963), and O'Kelley (1974).

In general, any ions in excessive concentration may be detrimental to

Table 3.4. Proposed role(s) of mineral elements (Evans and Sorger, 1966).

Element	Proposed role(s)
N	Constituent of amino acids, amides, proteins, nucleic acids, nucleotides and coenzymes, hexoseamines, etc.
P	Component of sugar phosphates, nucleic acids, nucleotides, coenzymes, phospholipids, phytic acid, etc. Plays key role in energy transfer.
K and other univalent cations	See text
S	Component of cysteine, cystine, methionine, and thus proteins. Constituent of lipoic acid, coenzyme A, thiaminepyrophosphate, glutathione, biotin, adenosine-5'-phosphosulfate and 3'-phosphoadenosine-5'-phosphosulfate, and other compounds.
Ca	A constituent of the middle lamella of cell walls as Ca-pectate. Required as a cofactor by some enzymes involved in the hydrolysis of ATP and phospholipids. An essential component of amylases from certain bacteria, fungi, and animals. Required by polygalacturonic-transeliminase from *Erwinia carotovora* and *Bacillus polymyxa*. Ca deficiency results in increased chromosome fragility. Possible involvement in binding of RNA to protein in chromosomes.
Mg	Required nonspecifically by large number of enzymes involved in phosphate transfer. A constituent of the chlorophyll molecule.
Fe	A constituent of cytochromes. A constituent of nonheme iron proteins, which are involved in photosynthesis, N_2 fixation, and respiratory-linked dehydrogenases.
Mn	Required for activity of some dehydrogenases, decarboxylases, kinases, oxidases, peroxidases, and nonspecifically by other divalent, cation-activated enzymes. Reportedly required by nitrite and hydroxylamine reductases, but not in higher plants. Required for photosynthetic evolution of O_2.
B	Indirect evidence for involvement of B in carbohydrate transport. Deficiency in *Lycopersicon esculentum* results in decreased RNA content. Borate forms complexes with certain carbohydrates, but natural borate complexes in plants have not been identified.

Table 3.4. (*Continued*)

Element	Proposed role(s)
Cu	An essential component of ascorbic acid oxidase, tyrosinase, laccase, monoamine oxidase, uricase, cytochrome oxidase, and galactose oxidase. Component of plastocyanin from *Spinacia oleracea*. Deficiency in algae causes decreased activity of cytochrome photooxidase.
Zn	Essential constituent of alcohol dehydrogenase, glutamic dehydrogenase, lactic dehydrogenase, carbonic anhydrase, alkaline phosphatase, carboxypeptidase B, and other enzymes.
Mo	A constituent of nitrate reductase of fungi, bacteria and higher plants, and of xanthine oxidase, aldehyde oxidase from animal sources, and essential for N_2 fixation.
Cl	Required for photosynthetic reactions involved in oxygen evolution.
Co	Essential for some microorganisms, free-living, N_2-fixing microorganisms, and nonleguminous and leguminous N_2-fixing symbionts. A constituent of vitamin B_{12} coenzymes.
Si	Appears to function as a structural component of diatoms.
V	Role unknown; required, however, by *Scenedesmus obliquus*.

growth or have a toxic effect. The ecological limitations of saline soils (Waisel, 1972; Gauch, 1972) and marine environments to halophytes are quite well known. There are a number of situations, for example the slack pile of mine dumps, in which there may be an unusually rich supply of certain elements such as Fe, Ni, Cu, Zn, etc. Although these elements at higher concentrations are toxic to most plants, some species appear to have developed a high tolerance and act as "accumulators" (Gauch, 1972). *Hybanthus floribundus* (Violaceae) can accumulate Ni to 1% of its dry weight (Severne and Brooks, 1972). Different species of lichens show individual patterns of uptake but can grow on mine tailings and accumulate high concentrations of Fe, Zn, and Cu (Noeske et al., 1970).

3.2.5 Site of Absorption and Circulation in the Plant

In aquatic organisms, such as algae and many of the simpler terrestrial plants, there may be no highly specialized organs of absorption and translocation; certain aquatic flowering plants do have glands, hydropotes, which show a transport function (Lüttge, 1973a; see Fig. 7.21). In vascular plants the roots are especially adapted for absorption and translocation as well as anchorage. While roots absorb water and minerals from the substrate, shoots, particularly leaves, absorb CO_2 from the atmosphere. Despite the evolution of specialized organs, the cells of all parts of the plant are capable of absorbing those materials available to them in the apoplast; this is particularly relevant for O_2 and CO_2 exchange, but also applies to H_2O and solutes generally.

The pathway for H_2O and dissolved minerals in vascular plants is: entry into the root hairs and epidermis, to the xylem via the symplast, and up the xylem to the leaves. Along the pathway there may be exchange or loss of both water and solutes. With turgid photosynthetic plants, there is a "loading" of the phloem, especially with sucrose from photosynthesis, and a movement via the apoplast to the adjacent phloem and downward back to the roots; with growing plants or plants with developing fruits there is, in addition, significant upward transport through the phloem to these structures. Nonetheless, the bulk of the circulation in plants is upward in the xylem and downward in the phloem; solutes carried upward which are not absorbed by cells in the shoot are either extruded there (e.g., by guttation, by salt glands, or leached) or are recirculated to the root, thus tending to suppress further absorption from the exterior (see Chaps. 12 and 13).

During circulation in the plant, selective lateral transport seems to occur all along the pathway so that the concentrations of various solutes in the xylem and phloem saps may be quite different (Table 3.5). Furthermore the xylem sap is not necessarily restricted to mineral solutes but may also carry sugars and other organic materials (see next section). Of course, the phloem is adapted for the conduction of organic solutes and

Table 3.5. Comparisons of phloem and xylem sap composition in two species of annual lupin (Pate, 1975).

	Lupinus albus		*Lupinus angustifolius*	
	Xylem sap (Tracheal)	Phloem sap (Fruit bleeding)	Xylem sap (Tracheal)	Phloem sap (Fruit bleeding)
mg ml^{-1}				
sucrose	—	154	—	171
amino acids	0.70	13	2.6	15
μg ml^{-1}				
potassium	90	1,540	180	1,820
sodium	60	120	50	101
magnesium	27	85	8	140
calcium	17	21	73	64
iron	1.8	9.8	1.0	7.0
manganese	0.6	1.4	0.4	0.6
zinc	0.4	5.8	0.7	5.5
copper	Trace	0.4	Trace	0.2
nitrate	10	—	31	Trace
pH	6.3	7.9	5.9	8.0

the phloem sap generally has a greater variety and higher concentration of solutes than the xylem sap. More information on the methods for obtaining phloem exudate and xylem sap is given in Chapters 11 and 12.

3.3 Organic Materials

The organic materials translocated in greatest amount are quite certainly the sugars; however, a number of other organic compounds are subject to short- or long-distance transport. The total concentration of organic and inorganic solutes in the phloem sap is -14 to -40 bars (Ziegler, 1975).

3.3.1 Sugars

In vascular plants the carbohydrates produced in photosynthesis are quite generally translocated from the leaves to other parts of the plant, e.g., roots, meristems, developing fruits, etc., in a "source-to-sink" relationship. The long-distance movement is generally in the phloem but in deciduous trees, such as maple in the early spring, sugar may occur in quantity as part of the xylem sap (Sauter et al., 1973).

Both sugars and sugar alcohols are translocated in the phloem with differences in composition being related to taxonomic groups. Ziegler (1975) lists three main types: (1) species with sucrose as the predominant sugar; (2) species with sucrose and oligosaccharides (raffinose); and (3) species which contain sucrose and some raffinose, but in addition a sugar alcohol such as mannitol or sorbitol. Zimmermann and Ziegler (1975) have compiled a survey of more than 500 species representing about 100 families of the Dicotyledonae. The sugars and derivatives for which the phloem exudates were examined were: sucrose, raffinose, stachyose, verbascose, ajugose, D-mannitol, sorbitol, dulcitol, and myoinositol.

It is very interesting to note that glucose and fructose are absent or in negligible amount in the phloem; this appears to be true for all reducing sugars (Ziegler, 1975). The reason for this is not clear but among other ideas it has been postulated that the reducing sugars are selectively excluded from sieve elements in the loading process (vein loading). Accumulation of sugars, especially sucrose, may occur against a concentration gradient and by various criteria appears to be an active process in higher plant tissue (Geiger, 1975; Lüttge and Schnepf, 1976). Only in the green alga *Chlorella* has an active hexose transport system been well characterized (Komor and Tanner, 1971), although hexoses do occur in exudation from glands such as the nectaries, and this is an active process (Lüttge and Schnepf, 1976; see Chap. 12).

Needless to say, the phloem sieve tubes in evolution have become especially adapted for translocation (see Sect. 11.4). Among other features in vascular plants, the tonoplast is lost and the plasmalemma is selective for transport of certain sugars. The sugars selected for transport are not consumed in the process or converted to starch showing a biochemical modification of the protoplast; one such adaptation is the absence of invertase (Eschrich and Heyser, 1975).

The evolution of a specialized living conducting system in the brown algae, the kelps, seems unique in the lower plants and apparently developed independently of phloem origin in vascular plants. The "sieve tubes" (or "trumpet" cells) apparently retain a nucleus and the tonoplast (Willenbrink, 1976); they translocate mannitol primarily (Parker, 1966) but also appreciable amounts of serine and glycine (Willenbrink, 1976).

3.3.2 Amino Acids and Other N-Compounds

Significant amounts of several amino acids and amines are found in the phloem sap (Ziegler, 1975) and undoubtedly constitute a long-distance source-to-sink relationship. Predominant are glutamine, glutamate, asparagine, aspartate, and serine; the content appears to increase or decrease as does sugar. In legumes, the N fixed in the nodules is apparently transported via the xylem primarily as asparagine, glutamine, and aspartate (Pate, 1976); this transport appears to be selective and active since it is against the concentration gradient and certain amino acids in the nodule, such as glutamate and glycine, are not exported in significant amounts: the selectivity presumably resides in the parenchyma leading to the xylem.

There is some evidence that several amino acids are absorbed quite readily by higher plant cells (Birt and Hird, 1958; Cseh and Böszörmenyi, 1964; Borstlap, 1974; Cockburn et al., 1975); this should receive further study but it is consistent with the idea that amino acids are mobile and can be readily exchanged between phloem and adjacent tissue.

In certain species other N-compounds of low molecular weight may be found in the phloem; Ziegler (1975) states that selectivity of the phloem to smaller N-compounds has not been shown, in contrast to the case with sugars.

Proteins, including some with enzymatic activity, are also found in phloem exudate but there is doubt that they are translocated (Ziegler, 1975); it is unlikely that they would pass through cell membranes.

3.3.3 Organic Acids

Although a number of organic acids are found in phloem sap they are present only in small amounts; Ziegler (1975) believes that the sink tissues are not dependent upon long-distance transfer of acids but rather utilize sugar.

Short- and medium-distance transport of various organic acids undoubtedly occurs either across membranes or via plasmodesmata (see Chap. 11). Many plants or plant parts are strongly acidic due to the accumulation of malic, citric, or other acids intermediate in the Krebs cycle. Oxalic acid is often found in cells as membrane-bound crystals of the Ca salt (Sect. 3.2.4.D).

A model involving organic acid translocation and NO_3^- uptake is discussed in Chapter 13.

3.3.4 Growth-Regulating Substances

Growth-regulating substances may have significant effects on membrane transport processes (Pitman, 1975; Ziegler, 1975). This is discussed more fully in Chapter 9.

Cytokinins and gibberellins are produced in the roots, but appear to move readily throughout the plant; they are found in exudates of both xylem (Pitman, 1975) and phloem (Ziegler, 1975).

Abscisic acid and the auxin indoleacetic acid, IAA, are produced in the shoot but can move to the roots; both are found in phloem sap (Ziegler, 1975). IAA is well known for having polar transport over medium distances, e.g., down an *Avena* coleoptile; since phloem transport may occur upward or downward, it seems clear that the polar transport must be in the parenchyma. The possible mechanisms for IAA transport have been reviewed by Goldsmith (1977). A passive diffusion model is developed using electrochemical principles; it is shown that at pH values of 6 and 7 the permeability of the anion, IAA^-, may be on the order of 10^{-3} or 10^{-2} that of the undissociated auxin, IAAH. If the ratio P_{IAA^-}/P_{IAAH} at the upper end of a cell differs significantly from the base of the cell, a polar flux of $IAA^- + IAAH$ can occur until the outer solution at the cell base contains 3.8 times the concentration of the apex.

Goldsmith (1977) also provides a summary of polar transport of other substances; this shows that the following have no polar movement or it is weak: abscisic acid, various kinetins, cyclic AMP, gibberellins, thiamine, and Cl^-.

In addition to the substances mentioned above, there are a number of other compounds, synthetic or naturally occurring, which move in plants. Compounds with auxin-like effects, α-naphthaleneacetic acid, 2,4 dichlorophenoxyacetic acid, 2,4-D, etc., seem to move much like IAA (Crafts and Crisp, 1971). The mobility of a number of herbicides, including 2,4-D, in the xylem and phloem has been compiled by Jacob et al. (1973). Herbicides may have an effect on phloem translocation and on lateral transport in and out of the phloem.

Fusicoccin, a phytotoxin produced by the wilt fungus, *Fusicoccum amygdali*, has many effects like auxins but appears to be more mobile (Ballio et al., 1968; Marrè, Colombo et al., 1974).

With respect to movement of other growth regulators, little is known.

It appears that phytochrome is immobile. It should be added, however, that phytochrome activity is associated with membrane transport of K^+ and Cl^- during organ bending movements (Marmé, 1977; also see Chap. 9). Evidence exists that other unknown substances, such as the flowering stimulus, must be translocated; until these substances are better identified, details of their transport must remain unknown.

3.3.5 Other Organic Materials Transported

As seen in the preceding sections, some organic substances such as amino acids have access, perhaps passive, to the sieve tube through the plasmalemma for long-distance transport, while in other cases, e.g., the various sugars, the transport is active and selective. Passage through membranes depends not only on molecular size and chemical properties, but also on the particular characteristics of membranes governing selectivity; the hypothesis of specific carriers is often invoked to explain such selectivity. In any case, there are no well-defined rules enabling us to predict with certainty whether a given substance will be translocated. Generally, the plasmalemma is quite impermeable to small highly charged ions and to large molecules, such as proteins, whether highly charged or not. Of course this does not preclude movement through the symplast via plasmodesmata or even across the plasmalemma if pinocytosis occurs.

The phloem exudate, which of course flows from living protoplasm, contains proteins, including enzymes, nucleic acids, adenosine phosphates, other organic phosphates, and some vitamins (Ziegler, 1975). It seems unlikely that the proteins, nucleic acids, and nucleotides are translocated, since they would not move readily between the sieve tubes and sink tissues. In fact, it has been shown that ATP content of the phloem is constant over a long distance (10 m in *Tilia*) whereas substances subject to a source-to-sink relationship become depleted due to lateral transport (Ziegler, 1975). The evidence suggests that ATP does not penetrate the plasmalemma (Lüttge, Schöch et al., 1974).

In the case of the vitamins, thiamine, pantothenic acid, niacin, and pyridoxine, there appears to be genuine long-distance transport. These are known to be required, at least in part, for culture of isolated roots which cannot synthesize them; in the intact plant the shoot apparently synthesizes these substances and exports them to the roots (Street, 1966; Ziegler, 1975). The opposite situation apparently occurs in tobacco with the alkaloid nicotine being synthesized in the root and exported to the shoot (Dawson, 1942; Street, 1966).

3.3.6 Coupling Effects

In approaching the problems of transport, the experimenter usually starts with the assumption that ions and molecules move independently. The evidence in many cases indicates otherwise, namely that the movement of one substance is dependent on the movement of another, i.e., there is transport coupling. For example, does the flow of water in transpiration sweep solutes with it, as appears to be the case in some experiments? (See Chap. 11.) Membrane transport also may show coupling of the movement of two substances with one another, and also coupling in some way with the energy-producing reactions of metabolism (see Chaps. 8 and 10).

Part II

Complications of Models by Cellular Structures: Chapters 4 to 7

In Chapter 2 we discussed a simple outside–inside model. It was composed of two compartments which were separated by a boundary layer or membrane. It was somewhat similar to the simple primitive initial "organisms" or coacervate droplets postulated in Chapter 1. We made no assumptions about the structure of the interior phase, we simply assumed that the interior phase, like the exterior phase, was a well-stirred aqueous solution of the particles whose transport was considered. We also made no assumptions about the structure of the boundary layer, although certain properties were ascribed to it. On one occasion it was assumed that the boundary layer was built up from lipid molecules (Sect. 1.1), and on another occasion ion exchange membranes having a matrix with fixed charges were mentioned (Sect. 2.2.2.2). Consideration of the structure and compartmentation of the interior phase will be given in Chapters 6 and 7. It is now important to account for the structure of the boundary layer.

Plant cells are usually surrounded by a more or less rigid cell wall. The external surface and boundary of the cytoplasm of the cell is the plasmalemma, which is more or less attached to the cell wall. Only the plasmalemma or plasma membrane or similar structures of cytoplasmic origin are called a membrane here. The rigid outer cell wall is distinct from the membrane in respect to its structure and function.

By the simple distinction between cell wall and cell membrane we can characterize the two structurally very different boundary layers of a plant cell. In a somewhat different sense we use the term membrane if we talk about artificial "membranes" in physicochemical model experiments. The term membrane in this case may be used for ion exchange membranes whose matrix is more comparable to the plant cell wall, but also for artificial lipid films which are much more like biological membranes in the strictest sense.

Chapter 4

The Cell Wall as a Phase for Transport

4.1 The Structural Basis for Transport in the Cell Wall Phase

The structure of the cell wall must be known before we can understand the role which the wall plays as a phase into and through which transport occurs in absorption by cells and in translocation across tissues.

With a few exceptions (among algae, fungi, and bacteria) plant cell protoplasts are enclosed by a relatively firm wall and, because of the osmotically active solutes they contain, may develop a turgor pressure of several atmospheres; without this pressure a plant wilts. Animal cells, on the other hand, lack such a restraint of hydraulic pressure and under usual circumstances regulate their osmotic content by ion pumps, e.g., an Na^+-K^+ pump in erythrocytes (Sect. 2.4). If erythrocytes are placed in water, rather than in the usual Ringer's solution, they will swell and burst (hemolysis) because in the absence externally of K^+ and Na^+ they cannot osmoregulate. The development in evolution of cell walls, a special feature of plant phylogeny, has permitted protoplasts a wide range of hydraulic pressures. For example halophytes growing on the coast may, in dry periods, be exposed to salinities well above that of ocean water (ca −30 bars) and the next day to rainwater (ca 0 bars); this is an extreme example, but plants are typically exposed to wide ranges of osmotic stress. As a consequence of this evolution the nature of active transport systems conducive to higher development in plants differs markedly from that in animals. Nonetheless, plant cells have retained the capacity to osmoregulate (Sect. 2.4).

4.1.1 Chemical Properties of the Cell Wall

Chemically the cell wall is built up largely from cellulose and the so-called hemicelluloses, pectins, and glycoproteins.

The cellulose molecules (Fig. 4.1a) are subject to a very strict spatial order. About 40 up to 300 cellulose molecules are joined together into so-called microfibrils in which there is formation of lattices, crystal-like areas, the so-called micelles. The diameter of the microfibrils is 3 to 30 nm. There is no certainty about the size of the micelle strands, but in a scheme by Frey-Wyssling (Fig. 4.2) their diameter is given as 6 nm. The fibrillar elements are also called the framework of the cell wall. The sizes and diameters of micelles and fibrils of the wall framework are not as important for transport as the sizes of the intermicellar and the interfibrillar spaces which are left open for diffusion (Fig. 4.2). They have a diameter of about 1 and 10 nm, respectively, as an order of magnitude and, therefore, are much larger than many solutes which are transported in plants (Table 4.1). The size of these spaces may be reduced to some extent by

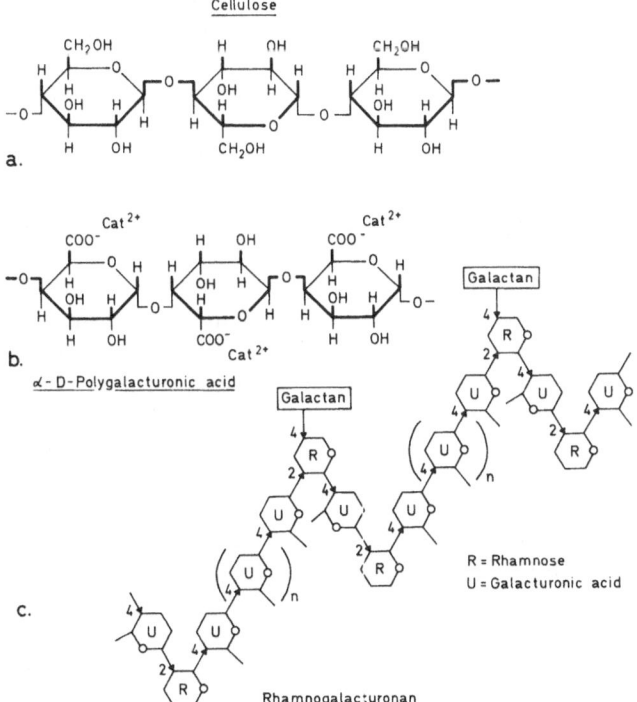

Fig. 4.1. Cellulose **a** and cell wall matrix substances; **b** α-D-polygalacturonic acid—the basic structure of the pectins, where the free carboxyl groups can form salts with divalent and monovalent cations but some carboxyl group are esterized to methoxy groups; **c** possible structure of rhamnogalacturonan, n an undetermined number, probably between 4 and 10. (Talmadge et al., 1973.)

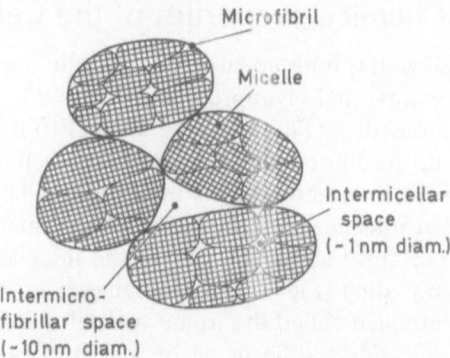

Fig. 4.2. Model of microfibrils in cell walls. (Modified after Robards, 1970, from Frey-Wyssling, 1959.)

matrix substances (Sect. 4.1.2). It will be shown below, however, that water and solutes indeed can move in these spaces.

Matrix substances of the cell wall, in contrast to cellulose, are amorphic. The so-called hemicelluloses are polysaccharides which, unlike cellulose, are soluble in concentrated alkali (Frey-Wyssling and Mühlethaler, 1965).

Particularly well known is pectin which chemically is closely related to α-D-polygalacturonic acid; pectin or pectic acid molecules are polygalacturonic acid chains on which numerous carboxyl groups are esterized to methoxy groups (Fig. 4.1b). In an example which is quoted by Rogers and Perkins (1968) about 37% of the COOH groups are esterized in this way. The free carboxyl groups constitute a considerable reservoir of negatively charged fixed ions in the cell wall. With bivalent cations, especially calcium, single chains of pectic acid can be cross-linked. Other acid

Table 4.1. Particle diameters [nm] of water, glucose, and some inorganic cations and anions. (Tabulation by Läuchli, 1976a.)

Particle	Molecule[a] or ionic crystal[b] diameter	Hydrated ionic diameter[d]	Particle	Molecule[a] or ionic crystal[b] diameter	Hydrated ionic diameter
H_2O	0.39		Mg^{2+}	0.13	0.92
Glucose	0.89		Ca^{2+}	0.20	0.88
Na^+	0.19	0.60	Cl^-	0.36	0.50[e]
K^+	0.27	0.54	NO_3^-	0.41[c]	
NH_4^+	0.30				

[a] From Beck and Schultz (1970).
[b] From Pauling (1960).
[c] Estimated from Wells (1945).
[d] Estimated from McFarlane and Berry (1974).
[e] From Adamson (1973).

polysaccharides and neutral polysaccharides found in the plant cell wall are rhamnogalacturonans and arabinogalactans, respectively (Northcote, 1972; Fig. 4.2c). Among the matrix substances of the cell walls there are also proteins. These are glycoproteins, in which the protein fraction is characterized by being particularly rich in the amino acid hydroxyproline. Several structures of cell wall proteins have been hypothesized. A possible configuration is shown in Figure 4.3, in which a highly branched arabinogalactan polymer is bound to a serine residue of the hydroxyproline-rich cell wall protein (cf. review by Läuchli, 1976a).

Free carboxyl groups of polysaccharides and proteins provide a large number of fixed negative charges in the cell wall. Free amino groups of proteins contribute a somewhat smaller amount of positive charges. The charged groups of the matrix substances lead to the formation of a Donnan system in the cell wall which must be highly important for ion transport in plant tissues.

Albersheim and coworkers have devoted a lot of work to the macromo-

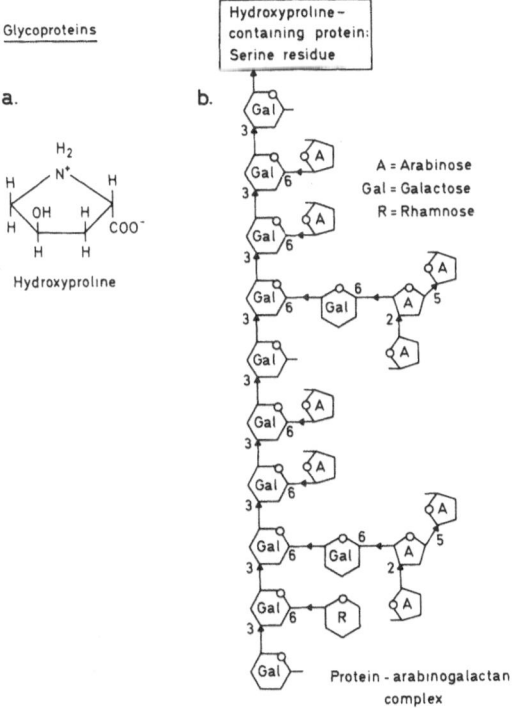

Fig. 4.3 a and b. Glycoproteins. a Hydroxyproline—the major amino acid of the cell wall glycoproteins; b possible structure of a glycoprotein: the glycoprotein is a protein-arabinogalactan complex. The linkage between the arabinogalactan and the hydroxyproline-containing proteins is caused by a covalent bond between galactose and a serine residue of the protein. (Keegstra et al., 1973; Läuchli, 1976a.)

lecular structure of primary cell walls. The primary cell wall of sycamore, according to Talmadge et al., (1973; from Läuchli, 1976a), is composed of:

> 23% cellulose
> 55% hemicellulose:
> > 34% pectins—10% neutral arabinan
> > — 8% neutral galactan
> > —16% acidic rhamnogalacturonan
> 21% xyloglucan
> 21% glycoprotein:
> > 10% hydroxyproline-rich protein
> > 9% oligo-arabinosides attached to hydroxyproline
> > 2% arabinogalactan attached to serine

A structural model of the primary cell wall of sycamore is shown in Figure 4.4. Associations between the various constituents suggest that the whole primary cell wall could be considered as one giant complex macromolecule. The consequences for transport are twofold. First, the matrix substances of the primary cell wall limit accessibility of the free spaces between the cellulose micelles and fibrils (Fig. 4.2). Second, they provide the chemical basis for transport along ion exchange sites in the Donnan system of the cell wall.

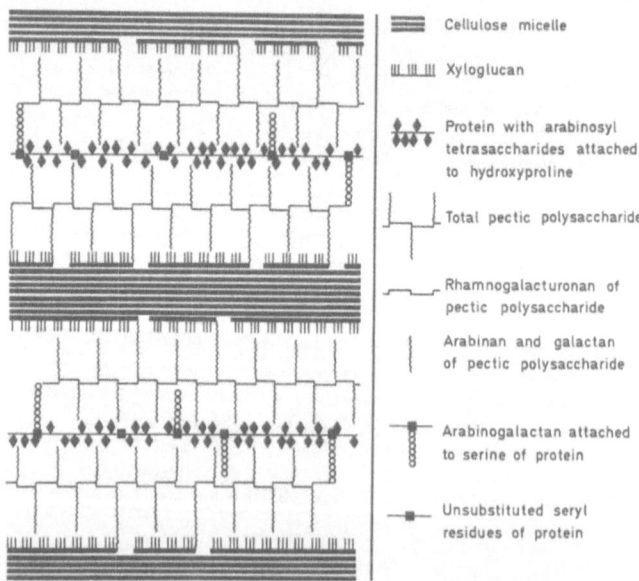

Fig. 4.4. Tentative model of the macromolecular structure of a primary cell wall in suspension-cultured sycamore cells. The components are presented in approximately proper proportions except for the distance between cellulose micelles, which is expanded to allow room for the interconnecting structures. (Keegstra et al., 1973; Läuchli, 1976a.)

4.1.2 Cytological Properties of the Cell Wall

Cytologically the cell walls are formed of a number of lamellae which are built one after another. When a new cell wall is formed a layer with a very high proportion of matrix substances, the middle lamella, is laid down first; this is followed by the primary wall layer, then, in many cells, various layers of secondary wall, and eventually tertiary wall thickenings. In the primary wall, the cellulose fibrils are nonordered (rather, criss-crossed), whereas these fibrils are in a rather regular parallel array within each of the secondary wall layers. These features are important for growth and for strength of plant cells. More important for transport, as already stressed above, are the fine pores in the cellulosic framework and the fixed electric charges of ionic groups of the matrix material. It is understandable that alterations of the fine pore structure must have important effects on transport. Such alterations can be brought about by imbibition and infiltration of cell walls with polymerizing material. This is called *incrustation* of the cell wall. Modifications are also due to materials layered onto cell walls. This is called *adcrustation* (Sitte and Rennier, 1963).

4.1.2.1 Cell Wall Incrustation (Lignin)

When certain organic molecules move into the spaces between the cellulose microfibrils and then react with each other to form a highly polymerized mass, which literally solidifies the framework of the wall, then we have a so-called incrustation. The importance of the incrustation of the cell wall with rigid amorphic material for the mechanical properties of larger land plants is quite obvious. It is disputable, however, what effects this process of incrustation has on transport within the cell wall. This depends, on one hand, on the chemical nature of the incrusting substances and, on the other, on the extent to which the spaces in between the cellulose micelles and fibrils are occupied by the incrusting material. The most important incrustation is that of lignin. Its monomers are various phenylpropane compounds derived from secondary metabolism, for instance, coniferyl alcohol, sinapyl alcohol, or p-coumaryl alcohol (Fig. 4.5a). When we look on the various chemical structures which have been suggested for the polymer lignin (e.g., Fig. 4.5b), we can see that lignin is not an entirely hydrophobic substance, but that there are a number of polar groups, especially nonesterified OH groups. Nevertheless, the hemicelluloses are much more polar than lignin and it is interesting to note that the wall layers with the largest relative amount of matrix substance are also most highly lignified. The lignin content in the region of the middle lamella can be as high as 90% of its total dry weight. In lignification of the middle lamella the physical characteristics of the matrix substance is altered considerably; the matrix substance is shrunken and

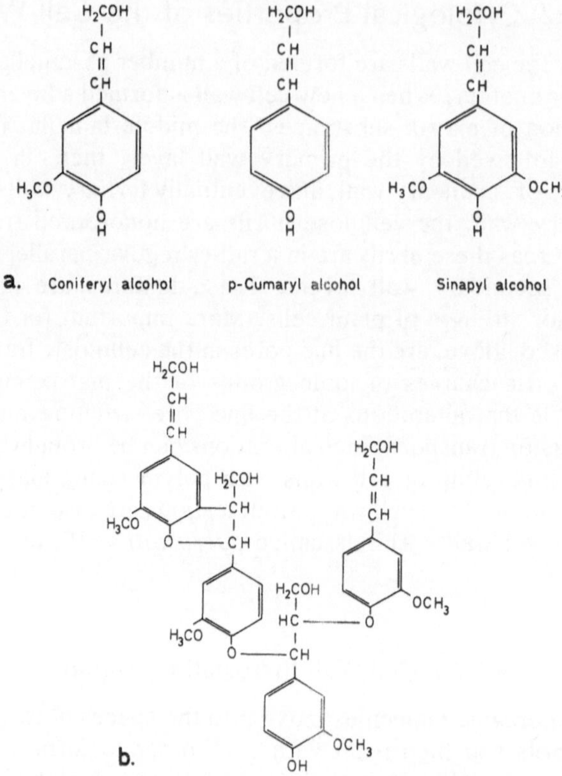

H$_2$COH
CH
CH
H$_3$CO
O
H
a. Coniferyl alcohol

H$_2$COH
CH
CH
O
H
p-Cumaryl alcohol

H$_2$COH
CH
CH
H$_3$CO   OCH$_3$
O
H
Sinapyl alcohol

H$_2$COH
CH
CH
H$_3$CO
O — CH

H$_2$COH
CH
H$_3$CO
O — CH

H$_2$COH
CH
CH
OCH$_3$
O

H$_2$COH
HC —
H$_3$CO
O — CH
OCH$_3$
OH

b.

Fig. 4.5. a Monomers and **b** polymer structure of lignin after Freudenberg, from Sitte, 1965.

pressed together (Sitte, 1965). One can assume that the movement of ions and the exchange at fixed charges in cell walls must be limited very much after lignification, if not totally prevented. This point of view is not shared entirely by all transport physiologists.

4.1.2.2 Cell Wall Adcrustation (Suberin, Cutin)

Other substances which participate in the cytological modification of the cell wall are macromolecules built from fatty acid residues, i.e., the suberins and the cutins. The degree of the hydrophilic character of these substances depends on the amount of nonesterified carboxyl and hydroxyl groups (Fig. 4.6). Suberin and cutin usually form adcrustations. This means they are layered onto the lamellae of the primary, secondary, or tertiary cell wall. In addition to adcrustations, cutin can also be found as an incrusting substance, especially in the wall layers of outer cells of all or most aerial organs of plants. The details of the anatomical and chemical structure of suberin and cutin lamellae determine to what extent suberiza-

Suberin (cutin)

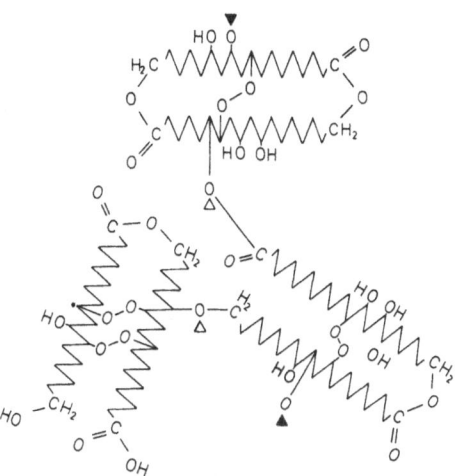

Fig. 4.6. Scheme of suberin-(cutin)-structure from Sitte (1965) after Heinen. The subunits each consist of two fatty acid chains. Three subunits are depicted, with *open triangles* showing connections between the subunits, and *solid triangles* indicating further possible sites for attachment of other subunits.

tion and cutinization cause a limitation of transport within and particularly across the cell wall (review, Läuchli, 1976a).

4.1.2.3 Cuticles

Cuticles at the surface of aerial organs of plants combine incrustation and adcrustation (cuticularization) with cutin. Investigations of Sitte and Rennier (1963) show that cuticular walls of leaves are built up of four layers: the cellulose wall, a layer with cellulose fibrils and cutin incrustation; and the actual cuticle adcrustation of a cutin and a wax layer, where the cutin is less lipophilic than the wax. Since cuticular transpiration does occur, cuticles cannot be entirely impermeable to hydrophilic particles such as H_2O. A detailed biophysical analysis of solute transport across cuticles and their permeabilities is available from the work of Schönherr (1974, 1976a,b). The water permeability of cuticles is largely determined by the wax layer. Extraction of cuticular waxes increases water permeability by a factor of $3-5 \cdot 10^2$ (Schönherr, 1976b). Cuticles contain a certain amount of fixed charges allowing ion transport and exchange in a Donnan phase (Schönherr, 1974, 1976a). This agrees with the observation that the cuticle of the adaxial epidermis of the scale-like leaves of onion bulbs is impermeable to hexose molecules, while potassium and chloride diffusion across the cell wall is reduced but not entirely blocked (Steinbrecher and Lüttge, 1969).

4.2 Special Cases of Cell Wall Incrustation and Adcrustation

The incrustation of cell walls with lignin adds considerable mechanical strength to the plant body. Phylogenetically, the capacity for lignification of cell walls was one of the most important prerequisites for the evolution of the larger land plants following the emergence from an aquatic environment. The consequences of lignification for transport in the cell wall phase appears, phylogenetically at least, to be of secondary nature. The function of adcrustation, by contrast, appears to be primarily related to transport. Cell walls with adcrustations are usually found in surface tissues. Cuticles at the external cell wall of the epidermis cells are particularly well developed when there is a need to limit loss of water (cuticular transpiration) especially as in xerophytes. Adcrustations are found in all secondary boundary tissues (e.g., periderms or bark) whose cells are eventually dead and air-filled. Adcrustations in cork cells are multilayered with intermittent suberin and wax layers (Sitte, 1962; cf., Läuchli, 1976a), and Sitte (1975) points out that this lamellar structure provides a particularly efficient insulation in plants as in man-made insulating materials.

In tissues which have a particular transport function, we often find at specific locations incrustations which are limited to particular parts of cell walls. We refer to the Casparian strips in the radial walls of the endodermal layers surrounding the stelar cylinders of roots and shoots and to similar wall formations in gland tissues.

4.2.1 The Casparian Strip in the Primary Endodermis

Figure 4.7 shows that the Casparian strip is located in a strategic position forming a band in the radial walls of endodermal cells. The Casparian strip shares many cytological properties with lignified and with polymerized lipophilic wall structures. Electronmicroscopy proves that the Casparian strip certainly is an incrustation. After what we have learned so far about such incrustations, we can assume that the mobility of water and ions and other low molecular substances in the cell wall—and hence radial transport—are greatly hindered by the Casparian strip. The schematic drawing in Figure 4.7 also shows that for a blockage of radial transport across the endodermis, in addition to the Casparian strip, a tight association of the endodermis cells is necessary. If the endodermis were interrupted by intercellular spaces, transport in the apoplastic free space could be possible despite the Casparian strip, for instance on the surface of the cell walls or in the intercellular spaces themselves, if they were wet. Observations show, however, that the endodermis is indeed free of intercellular spaces. Another possibility is that solutes could move at the internal surface of the walls, i.e., between the Casparian strip and the plasma-

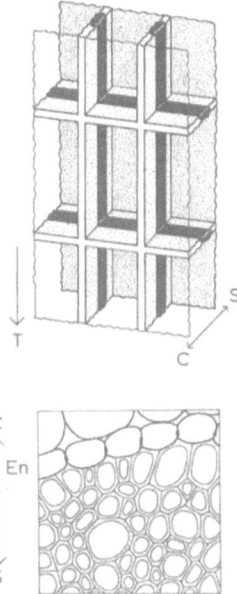

Fig. 4.7. Schematic three-dimensional diagram of a root endodermis cell with eight adjacent cells (*upper diagram*). The Casparian strip is indicated as a *boldly black structure*. The spatial relationships within the root are indicated by *symbols* and *arrows;* see also root cross section (lower diagram). *T*, root tip; *C*, cortex; *S*, central cylinder or stele; *En*, endodermis.

lemma of endodermal cells. This is unlikely because at the Casparian strip the contact between the plasmalemma and the cell wall is particularly intimate. This is seen in the plasmolysis of endodermal cells, when the plasmalemma often does not separate from the cell wall adjacent to the Casparian strip (see Falk and Sitte, 1960; Bonnett, 1968; Haas and Carothers, 1975).

It is most important for plants to subject transport processes to metabolic control for which the participation of the cytoplasm is an indispensible prerequisite. A blockage of transport in the cell wall phase by the incrustation of the Casparian strip means that, from a critical location onward, solutes moving to the stele of roots can use the cytoplasmic pathway only. Such a function of the specific incrustations of the Casparian strip appears to provide the only possible advantage in the phylogenetic selection of these strange structures.

The assumption that the Casparian strip indeed does effectively block radial transport of solutes has been supported recently by a great deal of evidence. Investigations combining experiments on uptake and washout of radioactively labeled ions (see Sects. 4.3 and 6.2.3.2) with microautoradiographic localization in root tissue, show that the endodermis blocks passive radial transport in the cell walls (e.g., Fig. 4.8; also see Krich-

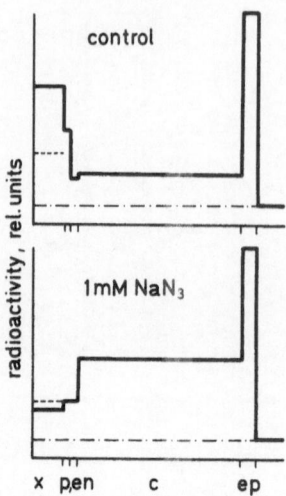

Fig. 4.8. Grain counts obtained from microautoradiographs showing the distribution of radioactivity in maize roots after the roots were transporting ^{35}S labeled SO_4^{2-} for 1–3 h in the absence and in the presence of 1 mM NaN_3 (azide) respectively. In the presence of the metabolic inhibitor the transport to the stele across the endodermis is highly reduced. (Weigl and Lüttge, 1962). *x*, xylem; *p*, pericycle; *en*, endodermis; *c*, cortex; *ep*, epidermis; the *solid line* represents radioactivity in the above-mentioned tissues; *line of short dashes*, radioactivity in the phloem; *line of dots and dashes*, average background.

baum et al., 1967). With electron-opaque elements (like La and Pb) used as tracers, blockage of radial cell wall transport by the endodermis can be made visible (Tanton and Crowdy, 1972; Nagahashi et al., 1974; Robards and Robb, 1974).

4.2.2 Secondary and Tertiary Endodermis

The Casparian strip as discussed in Section 4.2.1 is usually developed in the endodermis in its primary state, a short distance behind the root tip. Endodermal cells of monocotyledonous roots undergo further changes as depicted in Figure 4.9. In the secondary state a suberin lamella is deposited, and in the tertiary state many cellulose layers, often lignified, are added. This completely blocks transport in the cell wall phase (reviews Clarkson and Robards, 1975; Läuchli, 1976a).

4.2.3 Suberized Root Hypodermis

For the evaluation of radial transport in roots (see Sect. 12.2) it is essential to note briefly that suberization of the external cell layers of roots may very effectively prevent the movement of ions into the root tissue (*Iris* roots: Ziegler et al., 1963). Formation of a suberized hypodermis re-

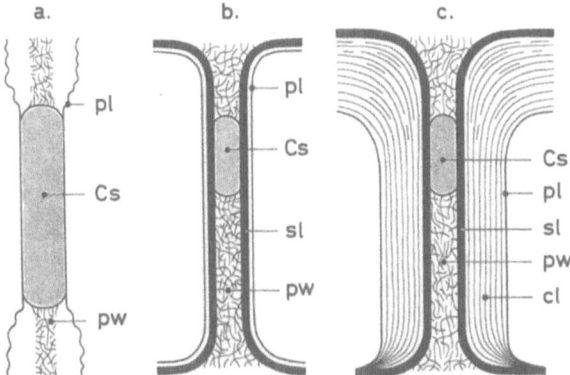

Fig. 4.9a-c. Wall development of an endodermal cell in a monocotyledonous root depicted in schematic drawings of cross sections through the root. **a** Primary state: Casparian strip (*Cs*) in the radial wall; *pw*, primary wall; *pl*, plasmalemma. **b** Secondary state: a suberin lamella (*sl*) is deposited over the whole inner wall surface and separates the plasmalemma (*pl*) from the Casparian strip (*Cs*). **c** Tertiary state: a thick cellulose layer (*cl*) covers the inside of the wall, mainly the inner tangential and the radial walls. The entire cell wall becomes lignified. Note that **b** and **c** are at a different scale than **a**. (Läuchli, 1976a.)

stricting ion uptake also has been observed in maize but not in barley roots (Ferguson and Clarkson, 1976a,b).

4.2.4 Cell Wall Incrustations in Glands

Lignin-like cell wall incrustations are observed in many glands: salt glands and salt hairs, nectar glands, and nectar trichomes. Like the Casparian strip of the endodermis, these incrustations are located in glands at strategic positions where they force uptake of solutes into the cytoplasm before the solutes can be excreted or secreted (Fig. 4.10; reviews: Frey-Wyssling, 1935; Schnepf, 1969; Lüttge, 1969, 1971a, 1975; Lüttge and Schnepf, 1976; Hill and Hill, 1976). Applying fluorescent staining techniques to the nectar trichomes of *Abutilon* flowers (see also Figs. 12.6 and 12.7) Gunning and Hughes (1976) clearly demonstrated that such incrustations may completely impregnate the cell wall and prevent cell wall transport of solutes.

4.3 Apoplastic Transport

The preceding discussions readily imply that the cell wall space may serve as a phase or pathway for transport. Movement in the cell wall is determined by the width of the interfibrillar and the intermicellar spaces and by the kind and concentration of charged ions fixed there. Transport in the cell

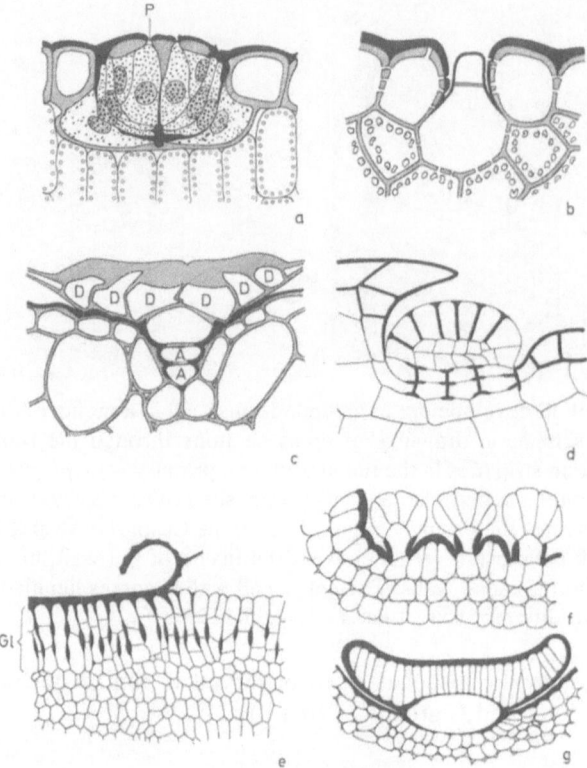

Fig. 4.10. Cell wall incrustations (indicated as *boldly black cell walls*) of glands, which are analogous to the Casparian strip of the endodermis of roots and stems. **a** Salt gland in the leaf epidermis of *Limonium vulgare; P,* secretion pores (after Ruhland, 1915; from Arisz et al., 1955); **b** Salt gland of *Spartina* (after Helder, 1964); **c** water "sucking" scale shaped hair of *Tillandsia usneoides. A,* absorption cells; *D,* cells of the disk-like surface of the scale (Dolzmann, 1965); **d** digestive gland from a pitcher of the carnivorous plant *Nepenthes compacta* (Stern, 1917); **e–g** extrafloral nectar glands (Frey-Wyssling, 1935); **e** *Hevea* where the tissue of gland cells *(Gl)* resembles closely packed gland hairs; **f** nectar trichomes of *Syringa sargentiana;* **g** scale-shaped nectaries of *Glaziova.* (See also Lüttge, 1969.)

wall phase can be separated kinetically from transport across the plasma boundary layer, the plasmalemma, i.e., from transport into the interior of the cell.

In order to show this, one can put plant cells or plant tissues from a particular solution into another of different composition which, for instance, contains a substance, S, not present in the external medium initially. Experimentally this can be done by transferring cells or tissues from a nonradioactive into a radioactively labeled solution. The time course of uptake of radioactively labeled particles S* into the cells is mea-

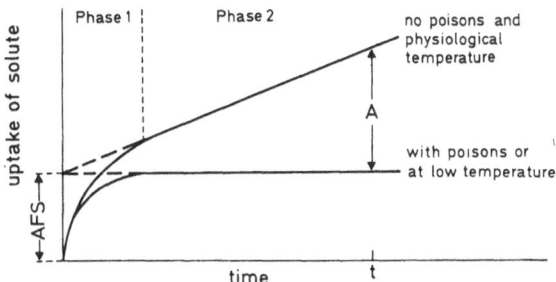

Fig. 4.11. Uptake of solutes into plant tissues. *Phase 1*, penetration of the "apparent free space" (AFS); *Phase 2*, accumulation into the cells. Phase 2 depends on metabolism and can be inhibited by low temperatures and poisons of metabolism. *A*, amount of solute accumulated at time *t*. *Dotted curve*, determination of AFS by extrapolation of the uptake curve of Phase 2. (After Briggs et al., 1961.)

sured. As shown in Figure 4.11 there are kinetically two clearly separated phases. First, there is a very rapid phase with a saturation half-time ($t_{1/2}$) on the order of a few minutes. Second, there is a slow phase which, after completion of the first phase, continues for several hours at a constant rate (linear time course). The rapid initial phase also can be seen in the time course of washout curves of radioactive particles from a tissue which has been loaded previously (Sect. 6.2.3.2, Fig. 6.11). Using inhibitors of metabolism, or low temperatures, one can further characterize the two phases shown in Figure 4.11. Only the slow phase is blocked by metabolic inhibitors and is obviously controlled by the living cytoplasm. In the first phase the uptake of solutes apparently takes place unhindered by a membrane barrier and is very rapid, i.e., as if a distinct part of the cell or tissue is freely open for movement of the solute particles. Therefore, this uptake has been termed movement into the *apparent free space* (AFS).

This implies that transport in the free space is a passive process, exclusively due to physical driving forces. It should be noted, though, that there may be a group of negatively charged binding sites in the free space, the existence of which depends on unimpaired metabolism (Ighe and Pettersson, 1974). These could be at the external surface of the plasmalemma. Thus metabolism may indirectly control part of the Donnan exchange in the free space by maintenance of structure.

The size of the AFS can be determined by linear extrapolation of the second phase of the uptake time course to the zero point of the time axis (Fig. 4.11). It can be expressed as a percentage of the total cell or tissue volume. But even in the initial phases of the research on AFS it became quite clear that there are serious experimental difficulties. In order to investigate the uptake of substances, one must remove a film of external solution adhering to the surface of the tissue. For important reasons—one of which is the unstirred layer effect discussed in Section 2.1.3 and Figure 2.3—this is not always entirely possible, either by careful blotting with absorbent paper or by short rinses with a nonradioactive wash solution.

The volume of the surface layer, or unstirred layer, on the outside must be taken into account in determinations of the AFS to estimate its volume. One can calculate the size of the surface layer on the basis of theoretical considerations or it can be determined experimentally, for instance, by removing the surface layer through centrifugation (Ingelsten and Hylmö, 1961), or by estimating it with microautographic procedures (Krichbaum et al., 1967). After the appropriate corrections the AFS volumes obtained are between 8% and 15% of the total tissue volume.

For charged particles, the AFS consists of two components because, as we have seen, some ions are free in the solution occupying the cell wall spaces and some are bound on fixed-charge sites in the cell wall. Accordingly, we distinguish between a *water free space* (WFS) and a *Donnan free space* (DFS); it follows that:

$$WFS + DFS = AFS \qquad (4.1)$$

The DFS depends on the nature and the density of the fixed charges in the cell wall; since the wall has many more negatively charged sites than positive ones, the DFS for cations is much larger than that for anions; in red beet tissue a DFS of 2% of the total tissue volume was found (Briggs et al., 1958a). More data on the relative sizes of WFS and DFS are given in Table 4.2; the bulk of the ions in the DFS are divalent cations if they are present in the external solution (Briggs et al., 1961).

Table 4.2. Properties of free space exchange in various plants and tissues. Total content in the free space can be calculated from concentrations of ions in solution and the free space properties. (See Briggs et al., 1961.) (Tabulation by Walker and Pitman, 1976.)

Plant cell or tissue	WFS Volume ($cm^3 \ g^{-1}$)	Fixed negative charges in DFS	
		Amount (μmol g_{FW}^{-1})	Concentration (mM)
Beetroot[a] slices (1 mm thick)	0.20	12	560
Barley roots[b]	0.24	2.0	—
Barley leaf[c] slices (0.9 mm)	0.21	3.5	300
Atriplex leaf slices[d]	—	16	—
Chara cell walls[e]	0.34	146	600
Chara cells (if wall is 5.8% of FW)	0.081	8.5	600

[a] Briggs et al. (1958b).
[b] Pitman (1965, 1972).
[c] Pitman et al. (1974).
[d] Osmond (1968).
[e] Dainty and Hope

The term *apparent free space* is a purely operational one. There has been controversy for some time over whether the AFS is solely at the surface and in the walls of cells and tissues, or whether it also extends into most or all of the cytoplasm. In the latter case the tonoplast would be the effective external diffusion barrier of plant cells. The relative size of the AFS has played a large role as an argument in this controversy. For very different and more physiological reasons (Sect. 6.2.3) most authors now agree that the AFS extends only to the plasmalemma, i.e., it is restricted to the cell walls. The more concrete term *apoplast* then appears to be more suitable than the operational term AFS. The apoplast of a cell or tissue is given by all spaces outside the plasmalemma barrier of living cells. According to this view the apoplast is identical with the cell wall space plus intercellular spaces. It should be noted though that with this definition the lumen of dead xylem tracheids and vessels also belongs to the apoplast (Sect. 11.2). (The analogous term for what is inside the plasmalemma barrier is "symplast"; see Sect. 11.3.)

Apoplastic transport in cell wall systems of tissues over medium distances can be best demonstrated in the electronmicroscope by using electron-opaque tracers (Sect. 4.2.1).

In many experiments on transport in plants the AFS is very disturbing, because one has a particular interest in transport across the plasma membrane into the cell interior onto which the diffusion and ion exchange in the cell wall is superimposed. Therefore a correction is required in order to measure the accumulation into the cell interior. In experiments with radioactively labeled particles, on which much of our knowledge is based, this correction is achieved kinetically. At the end of the experiments the tissue is transferred for a short period into an unlabeled ice-cold solution where it exchanges the particles in the WFS and DFS for nonlabeled particles. The low temperature increases the half-time for free-space exchange only slightly, but the exchange of particles from the cytoplasm and from the vacuole is greatly reduced (see also Sects. 2.5.4 and 6.2.3.2). In this way, one can obtain ion exchange in the AFS without an appreciable loss of radioactivity from the cell itself. After exchange of the AFS one can determine the amount of substance which has been taken up into the cell interior. The optimal time for AFS exchange, however, has to be newly determined for each tissue and for each species of particle under investigation (refer also to Cram, 1969a; Cram and Laties, 1971). The method of corrections of AFS which has been described here is much more often used than the extrapolation shown in Figure 4.11.

Chapter 5
The Membranes

5.1 The Historical Development of Membrane Research

5.1.1 The Danielli-Davson Model of the Membrane and the Concept of the "Unit Membrane"

The investigation of biological membranes looks back to a long history. Its early stages coincided with the beginning of studies on the behavior of long chain amphipolar molecules (e.g., fatty acids, lipids) in forming films on water–air boundary layers (Langmuir, 1917a, b, 1933). With fatty acid molecules the hydrophilic poles of their carboxyl groups intrude the aqueous phase. Their longer hydrocarbon axis stands perpendicular to the water surface, so that the hydrophobic or lipophilic end of the fatty acid chain sticks away from the aqueous phase. The first highlights in membrane research were provided by Gorter and Grendel in 1925 in their analysis of the lipids extracted from erythrocytes. In a regular and appressed order at a water–air interphase, the surface area occupied by erythrocyte lipids is just twice as large as the surface of the erythrocytes. Since erythrocytes are not compartmented by membranes internally, one could conclude from the experiments of Gorter and Grendel that the outer membranes of erythrocytes are formed by a bimolecular layer of lipid molecules. It was assumed that the external surfaces of the membrane on either side were formed by the hydrophilic poles of the lipid molecules and the interior of the membrane by their hydrophobic chains perpendicular to the membrane surface.

An important new piece of evidence for the concept of a membrane model came from measurements of surface tension (Danielli and Harvey,

Table 5.1. Surface tension of oil droplets and of cells.

	Surface tension [dyne · cm^{-1}]
Oil droplets in water	7–10
Erythrocytes, eggs of sea urchin, mollusks, and salamander	0.2–0.8
Oil droplets within cytoplasm	0.6

1935). The surface tension of erythrocytes, sea urchin eggs, mollusk eggs, and salamander eggs in water turned out to be very different from the surface tension of lipid droplets in water, e.g., of oil taken from mackerel eggs (Table 5.1). However, the surface tension of oil droplets within the cytoplasm was similar to that of the cell surface, but not to that of oil droplets in water. For an explanation of the lower surface tension of cell membranes and oil droplets in the cytoplasm, one has to assume that the lipid films of plasmatic boundary layers bind with surface active substances of the cytoplasm (Danielli and Harvey, 1935). Thus Danielli and Davson (1935) suggested a membrane model according to which the hydrophilic poles of the lipid molecules at the surface of the bimolecular lipid layer of the membrane were binding with globular protein via hydrogen bonds or salt bonds. In this way the membrane surfaces were thought to be coated with proteins. The membrane as a whole constituted a lipoprotein complex (Fig. 5.1a).

The lipoprotein nature of biological membranes was soon corroborated by additional evidence. It was observed that droplets of oil which were injected into the cytoplasm of sea urchin eggs were covered by a protein film (cf. Harvey, 1954). Modern membrane research with its highly sophisticated methodology has proven beyond doubt that biological membranes are lipoprotein complexes (see Sects. 5.2.2, 5.3.3). The arrangement of lipids and proteins assumed by the Danielli and Davson model also appeared to be confirmed by X-ray diffraction studies.

Very important support came from electronmicroscopy. When the first interpretable electronmicroscopic pictures of biological cells and tissues became available, all biological membranes appeared as two parallel dark lines with a bright layer in between. Similar pictures were obtained with artificial membranes (e.g., artificial myelin layers). In early electronmicroscopy, potassium permanganate and osmium tetroxide were predominantly used as fixation and contrasting agents. With great care, using model membranes, the attempt was made to prove that the molecules of the fixation and contrasting agents were bound to the hydrophilic ends of the lipid molecules. In biological membranes according to the Danielli-Davson model, these were supposed to be the sites where the protein molecules were attached. Thus with a preferential binding of the contrasting molecules at the hydrophilic poles of the membrane, the two dark lines shown

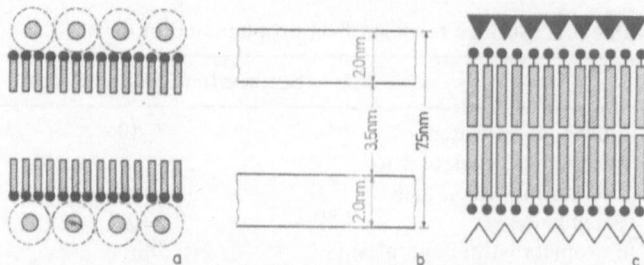

Fig. 5.1. Membrane model of Danielli and Davson (1935) and "unit membrane" of J.D. Robertson (1964). **a** Danielli-Davson membrane model. The lipid film formed by a lipid double layer is covered with protein on both faces. The interior of the membrane is formed by a hydrophobic lipid region (*hatched blocks* represent the hydrophobic tails of lipid molecules), the hydrophilic or polar groups of the lipid molecules (*black circles*) are oriented at the surface; **b** diagram of an electronmicroscopic picture of a membrane after fixation with $KMnO_4$ or OsO_4 with the average dimensions of the two electron absorbing lines and the central electron transparent line; **c** model of the "unit membrane" from Robertson (1964) corresponding to the Danielli-Davson model. By assumption of different kinds of protein on both faces the membrane becomes asymmetric.

by electronmicroscopy would correspond to the external zones of the Danielli-Davson membrane as shown in Figures 5.1a and 5.1b. In spite of much criticism (cf. Korn, 1966) important arguments could be made supporting such a binding of the fixing agents (Stoeckenius, 1959, 1960; Stoeckenius et al., 1960). In electronmicrographs all membranes looked very similar. One should perhaps say that all membranes appeared to react in a similar way to the fixation, i.e., that in this respect they had similar properties. This led J.D. Robertson to the concept of the "unit membrane" according to which all membranes are structurally analogous and ontogenetically or phylogenetically of similar origin (Robertson, 1964). A sketch of the "unit membrane" (from Robertson, Fig. 5.1c) is different from the original Danielli-Davson model only in respect to the assumption that different protein films line the two membrane faces, i.e., an assumption of asymmetry.

5.1.2 Early Theories of Membrane Permeation as Related to the Danielli-Davson Model

Along with the developments ending with the structural concepts of the Danielli-Davson model, hypotheses emerged which were concerned with the mechanism of permeation across biological membranes.

5.1.2.1 The Lipid Theory of Permeation

The oldest of the more famous membrane transport theories is the lipid theory of permeation (Overton, 1899; cf. Wartiovaara and Collander,

1960). This was based on a large body of data showing that highly lipophilic substances can generally permeate more rapidly than less lipophilic
substances. For example, with alcohols, membrane permeability increases with increasing length of the lipophilic carbon chain. The lipid
theory of permeation is based on the Nernst law of distribution of solutes
between two different solvents adjacent to one another. The ratio of the
concentration of the solutes in both solvents is dependent only on the affinities of the solutes and solvents; it is independent of solute concentration. At isothermal and isobaric conditions this is given by the partition
coefficient:

$$\frac{\text{Concentration in lipid phase}}{\text{Concentration in aqueous phase}} = \text{constant.} \qquad (5.1)$$

Thus the permeation from an outer phase across a continuous lipid membrane into an inner phase should be much easier for lipophilic solutes than
for hydrophilic ones (Fig. 5.1). When we compare this conclusion of the
lipid theory of permeation with Figures 5.1a and 5.1c, it is quite clear that
these ideas are in accord with the Danielli-Davson model.

5.1.2.2 The Ultrafilter Theory of Permeation

One difficulty with the lipid theory of permeation results from what we
know about permeability of water (Sect. 2.2.3.5). Water molecules are
"hydrophilic" particles and they permeate extremely readily. Furthermore, it is observed that the size of solute particles often is much more
important in permeation than indicated by the ratios

$$\frac{C_{\text{lipid phase}}}{C_{\text{water}}}$$

as found in Eq. (5.1) in the lipid theory of permeation. For a large number
of substances the volume of the molecules is more important for their permeation. Based on these results, Ruhland coined his ultrafilter theory of
permeation, which assumes a sieve, or filter effect of the membrane (Ruhland, 1912; Ruhland and Hoffmann, 1925).

This implied the postulate of fine pores crossing biological membranes.
The Danielli-Davson membrane model has therefore been amended by
adding water-filled pores which are lined by a protein layer (Fig. 5.2).
With this modification, the Danielli-Davson membrane model is in accord
with the ultrafilter theory of permeation. (The question of pores in biological membranes is further evaluated in Sect. 5.1.2.4).

5.1.2.3 The Lipid-Filter Theory of Permeation

The two theories, the lipid theory and the ultrafilter theory of permeation
are not mutually exclusive. Collander (Wartiovaara and Collander, 1960)

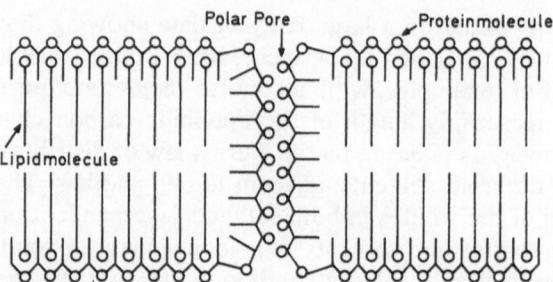

Fig. 5.2. Danielli-Davson membrane model with hydrophilic pores lined with protein. (Branton and Deamer, 1972.)

unified them, coining the term lipid-filter theory, and Höfler (1958, 1959, 1960, 1961) later talked about the two-pathway theory. In this the permeating particles, depending on their properties, should either prefer the lipid phase or the hydrophilic pores as a pathway of permeation.

5.1.2.4 Membrane Pores

We have seen above that the apparent sieve effect of membranes during permeation, in a trivial way, leads to the assumption of pores in the membranes. By experiments with hydrophilic molecules of known dimensions, and by taking into account the length and thickness of the molecules, and the degrees of freedom of rotation in narrow pores, one can arrive at an estimate of the pore diameter. For the membrane of beef erythrocytes a pore diameter of approximately 0.4 nm has been estimated.

With the model of porous membranes it is particularly easy to explain the meaning of the reflection coefficient, σ, of hydrophilic solute particles (Sect. 2.1.4). σ can be considered as a measure for the frictional resistance in the pores. $\sigma = 0$ means that there is no interaction between the solute particles and the membrane matrix, i.e., the friction is the same as in diffusion in free solution. The pore radius in this case must be much larger than the migrating particles. Conversely, when $\sigma = 1$, one can imagine that the particles are too large to pass through or even enter the pores.

Using the thermodynamics of irreversible processes, mentioned briefly before [see Eqs. (2.42)–(2.44)], one can arrive at formulations allowing calculation of the pore radii. σ is determined experimentally [see Eq. (2.44)] and is assumed to be a function of the size of the solute and solvent particles (of known parameters) and of the pore radius (parameter to be determined); (Passow, 1963). This interpretation of the reflection coefficient applies, of course, only to hydrophilic particles which move through hydrophilic pores. For lipophilic particles the reflection coefficient is always $\sigma < 1$ (Dainty and Ginzburg, 1964b,c). In this case σ depends on parameters other than the size of molecules and the pore radius (Sect. 5.1.2.1). The different meaning of the reflection coefficient for lipophilic and for hydrophilic substances is another expression of the two-pathway theory mentioned in Section 5.1.2.3.

Another argument in favor of the existence of pores in membranes is the phenomenon of "solvent drag". Water moves according to its concentration gradient given by Ψ_S [Eq. (2.7)] or $\Delta\pi$ [Eq. (2.8)]. We can call this a diffusion component of water transport. At the same time another driving force for water transport can be effective, namely a gradient of the hydrostatic pressure (ΔP), which is formed when $0 < \sigma < 1$. This moves water as a laminar streaming. The volume flow observed is a result of the two oppositely directed driving forces. Diffusion of water, however, and laminar streaming of water are two processes of somewhat different nature. If one investigates transport of water across a membrane by labeling with deuterium or tritium, one can find that water permeation with hydrostatic pressure as the driving force occurs with a different velocity than with the concentration gradient as the only driving force. This phenomenon can be explained by the assumption of pores. Passow (1963) has considered this process in detail. The diffusion through pores depends only on the area available for diffusion, i.e., it depends on the radius of the pores to the second power. The laminar streaming, however, according to hydrodynamic laws. depends on the radius of the pores to the fourth power. By the laminar streaming of water (solvent), molecules of the solute will be dragged along. This phenomenon, *"solvent drag,"* can scarcely be explained without the presence of pores. The friction in the membrane pores is larger for the solute molecules than for the water molecules. The solute molecules, at the same time, will also be slowed down by those particles of the solvent which move in the opposite direction. Applying the appropriate equations, "solvent drag" can be used to calculate the pore radius. For various animal membranes, values on the order of 0.4–0.5 nm were obtained in this way (Passow, 1963).

The counterpart to "solvent drag" is "solute drag" (Sitte, 1966), which is observed in electroosmosis (Sect. 2.5.2.2). "Solute drag" is "pulling along" of particles of the solvent (H_2O) by ions moving in an electrical field. The interactions between ions and water molecules and the pore wall are important in this case. The interactions between ions and water molecules are more pronounced with large ions having a large hydration shell than with small ions. One can imagine that ions driven by an electric field move in the pores like pistons, the larger the ions the more water they will move in front of them through the pores. This, of course, has a natural limit when the diameter of the hydrated ion is nearly that of the pore diameter. Fensom and coworkers observed with *Nitella* cells that the electroosmotic efficiency which is given by the amount of water molecules transported per cation is proportional to the size of the hydrated ions:

| Ion | | Ca²⁺ | Li⁺ | Na⁺ | K⁺ | H⁺ |

Ion Ca^{2+} Li^+ Na^+ K^+ H^+

Electroosmotic efficiency
(= number of H_2O molecules transported
per cation)   218   186   178   114   44

Such data permit estimates of relative pore sizes (Fensom and Dainty, 1963; Fensom et al., 1965, 1967; Fensom and Wanless, 1967). Thus electroosmosis is readily explicable only by the assumption of hydrophilic pores in the membrane.

Nevertheless, with all of these methods and arguments only circumstantial evidence is provided and no unequivocal proof of the presence of pores in membranes has been found. Passow has stressed very emphatically that no predictions about the structure of the pores can be made, and it is not justified on this basis to envisage pores as cylindrical aqueous channels perpendicular to the membrane surface. Gutknecht (1968) found that labeled water moved through the membranes of *Valonia ventricosa* by independent molecular diffusion, not by flow through pores (see Dainty, 1976). Membranes are not rigid structures and it is not surprising that such pores are not seen in electronmicrographs. An argument against the pore hypothesis is the high electrical resistance of artificial lipid films, $10^5–10^8\,\Omega$ cm² and of living lipoprotein membranes, $10^3–10^5\,\Omega$ cm².

Thus it should not be assumed that the pores, if they exist at all, are finite and temporary structures in the membrane. One may, perhaps, envisage them as parts of the membrane which open and close in a statistical way. But this means that strictly we cannot speak of pore "radii". The pore "radii" obtained by the methods mentioned above should be called "equivalent radii" (Solomon, 1961). Alternatively, the term pore radius could be discarded altogether and the membrane characterized by the parameters which were really measured in each particular case (Passow, 1963; see also Diamond and Wright, 1969).

5.1.3 Conclusions

To conclude this historical survey, we may say that between the mid 1930's and the late 1960's the Danielli-Davson membrane model, modified by the assumption of pores, was a hypothesis which covered all known membrane phenomena. The phase of membrane research described in Sections 5.1.1 and 5.1.2 right from the beginning was molecular biology, although its onset dates back to the end of the last century. Modern molecular biology has more or less questioned what was believed to have been achieved. Bünning (1975) has pointed out that the somewhat naive considerations in the lipid and ultrafilter theories of permeation regarding lipid solubility or particle size as the dominant parameters were steps backward. In the last quarter of the 19th Century Wilhelm Pfeffer already had much more sophisticated views (see frontispiece and quotation), and

Ernest Overton also anticipated the participation of active processes of the cytoplasm in permeation (Bünning, 1975: pages 42–48).

5.2 Modern Membrane Research

5.2.1 General Considerations

By new methodological developments, membrane research has entered a pace of such rapid expansion that a brief survey is difficult to give. Modern macromolecular chemistry allows much more detailed investigations of proteins and of lipoprotein complexes than was possible earlier. In addition, cell fractionation permits the investigation of specific membranes, especially of chloroplasts and mitochondria, by biochemists and biophysicists studying energy transfer processes bound to membranes. Chloroplast and mitochondrial membranes now can be readily isolated from plant cells, but also other membrane fractions have been obtained including ER and Golgi membranes, plasmalemma, and tonoplast. Electronmicroscopy with technically highly developed instruments and new methods of preparation (for instance, freeze etching) provides pictures of membranes of increasingly improved quality.

The Danielli-Davson model is no longer sufficient and must be modified or replaced. Thus far, however, there is no new final and complete general hypothesis on the structure of membranes. Various membrane models incorporate more or less of the salient features of the Danielli-Davson model. A multiplicity of membrane models is available not only because of the lack of agreement about details, but also because various membranes within the cell have very different functions and therefore must have different molecular structure. Thus, the hypothesis of the "unit membrane" loses its image, at least to the extent that similar structure and a close relation of all membranes is implied. One perhaps can still use it as a hypothesis of an "elementary membrane" (Sitte, 1966) which envisages all membranes as elementary morphological units of varying molecular fine structure.

The preceding statements are illustrated by a brief consideration of the inner membranes of mitochondria and chloroplasts. Invaginations of the inner membranes of these organelles form particular membrane systems, i.e., cristae, tubuli, and sacculi in the mitochondria, and thylakoids in the chloroplasts. Systems serving biological energy transfer are localized in these membranes. In the mitochondria they consist of enzymes and cofactors of the respiratory redox chain which are associated forming specific and distinctly oriented respiratory units (cf. Lehninger, 1964). The thylakoid membranes of the chloroplast contain the photosynthetic pigments and other redox components of the photosynthetic electron transport chain. The specific spatial arrangement and orientation of the various enzymes, pigments, and cofactors in the membrane matrix are coordinated so that the reactions occurring on them are also spatially ordered, and thus "vectorial reactions" become possible. According to the Mitchell hypothesis (Mitchell, 1961, 1962; Robertson, 1968) on both the cristae and the thylakoid membranes charge separation takes place consistent with the scheme:

$$
\begin{array}{c}
H_2O \\
\\
H^+ \qquad\qquad OH^-
\end{array}
$$

The H^+ and the OH^- ions after formation are separated by the membrane. The energy which is contained in this potential gradient can be used by the cell in various ways. (See Sects. 7.2.2, 7.2.3; Fig. 7.5.) These rather general remarks may permit a few conclusions on the principal properties which membranes can have:

1. Membranes are complex structures. They not only consist of a matrix of structural protein and structural lipids, but they can also contain highly specific enzyme, pigment, and cofactor molecules.
2. By chemical variability of the protein and lipid molecules constituting them, and, depending on their function, various membranes can have different structures.
3. Membrane structure provides a high degree of order in which specific molecules can form complex functional units.
4. Enzymatic catalysis and other processes in membranes can lead to spatially directed, i.e., to vectorial reactions; the membranes are asymmetric.

5.2.2 New Membrane Models Based on Chemical and Physical Evidence Obtained With Artificial and Biological Membranes

Let us return from the specialized membranes of mitochondria and chloroplasts to simpler membranes which function largely as barriers, but also as locking gates. Typical examples for this are the plasmalemma at the outer surface of the cytoplasm and the tonoplast at its inner surface enclosing the vacuolar sap. New discoveries of chemical and physical investigations with artificial membranes, and with biological membranes, require a modification of ideas developed in the earlier phase of membrane research.

Roentgenological and electronmicroscopical methods have revealed that the lipid double layer is not the only energetically possible configuration of artificial lipid membranes at polar phase boundaries. Under particular conditions, lipid molecules can also be ordered in hexagonal patterns or lipid globuli can be formed. In this respect surface-active or "membrane-active" substances, e.g., cardiac glucosides (e.g., ouabain), detergents, and phospholipids play an important role. For instance these substances influence the arrangement of single molecules in artificial lipid films when they are added, in small concentration, to mixtures of lipids or lipid extracts from biological material. Artificial membranes of varying catalytic activity can be made in the Langmuir trough with appropriate lipids on top of an aqueous subphase containing appropriate proteins (see Hoelzl-Wallach and Fischer, 1971).

The effect of membrane-active substances on biological membranes can be investigated readily by aid of hemolysis. Among all biological objects, erythrocytes most closely represent the simple outside-inside model (Sect. 1.1, Fig. 1.2), because they are not compartmented, and have only one unit membrane as their outer boundary. Membrane-active substances alter membrane permeability bringing about volume changes and, in case of very drastic effects on membrane structure, a rupture of the erythrocytes may occur (hemolysis). The volume changes can be recorded by the aid of light scattering of erythrocyte suspensions, and thus the effectiveness of various membrane-active substances can be assessed. Among plant objects the tissue of red beet has proven to be well suited for similar studies. Although its cells are far more complex than the erythrocytes, they have an important characteristic. If the semipermeability is disturbed, they lose the red pigment (betacyanin) which is localized in aqueous solution in the vacuoles. This can be measured colorimetrically directly in the external solution. With other tissue similar experiments can be done, although detection of the substances leaking from the tissue in these cases needs special analytical efforts (Hendrix and Higinbotham, 1973).

Even more important for the discussion of the structure of biological membranes appears to be the great progress of protein chemistry which provides new knowledge about membrane proteins. A large proportion of the protein isolated from biological membranes at physiological pH values is hydrophobic. In mitochondria this is 50%–70% of the total membrane protein. From this it can be concluded that not all of the membrane protein is hydrophilically bound at the exterior surfaces of the lipid double-layers as presumed by the Danielli-Davson model. Furthermore, it is observed that in certain biological membranes approximately a fourth to a third of the total membrane protein is present in the α-helix configuration. This tertiary structure is not stable on the polar phase boundary. Under the conditions of the Danielli-Davson model, numerous nonpolar amino acid residues would be associated with the aqueous phase and this is thermodynamically unfavorable. One has to assume that the α-helix

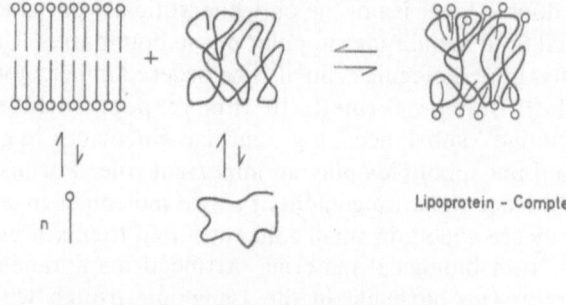

Lipidmolecule   Proteinmacromolecule

Fig. 5.3. Steps in the responses of lipid and protein molecules to form a stable membrane. (Benson, 1966.)

configuration is stabilized by hydrophobic forces in the interior of the lipid membrane (Lenard and Singer, 1966; Wallach and Zahler, 1966). Thus between the proteins and the lipids of the membrane more intimate interactions must be possible than those envisaged by the Danielli and Davson model:

1. There are hydrophilic *and* hydrophobic interactions between the proteins and the lipids in the membranes.
2. Protein must also be in the interior of the membranes.

The ideas of Benson and Singer (Benson, 1966) on the equilibrium reactions taking part in the assembly of lipid-protein membranes will elucidate this further, as shown in Figure 5.3. Still other ways to depict these new ideas about membranes are given by Figure 5.4. Membrane models

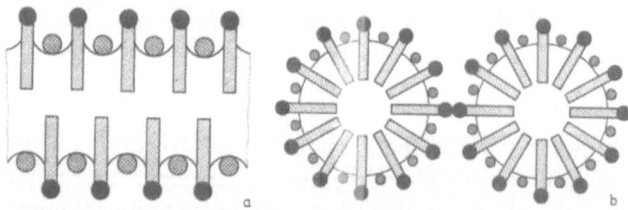

Fig. 5.4. More recent membrane models. **a** Membrane model drawn according to some ideas of Benson and co-workers. The interior of the membrane is formed by hydrophobic protein which interacts with the hydrophobic tails of lipid molecules (*shaded blocks*). The polar groups of the lipids (*black circles*) and proteins (*hatched circles*) are localized at the surface; b membrane model drawn according to some ideas of Sjöstrand and Elfvin (1964). The interior of globular membrane subunits is formed by hydrophobic protein which interacts with the hydrophobic tails of lipid molecules (*shaded blocks*). The polar groups of the lipids (*black circles*) and proteins (*hatched circles*) are localized at the surface of the globular subunits. In this way hydrophilic regions are formed which extend across the membrane. (Lüttge, 1969.)

with globular protein are based on electronmicroscopical observations of globular subunits in the membrane.

A synoptic view of various membrane models is given in Figure 5.5. In Figure 5.5k is incorporated most of the information presently available. The membrane appears as a mosaic of purely lipid regions, purely protein-aceous regions, and regions of lipid–protein interactions. There is some mobility of the components within the membranes, so that we consider membranes as dynamic fluid mosaic structures.

These molecular models should help our imagination. By contrast to the Danielli-Davson model, however, which in its time was a general hypothesis, there is no comprehensive hypothesis at present (for reviews see Hoelzl-Wallach and Knüfermann, 1973; Kotyk and Janáček, 1977). The new developments have important consequences for the discussion of membrane transport phenomena.

5.2.3 Transport Mechanisms as Related to Modern Membrane Models

Membrane models, which assume participation of proteins in the internal structure of the membrane, have advantages for explanation of membrane transport in two respects. They facilitate understanding of permeation through "pores" and they allow explanations of phenomena of catalyzed and active transport.

5.2.3.1 Membrane Pores

The discussion of the circumstantial evidence for the existence of aqueous pores in biological membranes (Sect. 5.1.2.4) has shown that the general idea emerging from the colloquial use of the term "pore," i.e., water-filled channels in the membrane, is certainly not valid. We have to discard this simple image. However, protein extending from one face to the other can provide regions of hydrophilic material continuous across the membrane. Wallach and Zahler (1966) think it is possible that the α-helix type proteins mentioned in Section 5.2.2 constitute rod-like aggregates extending perpendicularly across the whole membrane. At the exterior surface of the protein rods, the lipophilic residues of the proteins would have hydrophobic interactions with the lipids of the membrane interior. Polar groups of the proteins, however, could form the center of the rods (Fig. 5.5g). Such hydrophilic regions extending across the membrane would be pores in a functional sense because they could serve as pathways for hydrophilic particles. They are not spaces left empty by the molecules constituting the membrane. The transport of low molecular weight particles in these hydrophilic regions may occur by interaction with the polar groups of the proteins.

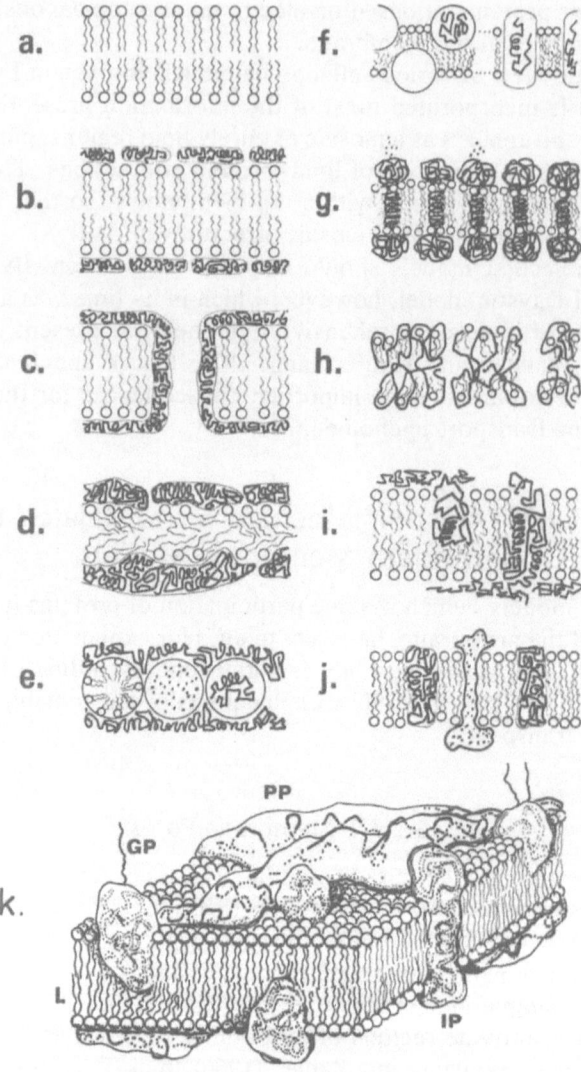

Fig. 5.5. A synopsis of various historical and modern models of cell membrane structure. An attempt was made to use the same graphical means for all models, *a circle with two wavy lines* representing a phospholipid, *heavy twisted line* a polypeptide chain, *dotted areas* a glycoprotein. **a** Gorter and Grendel, 1925; **b** Davson and Danielli, 1943 (see 1952); **c** Stein and Danielli, 1956; **d** Hybl and Dorset, 1970; **e** Lucy and Glauert, 1964; **f** Vanderkooi and Green, 1970; **g** Zahler, 1969; **h** Benson, 1968; **i** Singer and Nicolson, 1972; **j** Green et al., 1972; **k** a comprehensive view of a typical plasma membrane, incorporating most of the recently available information. *L*, Lipid continuum; *IP*, integral proteins; *PP*, peripheral proteins; *GP*, glycoproteins. The number of polar heads of phospholipids should be roughly 10 times greater than the number of integral proteins. (Figure and legend from Kotyk and Janáček, 1977, with kind permission.)

5.2.3.2 Molecular Carrier Models

Carriers have already been mentioned repeatedly (e.g., as a prerequisite for facilitated diffusion, exchange diffusion, and countertransport, Sect. 2.5.2.3). In the present section some molecular models of carrier mechanisms are considered. Principally, one can distinguish two types of carriers, diffusible carriers and carriers fixed or bound in the membrane.

The most simple case of a transport catalysis by a diffusible carrier is depicted in Figure 5.6. The carrier–substrate complex can diffuse freely within the membrane. As we have learned from the discussion of facilitated diffusion, this allows passive transport of S across the membrane. If initially the concentration S_0 at the external face of the membrane is much larger than the concentration S_i there is facilitation of complex formation at the outer face and of dissociation at the inner face. With an equalization of the concentration between the two sides of the membrane net transport stops. We have also indicated, however, that carrier molecules can serve well for the explanation of active transport. Let us assume that the metabolic source of energy is ATP and extend the scheme depicted in Figure 5.6 to derive the model shown in Figure 5.7. Now the energy-rich, or activated, carrier R–P can react with S, and again the carrier and the carrier substrate compounds are freely mobile by diffusion within the membrane. The transport of S in the direction of outside to inside can continue as long as ATP is available. In living cells the availability of ATP for membrane transport depends on metabolic activity.

Models such as that of Figure 5.7 can be tested experimentally. For instance, take as a membrane a piece of filter paper and allow it to imbibe a mixture of lipids. As a carrier one can dissolve in this lipid ''membrane''-phase cholesterol-3'-phosphate representing the activated carrier R–P in our scheme. Cholesterol-3'-phosphate can react with a cation corresponding to the substrate S and form a carrier substrate complex, S–R–P. If the aqueous phase on the right side of the membrane contains a phosphatase which can split the cholesterol phosphate, then a cation transport can occur from left to right until the whole cholesterol-phosphate present initially is used up. It can be imagined that such a

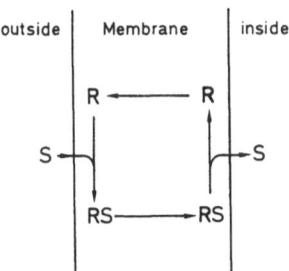

Fig. 5.6. Simple carrier cycle. R, carrier; S, substrate. The carrier R and the carrier-substrate complex RS are assumed to diffuse within the membrane phase.

cycle will go on in living cells indefinitely as long as the carrier is steadily activated by rephosphorylation (Keller, 1960; Netter, 1961).

Another example may show how one can proceed from the initially
quite simple assumption of diffusible carriers to much more complex
models. Let us first keep to the scheme depicted in Figure 5.7. The carrier
R shall be assumed to be a diglyceride molecule. By a diglyceride–kinase
reaction the diglyceride will be phosphorylated to form phosphatidic acid
corresponding to R–P. This then can bind on the left membrane phase
with a cation, S, to form S–R–P and diffuse in the membrane. On the
right side of the membrane a phosphatase splits the S–R–P complex; the
cation and inorganic phosphate go to the inner phase, and R can return
into the cycle. The localization of the kinase reaction in respect to both
sides of the membrane is not important when R, R–P, and S–R–P in the
membrane can diffuse unhindered.

Hokin and Hokin (1959, 1961, 1963) described this model and have
worked it out in more detail experimentally by using, among other materials, the salt gland tissue of sea gulls. They analyzed the enzymes participating in the phosphatidic acid cycle, the turnover of the phosphate, of
the phosphatides, and other important parameters. For stoichiometric
reasons they eventually arrived at the conclusion that, at least in the cases
which they investigated, the model shown in Figure 5.7 is too simple. According to the model, with the utilization of 1 mol ATP, a maximal transport of two cation equivalents should occur, but they found much higher
values.

We now know that macromolecules are subject to steric conformational changes when they react with various ligands (allosteric effect, see
also Sect. 6.2.2.2 and Fig. 6.9). In the process of their work, Hokin and
Hokin discarded the idea of the diffusible carrier. Instead they assumed

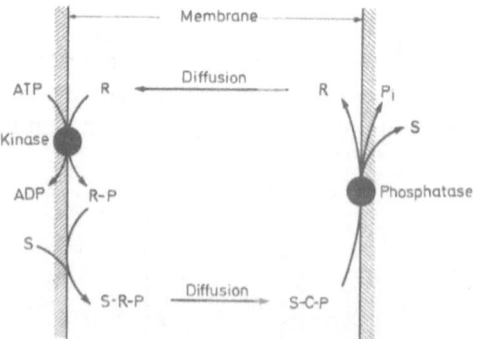

Fig. 5.7. Energetic coupling of a carrier cycle with ATP as a source of energy. P_i,
inorganic phosphate; R, carrier; R-P, phosphorylated carrier; S-R-P, carrier-
substrate complex. In the model of Keller R is cholesterol and R-P is
cholesterol-3'-phosphate; in the model of Hokin and Hokin R is diglyceride, R-P
is phosphatidic acid.

that in the living membrane the phosphatidic acid carrier is bound to a lipoprotein complex for which it is just a prosthetic group. By the various processes taking place at this prosthetic group during the phosphatidic acid cycle, changes are assumed to occur in the configuration of the macromolecular part of the transport site by which, first, specific cation binding sites become available, and second, spatial changes occur in the orientation of these binding sites in relation to the two sides of the membrane.

According to Hokin and Hokin, these two principles can explain an ATP-utilizing coupled Na^+–K^+ carrier transport as follows: Let us start with the phosphatidic acid bound to the carrier lipoprotein on the right hand side of the membrane. By the phosphatase reaction, the phosphatidic acid will be changed to the diglyceride. By the change of configuration associated with this reaction, specific Na^+ binding sites can become available to the Na^+ ions in the adjacent solution. If the Na^+ ions now occupy these sites a second change of configuration is supposed to turn the Na^+ binding sites, together with the Na^+ bound, by 180°, i.e., to the left side of the membrane. In this way Na^+ is transported from right to left. If now the diglycerokinase localized there changes the diglyceride to phosphatidic acid, another change of configuration is possible which affects the Na^+ binding sites in a way such that the Na^+ becomes liberated and a binding specificity for K^+ will be formed. The binding of K^+ will now cause another change of configuration which leads to turning the active sites back by 180° to the right side of the membrane. The phosphatidic acid phosphatase now can act again and K^+ will be liberated on the right side of the membrane. Then the cycle can start again.

This is an imaginative model which Hokin and Hokin used to build a clear hypothesis. Here it was used for the purpose of showing the principles of how carrier mechanisms can function without the assumption of diffusible low molecular weight carriers in the membrane. Symbolically, for instance, one can envisage the function of carriers bound in the membrane by the turning door mechanism which has been considered for some time (Fig. 5.8).

We have concluded that the Danielli-Davson model, modified by the assumption of pores, can explain passive membrane transport of low molecular weight solutes according to the lipid-filter theory. Explanations of active transport can be in agreement with the Danielli-Davson model if small lipid-soluble carrier molecules are assumed which diffuse within the membrane and can react with energy-providing metabolites and with the transported solute particles. Binding of the solute particles to such a carrier would make the particles more lipid-soluble and allow their movement across the lipid phase of the membrane. The alternative of a metabolically controlled movement in aqueous pores is more difficult to imagine. This could be possible, however, assuming metabolically dependent regulation of the width of the pores. Some time ago, it was as-

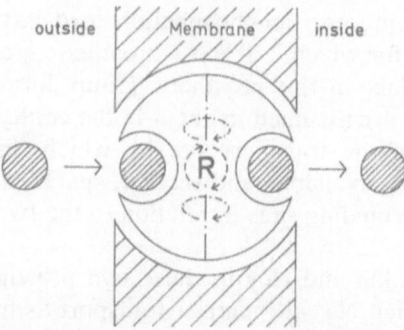

Fig. 5.8. Scheme of the "turning door mechanism". The transported particles are drawn as *hatched circles*. The carrier *R* is assumed to be able to turn either within the plane of the drawing (*dashed arrow*) or perpendicularly to it (*arrow with points and dashes*). In this way the carrier *R* can mediate movement of particles across the membrane.

sumed that the protein lining the pore walls of the Danielli-Davson model (Figs. 5.2 and 5.5c) is a contractile protein and that ATP plays a role similar to that in muscular contraction. When the protein of the pores contracts, substances bound to this protein prior to contraction were assumed to be liberated and thus moved through the pores (Goldacre hypothesis). Carrier mechanisms based on steric conformational changes of specific macromolecules integrated into the membrane matrix are much more easily understood on the basis of the more modern membrane models than on the basis of the Danielli-Davson model. Only protein macromolecules appear to provide the wide variety of structural possibilities needed to explain a multiplicity of specific transport processes which is presumably mediated by carriers. Prosthetic groups, or also small peptides bound to membrane proteins, could participate and add to variability. Shamoo and MacLennan (1974) think that "it seems appropriate to frame the hypothesis that transport mediators whether carriers or channels, cyclic or otherwise, are a universal property of membrane-bound transport proteins. These transport mediators are likely a covalently bonded integral part of the membrane-bound protein and are likely to be ion-specific."

5.3 Molecular Characterization of Carriers

Much circumstantial evidence has been provided suggesting the operation of enzyme-like carrier mechanisms in biological membranes, e.g., kinetic data (applicability of Michaelis-Menten kinetics; Sect. 2.5.2.3; Chap. 6), the observation of specific exchange processes (Sect. 6.1.2.2) and artificial models (Fig. 5.7 in Sect. 5.2.3.2). In the present section we aim to assess the question of whether we can physically grasp carrier molecules of biological membranes. Modern methods of molecular biology should provide approaches for this. These include genetics, induction phenomena, isolation, and reconstitution. If transport processes are regulated genetically, if they can be induced or repressed, they must have a molecular basis. The most rigorous proof for the real existence of carriers, of course, is isolation, characterization of molecular properties, and reconstitution of transport capacity of membranes by readdition of carrier molecules isolated from them. Auspicious achievements of molecular biology have long been limited to micro-organisms, but modest progress with eukaryotic plant cells is gradually being made.

5.3.1 Induction and Repression, Genetic Control, and Isolation of Carriers in Micro-Organisms

Mutants of *Salmonella typhimurium* have been isolated in which the ability for active SO_4^{2-} uptake, characteristic of the wild type, has been lost by gene repression. After derepression the mechanism of SO_4^{2-} transport became intact again. By the method of osmotic shock a certain protein fraction can be isolated from *Salmonella*, which appears to be involved in SO_4^{2-} uptake by the cells. From *Salmonella* with an intact SO_4^{2-} transport system a protein was obtained which had a high affinity for binding of sulfate. No SO_4^{2-} binding protein was found in the defective mutants. The SO_4^{2-} binding protein from SO_4^{2-} transporting cells was purified and even crystallized. It has a molecular weight of about 70,000 and very specifically binds about one mol SO_4^{2-} per unit of molecular weight. This binding is about 10^5 times as strong as the SO_4^{2-} binding to a Dowex-1 ion exchange resin. A number of facts suggest that this binding protein is localized on the cell surface; e.g., its ready separation from the cell by osmotic shock. Furthermore, antibodies for this protein bind to the cell surface. From this evidence it may be concluded that the SO_4^{2-} binding protein in *Salmonella* is an important part of the SO_4^{2-} transporting system. Nevertheless, the SO_4^{2-} carrier probably has a still more complex structure. This conjecture results, for one thing, from the fact that the SO_4^{2-} binding protein can be separated from the cell so easily by osmotic shock, suggesting that it cannot be bound very strongly in the membrane matrix. There are also mutants which do not take up SO_4^{2-} but which are able to bind SO_4^{2-}, showing that the molecular organization of the SO_4^{2-} carrier in *Salmonella* must be complex. These investigations on *Salmonella*, reviewed from the work of Pardee (Pardee, 1967, 1968; Pardee and Prestidge, 1966), clearly demonstrate genetic control of carriers and the possibility of isolating them.

One of the most outstanding examples for genetic control and inducibility of a transport mechanism is β-galactoside uptake by cells of *Esche-*

richia coli. By the addition of lactose (= gluco-β-galactoside) or other β-galactosides to the medium of *E. coli* cells, the ability to transport and to metabolize lactose can be induced. This is controlled by the lac operon. A distinction between uptake and metabolism of lactose can be made by aid of mutants and using β-galactosides which are not metabolized. It turns out that a specific gene, the z-gene of the lac operon is coding the β-galactoside permease (or carrier) of *E. coli*.

Fox and Kennedy (1965) have worked on isolation and characterization of the β-galactoside carrier from *E. coli*. They found an active membrane component M, which was more firmly bound to the membrane matrix than the SO_4^{2-} binding protein of *Salmonella*. Purification of the component M showed that it is a lipoprotein complex with high affinity for β-galactosides which, however, is not identical to β-galactosidase. After induction of the β-galactoside-permease with isopropyl-β-D-thiogalactoside, and subsequent addition of methyl-β-D-thiogalactoside as a transportable substrate, the incorporation of inorganic phosphate (P_i) into the phospholipid fraction of the cells was increased (Nikaido, 1962). A comparison with the model of Figure 5.7 shows that this is a necessary consequence of a cycle in which a lipophilic carrier is continuously reactivated by phosphorylation. ATP utilized as an energy source in carrier phosphorylation has to be steadily resupplied by synthesis from ADP and inorganic phosphate explaining the P_i incorporation.

Fox and Kennedy (1965) have developed a model according to which the membrane component M can mediate both passive facilitated diffusion and active transport of β-galactosides (Fig. 5.9). The carrier mediates passive uptake when the *E. coli* cells have an active β-galactosidase within the cytoplasm which continuously keeps the internal β-galactoside pool at a low level. This drives β-galactoside uptake downhill without active transport. In the case of metabolically driven β-galactoside uptake, by consumption of energy-rich phosphate compounds, the carrier M is supposed to be changed to the configuration M_i in which it has only very low affinity for β-galactosides. When M_i moves in the membrane to the outer face its configuration is supposed to be transformed into the active form, M, in which the carrier has high affinity and an active galactoside transport from the outside to the inside becomes possible.

In the realm of micro-organisms there is a wealth of other examples. One more case of isolation and molecular characterization of a carrier system is that of glucose transport of bacteria described in Section 10.2.3.1 (Fig. 10.3). Inducible and genetically controlled permeases, or carriers, for amino acid uptake by bacteria have also been described. (Review Kaback, 1970a,b; see C.W. Slayman, 1973, for an extensive and comprehensive tabulation of transport mutants of micro-organisms.) As compared with eukaryotic cells prokaryotic micro-organisms represent a special case additionally because their energy transfer systems and redox chains reside in the outer membrane. Strictly speaking, all vectorial reac-

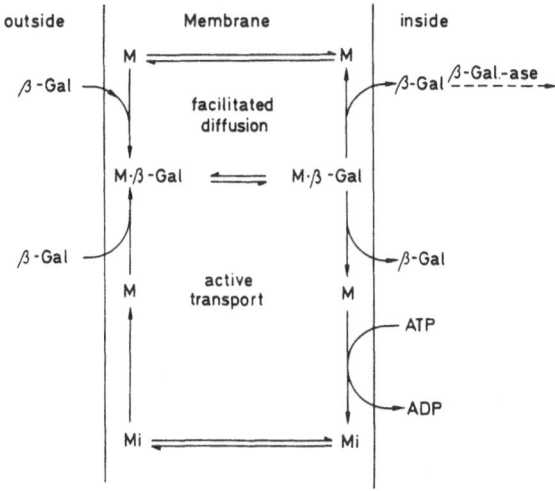

Fig. 5.9. Transport of β-galactosides in cells of *Escherichia coli* according to the ideas of Fox and Kennedy (1965). The carrier or membrane component *M* mediates facilitated diffusion or, in the case of supply of metabolic energy, mediates active transport of β-galactosides across the membrane. *M*, active, M_i, inactive form of the carrier.

tions—for instance the processes which lead to charge separation across membranes during energy transfer (Sect. 5.2.1)—are also transport processes because the formation of intermediary products is spatially and vectorially determined. It will be shown in Chapter 7 that a large number of carriers can also be identified in the inner membranes of mitochondria and chloroplasts of eukaryotic plant cells.

5.3.2 Genetic Control and Induction of Transport in Eukaryotic Plant Cells

The most famous example of genetic control of transport in higher plants dates back to 1943, when Weiss (1943) described two soybean varieties of *Glycine max,* one of which was efficient in Fe absorption and the other not. This was due to a single gene mutation. A considerable number of other examples of genetic and varietal differences of transport processes in higher plants have been observed. They were reviewed and tabulated by Epstein and Jeffries (1964), Epstein (1972), and Läuchli (1976b). Most of these examples refer to genetic control of micronutrient transport. In this case varietal differences and mutants can be most readily recognized by detection of micronutrient deficiencies. Examples showing genetic control of macronutrients are more scarce. In a fair number of species, however, varietal differences have been observed in salinity adaptation suggesting genetic control of Na^+ and Cl^- transport and K^+-Na^+ selectivity (see reviews cited above; see also Table 6.6). This may have much practical importance in breeding salt-resistant crops for agriculture on saline soils in coastal regions and in the arid zones (Epstein, 1977; Epstein and Norlyn, 1977).

The best example for an induction of transport by the transportable molecules in eukaryotic green plant cells is that of hexose transport in the unicellular alga *Chlorella* depicted in Figure 5.10. (For other aspects of the hexose transport system of *Chlorella* see Sect. 8.2.4.2). When cells of *Chlorella vulgaris* (but not of other *Chlorella* species and the related Chlorococcales *Scenedesmus* and *Ankistrodesmus*) are kept for some time in a medium without hexoses, they lose the ability to take up hexoses. When transportable sugars—or sugar derivatives which are transportable but not metabolizable by the cells—are added, there is a lag phase of about 20 min after which uptake commences. Obviously the transport mechanism has to be induced during the lag phase. Cells continuously kept in hexose solution do not show a lag phase of hexose uptake. The substrate specificity of induction turns out to be similar to the substrate specificity of transport when various hexoses and hexose derivatives are compared. Induction seems to involve a proteinaceous membrane component. Inhibitors of RNA synthesis prevent induction. A membrane fraction (protein?) becomes labeled when ^{14}C-phenylalanine is administered to induced-transport cells but not with noninduced cells. The most important evidence in favor of the involvement of a proteinaceous enzyme-like carrier is the turnover of the induced factor (Fig. 5.10). The half-life of the factor is about 4–6 h. (Reviewed from Tanner and Kandler, 1967; Tanner, 1969; Haass and Tanner, 1974; Tanner et al., 1970, 1974.)

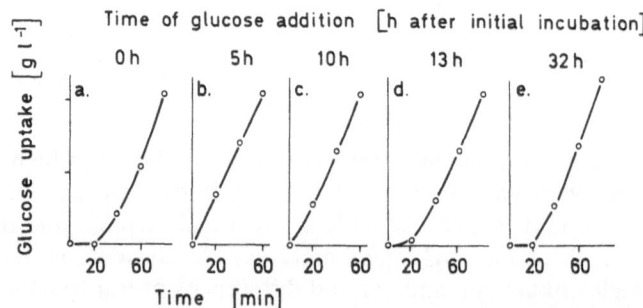

Fig. 5.10. Induction of hexose uptake in *Chlorella vulgaris* by glucose and turnover of the hexose uptake system. At time 0 h, glucose was added to five parallel samples, and the time course of glucose uptake was measured immediately in sample **a**. The other samples completely consumed the glucose, added at time 0 h, within 1.5 h; they received additional glucose after 5, 10, 13 and 32 h, respectively. The time course of glucose uptake in samples **b** to **e** was measured after the second addition of glucose, and the lag phase becomes discernible again at 13 h. (After Tanner et al., 1970, from Lüttge, 1973; Läuchli, 1976b.)

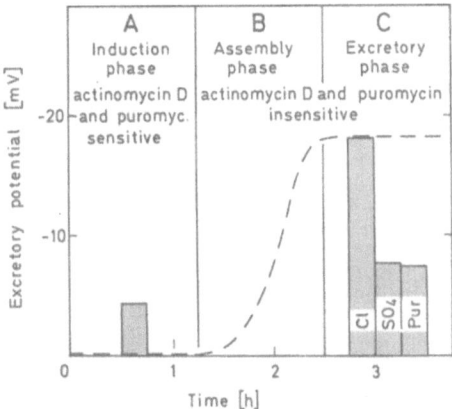

Fig. 5.11. Development of excretory activity in low-salt-adapted leaf disks of *Limonium* in response to a load of 100 mM NaCl at 25° C. *Broken line*, gland activity [excretory potential Ψ (mV)] (from Fig. 9 of Hill and Hill, 1973a). *Bars* represent relative Cl⁻-ATPase activities compiled from Figures 4, 5, 7 and 8 of Hill and Hill (1973b). These ATPase data were normalized for the ATPase activity of membrane preparations (microsomes) of induced leaves tested in the presence of Cl⁻ (*large bar in phase C*; 0.014 to 0.14 μmol $P_i g_{protein}^{-1}$ s⁻¹, depending on the leaf sample and preparation). *Bar in phase A*, ATPase activity of noninduced leaves; *bar-SO₄ in phase C*, ATPase activity of induced leaves tested in the presence of SO_4^{2-} instead of Cl⁻; *bar Pur in phase C*, ATPase activity of leaves treated with NaCl in the presence of puromycin. (Lüttge, 1975; and Läuchli, 1976b.)

An example of a substrate-induced transport process in an Angiosperm is active Cl⁻ excretion by the salt glands in the leaves of the halophyte *Limonium vulgare*. Development of excretory activity in plants adapted to low NaCl concentration after transfer to a medium with 100 mM NaCl is shown in Figure 5.11 to be related to Cl⁻-activated ATPase. Inhibitors of transcription of DNA and translation of m-RNA inhibit development of excretory activity when added during the lag phase in which induction occurs. The induced factor obviously is a Cl⁻-stimulated ATPase. (Reviewed from Shachar-Hill, and Hill, 1970; Hill and Hill, 1973a, b.) This induction process may have ecological significance allowing the plant to adapt to varying salinity of its habitat, i.e., to periods of flooding of salt marshes by seawater alternating with periods of rain.

Nitrate reductase, NR, the enzyme reducing NO_3^- to NO_2^-, is an inducible enzyme in higher plants. In roots of NO_3^- and Cl⁻ starved maize seedlings, NO_3^- uptake but not Cl⁻ uptake proceeds with an initial lag phase (Fig. 5.12, W.A. Jackson et al., 1973). Whether NO_3^- uptake as well as NO_3^- reduction is an inducible process has often been debated. NO_3^- reduction may be responsible for driving passive NO_3^- uptake by maintaining a downhill gradient of NO_3^- concentration. There are even indications that there is a very close relation between NO_3^- transport activity

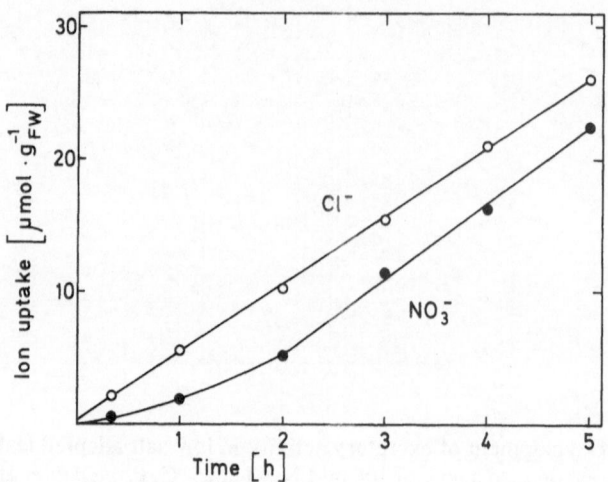

Fig. 5.12. Time course of NO_3^- and Cl^- uptake to excised corn roots exposed to 0.5 mM solutions of $Ca(NO_3)_2$ and $CaCl_2$, respectively. (After Jackson et al., 1973; from Läuchli, 1976b.)

and NR activity (Ullrich-Eberius, 1973; Eisele and Ullrich, 1975; Butz and Jackson, 1977). On the other hand, evidence has been reported that NR and the NO_3^- uptake system are different and independent entities; NO_3^- can be taken up by cells devoid of active NR (Heimer et al., 1969; Heimer and Filner, 1971; Schloemer and Garrett, 1974; Rao and Rains, 1976; see also Vennesland and Guerrero, 1978). Neyra and Hageman (1975) investigated NO_3^- levels and fluxes in excised corn roots, and they suggest that NO_3^- in the medium first induces the NO_3^- uptake system at the plasmalemma, before NO_3^- levels in the cytoplasm rise to an extent allowing induction of NR. In this respect it is noteworthy that nitrite (NO_2^-) can also induce the NO_3^- uptake system (Heimer, 1975) and that in higher plants there seems to be a division of labor to some extent, with predominant NO_3^- reduction in the shoots and uptake by the roots (Sect. 13.3.2, Fig. 13.6).

Inhibitors of protein synthesis at the transcriptional and translational level have been used widely in transport studies (review: van Stevenick, 1976b). Inhibitions observed have been interpreted as showing involvement of proteins as carrier molecules. However, care is needed if inhibitor data are used alone and not in connection with other evidence such as the induction phenomena discussed above. Inhibitors of protein synthesis may affect transport rather indirectly. (For examples, see Sect. 12.2.5.)

A very peculiar but highly interesting case is the transfer of information from some gram-negative bacteria (e.g., *Pseudomonas tolaasii*) to barley roots inducing an uptake capacity for choline sulfate (Nissen, 1968, 1971a, 1973). This induction occurs in at least three steps or phases as shown in Figure 5.13. In phase I an uptake mechanism is induced in the bacteria. For this induction choline sulfate is highly specific. Sulfate uptake in bac-

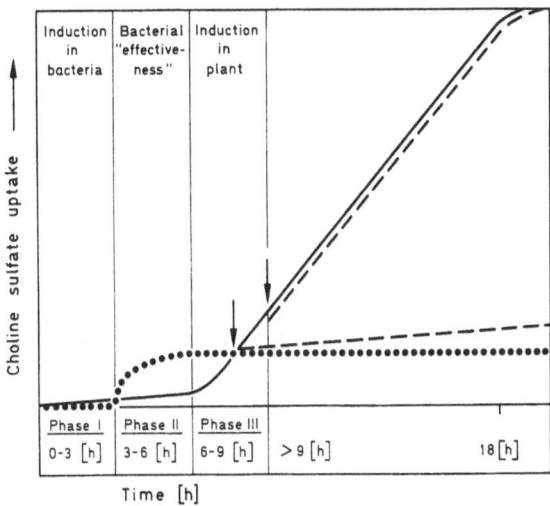

Fig. 5.13. Induction processes and time course of choline sulfate uptake by bacteria and higher plant. Bacteria: gram-negative bacteria, e.g., *Agrobacterium, Pseudomonas, Rhizobium;* higher plant: roots of barley. *Solid line,* plant with bacteria. *Broken line,* plant without bacteria (bacteria removed at time indicated by *arrow*). *Dots, bacteria. Phase I,* induction of a specific permease in bacteria; *Phase II,* bacterial ''effectiveness'' and contact between effective bacteria and plant tissue. *Phase III,* induction of a specific permease in plant tissue. (After Nissen, 1971a; from Läuchli, 1976b.)

teria is not dependent on Ca²⁺. In phase II the bacteria obtain the effectiveness to induce the uptake system in the roots, a property which they do not have at completion of phase I. To gain effectiveness in phase II either choline sulfate or choline must be present in the medium. In phase III the effective bacteria can induce the uptake system in the roots. The uptake mechanism of the plant is distinct from that of the bacteria, for example by its requirement for Ca²⁺. Furthermore, after completion of the induction of the root, the bacteria can be removed without affecting uptake by the roots. It is not clear how the apparent transfer of information between the bacteria and the roots is brought about. The phenomenon may be of ecological significance for associations of bacteria and plant roots in the rhizosphere; the effective bacteria include genera which frequently occur in the rhizosphere. (For review, see Läuchli, 1976b.)

5.3.3 Isolation of Membrane ATPases From Higher Plant Cells

Membrane ATPases have been known for a long time in animal cells (Lowe, 1968). They hydrolyze ATP, and the energy available from this reaction can drive an active K⁺–Na⁺ exchange between the two compart-

ments separated by the membrane. Thus membrane ATPases are important parts of an active alkali ion transport mechanism (see Fig. 10.2).
Membrane ATPases are activated by alkali ions, or by Mg^{2+} and, to a
lesser extent, Ca^{2+}. In animal cells they are specifically inhibited by the
cardiac glycoside, ouabain (strophanthin). An inhibition of K^+–Na^+
transport mechanisms in a few cases has been observed also in plant cells
(e.g., Simonis and Urbach, 1963; Cram, 1968b; Raven, 1967a,b, 1968a,b;
Thomas, 1970). However, usually plant systems are insensitive to ouabain (see Hodges, 1976); nevertheless, considerable evidence has accumulated in the last 5 years showing that membrane ATPases can be isolated from plant cells and may be involved in ion transport across plant
membranes (review, Hodges, 1976).

Cytochemical methods in the electron microscope reveal positive
staining for ATPases in a large number of membranes of plant cells, e.g.,
plasmalemma, endoplasmic reticulum, and tonoplast (Winter-Sluiter et al.,
1977; Malone et al., 1977). Hodges and his coworkers and Kylin with his
group have pioneered in research isolating and characterizing membrane
fractions with ATPase activity from plant cells, and progress has also
been made in many other laboratories. This, together with methods of isolation and purification of membrane fractions and the properties of their
ATPases, was comprehensively reviewed by Hodges (1976); it is apparent
that a number of different ATPases are associated with membranes and
these are unlike those of animal origin.

One of the greatest problems in this work is the identification of the
membrane fractions obtained after grinding plant tissue and purification
by centrifugation in a density gradient. Especially this is true for the plasmalemma and tonoplast, i.e., the two membranes of utmost interest for
ion uptake and accumulation in plant cells (Chap. 6); there are no typical
intrinsic markers for identification (Flowers and Hall, 1976; Hendriks,
1977). For instance, with mitochondrial membranes such markers would
be the components of the respiratory electron transport chain. For some
time it appeared that the plasmalemma of plant cells could be stained specifically with a mixture of periodic acid, phosphotungstic acid, and
chromic acid, but later ambiguities have been revealed, e.g., the plasmalemma not staining in some cases and the tonoplast also staining under
certain conditions (Hall and Baker, 1975; Hall and Flowers, 1976; Hendriks, 1976). The detection of intrinsic markers that have been allocated
to the plasmalemma in all cases can be traced back to an identification by
the phosphotungstic acid/chromic acid-staining method and thus the usefulness of the work is contingent on the reliability of this method. Also the
possibility cannot be excluded that ATPases may become associated with
membranes during grinding and purification. Hendriks (1977) has
suggested a way of labeling the plasmalemma proteins externally by
lactoperoxidase-mediated iodination prior to grinding; this may lead to
some progress in confirming identification of plasmalemma fractions.

The best evidence for an involvement of membrane ATPases in ion transport of plants seems to come from a different approach, in which the properties of transport and of ATPases are subject to a detailed comparison. Figure 5.14 shows that in four cereal species there are similar kinetics of K^+ or Rb^+ uptake and ATPase activity in relation to KCl or RbCl concentration of the medium. Among the four cereals the relative rates of uptake correspond to the relative ATPase activities. Figure 5.15 gives a more detailed kinetic comparison of K^+ influx into oat roots and the ATPase from a preparation which most likely was a plasmalemma fraction. Sze and Hodges (1977) have carefully compared the selectivity and Ca^{2+} dependence of Li^+, Na^+, K^+, Rb^+, and Cs^+ uptake by oat roots and, respectively, of the properties of ATPase. Ca^{2+} inhibited the uptake of all these ions and altered the selectivity of uptake. The K^+, Rb^+, and Ca^{2+} influxes and the relative stimulation by these ions of the ATPase of the plasmalemma fraction were closely correlated. Na^+ and Li^+ are presumably transported by a different system. Sze and Hodges (1977) conclude that

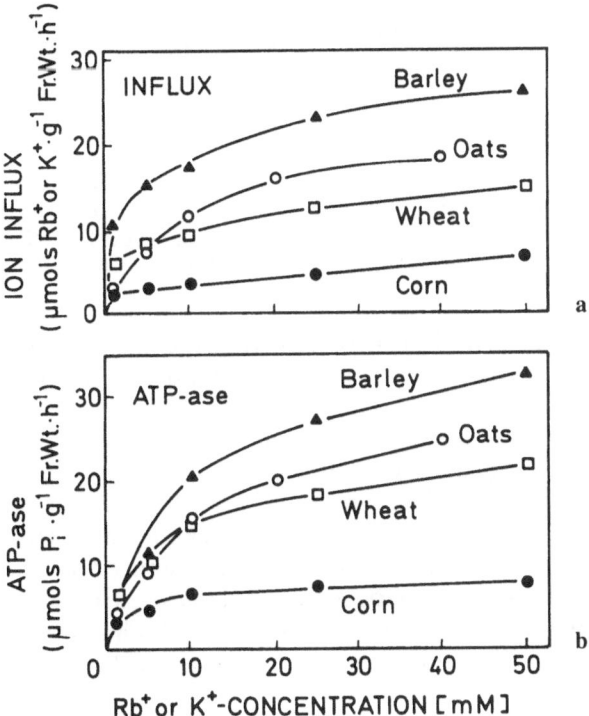

Fig. 5.14. a The effect of KCl or RbCl concentrations on the influx of K^+ into oat roots and the influx of Rb^+ into roots of barley, corn, and wheat; **b** the effect of KCl or RbCl concentrations on the ATPase activity that is stimulated by these ions (the activity in the presence of Mg^{2+} was subtracted) in membranes isolated from roots of the four plant species. (Fisher et al., 1970: and Hodges, 1976.)

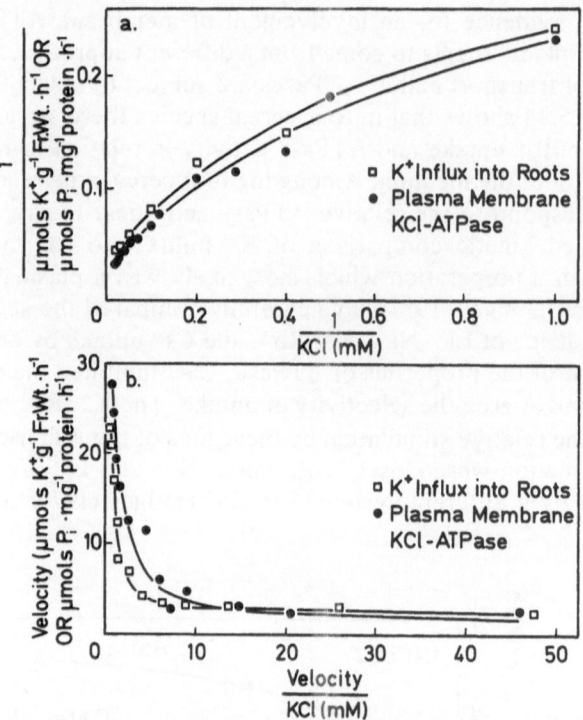

Fig. 5.15. a Lineweaver-Burk plot of K^+-stimulated ATPase of plasma membranes of oat roots and $^{42}K^+$ influx into excised oat roots; **b** Eadie, Hofstee plot of K^+-stimulated ATPase of plasma membranes of oat roots and $^{42}K^+$ influx into excised oat roots. (Adapted from Leonard and Hodges, 1973; from Hodges, 1976.)

ion uptake specificity in part results from specificity of the ATPase, but other factors such as specific carriers and differential diffusion rates can also be involved.

The membrane-bound vesicles of microsomal membrane fractions also can take up ions passively in vitro (Sze and Hodges, 1976) and actively (Gross and Marmé, 1978). Gross and Marmé observed ATP-dependent Ca^{2+} uptake into membrane vesicles obtained from several higher plants and an alga. They could not unequivocally identify their membrane fraction, but an origin from mitochondria, etioplasts, ER, and Golgi vesicles was excluded. (Hodges believes that the vesicles he isolated are from the plasmalemma.) The membranes contained an ATPase and an adenylate kinase.

A promising recent development is the isolation of intact vacuoles surrounded by the tonoplast membrane (Wagner and Siegelman, 1975); quite pure tonoplast membranes can be obtained in this way and they have been shown to contain ATPase (Lin et al., 1977) presumably active in ion transport. The isolated vacuoles apparently retain their content well, during isolation, for organic materials (Buser and Matile, 1977) and for inorganic ions (Lin et al., 1977; also see Sect. 2.2.3.2).

Reconstitution of an ability to transport Na^+ and K^+ by addition of purified and active $(Na^+ + K^+)$-ATPase to membrane vesicles was achieved with animal systems. Sze and Hodges (1976) summarized the literature: "Na^+ and K^+ transport in vesicles prepared from egg lecithin can be coupled to ATP via the purified $(Na^+ + K^+)$-ATPase of the plasma membrane of canine brain gray matter and the rectal gland of *Squalus acanthias*" (see Goldin and Tong, 1974; Hilden and Hokin, 1974).

Chapter 6

The Simplified Cell Models of Transport Physiology

In Chapter 2 we considered a simple two-compartmental model with out-side and inside phases. We have seen how useful it would be for investigations of basic problems in transport. However, living cells have very many compartments separated by membranes across which transport of a multiplicity of solutes occurs. One can try to understand each of these transport processes and find out to which extent they are co-operating or unco-operative.

Another approach is not based on the obvious multiplicity of compartments and transported solutes, but rather starts by simply measuring the uptake of a given solute from the outside. From this approach several models have emerged.

6.1 The Model With Two Compartments, Outer and Inner

6.1.1 The Outer Diffusion Barrier of Plant Cells

Models which consider the plant cell as a simple osmotic system (see Fig. 2.4) assume that the total of the cytoplasm including the plasmalemma and the tonoplast acts as semipermeable barrier.

If one tries to differentiate further, the question arises as to which of the two membranes, plasmalemma or tonoplast, is the more important barrier to the external medium. Of course, the most suggestive assumption is that this is the external plasma surface, the plasmalemma. This is simply expected on the basis of the necessity of an effective separation of metabolism, occurring within the cytoplasm, from the external medium. On the other hand, experimental findings have suggested that part of the cytoplasm, or all of the cytoplasm, belongs to the apparent free space (AFS) and that the tonoplast is the first barrier bordering the AFS. This can depend very much on the state of the plant material used and on the experimental conditions, especially on the external solute concentration. This was a serious problem some time ago in discussions of ionic relations of plant cells. However, more recent measurements of electropotential differences across the plasmalemma and across the tonoplast clearly demonstrate that under physiological conditions the plasmalemma is the decisive external barrier of the cell (Sect. 2.2.3.2, Table 2.2).

If this is so, transport processes must permit uptake and release of substances at the plasmalemma. Kinetic investigations and experiments concerned with selectivity and counterion effects during ion uptake have revealed distinct mechanisms of uptake. These mechanisms have been assumed by some authors to be exclusively localized in the plasmalemma. We now deal with the results of such investigations in detail before we can go to more complex cell models.

6.1.2 Dual-Isotherm Michaelis-Menten Kinetics of Ion Uptake

6.1.2.1 The Kinetic and Qualitative Characterization of System 1 and System 2 of Ion Uptake

In 1937, in a journal receiving only very limited distribution, van den Honert reported experiments showing that the rate of absorption of phosphate by sugar cane plants followed a hyperbolic relation with increasing external phosphate concentration. In 1952 Epstein and Hagen studied the effect of external cation concentration on rate of uptake. Generally ion uptake by plant tissue with increasing external concentration approaches a saturation or maximal rate.

The curves describing the relationship between external concentration and rate of uptake are called absorption isotherms because they are obtained at constant temperature generally using tracer-labeled ions but varied concentration. The uptake into the AFS in these experiments is eliminated by rinsing in nonlabeled solution (see Sect. 4.3). An analysis of ion uptake isotherms showed that they obey the Michaelis-Menten equation [Eq. (2.45) Sect. 2.5.2.3]. Thus, formally, ion uptake kinetics are identical with the kinetics of some enzymes. Various authors have investigated different ranges of concentration. Initially, the salt concentrations chosen most often were between 1 and 50 mM. Somewhat later K^+ uptake rates by barley roots were investigated using a very wide range of concentrations (10^{-6} to 10^{-2} M), and it was discovered that there are two systems participating in ion uptake; these show saturation in two different concentration ranges (Fried and Noggle, 1958):

1. System 1 with a low Michaelis constant, i.e., with high affinity for the ions, and with a low maximal rate, and
2. System 2 with a high Michaelis constant, i.e., with low affinity, and with a high maximal rate.

Therefore there appear to be two inflections, or there is a dual isotherm of ion uptake. These kinetics were investigated over many years especially by Emanuel Epstein and his coworkers. It turned out later on that the system 2 isotherm has a characteristic fine structure and is separable into multiple inflections. Figure 6.1 shows such a curve. Meanwhile similar curves have been obtained by many authors for various plant tissues (roots, leaves, storage tissues, unicellular algae) and for several different ions (see Table 6.2 in Epstein, 1972; and Table 3.4 in Epstein, 1976). It is quite striking that although the maximal rate can differ, more or less depending on the tissue and on the ion species, the saturation concentrations and the critical concentrations where the isotherms have inflections are very similar (Epstein, 1966, 1972; Lüttge, 1968, 1969).

The two systems of ion uptake are not only characterized kinetically, they can also be characterized qualitatively, especially by their selectivity and also by counterion effects. These qualitative properties have given rise to the idea that system 2, in spite of its multiple inflections, can be regarded as relatively homogeneous and as a whole quite clearly distinguished from system 1.

An enumeration of the more important properties of both ion uptake systems is given in Table 6.1. Largely, results with roots are cited. In the first row we find some typical values of Michaelis constants. The ion for

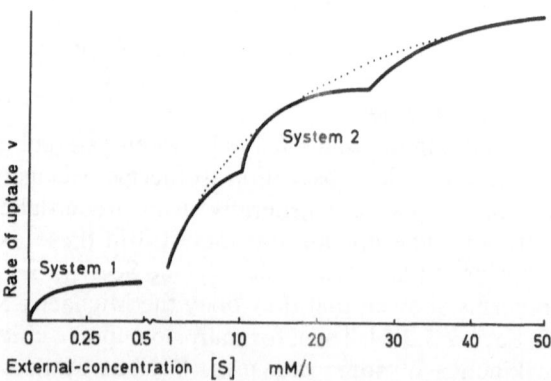

Fig. 6.1. Concentration-dependence of ion uptake by plant cells: dual-isotherm Michaelis-Menten kinetics (Epstein and co-workers, and other laboratories since 1952). Basically there appear to be two systems but system 2 may show multiple saturation points.

Table 6.1. Properties of system 1 and system 2 of ion uptake (see also Table 6.1, p. 129 in Epstein, 1972). (For K_M see Eq. [2.46])

		System 1	System 2
K_M [mM/l] for	K^+Cl^-, barley roots[a]	0.02	11.5 (up to 50 mM)
	K^+Cl^-, maize roots[b]	0.09	1 (up to 15 mM)
	Rb^+Cl^-, maize roots[b]	0.10	4 (up to 10 mM)
mechanisms cation uptake		$[K^+, Rb^+, Cs^+], [Na^+]$ $[Ca^{2+}, Sr^{2+}, Ba^{2+}], [Mg^{2+}]$	$[K^+, Rb^+, Cs^+, Na^+]$ $[Ca^{2+}, Sr^{2+}, Ba^{2+}], [Mg^{2+}]$
mechanisms anion uptake		$[Cl^-, Br^-], [NO_3^-]$ $[SO_4^{2-}]$	$[Cl^-, Br^-, NO_3^-]$ $[SO_4^{2-}]$
counterion effects: various anions as counterion		$V_{K^+(KCl)} \cong V_{K^+(K_2SO_4)}$ $V_{Na^+(NaCl)} \cong V_{Na^+(NaF)} > V_{Na^+(Na_2SO_4)}$	$V_{K^+(KCl)} > V_{K^+(K_2SO_4)}$ $V_{Na^+(NaCl)} > V_{Na^+(Na_2SO_4)}$
various cations as counterion		$V_{Cl^-(KCl)} \cong V_{Cl^-(CaCl_2)}$	$V_{Cl^-(KCl)} > V_{Cl^-(CaCl_2)}$

[a] From Epstein et al. (1963).
[b] Calculated from data of Torii and Laties (1966a).

which uptake was investigated is given by bold letters and thus can be distinguished from the counter ion. For system 2 the concentration range in which the Michaelis constant was determined is given in brackets. In the second and the third rows anion and cation uptake mechanisms are listed. Ions which compete with each other in uptake are included within the same brackets. Ions which are in separate brackets are transported more or less independently of each other. For instance, the addition of Br^- to a Cl^- solution inhibits Cl^- uptake in both concentration ranges, whereas the addition of SO_4^{2-} has no effect on Cl^- uptake. K^+, Rb^+, and Cs^+ compete with each other, but in the low concentration range Na^+ is obviously transported independently of the other three alkali cations; in the high concentration range, however, Na^+ competes with the other alkali ions. This indicates that there are a total of three mechanisms for alkali ion uptake. As a consequence, in the low concentration range there is a high selectivity. Conversely, the Na^+–K^+ selectivity in the high concentration range is low.

6.1.2.2 The Mechanism of System 1 and System 2 of Ion Uptake

After this description of the two ion uptake systems the question arises as to whether we can make any conclusions about the mechanism. All properties of both systems can be interpreted as circumstantial evidence for the existence of carrier mechanisms. We have seen in Section 2.5.2.3 that carrier mechanisms obey Michaelis-Menten kinetics. This criterion is fulfilled by both systems. Epstein and coworkers have used this as a base for their enzyme-kinetic hypothesis of ion transport and carrier function (review Epstein, 1976). The two uptake systems have been interpreted as representing two different carrier mechanisms. In addition, it has been assumed that the system 2 carrier has various active sites by which the fine structure, i.e., each of the kinetically separate parts of the system could be explained. Ion antagonisms and the phenomenon of selectivity can also be readily explained on the basis of the carrier hypothesis as competition for the active binding sides on the carriers. These phenomena therefore add to the support for the carrier hypothesis.

In Section 2.5.2.3 it has been shown that carriers may mediate exchange of solutes between two compartments, if the solutes have appropriate affinities for the carrier binding sites on both membrane faces (Fig. 2.18). These conclusions were based mainly on work with erythrocytes. However, such specific exchange also can be observed in investigations of ion transport systems in plant cells. Figure 6.2 shows a relevant experiment. To demonstrate a specific exchange the tissue is first loaded by long-term uptake with radioactive ions. Then the rate of exchange of the tracer in the tissue against nonradioactive ions in the external solution is measured over a range of external ion concentrations. The efflux

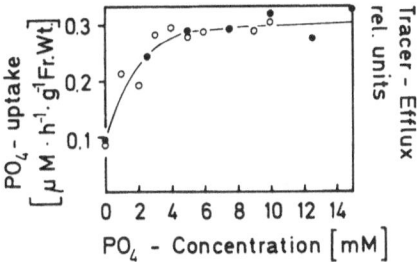

Fig. 6.2. Phosphate uptake (tracer influx) and phosphate release (tracer efflux) as related to phosphate concentration in the external solution. *Closed circles*, tracer influx; μmol h^{-1} g^{-1} Fr.Wt. *Open circles*, tracer efflux; relative units which were normalized so that it becomes clear that the influx and efflux isotherms are kinetically identical. (After Figures 1b and 2 from Weigl, 1968.)

isotherms so obtained exactly correspond to the uptake isotherms. The exchange is very specific; for instance $H_2PO_4^-$ exchanges only with $H_2PO_4^-$, Cl^- exchanges only with Cl^- (Weigl, 1968; Migliaccio and Weigl, 1973).

As much as these phenomena on one hand suggest participation of carriers in ion uptake, on the other hand, objections to the value of such circumstantial evidence are possible. Again and again it must be stressed that from the pure formalism of kinetic analysis no clear conclusions can be drawn about molecular interactions and, in this case, the molecular mechanisms of carriers (see also Sect. 5.3). This becomes particularly clear by pointing out that the mathematical form of the Michaelis-Menten isotherm [Eq. (2.45)] also is identical to the Langmuir isotherm which describes purely physical adsorption processes (Netter, 1959). Göring (1976) has made "models" of multicellular higher plant tissues from layers of living yeast cells embedded in gelatin. He argues that diffusion in the free space and unstirred-layer effects can lead to curves like those of apparent dual isotherms.

In the qualitative characterization of properties of system 1 and 2 we have mentioned above $Na^+–K^+$ specificites, and it was suggested that this lends support to the existence of different carrier entities (Table 6.1, Sect. 6.1.2.1). However, it has been emphasized that this selectivity could be due to the participation of alkali ion-dependent ATPases in ion uptake, and that the preferential K^+ accumulation in plant cells is brought about by active Na^+ extrusion (see for instance Dodd et al., 1966; Pitman and Saddler, 1967; Pitman et al., 1968; Cram, 1968b; see also Figs. 6.15 and 6.16). Jeschke (1972) and Jeschke and Stelter (1973) have shown that K^+ in the external medium enhances Na^+ efflux from root cortex cells and at the same time inhibits Na^+ transport into the root xylem. Nevertheless, of course, alkali ion-transporting ATPases also have to be considered as possible carrier mechanisms (Sect. 5.3.3).

A different type of criticism suggests that the phenomena observed
are not exclusively due to metabolism-dependent carrier mechanisms.
Quite a number of interpretations assume that selectivity is brought about
by an interaction of passive and active fluxes. Investigations of Pitman,
1970, suggest that protons may play an important role. H$^+$ ions are extruded from the cells and in exchange external cations, e.g., Na$^+$–K$^+$, may
be taken up thus maintaining electroneutrality. The pH value in the apoplast must change. H-ions can exchange for cations, especially for Ca^{2+},
at the fixed charge sites of the Donnan free space. Thus the pH value in
the immediate vicinity of the plasmalemma is lowered, and this could increase passive membrane permeability. Therefore, with increasing external concentration, and with increasing K$^+$ uptake, the K$^+$–Na$^+$ selectivity could be decreased. Figure 6.3 shows that indeed the critical external
ion concentration in which marked changes of proton extrusion rates become apparent during ion uptake by roots is identical with the concentration at which the system 2 isotherm becomes clearly separated from the
system 1 isotherm. This suggests that the low K$^+$–Na$^+$ selectivity in the
high concentration range may result from changes in passive permeability. It is also notable that there is a large drop of membrane electropotential due to increased external ion concentration, which begins at about the
same concentration at which an increased H$^+$ efflux and a separation of
system 1 and 2 isotherms become discernible (Fig. 6.3; Pitman, Mertz et
al., 1970). The membrane potential is an essential element in an attempt
by Gerson and Poole (1971) to provide a "unary" interpretation of dual
isotherms.

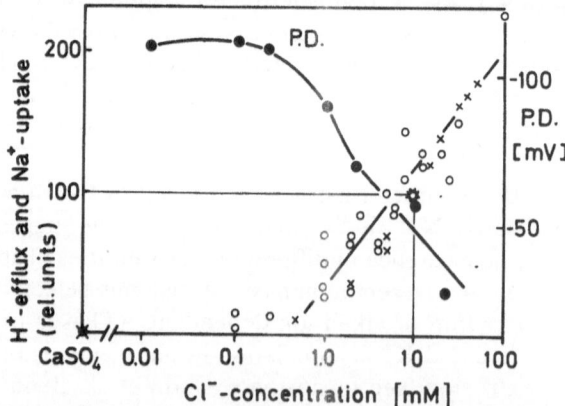

Fig. 6.3. Correlation between H$^+$-efflux, Na$^+$ uptake and membrane potential difference (PD) at increasing Cl$^-$ concentration (zero concentration = pure CaSO$_4$-
solution) after Pitman (1970) and Pitman et al. (1970). *Open circles,* H$^+$-efflux
measurements of Pitman normalized for H$^+$-efflux in 10 mM KCl = 100 (*aster-
isk*). *Crosses,* Na$^+$ uptake according to measurements of Rains and Epstein (1967);
closed circles, PD.

If physical effects on the membrane play a more important role, this perhaps gives an explanation of the strange fact mentioned above, that the saturation concentration of the particular isotherms is so similar for the various tissues and ion species.

Another possibility is that protons affect a carrier. For example Komor and Tanner (1974) have investigated the isotherms of hexose uptake by *Chlorella* cells at different pH. Figure 6.4 shows that there is a shift from a system 1-type mechanism with a $K_M = 0.3$ mM to a system 2-type mechanism with a $K_M = 30$ mM glucose as the pH of the medium is increased. At neutral pH glucose uptake by *Chlorella* shows the dual-isotherm Michaelis-Menten kinetics just as in Figure 6.1 (without the multiple inflections of system 2). H^+-carrier interactions, or protonation of carriers, may be essential in a basic mechanism of energy coupling of active transport (Sect. 10.2.3.2). H^+-cation exchange mechanisms also are important in regulation of charge balance and cytoplasmic pH control.

At this stage we can summarize as follows: ion uptake by plant tissues has kinetic properties which deductively one can regard as a necessary consequence of the operation of carrier mechanisms. On the other hand,

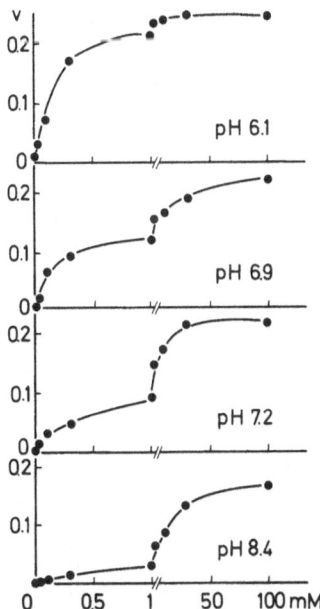

Fig. 6.4. Kinetics of hexose uptake by *Chlorella vulgaris* cells at different pH. Cells (36 μl packed cells) induced for hexose uptake (see Sect. 5.3.2) were incubated in 2.2 ml 40 mM Na-phosphate buffer of the pH as indicated. (^{3}H)-6-deoxyglucose was then added at different concentrations (0.01–100 mM). Samples were taken in half-min intervals, filtered and the amount of 6-deoxyglucose determined. v is given as mmol h^{-1} cm^{-3} packed cells. 0.2 mmol h^{-1} cm^{-3} correspond to 11.2 pmol cm^{-2} s^{-1}. (Komor and Tanner, 1975.)

these are not enough for the inductive conclusion that unequivocal proof exists for the occurrence of carrier mechanisms, at least in the sense of a mobile carrier. Undoubtedly the pathway of solutes is restricted if viewed in terms of other possible models which may conform to similar kinetics (see Chap. 5).

6.1.2.3 The Problem of Cytological Localization of Systems 1 and 2 of Ion Uptake

When the two systems of ion uptake were discovered, and also during the whole phase of the first quantitative and qualitative characterization, it was more or less tacitly implied that both systems 1 and 2 are located beside each other in the plasmalemma. It appears that, after a different scheme was suggested by other authors, placing system 1 at the plasmalemma and system 2 at the tonoplast, Epstein and coworkers (Läuchli, 1972) systematically began to develop evidence for a simple outside-inside model with both system 1 and system 2 in the plasmalemma. This led to an interesting controversy, considered in the following section.

6.2 The Model With Three Compartments: Outside-Cytoplasm-Vacuole

6.2.1 The Torii-Laties Hypothesis

According to the Torii-Laties (1966a) hypothesis, the two systems of ion uptake occur on two different membranes in the cell, namely in the plasmalemma and in the tonoplast, i.e., they are in series with each other. This is possible only if the system which determines the rate of ion uptake by the whole cell in the low concentration range, i.e., system 1, is located at the external barrier, i.e., the plasmalemma. With the two in series the system which determines ion uptake rate at high concentrations, i.e., system 2, can be located only at a barrier farther within the cell, i.e., at the tonoplast. Otherwise, in ion uptake experiments with intact cells and tissues only the second mechanism would be observed. According to the Torii-Laties hypothesis, the reason why system 2 can become rate-limiting for ion uptake into the cell in the high concentration range is that there is a more rapid passive diffusion of ions through the plasmalemma into the cytoplasm than into the vacuole. In the range of high external concentrations the ion concentration in the cytoplasm is thought to increase more rapidly than would be possible by the maximal uptake rate of system 1 according to Epstein. This latter assumption, which follows logically from the assumption of the localization of system 1 and 2 respectively at two membranes in series, has become a critical issue in the controversy between the protagonists of the outside-inside model and those of the outside-cytoplasm-vacuole model. We will come back to this problem later on, but first, we will consider the initial experimental basis for the Torii-Laties hypothesis.

6.2.1.1 Ion Uptake by Vacuolated and Nonvacuolated Root Tissue

If system 2 is located in the tonoplast it should not be observed in experiments with cells which have no vacuoles. As a largely nonvacuolated tissue Torii and Laties chose the tip region of maize roots and compared it with vacuolated root tissue further from the apex. The isotherms obtained in such experiments are shown in Figure 6.5. Vacuolated and nonvacuolated tissues have hyperbolic system 1 isotherms which obey Michaelis-Menten kinetics. A typical system 2 isotherm, however, is found only with vacuolated tissue. The isotherm of ion uptake of nonvacuolated tissue in this concentration range rises with increasing concentration linearly or parabolically (exponentially). Such a linear or exponential curve is interpreted as a diffusion isotherm (Torii and Laties, 1966a). By this experiment an important premise of the Torii-Laties hypothesis is verified. The rate of passive permeation of ions into the cytoplasm, which at high external ion concentration surmounts a maximal rate of system 1, becomes apparent as a linear isotherm when the vacuole and tonoplast are absent or greatly reduced. The isotherms obtained by Torii and Laties whith nonvacuolated root tissue, were linear up to 50 mM externally without a slight indication of a reduction in the increase of rate at higher concentrations. On the contrary, usually there was a tendency for even more rapid parabolic increase of the isotherm. Taking the so-called linear isotherm as the initial slope of a carrier system based on Michaelis-Menten kinetics we would obtain K_M values of 100 to 200 mM, which are far beyond the range of physiological concentrations. There would be such

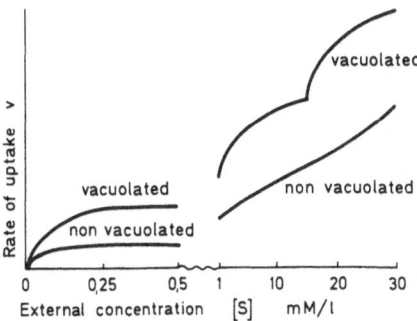

Fig. 6.5. Principle experimental basis of the Torii-Laties hypothesis: Ion uptake by maize root tissue with vacuolated and nonvacuolated cells respectively (Torii and Laties, 1966a).

a low affinity for the transported particles that one could not consider this as a specific carrier.

6.2.1.2 Why Dual Uptake Mechanisms?

Why can we observe two mechanisms when we investigate ion uptake rate as related to dependence on the external concentration? This appears to be the crucial question in discussing the two models considered so far, and which are schematically symbolized in Figure 6.17a and d. We want to designate these two models according to the assumed compartmentation of the inner phase in the following simplifying way as one-compartment and two-compartment cell models, respectively.

Let us try to give an answer to the question by using the one-compartment cell model. The external concentration then acts directly on both uptake systems in the plasmalemma. A descriptive comparison is given when we consider an experiment *in vitro* where we have two enzymes with different affinities for a single substrate. At low substrate concentration, only the enzyme with high affinity is effective, whereas at high substrate concentration, both enzymes will operate. To describe the kinetics of the low-affinity system precisely, a correction is needed which accounts for the high-affinity system operating in parallel.

Conversely, in the two-compartment model, only the rate of system 1 is determined by the external concentration. The rate of system 2 is determined by the ion concentration in the cytoplasm. In the high concentration range, therefore, a correction of the transport rates observed by the contribution of system 1 should not be necessary. In this range the concentration in the cytoplasm should be not so much determined by system 1, but more by the rapid formation of a diffusional equilibrium between the cytoplasm and the external solution.

The kinetics of this equilibration or filling of the cytoplasm has become an important argument in the controversy about the two models. The tissue used for investigations of isotherms usually contains very low levels of salt; often the roots used are grown only in $1-5 \times 10^{-4}$ M $CaSO_4$ solution. If it is possible to show that the time after the transfer of the tissue into a solution of high ion concentration is important for the observed kinetics, i.e., whether both uptake systems, or system 2 alone, determine ion uptake kinetically, then the two systems must be in series with each other and separated by a distinct compartment. This would mean the two systems cannot operate in parallel at the same membrane, unless perhaps they are subject to adaptive changes as external ion concentration increases. (See also Sect. 6.2.2.2 where the possibility of multiphasic uptake systems is discussed.)

Linear transformations of the Michaelis-Menten-Equation [Eq. (2.45)] provide the simplest way of checking on whether or not these kinetics apply. Most frequently the derivation of Lineweaver-Burk (1934) is used.

$$V_{max} \cdot \frac{1}{v} = K_M \cdot \frac{1}{[S]} + 1. \tag{6.1}$$

A straight line should be obtained by plotting $\frac{1}{v}$ against $\frac{1}{[S]}$.

The first isotherms found in the high concentration range yielded good straight lines in this way (Epstein and Hagen, 1952). At that time the division into systems 1 and 2 was not known. Later on, however, Epstein and coworkers showed that system 2 gave a straight line only if the rate of system 1 (v_1) were subtracted from the observed rate of uptake v, and therefore Epstein formulated the following equation:

$$v = v_1 + v_2 = \frac{V_{max\,1} \cdot [S]}{K_{M1} + [S]} + \frac{V_{max\,2} \cdot [S]}{K_{M2} + [S]} \tag{6.2}$$

where the indices 1 and 2 mark the uptake systems 1 and 2. To obtain straight lines for system 2 in a Lineweaver-Burk plot, the actual rate of system 2 (v_2) must be calculated from the observed rate (v) and the parameters of system 1 obtained at low concentrations in the following way:

$$v_2 = v - \frac{V_{max\,1} \cdot [S]}{K_{M1} + [S]} \tag{6.3}$$

and one must plot $\frac{1}{v_2}$ against $\frac{1}{[S]}$.

How do we explain the discrepancy between the original results and the later results by Epstein? Laties (1969) has suggested that the subtraction in Eq. (6.3) is necessary only when the times over which v has been measured were short, i.e., on the order of 10 to 20 min. In this case, the cytoplasm, according to Laties, is relatively empty during the entire period of uptake and the contribution of system 1 to ion uptake in the high concentration range is considerable. After longer times of uptake (a few hours) the cytoplasm should be much closer to equilibrium with the external medium. Then a correction according to Eqs. (6.2) and (6.3) is not necessary because the contribution of v_1 in the high concentration range becomes negligible. In terms of numbers this can be made clear when one compares the K^+ uptake rates at 10 mM KCl obtained during long-time experiments with those obtained in 10-min experiments. Rates representative for various plant tissues in longer uptake periods are on the order of 1 to 6 μmol h^{-1} g^{-1} fresh weight. K^+ uptake by barley roots from 10 mM KCl in 10-min experiments, however, is 18 μmol h^{-1} g^{-1} or higher. A correction for $V_{max\,1}$ brings this value back to the rate observed in long-time experiments.

The importance of ion concentration in the cytoplasm for the kinetics of ion uptake into the vacuoles can be shown by experiments with tissue having lower and higher ion concentrations in the cytoplasm, respectively; such tissues can be prepared by appropriate pretreatments. Ion uptake into the vacuoles of beet-root tissue with high ion concentration in the cytoplasm in the low concentration range (lower than 1 mM) is almost entirely independent of the ion concentration in the external medium. The change from low to high cytoplasmic concentration can be followed by kinetic investigations with tissue disks of red beet. The time required for filling of the cytoplasm is inversely related to the external ion concentration (Osmond and Laties, 1969). The kinetics of the filling of the cytoplasm depends on the temperature. In this way, one can also show the importance of the ion concentration in the cytoplasm. Thus loading time and temperature affect ion uptake kinetics (Robinson and Laties, 1975).

Changes in cytoplasmic ion concentrations may be responsible for certain phenomena of K^+–Na^+ selectivity. It has been found that K^+–Na^+ selectivity changes during a few hours after transfer of barley root tissue into an Na^+–K^+ solution (Pitman, 1967; Pitman et al., 1968). It is conceivable that an Na^+ extrusion pump at the plasmalemma is induced with increased Na^+ concentration in the cytoplasm and this influences the selectivity. Johansen and Loneragan (1975) confirmed this in experiments with intact barley plants. KCl pretreated plants have an anion-independent and an anion-controlled component of K^+ uptake. During pretreatment with KCl the selectivity for K^+ over Na^+ is increased. Contrasting results were obtained by Leigh and Wyn Jones (1973), who investigated the effect of internal ion concentrations on the ion uptake isotherms of excised maize root segments. The Na-extrusion pump seems to be active always and does not need to be activated by salt loading. In the external concentration range 0.1–0.3 mM K^+ and Na^+ influx is higher in nonloaded than in loaded roots; in the range 1–30 mM K^+ uptake is lower in nonloaded roots than in loaded roots but Na^+ uptake is larger in nonloaded roots.

6.2.1.3 The Synthesis and Compartmentation of Organic Acids in Relation to Ion Uptake

From Table 6.1 it becomes clear that under certain circumstances the kind of the counterion may influence the uptake of a given ion. One can distinguish between rapidly accumulated ions, for instance, K^+ and Cl^-, and slowly accumulated ions, e.g., Ca^{2+} and SO_4^{2-}. A slowly penetrating ion can slow down the uptake of a rapidly penetrating counterion, especially in the concentration range of system 2; Ca^{2+} reduces Cl^- uptake and SO_4^{2-} reduces K^+ uptake. If an ion is taken up more rapidly than its counterion and there is no equalization of charge, e.g., by an ion exchange, then we observe an electrogenic mechanism (Sect. 2.2.2.4). The membrane potential brought about by such a process then affects ion movement.

Different plants can balance the charge of surplus cation uptake more or less well by the synthesis of organic acids and the exchange of H^+ for cations taken up. It is well known that plant tissues can accumulate a stoichiometric amount of organic acid anions when cation uptake exceeds

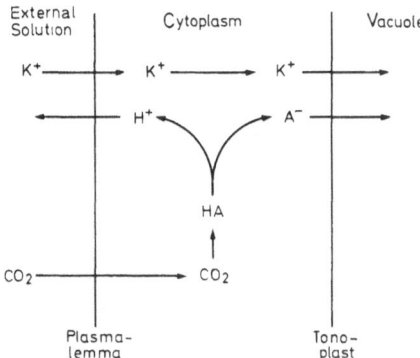

Fig. 6.6. Correlations between acid synthesis and uptake of surplus cation equivalents according to the suggestions of Torii and Laties (1966b). K^+, cation; HA, organic acid; A^-, organic acid anion; H^+, proton. (Lüttge, 1969.)

anion uptake. Conversely, the level of organic acid anions in tissues may decrease if more anions are taken up than cations. Figure 6.6 shows how one can envisage the relations between acid synthesis as a consequence of CO_2 dark fixation and a surplus cation uptake. Release of protons, i.e., an acidification of the external medium, has frequently been observed during uptake of surplus cations. It has been argued that concomitantly cytoplasmic pH is changed, and that the activity of several enzymes is influenced in a way leading to acid synthesis (Hiatt, 1967). However, it is unlikely that cytoplasmic pH really changes very much. With a stoichiometric cation-H^+ exchange and organic acid anion accumulation there is not much reason to assume this. In fact a "pH-stat" mechanism proposed by Davies (1973a,b) seems to operate with two enzymes involved in CO_2-dark fixation and organic acid metabolism as follows:

$$PEP \xrightarrow[CO_2 \text{ or } HCO_3^-]{\substack{\text{high pH} \\ \text{PEP-carboxylase}}} OAA \longrightarrow \underset{\text{(strong acid)}}{\text{malate}} \xrightarrow[CO_2]{\substack{\text{low pH} \\ \text{malic enzyme}}} \text{pyruvate}$$

(Raven and Smith, 1973; Smith and Raven, 1976).

With a K^+–H^+ exchange at the plasmalemma alone, no organic acid anions are removed from the reaction equilibria in the cytoplasmic phase. This occurs only when the anions (A^-) accompany the cations (C^+) into the vacuole (Fig. 6.6). Synthesis regulated by this acid transport therefore, in particular, should be observed in vacuolated root tissue and not so much in root tips. Furthermore, the effect should be more pronounced in the range of system 2, thought to be localized at the tonoplast, than in the concentration range of system 1. Data obtained by Torii and Laties (1966b) are consistent with these predictions (Table 6.2). The fixation of externally applied $^{14}CO_2$ in an ion-free solution into the fraction of organic

Table 6.2. Labeling of organic acids in vacuolated root tips and in non-vacuolated root tissue of maize after $^{14}CO_2$ fixation in the presence of various salts at two different concentrations in the medium. (After Torii and Laties 1966b.)

Concentration mEq/l	Salt	Non-vacuolated	Vacuolated
	H_2O	100	100
0.2	KCl	83	107
	K_2SO_4	164	254
	$CaCl_2$	117	86
	$CaSO_4$	127	92
20	KCl	102	118
	K_2SO_4	108	**460**
	$CaCl_2$	100	**27**
	$CaSO_4$	104	60

acids was taken as 100 and the effect of other conditions on fixation was measured (Table 6.2). The largest deviation from 100 (bold numbers in the table) result, as expected, when the following three conditions are given at the same time: (1) vacuolated tissue, (2) high external ion concentration, (3) combination of a readily penetrating anion with a slowly penetrating cation, or vice versa, in the external medium.

Accumulation of malate in the vacuole also follows from quantitative assessment of malate levels in carrot and barley root cells. Malate at 35 μmol g^{-1} must be at least partly in the vacuole since it would be at an impossibly high concentration (i.e., about 700 mM) if exclusively in the cytoplasm (Cram and Laties, 1974).

The correlations between acid synthesis and ion transport in the two-compartment cell system further can be demonstrated with pulse-chase experiments. A $^{14}CO_2$ pulse of 30 min duration is given to the tissue of red beet; then the tissue is transferred to nonlabeled CO_2 and the decrease in the specific activity (radioactivity per mol) of malate in the tissue is followed with time. Malate is the most abundant of the acids formed. The specific activity of malate decreases rapidly when, at the same time, no surplus cation uptake is possible. When there is concomitantly high cation uptake but low anion uptake, the radioactivity of the labeled malate remains high for a longer time, i.e., the turnover of malate is decreased (Fig. 6.7). A compartmental analysis, by aid of the kinetic method discussed in Section 6.2.3.2, shows that under these conditions there is more malate transported into the vacuole and it is thus eliminated from turnover in the cytoplasm (Osmond and Laties, 1969). In roots, in similar experiments, malate levels of vacuolated tissues remain high and constant for a fair amount of time when an external ion solution is replaced by a salt-free medium, but in root tips under similar conditions malate is rapidly degraded (Jacoby and Laties, 1971).

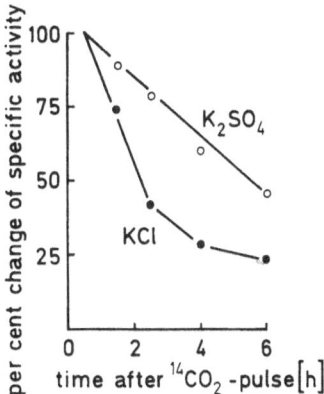

Fig. 6.7. Turnover of labeled malate in aged red beet tissue disks after a 30-min pulse of $^{14}CO_2$ under different conditions of ion uptake: in K_2SO_4 (20 mN) there is rapid cation uptake, slow anion uptake; in KCl (20 mN) there is equally rapid uptake of cations and anions. (Osmond and Laties, 1969.)

Double labeling experiments also show that malate may be compartmented within the cell in a different way than other acids of the Krebs cycle. The turnover of intracellular malate taken up from the external solution into the cells as ^{14}C labeled malate is very slow, as compared with the turnover of malate formed via the Krebs cycle in the mitochondria and traced by introducing tritium-labeled acetic acid. Malate taken up from the external medium therefore must be separated by a compartmental barrier from the Krebs cycle (Steer and Beevers, 1967); whether this barrier is the mitochondrial envelope or the tonoplast is not clear.

In general these experiments show that coupling of acid metabolism and ion transport offers interesting possibilities for metabolic control of ion relations in plant cells.

Acid anions other than malate, e.g., citrate, oxalate, and notably sulfate may be accumulated by plants; as in the case of malate the evidence strongly suggests that most of this accumulation is in the vacuole.

Patterns of charge balance of various kinds of cations and organic and inorganic anions in the vacuoles of various plant species can be of taxonomic significance, and this has led to the definition of physiological types of plants ("physiotypes": Albert and Kinzel, 1973). In the case of oxalate a relatively large proportion is stored not in solution but rather in the form of Ca-oxalate crystals (raphides, druses, etc.) apparently inert. Such crystals are enclosed in a membrane (Schötz et al., 1970); this membrane might serve a special transport system but this is not known. In the case of sulfate accumulation, in some rare cases such as the membrane-surrounded statoliths of the geotropically responsive *Chara* rhizoids, SO_4^{2-} is accumulated together with Ba^{2+} and forms an insoluble precipitate as $BaSO_4$ (Schröter et al., 1975). Also free acids occur in the vacuoles. A special case is the diurnal rhythm of malic acid levels in the vacuoles of plants performing Crassulacean acid metabolism (CAM), where malic acid at the end of the light phase may be around 15–20 mM and rise to 200 mM, or occasionally more, towards the end of the dark phase (e.g., Fig. 3 in Lüttge and Ball, 1977). Free sulfuric acid is stored in the vacuoles of several species of the brown marine alga *Desmarêstia*. The pH of the cell sap may be lower than 1.0 and reach a concentration of 0.44 N H_2SO_4 (Eppley and Bovell, 1958). A hydrogen ion gradient of about 10^{-6} occurs across the tonoplast; since the cytoplasm is presumed to have a "pH-stat" controlling pH, this suggests that the tonoplast must have a powerful H^+ ion or H_2SO_4 pump.

6.2.1.4 Some Further Evidence for the Torii-Laties Hypothesis

A few more experimental findings in favor of the Torii-Laties hypothesis follow.

A. *Differences in Ion Uptake Isotherms due to Aging.* Freshly isolated disks of storage tissue organs, and also freshly isolated root segments, differ in the capacity and kinetics of ion uptake as compared to tissue which has been washed or aged for some time (10 to 20 h) in a dilute $CaSO_4$ solution. Van Steveninck (1975, 1976b) calls such changes of transport mechanisms, and also the biochemical changes which are observed, "adaptive aging". Changes of transport mechanisms during "adaptive aging" can be particularly pronounced for the system 1 isotherm. Flux studies (see Sect. 6.2.3.2) show that during aging, the influx at the plasmalemma is particularly altered. These results support the Torii-Laties hypothesis (Laties, 1967, 1969; Lüttge, 1968).

B. *Experimental Alterations of Membrane Permeability.* Using various so-called membrane-active substances (Sect. 5.2.2) one can alter membrane permeability. For instance, the polybase poly-1-lysine influences ion permeability of chloroplast and mitochondrial membranes. Poly-1-lysine also affects energy transfer reactions. By a suitable choice of the experimental conditions, especially of poly-1-lysine concentration and duration of applications, one can selectively increase ion permeability of the plasmalemma. This appears to interfere in particular with ion uptake by system 1 (Osmond and Laties, 1970).

C. *Coupling of Short-Distance and Long-Distance Transport.* The two-compartment model also offers a useful basis for the explanation of coupling of long-distance transport with membrane transport mechanisms. Observations on ion transport from an external medium across the root into the pathways of long-distance transport in the xylem may perhaps be explained on the basis of the Torii-Laties hypothesis, and, conversely, such studies can be used for its further evaluation.

This will be explicitly discussed later in Chapter 12.

6.2.2 Further Advancement

6.2.2.1 Test of the Models by Computer Simulation

The two models which we have discussed can also be described mathematically. In Chapter 2 we have learned about various equations which under certain circumstances are valid for the outside-inside model. In the following section (6.2.3) we will indicate how the more complicated model, outside-cytoplasm-vacuole, can be described by mathematical formulations. A considerable number of parameters which are mutually interdependent have to be taken into account: i.e., the fluxes at the plasmalemma and the tonoplast in opposite directions and the concentrations in the external medium, the cytoplasm, and the vacuole (Fig. 6.12). The number of parameters increases exponentially with increasing complexity of the cell model used (Sect. 6.4). With a suitable mathematical formulation it is possible to test a model by aid of computers. As in real experiments, certain parameters are kept constant and other parameters are varied. The resulting behavior of the model in the computer simulation can be compared with the real behavior of the object in actual experiments. This is perhaps the most rigorous quantitative test to which one can submit a model.

The outside-cytoplasm-vacuole model, with system 1 at the plasmalemma and system 2 at the tonoplast, was submitted to such a test by Pitman (1969).

He arrived at the following conclusions:

1. *Anion uptake:* An active transport mechanism must be operative at the plasmalemma in the range of low and high external concentrations because the rate of the active transport at the plasmalemma observed in the range of high external concentrations is larger than the maximum velocity of system 1. Additionally, a transport mechanism at the tonoplast is operative.
2. *Cation uptake:* To explain cation uptake system 1 must be located in the plasmalemma. In contrast to anion uptake, it is not necessary to assume that an active cation transport by system 2 is operative in the plasmalemma.
3. *Coupling:* The active transport mechanisms at the plasmalemma and at the tonoplast must be coupled with each other. This coupling can be brought about by a common source of energy or, in the simplest case, by dependence of both pumps on the ion concentration in the cytoplasm.

   It is clear that this system requires a two-compartment cell model. It needs both compartmental barriers, plasmalemma and tonoplast, and it needs active transport mechanisms at each of these membranes. It combines properties of the simple outside-inside model (anion-transport) with those of the Torii-Laties model (cation transport). According to Pitman's own wording, his results support a model which forms a compromise between the one-compartment cell model with the parallel localization of active transport mechanisms of system 1 and system 2 at the plasmalemma (Epstein and coworkers) and the two-compartment cell model with a serial arrangement of systems 1 and 2 at the plasmalemma and the tonoplast, respectively (Fig. 6.17). One may also conclude that Pitman's computer tests support a model which goes beyond the two other models (see Fig. 6.17f,g).

6.2.2.2 Multiphasic Uptake Systems and Co-operative Enzyme Kinetics

Nissen (1971, 1973a,b,c, 1974; Holmern et al., 1974; Vange et al., 1974) has posed the question as to whether the formulation of the dual Michaelis-Menten Eq. (6.2) is indeed adequate to describe ion uptake isotherms by plant cells. Also using computers he has reanalyzed all isotherms which he could obtain from the literature or by personal communications, and he also performed experiments. His data analysis, based on Lineweaver-Burk type $1/v$ vs $1/S$ plots (Fig. 6.8), reveals numerous discontinuities of isotherms such as those already described by Epstein and others for the concentration range between 1 and 50 mM (Fig. 6.1). Discontinuities, however, were also found at concentrations below 1 mM. There were discontinuities not only in isotherms of vacuolated tissue but also in root tip tissue. Nissen suggests that these discontinuities could be brought about by concentration-dependent alterations of the uptake mechanism. According to this concept, depending on the ion concentration in the external medium, the uptake system switches to different states which are characterized by different values of V_{max} and K_M. Nissen calls this a multiphasic system. He concludes that a multiphasic system is localized at both the plasmalemma and the tonoplast. Both systems act over a wide range of concentrations. In contrast to the Torii-Laties hypothesis, Nissen's concept does not need the assumption of a high diffusional component of ion uptake at the plasmalemma at high external concentrations to put the tonoplast system into action. By changes of phase in the plasmalemma system, an increased external concentration can lead to an increased ion uptake. A tonoplast system, in which V_{max} differs from the plasmalemma system, can become rate-determining only at extremely high external concentrations (about 10^{-2} M).

Nissen's ideas have found support from other workers (e.g., Linask and Laties, 1973; Glover et al., 1975, who worked with a micro-organism: Bacillus subtilis; Joseph and van Hai, 1976; Dogor and van Hai, 1977) but also have met much reservation. Epstein in his review points out that Nissen remains very unspecific as to what he thinks about the molecular mechanism of multiphasic states of uptake systems (Epstein, 1976, p. 88).

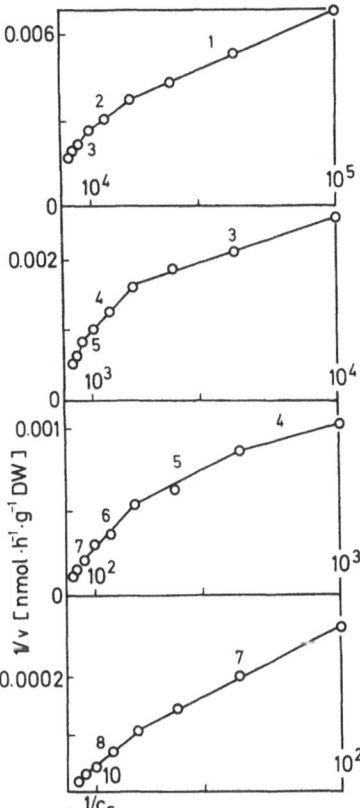

Fig. 6.8. Data on SO_4^{2-} uptake by barley roots obtained over a wide range of external concentrations and plotted as numerous short segments in the Lineweaver-Burk form (after Fig. 9B from Nissen, 1971b). v is rate and c_s is the external substrate concentration. DW = dry weight.

Although this is not accepted explicitly by Nissen, his ideas in many respects are akin to cooperative enzyme kinetics as discussed by Koshland and coworkers (Koshland, 1970; Levitzky and Koshland, 1969; Teipel and Koshland, 1969). These kinetics are based on multisubunit enzymes in which the binding of ligands (e.g., substrates) to subunits induces conformational changes of the protein molecule, thus affecting reactivity in other subunits so that the affinity may increase or decrease. Thus positive or negative co-operativity is observed. Figure 6.9 demonstrates, how such a system could function in membrane transport. The membrane ATPases of plant cells may very well constitute the molecular basis for such a mechanism. Involvement of ATPases of the plasmalemma and tonoplast in ion transport at these membranes is highly likely (Chapters 5, 8 and 10; Hodges, 1976). The complex kinetics of alkali cation stimulation of the plasmalemma ATPase were described by Leonard and Hodges

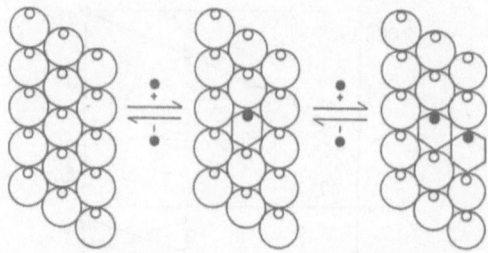

Fig. 6.9. Surface view of a membrane consisting of an array of subunits. Each subunit has a binding site for a ligand and can exist in two states, an R-state (hexagonal, with ligand attached) and an S-state (circle, having a binding site unoccupied). Ligands may be particles transported across the membrane, effectors like hormones, or substrates metabolized. Ligands can be activators or inhibitors. Co-operativity is due to R ⇌ S transitions depending on the state of the neighboring subunits; furthermore the affinity of both states (R, S) for the ligands may be different. Adapted from Hoelzl-Wallach and Knüfermann (1973) representing the ideas of Changeux and co-workers (Changeux et al., 1967, 1970; Changeux and Thiery, 1968; Blumenthal et al., 1971) on co-operativity in membrane arrays.

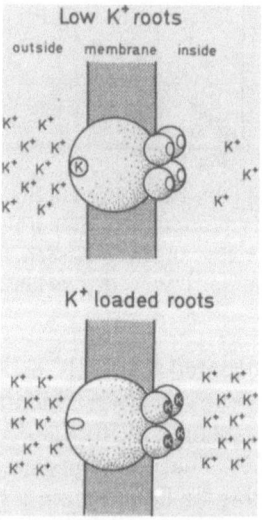

Fig. 6.10. Model of allosteric control system for K^+ influx into barley root cells. The large centrally located structure represents the K^+ carrier. On the outer surface, a single K^+-binding site is represented by the *small circle*. On the inner surface, four allosteric binding sites located on four subunits are represented by *four circles*. In low K^+ roots, the allosteric sites are vacant, and K^+ can be bound to the external site giving high initial rates of influx. When the internal concentration of K^+ is high (K^+ loaded roots), the allosteric sites are saturated, and the conformation of the external binding site is modified resulting in lowered affinity for K^+. (Glass, 1976a.)

(1973) as negatively cooperative, suggesting that the ATPase might consist of subunits and act in transport in a way akin to that depicted in Figure 6.9 (Hodges, 1973, 1976). Leonard and Hotchkiss (1976) note that

> "at present, hypotheses which invoke discrete phase changes in a multisubunit carrier (Nissen) or a single carrier with binding sites with different affinites (Epstein) are among several feasible interpretations of the complex kinetics for ATPase and transport."

An interesting allosteric carrier model is also suggested by Glass (1975, 1976) who investigated the feedback effect of cytoplasmic K^+ in the concentration range of 0.01 to 0.32 mM ("low concentration range"). K^+ uptake is related to K^+-concentration in the medium in the form of a rectangular hyperbola (Michaelis-Menten isotherm). K^+ uptake is related to $[K^+]_i$ in a sigmoidal fashion indicating an allosteric mechanism with a negative feedback of $[K^+]_i$ on K^+ uptake. Calculations suggest there is one binding site for K^+ at the outer face of the membrane and there are four allosteric sites at the inside (Figure 6.10).

6.2.3 Kinetic and Electrochemical Compartmental Analysis

In the application of Michaelis-Menten kinetics to interpret absorption isotherms, only rates are observed and the procedure provides no measure of driving forces or other thermodynamic relations giving information on the work done in transport uphill. The model with the three compartments, outside-cytoplasm-vacuole (two-compartment cell model), can be evaluated in a very different way and independently of isotherm kinetics by compartmental analysis (review Walker and Pitman, 1976). It has been mentioned before (Sect. 2.2.3.1) that only with the large coenocytic cells of some algae can a rather exact compartmental analysis be made directly by isolating and assaying the cytoplasm and vacuole; using cells of higher plants this is not practical, and indirect kinetic methods are required. At the same time electrical parameters can be measured to assess energy gradients.

6.2.3.1 Direct Compartmental Analysis: Coenocytic Algal Cells

The aim of compartmental analysis is not merely to determine ion concentrations in the particular compartments (cytoplasm and vacuole), but rather to identify passive or active ion fluxes at each membrane, plasmalemma and tonoplast; in other words, to determine what underlies the ion distribution. If one wants to apply the Nernst criterion [Eq. (2.19)] or the Ussing-Teorell criterion [Eq. (2.40)] in the two-compartment cell model, the membrane potentials at the plasmalemma (E_{co}) and at the tonoplast

Table 6.3. Ion concentrations [mM] in compartments of various fresh and brackish water algae, in higher plants, and in a marine alga. From data compiled by MacRobbie (1970a), Pierce and Higinbotham (1970), and Tazawa et al. (1974); consult these references for further details and literature. Italic numbers refer to analysis of the stationary cytoplasm (S) in *Characeae* which contains the chloroplasts and to chloroplast analyses (Chl), respectively.

Species	External solution			Cytoplasm			Vacuole		
Fresh and brackish water algae:	K^+	Na^+	Cl^-	K^+	Na^+	Cl^-	K^+	Na^+	Cl^-
Nitella translucens	0.1	1.0	1.3	119 *S:150*	14 *55*	65–87 *240*	75	65	150–170
Nitella flexilis[b]	0.1	0.2	1.3	125 *S:110*	5 *26*	36 *136*	80	28	136
Nitella flexilis	0.075	1.0	1.5	92	30	—	66	39	—
Nitella flexilis[a]	0.07–0.09	0.7–0.9	0.5–0.9	78	2	27	73	44	179
Tolypella intricata	0.4	1.0	1.4	87–97	4–22	23–31	90–119	3–39	110–136
Hydrodictyon africanum	0.1	1.0	1.3	93 *Chl.:340*	51 *36*	58 *340*	40	17	38
Lamprothamnium succinctum	6	289	337	137	47	86	250	136	373

Table 6.3. Continued.

Higher plants:	K^+	Na^+	Cl^-	K^+	Na^+	Cl^-	K^+	Na^+	Cl^-
Beta vulgaris (beet)	5	—	—	85–86	—	—	85–205	—	—
Hordeum vulgare (root)	2.5	7.5	—	102	70	16	74	29	150
Daucus carota (carrot)	—	—	0.5–95	169	—	—	55	—	—
Pisum sativum (epicotyl)	1	—	—	170	—	—	78	—	—
Avena sativa (coleoptile)	10	10	10	150–205	12–17	76	160–190	24–29	65
Range for fresh and brackish water algae (without *Lamprothamnium*) and higher plants	—	—	—	53–205	12–70	16–87	40–205	3–65	38–170
Sea water alga: *Valonia ventricosa*	10–13	470–510	520–600	434	40	138	625	44	643

[a] Results from a recent publication (Tazawa et al., 1974) where values given in the cytoplasm column refer to total protoplasm in contrast to the other Characeae data which refer to the cytosol. According to calculations by Tazawa et al., 1974, these new analyses do not seem to corroborate the results of older experiments.

[b] Kishimoto and Tazawa, 1965a, which suggested higher concentrations of Na^+ and Cl^- in the chloroplasts than in the cytosol.

Table 6.4. Fluxes with an active component in cells of freshwater and marine algae and in a higher plant. o = external solution, c = cytoplasm, v = vacuole. The sequence of the symbols indicates the direction of active fluxes. From MacRobbie (1970a) and Pierce and Higinbotham (1970).

	K^+	Na^+	Cl^-
Fresh and brackish water algae:			
Nitella translucens	o–c	c–o,c–v	o–c(c–v?)
Nitella flexilis	—	c–o,c–v	o–c,c–v
Tolypella intricata	—	c–o	o–c
Chara corallina	—	c–o	o–c,c–v
Hydrodictyon africanum	o–c	c–o	o–c
Lamprothamnium succinctum	—	c–o	o–c
Nitellopsis obtusa	—	c–o	o–c
Nitella clavata	o–c	c–o	o–c
Higher Plant:			
Avena sativa	o–c,c–v	c–o,c–v	o–c,c–v
Marine algae:			
Valonia ventricosa	c–o,c–v	c–o,c–v	(o–c?)
Chaetomorpha darwinii	o–c,c–v	c–o,c–v	—
Acetabularia mediterranea	(c–o?)	c–o	o–c

(E_{vc}) (see Table 2.2), and the ion concentrations in the cytoplasm and vacuole must be known. With large coenocytic algal cells reasonably reliable values can be obtained for the ion concentrations in the cell sap (i.e., for the vacuolar content) and the cytoplasm. Depending on the cell material chosen one can let the cell sap flow out of the cell or collect it for analysis by aid of a microcapillary. The cytoplasm can be pressed out of the cell or collected by gentle centrifugation and analyzed for ion concentrations; however, any contamination with the cell sap leads to a more or less reduced reliability of results. Some data on ion concentrations in various compartments of algal cells and on active ion fluxes at the boundaries of compartments are presented in Tables 6.3 and 6.4.

6.2.3.2 Indirect Compartmental Analysis: Isotope Exchange Kinetics

The measurement of the kinetics of isotope exchange between a radioactively loaded tissue and a nonlabeled external solution offers an indirect way for compartmental analysis, i.e., for the analysis of concentrations in particular compartments and of fluxes at the boundaries of such compartments.

A typical experiment is shown in Figure 6.11. For such investigations the plant tissue is immersed in a radioactively labeled ion solution where it takes up ions from the solution until it is in a pseudo steady state. In this situation influx and efflux are equal and constant with time, so that the ion

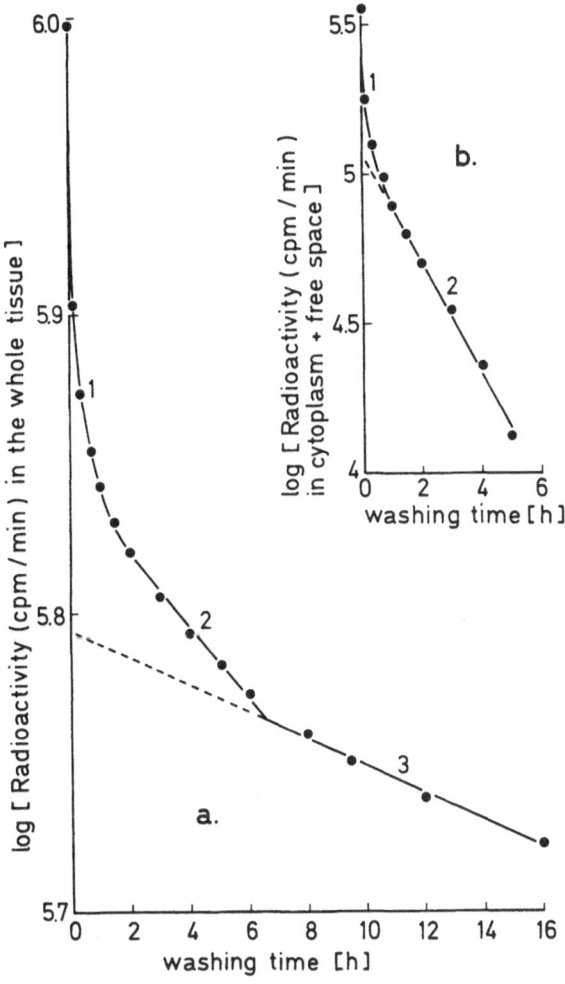

Fig. 6.11. Example of an efflux (or washout) experiment performed according to Pitman (1963). **a** Washout of labeled ions (K^+ labeled with $^{86}Rb^+$) from the tissue of excised maize roots after a 15 h loading period in 0.2 mM K(^{86}Rb)Cl-solution into 0.2 mM nonlabeled KCl. *1*, exchange of radioactivity in the apoplastic free space, with the external nonlabeled KCl; *2*, dominating exchange of the cytoplasm; *3*, efflux from the vacuole; **b** phases 1 and 2 after subtraction of vacuolar radioactivity obtained by extrapolation of phase 3 (*dotted line in a*). (Lüttge 1969.)

content for the duration of the exchange experiment does not change. After radioactive loading, one transfers the tissue into a nonlabeled ion solution which otherwise has the same concentration and composition as the solution used for labeling, and then follows the kinetics of the exit of isotopes, i.e., of washing out.[1]

---

[1] Note that in this respect this experiment is different from that shown in Figure 6.2.

In Figure 6.11 three phases of washout can be detected which differ in the half-times of washout, $t_{1/2}$. The $t_{1/2}$ of phase 1 is on the order of fractions of minutes up to a few minutes; the $t_{1/2}$ of phase 2 is a few hours. Loading and washout in this experiment were performed at a temperature just above freezing ($2°-4°C$). This temperature tends to increase $t_{1/2}$ of phase 2 as compared with physiological temperatures, so that phase 2 can be more easily separated from the temperature-independent phase 1. The $t_{1/2}$ of phase 3 extends over several days to several months. These three phases represent ion exchange from several compartments arranged in series within the plant cell. Tentatively each phase has been attributed to particular cell compartments namely, phase 1 to AFS, phase 2 to the cytoplasm, and phase 3 to the vacuole.

Since efflux kinetics alone do not allow an unambiguous identification of each compartment it was necessary to verify these correlations independently. In this context experiments with coenocytic algal cells subject to direct compartmental analysis again were very important. Indeed isotope exchange kinetics were first investigated in giant algal cells and verified by comparison with direct analysis (MacRobbie and Dainty, 1958) before the method was applied to the tissues of higher plants (Pitman, 1963).

Cram (1968b) has tried to characterize the three compartments in isolated carrot-root tissue more closely, and to verify the serial model of AFS-cytoplasm-vacuole, as follows:

First by experiments with killed tissue it could easily be shown that phases 2 and 3 represent exchange with living compartments. Working with sterile tissue, contaminating microorganisms as possible compartments could be excluded. Thus only the more important cell compartments, i.e., the cytoplasm and the vacuole, remained as the explanation of phases 2 and 3.

For an evaluation of the serial model which is the basis of interpretation of efflux kinetics (see also Fig. 6.12) it is important to show that the ion exchange from the vacuole with the external solution can occur only via the cytoplasm. A priori this is not self-evident. It is quite possible to imagine transport processes which lead to ion fluxes between the vacuole and the external medium without requiring an equilibration in the cytoplasm (see Sect. 6.4). As a test of the series model Cram (1968b) transferred $^{36}Cl^-$-loaded carrot-root tissue during washout from the usual nonlabeled ion solution (see above and Fig. 6.11) into distilled water. The rate of efflux was observed to decrease to a value near zero. This effect could mirror events at different membrane barriers. However, consider the state of a labeled tissue which has been washed in nonactive salt solution long enough for the two compartments with the shorter half-lives, i.e., compartments 1 and 2, to have fully exchanged. Then the radioactivities of the wall and cytoplasm will be equal to zero if the cytoplasm, compartment 2, and the vacuole, compartment 3, are not connected with each other by membrane fluxes. However, if the cytoplasm and the vacuole are

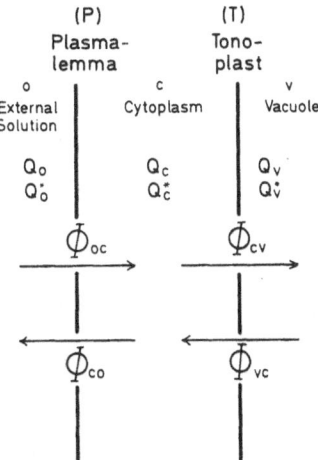

Fig. 6.12. Serial two-compartment cell model (not including the wall). Individual fluxes between the external solution and the cytoplasm, and between the cytoplasm and the vacuole. Q, amount of ions; Q^*, radioactivity. External solution: index o, cytoplasm: index c, vacuole: index v.

in series with each other the specific radioactivity in the cytoplasm will have a distinct level which is below the level of specific radioactivity in the vacuole, and there should be some radioactivity in the AFS. Since the AFS exchanges very rapidly, after an initial burst of washout of label from the water free space, it should show negligible tracer efflux on transfer to distilled H_2O. In the absence of Cl^- externally the exchange between the cytoplasm and the AFS and external solution can no longer take place. As expected on the basis of the serial model, Cram found a transient increase in efflux from the carrot tissue on transfer to distilled H_2O which was not followed by any further release of tracer-labeled ions (Fig. 6.13). However, fluxes of $^{36}Cl^-$ between the cytoplasm and vacuoles could continue without loss to the medium. This would lead to increased specific radioactivity of the cytoplasm as compared with the situation immediately before transfer to the distilled H_2O. Indeed, at a later time when unlabeled Cl^- was added again to the outer solution, there was another burst of $^{36}Cl^-$ efflux from the tissue. This was logically attributed to the increased specific radioactivity of Cl^- in the cytoplasmic phase. The bulk of the $^{36}Cl^-$ still remained in the tissue and quite certainly in the vacuole. The isotope efflux was expected to decrease again more or less rapidly to about the same value which it had before transfer from salt solution to the distilled H_2O. After the transient burst of washout of label on retransfer of the tissue to KCl this was really observed (Fig. 6.13). The kinetic analysis of this peak gives a $t_{1/2}$ of isotope exchange which corresponds with the $t_{1/2}$ of phase 2 of the normal washout curve (e.g., Fig.

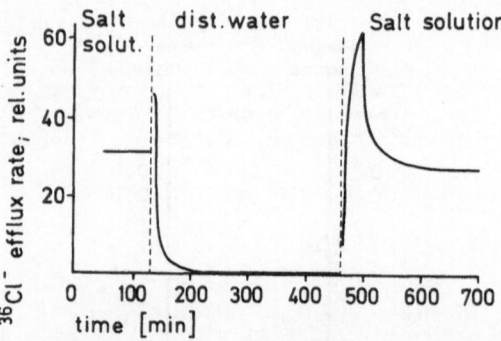

Fig. 6.13. Transient changes of ^{36}Cl$^-$ efflux from ^{36}Cl-loaded carrot root disks transferred from nonlabeled salt solution to distilled water and back to nonlabeled salt solution. After loading of the tissue with ^{36}Cl the apoplastic free space and the cytoplasm (phases 1 and 2 in Fig. 6.11) were washed out in nonlabeled salt solution overnight prior to the measurements shown here so that this experiment was performed during phase 3 (see Fig. 6.11). After transfer to water: rapid ^{36}Cl$^-$ efflux from the free space, followed by zero efflux. After re-transferring the tissue into the salt solution: initially increased efflux from the cytoplasm followed by an efflux similar to the initial rate in salt solution at the beginning of the experiment. (Simplified from Cram, 1968b.)

6.11). The results depicted in Figure 6.13 therefore agree with the expectations and confirm the series model shown in Figure 6.12.

By a mathematical analysis of washout curves ion concentrations in particular compartments and fluxes at the compartmental barriers can be calculated. The symbols which are used in the following are explained in Figure 6.12. The radioactivity in the cytoplasm (Q_c) and in the vacuole (Q_v) changes during washout due to the fluxes at the plasmalemma and at the tonoplast; these are referred to as

$$\frac{dQ_c^*}{dt} \text{ and } \frac{dQ_v^*}{dt}.$$

The most important equations resulting from this are:

$$\frac{dQ_c^*}{dt} = (\Phi_{oc} \cdot s_o + \Phi_{vc} \cdot s_v) - s_c(\Phi_{co} + \Phi_{cv}) \qquad (6.4)$$

$$\frac{dQ_v^*}{dt} = \Phi_{vc} \cdot s_v - \Phi_{cv} \cdot s_c \qquad (6.5)$$

(Pitman, 1963; Pierce and Higinbotham, 1970). The specific activities are:

$$s_o = \frac{Q_o^*}{Q_o}; \; s_c = \frac{Q_c^*}{Q_c}; \; s_v = \frac{Q_v^*}{Q_v}. \qquad (6.6)$$

The fluxes are mutually interdependent:

$$\Phi_{net} = \Phi_{oc} - \Phi_{co} = \Phi_{cv} - \Phi_{vc} \qquad (6.7)$$

[see also Eq. (2.23)]. By means of a number of other derivations then, the values for the individual fluxes can be obtained. The extrapolation of phase 3 to the origin of a washout curve (zero on the abscissa on Fig. 6.11) gives the radioactivity in the vacuole (Q_v^*). If in addition one determines the ion concentration in the tissue (see Sect. 2.2.3.1), the specific activity in the vacuole (s_v) can be calculated initially using the assumption that the total ion concentration measured is close to that in the vacuole. This means that the ion capacity of the cytoplasm must be considerably lower than that of the vacuole (Q_c considerably lower than Q_v).[2]

Thus this first approximation is valid only for cells with large vacuoles and thin cytoplasmic layers along the cell wall. The apparent content of labeled isotope in the cytoplasm is given by extrapolation of phase 2, after subtraction of the vacuolar content which has been obtained by extrapolation of phase 3 (Fig. 6.11b). Further parameters obtained from such experiments are the net uptake rates during pretreatment with the labeled solution and the rate constant of washout of the cytoplasmic phase. The derivation and the description of the equations necessary for flux and concentration calculations will not be given here (see especially Pitman, 1963; Cram, 1968a; Pierce and Higinbotham, 1970; Hope, 1971; MacRobbie, 1971b; Walker and Pitman, 1976).

By analysis of individual fluxes at varied external concentrations one can, so to speak, obtain absorption isotherms. Such isotherms for maize roots based on values of Φ_{oc} and Φ_{cv} of ^{86}Rb-labeled KCl solution[3] are shown in Figure 6.14; included for comparison is a Cl^- uptake isotherm taken from a paper of Torii and Laties (1966a). These fluxes were measured at 2°C. Loading of the root tissue with labeled ions took 15 hours. Before loading with label the roots were pretreated for 7 h with nonlabeled KCl solution. The washout period was also 15 h. Periodically influx of label and the K^+ content of parallel control tissue were measured during the 22–37-h period. Initial uptake rates were the same throughout and

---

[2] In mature root tissue the volume of the cytoplasm is estimated to be less than 5% of the tissue, the vacuole 80%–90%.

[3] Rb^+ behaves much like K^+ in transport and ^{86}Rb has been used frequently as a radioactive tracer for K^+ movement. Under appropriate conditions this procedure appears to give valid results (Jeschke, 1970b; Marschner and Schimansky, 1971). However, discrimination in transport between Rb^+ and K^+ may occur (Ramani and Kannan, 1976; Jacoby and Nissen, 1977). Rb^+ tends to inhibit K^+, particularly at K^+/Rb^+ ratios above about 10 (unpublished work of D. L. Hendrix in the laboratory of NH). For use of ^{86}Rb as a tracer for K^+ the specific activity should be high and the chemical amount of Rb^+ kept at a minimum relative to K^+.

two important requirements for interpretation of washout experiments
were confirmed, namely:

1. that the ion content of the tissue does not change during the washout
   period, i.e., $\Phi_{net} = 0$;
2. that Φ_{oc} remains constant during washout.

The Cl$^-$ uptake isotherm used for comparison in Figure 6.14 was obtained
at 0°C. At this temperature the ion uptake isotherms of vacuolated and
nonvacuolated tissues were identical. In the concentration range of
system 2 we find with both tissues a linearly, or parabolically, rising
isotherm (cf., however, Sect. 6.2.1.1; Fig. 6.5). One must assume that
mechanism 2 is inhibited so much by the low temperature that only the
diffusion component of ion uptake becomes apparent (Sect. 6.2.1.2) since
its temperature sensitivity is low. The essential point in Figure 6.14 is that
it shows a clear coincidence between Φ_{oc} and the Cl$^-$ uptake isotherms in
the low concentration range. This is in agreement with the hypothesis that
the system 1 isotherm reflects ion uptake at the plasmalemma. In the high
concentration range all three isotherms have a similar form. If the hyperbolic system 2 isotherm is suppressed by the low temperature this result is
as expected and does not contradict the Torii-Laties hypothesis.

Cram and Laties (1971) have investigated influx at the plasmalemma
and into the vacuole of barley roots over a wide range of concentrations.
At low external concentrations (0.02–1.0 mM) the two fluxes are equal. In
the range of higher concentrations (experiments have been extended up to
80 mM) influx at the plasmalemma increases linearly and is many-fold the

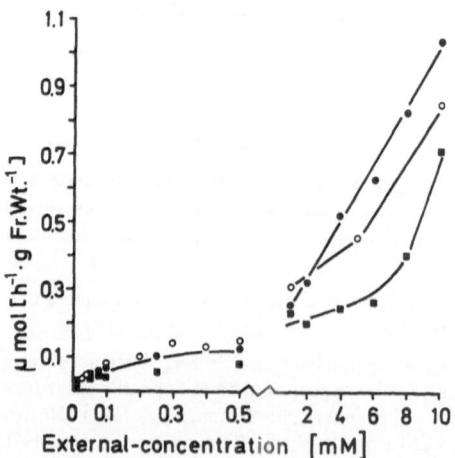

Fig. 6.14. Dependence of K$^+$ tracer fluxes Φ_{oc} (*closed circles*) and Φ_{cv} (*squares*)
of maize root segments at 2°C on external concentration during loading and wash-
out of label (^{86}Rb) (Lüttge, 1973) in comparison with the Cl$^-$-uptake isotherm at
0°C (*open circles*). Torii and Laties, 1966a.

rate into the vacuole. The influx into the vacuole is saturated at 20–30 mM and then does not increase further with increasing external concentration. In agreement with the Torii-Laties hypothesis, these experiments show that the flux which attains saturation at high external ion concentration cannot be located at the plasmalemma, but must be an influx into a compartment within the cell, i.e., the vacuole.

In studies with labeled solutions the tissue is often rinsed for a few minutes in nonlabeled uptake solution and at low temperature (0°C) after uptake to remove label from the AFS before the tissue samples are counted. Figure 6.11 shows, however, that even at low temperature there is efflux of label not only from the free space but also from the cytoplasm. Thus short rinsing (on the order of 2 min) at low temperature will selectively wash out the free space; long rinsing (on the order of 30 min) will also wash out appreciable amounts of label from the cytoplasm. Thus experiments using short uptake periods can reflect Φ_{oc} reasonably well, whereas with long uptake periods much of the label will enter the vacuole. Therefore, separate measurements for Φ_{oc} and Φ_{cv}, respectively, can be obtained by the appropriate combinations of short or long uptake periods with short or long rinses (Cram, 1969a, 1973; Cram and Laties, 1971; Pitman et al., 1974a; review: Walker and Pitman, 1976).

6.2.3.3 Electrochemical Compartmental Analysis Distinguishing Active and Passive Fluxes at Plasmalemma and Tonoplast

If one has measured membrane potentials, ion fluxes, and ion concentrations, then the Nernst criterion [Eq. (2.19)] and the Ussing-Teorell criterion [Eq. (2.40)] can be applied to both the plasmalemma and tonoplast. This provides a test for active transport at these boundaries. Values for algal cells are summarized in Table 6.4. Figure 6.15 shows the result of a compartmental analysis of oat coleoptiles. The multitude of data obtained for three more important ions, K^+, Na^+, and Cl^-, in different organisms can be summarized in the following way:

Sodium, *plasmalemma:*
    In most cells investigated, active efflux occurs, but never active influx.
tonoplast:
    In most cells investigated, there is active influx.
Chloride, *plasmalemma:*
    In all cells investigated, there appears to be active influx.
tonoplast:
    In most cells investigated, there is active influx.
Potassium, *plasmalemma:*
    Active influx occurs in some of the examples listed in Table 6.4, but

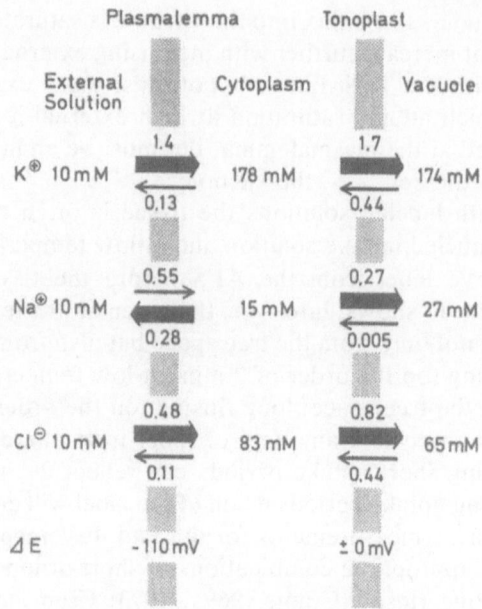

Fig. 6.15. Summary of electrochemical data on oat (*Avena*) coleoptile cells: K$^+$, Na$^+$, and Cl$^-$ compartmentation at 10 mM external concentrations. Membrane potentials and flux sizes (*arrows*) in pmols^{-1} cm^{-2}. *Heavy arrows*, active fluxes; *thin arrows*, passive fluxes. (Pierce and Higinbotham, 1970.)

the general situation is not very clear; see the many uncertainties in Table 6.4. In some experiments there is clear electrochemical evidence for downhill plasmalemma influx (Lüttge, Higinbothan and Pallaghy, 1972; Fischer et al., 1975). It is possible that efflux of K$^+$ can be active (see also Etherton, 1963). Efflux seems clearly to be active in *Valonia*.

tonoplast:

There may be active or passive influx as for the plasmalemma.

From this the basic types depicted in Figure 6.16 can be derived.

It should be pointed out that the conclusions of active transport in Table 6.4 may be based on rather strong or rather weak evidence. In Table 6.5 calculations of the driving force, E_K^D for K$^+$ movement, are provided ($E_K^D = E_M - E_K$). For *Nitella translucens* and *Valonia ventricosa* E_K^D values are quite high and it seems quite unlikely that experimental errors in measurement of electropotential or estimates of activities in compartments could account for the results. In the other cases, in which E_K^D values are only a few millivolts, it seems likely that K$^+$ may be moving passively; this probably is true also for K$^+$ transport in some of the cases cited in Table 6.4, e.g., *Avena sativa* (see Fig. 6.15). E_{Kco}^D for *Avena* is -37 mV suggesting a driving inward force for K$^+$ influx; however, the E_{Kco}

Table 6.5. The Nernst potentials in millivolts for K, E_K^{co}, between the outer solution and the cytoplasm, compared with the measured potentials, E_M^{co}, and between cytoplasm and vacuole, E_K^{vc}, compared with measured potential, E_M^{vc}. An activity ratio between ionic solutions inside and outside of 0.75 was assumed for *Nitella* and *Chara* whereas for *Valonia* the ionic activities were taken to be the same in and out. (From Higinbotham, 1973a.)

	Driving force			Driving force		
	E_K^{co}	E_M^{co}	$E_M^{co} - E_K^{co} = E_{co}^D$	E_K^{vc}	E_M^{vc}	$E_M^{vc} - E_K^{vc} = E_{vc}^D$
Nitella flexilis	−179	−170	+9(out)	+11	+15	+4(out)
Nitella translucens	−171	−140	+31(out)	+12	+18	+6(out)
Chara australis	−178	−173	+5(out)	+22	+18	−4(in)
Valonia ventricosa	−92	−71	+21(out)	−9	+88	+97(out)

values for Eq. (2.19) are based on a flux equilibrium whereas in the experiments cited in Figure 6.15 there was continuous K⁺ influx. The Ussing-Teorell flux equation [Eq. (2.40)] applies to such nonequilibrium conditions and was used as the criterion for the conclusions on active transport of K⁺ shown in Figure 6.15. This work was published prior to the discovery by Pitman et al. (1977) that PD measurements made shortly after cutting segments of tissue were quite low but with aging the cell PD in-

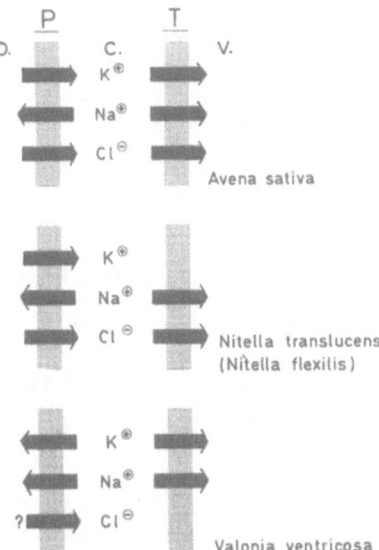

Fig. 6.16. Active ion fluxes in cells of a higher plant (*Avena*) in a fresh water alga (*Nitella*) and in a sea water alga (*Valonia*). See also Table 6.4. Abbreviations *P*, *T*, *o*, *c*, *v* as in Figure 6.12.

creased. Subsequent measurements of *Avena* cell potentials have yielded values up to about -135 mV; using these values the Ussing criterion [Eq. (2.40)] for K^+ (but not for Na^+ and Cl^-) indicates a passive movement for K^+ across the plasmalemma.

In contrast to the situation for K^+ the data on active transport of Na^+ and Cl^-, summarized in Table 6.4, is quite consistent and strongly based. However, we cannot explain results such as those of Macklon (1975a) indicating that in onion roots there is an active influx for Na^+ at the plasmalemma; the PD found in these cells, -32 mV, is quite low compared to most measurements.

By means of such analyses one arrives at an explanation of K^+–Na^+ selectivity in a very different way than by interpretation of isotherms (Table 6.1; Sect. 6.1.2.1), and some of the comments at the end of Section 6.2.1.2 appear to be confirmed. Preferential accumulation of K^+ over Na^+ in plant cells (as in animal cells) to a fair extent must be a result of active Na^+ extrusion. The effectiveness of this Na^+ exclusion becomes particularly clear in marine algae, which live in a medium of high Na^+ concentration and yet do not contain more Na^+ than freshwater algae and higher plants (Table 6.3). However, it also emerges from this approach, that the dangerous effects of high Na^+ levels on the proteins and enzymes in the cytoplasm are banished not only by active Na^+ export across the plasmalemma but also by active Na^+ removal into the vacuole. Kylin (private communication) and Kylin and Hansson (1971) have assessed this in an ingenious way working with four cultivars of sugar beet in which they investigated alkali transporting ATPases of the plasmalemma and tonoplast. The species from which sugar beet has been bred initially is a halophyte, *Beta maritima*. Thus sugar beet displays some resistance to NaCl salinity, but to a different extent depending on the cultivar or variety. Kylin's

Table 6.6. Na-transport and membrane ATPases in sugar beet cultivars (Kylin, private communication; and Kylin and Hansson, 1971).

Abbreviation of variety:	ADA	BEDA, EVA	FIA
$[Na^+ + K^+]$ Cl content:	low salt, high K^+ conc.	intermediate	high salt, high Na^+ conc.
Types of ATPases (according to ion requirements):	single type of Na-K-ATPase, requiring much K^+	two types of ATPase	single type of Na-K-ATPase, other requirements
Localization and function of ATPases:	ATPase at plasmalemma, K^+ uptake, Na^+ extrusion	ATPases at plasmalemma and at tonoplast removing Na^+ from cytoplasm in two directions	ATPase at tonoplast, Na^+ transport out of the cytoplasm into the vacuole

work shows that there are gradations among cultivars, and those having ATPases both at the plasmalemma and at the tonoplast removing Na^+ from the cytoplasm appear to be most viable and to have optimal protection (Table 6.6).

6.3 Survey of Models With Active Transport at the Plasmalemma Only and at Both Plasmalemma and Tonoplast

Figure 6.17 schematically summarizes the models discussed thus far in this chapter: (a) and (b) represent schemes consistent with the ideas of Epstein and coworkers. Systems 1 and 2 of ion uptake are localized at the plasma membrane either as two separate carrier entities (a) or as one carrier entity with two sets of active sites (b). (c) represents a scheme based on ideas of Gerson and Poole (1971, 1972), the salient feature of which is that the duality of the absorption isotherm is explicable by changes of membrane transport due to exchange sites in the AFS (Sect. 4.3), and

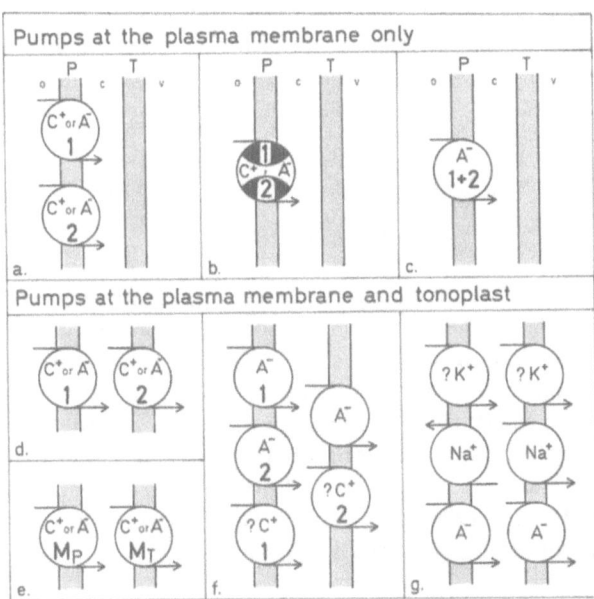

Fig. 6.17. Models with active transport at the plasmalemma only and at both plasmalemma and tonoplast. C^+ and A^- indicate whether the mechanism is thought to operate for cations and anions, respectively. Numbers *1* and *2* refer to system 1 and system 2 of the dual mechanism of ion uptake. *M*, multiphasic ion transport system. *Arrows* indicate direction of pumps. Abbreviations *P, T, o, c, v* as in Figure 6.12. (Sect. 6.3 provides a detailed explanation of this figure.)

which assumes an active anion pump at the plasma membrane effective over the entire concentration range of systems 1 and 2. (d) shows the model of Laties and coworkers, in which system 1 and system 2 are envisaged to be necessarily two separate entities and considered to operate at the plasma membrane and tonoplast in series. (e) gives Nissen's scheme of multiphasic plasma membrane and multiphasic tonoplast systems, the former operating at usual ion concentrations (≤ 10 mM) the latter at higher concentrations. (f) depicts the situation suggested by Pitman's computer simulation with anion pumping at the plasma membrane in the concentration ranges of system 1 and system 2 and with an additional anion pump at the tonoplast (the situation with cation pumping is less clear). (g) shows ion pumps as suggested by electrochemical evidence for most higher plant cells, with active inward pumping of K^+ or passive distribution of this ion, active pumping of Na^+ out of the cytoplasm into the vacuole and also to the outside, and active inward pumping of anions.

It is suggested that as an interesting exercise the reader should assess these models for similarities and discrepancies. He should bear in mind that (a), (b), (d), (e), and (f) were developed by analyzing tracer uptake kinetics only. However, (c) incorporates electrochemical data, and (d) is based on detailed electrochemical evaluations. Data based on kinetics only do not provide information on energy relations; electrochemical potential gradients do give an indication of forces involved and the amount of energy required to transfer the ions being moved against a gradient.

6.4 Models With Two Cytoplasmic Compartments

Various kinetic and electrochemical results apparently are not consistent with the two-compartment serial cell model (Fig. 6.12). An attempt has been made to explain these discrepancies by the assumption that two cytoplasmic compartments play a role in ion uptake and distribution in the cells. This led to the development of the so-called Y-model of Characean cells in which plastids in the nonmobile cytoplasm along the cell walls may play an important role as an additional and separate compartment (Walker and Pitman, 1976; Fig. 6.18a). The endoplasmic reticulum may also be important compartmentally (Fig. 6.18b).

6.4.1 "Unexpected Kinetics" of Ion Uptake and Ion Exchange in Cells of Higher Plants

One puzzling observation, which has been made in various laboratories and with various plant tissues, is the shoulder of the ion exchange curve (Lüttge and Pallaghy, 1972; see discussions in Anderson, 1973). During

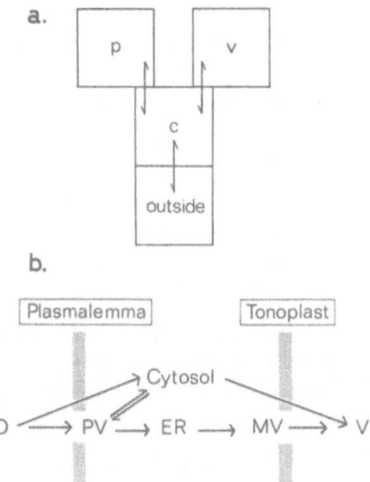

Fig. 6.18. Cell models with two compartments of the cytoplasm. **a** four-compartment Y-model (i.e., cell with three compartments) in which the cytoplasm (*c*) can exchange with the vacuole (*v*) or "plastids" (*p*). Walker and Pitman (1976); **b** pathways of ion transport during uptake into the vacuole of *Nitella* cells according to the ideas of MacRobbie (1970a). *O*, external solution; *PV*, pinocytotic vesicles; *ER*, cisternae of the endoplasmic reticulum; *MV*, micro-vesicles budding off from the ER, *V*, central vacuole. The compartments PV-ER-MV could form a pathway for "fast transport".

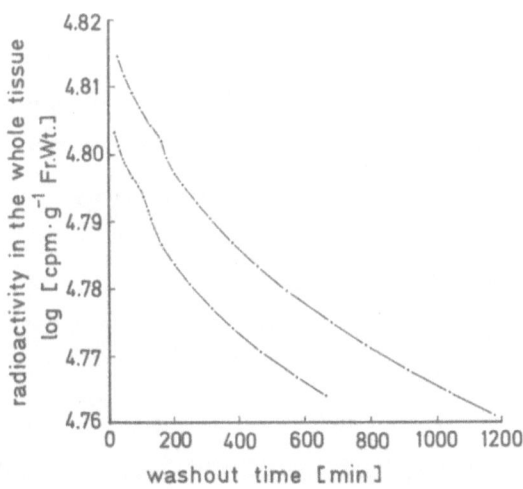

Fig. 6.19. Shoulder of the washout curve (^{42}K$^+$-efflux) of isolated root cortex of *Vicia faba* (2 separate experiments). K$^+$ concentration of uptake and washout solution 1 mM, temperature 23°C. (Lüttge and Pallaghy, 1972.)

washout in these cases radioactivity does not decrease continuously in three phases (Fig. 6.11). Rather, the washout curve shows a shoulder 100–180 min after the beginning of the washout (Fig. 6.19). Similar discontinuous kinetics are found also when tracer influx is studied in a *Vicia faba* root-cortex preparation which has been totally equilibrated in nonlabeled solution prior to the tracer experiments (Fig. 6.20). A satisfying mathematical compartmentation model explaining these unexpected ion flux kinetics has not been found yet. Certainly such a model must incorporate more than one cytoplasmic compartment.

With washout curves of Cl^- of carrot-root tissue, Cram (1968b) also obtained results not in agreement with a two-compartment cell model (cytoplasm-vacuole). The tonoplast fluxes Φ_{cv} and Φ_{vc} decreased when Q_v increased. Φ_{vc} therefore cannot be merely passive efflux from the vacuole because then it should increase with increasing Q_v. The nature of the relation between Q_v and Φ_{vc} is not clear. The plasmalemma fluxes reacted in different ways on increase in Q_v; Φ_{oc} decreased somewhat, which is in agreement with the active nature of this flux; Φ_{co} increased, and this is in agreement with the passive nature of this flux. All these results would be in agreement with the model of Figure 6.12 if Q_c also had increased with rising Q_v. This, however, was not the case; Q_c behaved in the opposite way and decreased. This unexpected discrepancy is resolved if one assumes two cytoplasmic compartments, namely a small compartment whose concentration changes in the same way as Q_v and a larger compartment whose ion content becomes lower as Φ_{cv} and Φ_{vc} decrease and when Q_v increases.

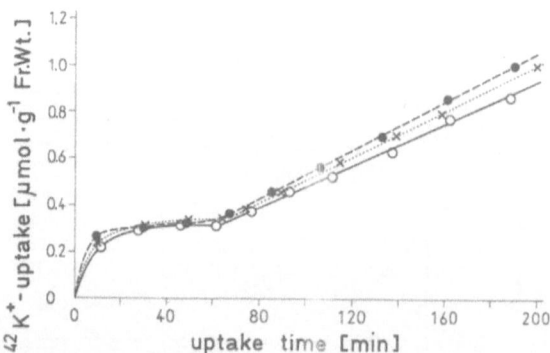

Fig. 6.20. $^{42}K^+$-influx in isolated root cortex of *Vicia faba* (3 separate experiments). The tissue was equilibrated for 24 h in nonlabeled salt solution and tracer ($^{42}K^+$) was added at time zero. K^+ concentration of nonlabeled and labeled solutions was 1 mM. (Lüttge and Pallaghy, 1972.)

6.4.2 Kinetic Investigations With Internodal Cells of Characeae

The possibility of comparing kinetic evidence with the results of direct analyses (see Sect. 2.2.3.1 and Sect. 6.2.3.1) has made Characean internodal cells a particularly well-suited object for compartmental studies. Therefore it is no surprise that they were used in the initial development of flux analysis (Sect. 6.2.3.1), and that now the dispute about cell models with three compartments (Fig. 6.18) again is concentrated on work with Characean cells (review Walker and Pitman, 1976).

On the basis of kinetic investigations with *Nitella* cells MacRobbie (1969, 1970b, 1971; Costerton and MacRobbie, 1970) has developed the model shown in Figure 6.18b. According to the serial two-compartment cell model, the only rate constant for exchange of ions between the cytoplasm and the vacuole must be:

$$k = \frac{\Phi_{co} + \Phi_{cv}}{Q_c} \qquad (6.8)$$

The constant k must not be a function of time and measurements of influx and efflux must yield the same values for k. These predictions of the two-compartment cell model initially appeared to be confirmed by experiment, but more detailed investigations of the time-dependence of k later on showed discrepancies when short uptake periods (below 15 min) were used. Transport of radioactively labeled ions from the external medium into the vacuole was more rapid than had been expected on the basis of the two-compartment serial model. According to this model, ions transported into the vacuole from the external medium must first mix with the ions already present in the cytoplasm and therefore the transport of labeled ions into the vacuole should show an initial lag phase; i.e., Φ_{cv} should be a function of Q_c. This, however, was not the case. Therefore two pathways for transport into the vacuole were assumed:

1. A rapid transport, during which a complete mixing of the newly absorbed ions with those ions already present in the cytoplasm is not necessary. This transport pathway can explain the deviation of the system from the two-compartment cell model which was observed during short periods of uptake.
2. A pathway of slow transport, in which an equilibration with the cytoplasm does occur. This transport pathway explains the agreement of the system with the serial two-compartment model observed during long experimental periods. The fast-transport pathway appears to be available in *Nitella* for Cl^- and Na^+ ions but only to a limited extent for K^+, whereas in *Tolypella* K^+ ions may also use this fast pathway for transport into the vacuole.

In this investigation ion transport into the vacuole was determined by direct analysis of cell sap which was clearly not contaminated by cytoplasm. The results suggest that the cytoplasm does not behave as a homogeneous phase during ion transport. With respect to which of the cytological compartments are responsible for the two kinetically determined transport pathways another experimental result of MacRobbie appears to be important. The contribution of the rapid phase relative to the total influx into a cell was always the same, irrespective of the absolute amounts of the total influx; in other words, the transport utilizing the fast pathway was directly proportional to the total influx. MacRobbie assumed that chloroplasts and mitochondria are not likely to be important as transport phases when transit of the cytoplasm to the tonoplast is concerned. Conversely, the endoplasmic reticulum (ER) appears to offer an explanation due to its particular localization in the cell and because of its dynamics. MacRobbie (Fig. 6.18b) postulated that ion uptake into the cytoplasm can occur by invagination of membrane vesicles at the plasmalemma (pinocytosis, see Chap. 7.1). These vesicles can merge with the cisternae (membrane-bound interior spaces) of the endoplasmic reticulum after they have lost part of their ion content to the cytoplasm. An increased influx thus would be correlated with an increased formation of pinocytotic vesicles; the membrane area available for the exchange between the vesicles and the cytoplasm would be proportional to influx, which would explain the constancy of the sizes of both phases relative to each other. Ions transported into the interior of the ER cisternae are supposed to move rapidly to the vacuole. Vesicles budding off from the ER profiles in the vicinity of the tonoplast and an incorporation of these vesicles into the vacuole could then be responsible for ion transport to the vacuole.

As experimental support for ion transport in vesicles, MacRobbie finds that uptake of ions in *Nitella* cells tends to be quantized:

$$\frac{\Phi_{ov}}{\Phi_T} = 0.22 \times n; \qquad (6.9)$$

where Φ_{ov} is the influx from the external medium into the vacuole, Φ_T the total influx into the cell, and 0.22 and n are empirical constants. In experiments with different individual *Nitella* cells values of n = 1, 2, and 3 were obtained. On first glance it seems to be quite simple to explain such a quantization by the formation of vesicles of different sizes participating in ion uptake. The size of the vesicles then should be specific for each particular cell. If only ion uptake at the plasmalemma were quantized in this way, then correctly $1 - \Phi_{ov}/\Phi_T$ should be quantized and not Φ_{ov}/Φ_T. Thus both plasmalemma influx and also the transport of ions into the vacuole appear to be quantized, and are assumed to be mediated by vesicular transport. One can therefore summarize the model of MacRobbie by the diagram as shown in Figure 6.18b.

The only cytological evidence for this model was the topographic distribution of the ER profiles in *Nitella* cells. However, recently there is support from cytological work with localization techniques suggesting that ion transport does in fact occur within the cisternae of the ER (Stelzer et al., 1975). Other support comes from work with the alga *Acetabularia* showing discrepancies between efflux kinetics and direct compartmental analysis, which also can be explained by assuming a "direct pathway" between the medium and the vacuole (Gradmann, 1975).

Nevertheless, the model remains very speculative and has been criticized severely. Criticism has focused on the question of quantization and on the problem of whether the vacuolar sap analyzed by MacRobbie appropriately represented the composition of the sap of cells *in vivo*. With respect to quantization it is puzzling as to why, in Eq. (6.9), n always turns out to be a whole number like 1, 2 or 3. Walker and Pitman (1976) have pointed out that it remains unclear whether this difference in size or capacity of vesicles is of genetic or of physiological origin. Findlay et al. (1971) have argued that the quantization is a product of a subjective interpretation of data by MacRobbie, which is statistically not justified.

As far as the analysis of the isolated vacuolar sap is concerned, it appears to be clear that it is not directly contaminated with cytoplasm. Yet it might be artificially modified by the occurrence of action potentials triggered as the internodal cells are cut open to collect the sap. Action potentials in charophyte cells occur at both the plasmalemma and the tonoplast (Hope and Findlay, 1964). It is likely that they cause a transient increase in tonoplast permeability leading to a rapid transport of ions from the cytoplasm into the vacuole (Walker and Pitman, 1976). MacRobbie (1975) concedes that the rapid uptake in her experiments could be an artifact due to such action potentials. However, she maintains that the kinetics of the slower phase, the absence of a discrimination between Cl^- and Br^- influx, and the link between influx and transfer to the vacuole still support the idea that ion transfer to the vacuole is via the creation and discharge of salt-filled vesicles, rather than processes of independent ion movements across the tonoplast.

Walker and Pitman (1976) have compared the Y-model (Fig. 6.18a) with MacRobbie's model (Fig. 6.18b). Clearly quantization of transport cannot be accommodated by the Y-model. However, with the reservations against accepting quantization as a fact, they conclude that thus far the Y-model with the vacuole (v) and with p representing plastids and c the "water phase" of the cytoplasm (i.e., better presumably the "cytosol") is the preferable model. Although vesicle transport cannot be excluded, they feel "it seems best to assume a molecular mechanism until it is disproved".

We have devoted much space to the argument of MacRobbie, to the description of her model, and to the controversy which has arisen about it. The reason for this is that we feel that with her attempt to relate the complex real plasma structure with mathematical-kinetic compartment models she has promoted a very important idea. This idea will certainly be of much relevance for the further development of transport models.

6.5 Conclusions and Outlook

The models discussed here owe their existence largely to a formalism which correlates kinetic results with various distinct simplifications which we can make of cell structure. By this kinetic formalism we obtain information on the number of compartments which play a role in transport and the distribution of solutes of interest, on the capacity of the compartments, and on the size and mutual interdependence of individual fluxes at the boundaries of the compartments. A determination of the identity of compartments characterized kinetically with visible structures in the cells, however, is not directly possible. Only in certain cases can this be accomplished by additional nonkinetic evidence. Examples are the experiments with nonvacuolated root tips and vacuolated root tissues, and the direct analyses with large coenocytic algal cells. In these cases we can be reasonably sure that the correlation of the kinetic data found with the cytoplasmic and vacuolar compartments are firmly based. However, our discussion about the models with more than one cytoplasmic compartment shows that even then we deal with only more or less good approximations.

In this chapter we have mainly discussed transport and distribution of ions which are not subject to significant metabolic alteration, i.e., Na^+, K^+, and Cl^-. In this respect these ions may seem less significant than those, such as NO_3^-, $H_2PO_4^-$, and SO_4^{2-}, which are altered or biochemically incorporated into various other forms. However, the nonmetabolized inorganic ions provide excellent material for establishing the basic principles governing the simplest transport and compartment models which we know. Even these turn out to be, in the framework of the complexity of cellular structure, extraordinarily difficult. Thus we now can only approximate the real models and equations governing the processes in real life. Nonetheless these processes are very important aspects of plant growth and development, and of ecological adaptation.

The transport of inorganic and of organic ions, which are subject to alterations in biochemical pathways, has not been dealt with in detail in this chapter. They must conform to the general principles found here, but superimposed would be the modifying effect caused by substrate-product relationships like those cited in Chapter 1 (Fig. 1.2). In this case substance A may change to B, or even to C, D, E, etc., perhaps in the same compartment or in different compartments. The biochemists speak of "pools" of species A, B, etc., and in many cases have established approximately whether these lie in the cytosol, the chloroplasts, the mitochondria, etc.; in very few cases are the electrochemical gradients clearly defined. Again we deal with a series of approximations to the "true" world. The information obtained has been found kinetically by dealing with rates of change of tracer-labeled substrates and of specific activities of products in various pools.

At this stage, we must state that on compartmentation we have two kinds of empirical observations of a very different nature:

1. the structurally visible and clearly manifold compartmentation of the cell by an apparently confusing multiplicity of membrane barriers, and
2. the kinetic evidence that a cell during transport does not behave as one space, but as a compartmented structure.

Since the whole work on compartmental models is occasionally dismissed as esoteric speculation, it appears important to stress that the conclusion summarized in the second point is a clear fact, notwithstanding the more subtle ambiguities in kinetic approaches. Speculation comes in only when attempts are made to correlate the kinetic compartments with real cytological structures or, in other words, in the attempt to make a synthesis between the two statements above. This becomes particularly difficult when more than one cytoplasmic compartment has to be accounted for (Sect. 6.4). For a general consideration a cell model suggested by Schnepf (1966) may be very helpful. This model is based on the assumption that all membranes separate an aqueous phase from a cytoplasmic phase. One distinguishes then basically only two principal phases, an aqueous mixing phase, and a nucleo-cytoplasmic mixing phase. The aqueous mixing phase consists of the external solution and the content of all compartments which are surrounded by a simple unit membrane within the space bordered by the plasmalemma. Thus, the aqueous mixing phase comprises the large central vacuole, small and tiny vesicles within the cytoplasm, the cisternae of the ER, and the dictyosomes, etc. A particular situation obtains with mitochondria and plastids which are surrounded by a double unit membrane. According to Schnepf's cell model, the space between the two unit membranes of a plastid or mitochondrion must belong to the aqueous mixing phase (see Fig. 6.21).

Our evaluations made above regarding the role of two or more cytoplasmic phases, on the basis of Schnepf's cell model, is semantically not quite correct. Ions, which after uptake from the external medium, are mixed within the cytosol and equally distributed within the plasma matrix are in the nucleo-cytoplasmic mixing phase. Ions which intrude into a compartment embedded in the plasma matrix are either in the aqueous mixing phase (if they only have passed one unit membrane) or in the nucleo-cytoplasmic mixing phase (if they have passed two unit membranes and for instance have intruded into the interior of a mitochondrion or plastid).

Now it is most important to mention that all these cytological compartments are very dynamic structures, and that they can continuously change their shape and size by membrane flow, may divide and also may merge one into another. These dynamics of membrane-bound compartments are clearly seen in moving pictures taken by phase contrast light microscopy. Several other investigations have shown also, for example,

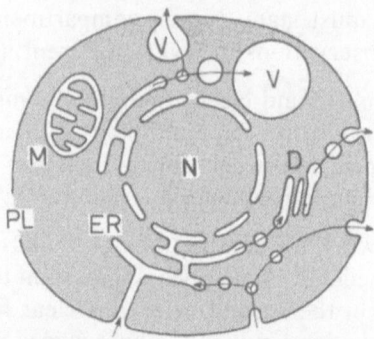

Fig. 6.21. Cell compartmentation model according to Schnepf. Nucleocyto-plasmic mixing phase is shaded: interior of the nucleus (*N*), mitochondrion (*M*), and cytosol. Aqueous mixing phase is white: vacuole (*V*), cisternae of the endo-plasmic reticulum (*ER*) and of the dictyosomes (*D*). The arrows indicate dy-namics of the compartments (vesicle intrusion and extrusion at the plasma-lemma; movement of vesicles in the cytosol; relations between ER, dictyosomes and vacuome). *PL*, plasmalemma. Simplified from Schnepf (1966).

that a particularly close relation exists between the endoplasmic retic-ulum and the cell vacuoles. Small vesicles budding off from the endo-plasmic reticulum can fuse with the vacuole (Chap. 7.1.3). During this process two compartments with aqueous content merge with each other. This, also, is transport, just as postulated in MacRobbie's model (Fig. 6.18b). In her discussion of the three-compartment cell model she specu-lated that the endoplasmic reticulum is identical with the kinetically ob-served additional cytoplasmic compartment. Because of the character-istic dynamics of the endoplasmic reticulum this appears to be not so bad a guess. At this point we are again at the utmost border of knowledge which we can attain, for the moment, without violating the warning of Hope and Walker (1975; p. 161) to "neither despair nor let the chain of hypotheses get too long".

This sixth chapter was based very much on indirect kinetic ap-proaches. We have developed the concepts of transport and compartmen-tation beginning with the very simple cell model of an outer and one inner phase only. We have subsequently considered increasingly complex models by more or less following the history of progress of research in this area. The following, seventh, chapter is closely related to the sixth chapter. It also deals with transport and compartmentation, but it does so from a very different point of view in that it starts with observations of real cytological compartments and the biochemical events localized in them.

Chapter 7

Correlations Between the Fine Structure of the Cytoplasm and Transport Functions: Further Complications of the Model

The kinetic models described at the end of the preceding chapter are very speculative. The experimental data are not sufficient yet to support a wide acceptance of the ideas behind these models. However, they reveal the unhappiness of transport physiologists who compare their simplified kinetic models with the reality of the cell as it is seen in the electron microscope, showing a multitude of tiny spaces and compartments (see Fig. 6.21). In the present chapter we want to collect some evidence for the occurrence of transport of substances in and by such compartments which has been gained by approaches different from those of the kinetic models.

7.1 Observations of Transport of Matter in Membrane-Bound Vesicles

The occurrence of small vesicles apparently formed by invaginations of the plasmalemma has often been observed in the electron microscope. An invagination into the interior of the cell could serve to take up substances (pinocytosis); and evagination to the exterior of the cell could serve to extrude substances. In the interior of the cytoplasm, as well, vesicles may be observed which can merge with other vesicles, with the central vacuoles or with the plasmalemma, and therefore serve a transport function within or across the cytoplasm (summaries, Schnepf, 1966, 1968, 1969).

7.1.1 Exocytosis

Exocytosis is extrusion of vesicle content to the exterior. An example, well investigated in many respects, is the extrusion of wall material by Golgi vesicles.

Numerous electron microscopic and electron microautoradiographic investigations have made it clear that the Golgi vesicles which bud off from the cisternae of dictyosomes excrete highly polymerized substances. In animal cells proteins and polysaccharides are transported to the exterior in this way. Plant cell Golgi vesicles excrete various polysaccharides, especially acid polysaccharides, pectins, and slimes of polysaccharide nature. Also little scale-like structures, which one often finds at the surface of some algal cells, are extruded by Golgi vesicles. Thus dictyosomes are observed to occur especially at the sites of extensive cell wall growth, for instance in the tip of root hairs and rhizoids, during the formation of new cell walls, after cell divisions, and during the wall thickening of tracheid walls; Golgi vesicles transport the material to be built into the cell wall. Also during secretion of the trapping slime of the leaf tentacles of the carnivorous plant, *Drosophyllum,* a similar function of the dictyosomes has been observed (Sievers, 1965; Mollenhauer and Morré, 1966; Morré and Mollenhauer, 1976).

Figure 7.1 shows the diagram of a dictyosome. Golgi vesicles bud off on one side of the disk-shaped dictyosome from the membrane-bound cisternae. During this process the dictyosome continuously loses membrane material. If the cisternae of dictyosomes are not to be exhausted quickly, a regeneration must occur concomitantly. This in fact happens at the face of the dictyosomes opposite to the face where the vesicles bud off. Having an excretion face and a regeneration face, the dictyosomes are thus structurally and functionally asymmetrical.

In addition to transport of polysaccharides there is synthesis of polysaccharides, or, at least, the polymerization from precursors within the

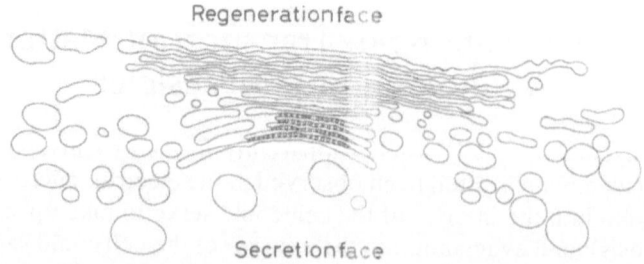

Regenerationface

Secretionface

Fig. 7.1. Cross section of a dictyosome with a clear asymmetry between a regeneration and a secretion face. Drawn after an original electronmicrograph of Whaley (Fig. 2 in Schnepf, 1969). About ×32,000.

membrane-bound systems of the dictyosomes. Thus during dictyosome activity we must distinguish three basic processes:

1. a biosynthesis of the secreted material,
2. a transport of secretion product in the vesicles, and
3. a cyclic flow of membrane material released at the secretion face and regained at the regeneration face.

A very good and broad description of all these processes with numerous examples is found in Schnepf (1969). He has summarized all the individual steps enumerated above in a schematical drawing. In doing this he also considered observations which suggest the participation of the membrane systems of the endoplasmic reticulum (ER) in the membrane flow and in the biosynthesis of secretion products. This scheme, simplified, is depicted in Figure 7.2 where the cyclic character of the whole process becomes very clear. As in a typical metabolic cycle, at a certain stage substances are injected (precursors for the secretion product) and at other stages substances are ejected (secretion or excretion); enzyme reactions, transport processes of substances, and flows of membrane material all participate in this cycle.

For the biosynthesis of the secreted material numerous steps are necessary, all of which probably can occur in the cisternae of the dictyosomes, in the Golgi vesicles already budded off, and also in the ER cis-

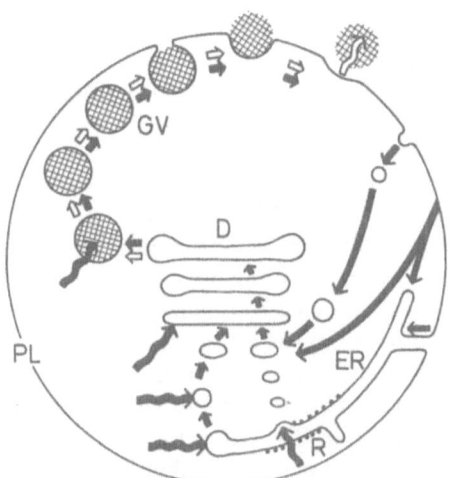

Fig. 7.2. Synthesis of secretion product, secretion and flow of membrane material during dictyosome activity. *Open straight arrows,* transport of secretion product; *open wavy arrow,* extrusion of secretion product; *dark wavy arrows,* injection of precursors of the secretion product; *dark straight arrows,* flow of membrane material. *D,* dictyosome; *ER,* endoplasmic reticulum; *GV,* Golgi vesicle; *PL,* plasmalemma; *R,* ribosomes. (Simplified from Schnepf, 1969.)

ternae where the ER membrane has ribosomes (= rough ER), as well as in the vesicles which have been budding off from the ER and which merge with the dictyosomes. Precursors of the secretion product can be injected into the cycle by uptake into the cisternae or vesicles from the cytoplasm at various stages during the cycle. By movement within the cell Golgi vesicles budded off from the dictyosomes approach the vicinity of the plasmalemma. During this process the structure of the membrane surrounding the vesicles changes so that it becomes very similar to the structure of the plasmalemma. Then the Golgi vesicles merge with the plasmalemma and the secretion product is extruded.

With the transport of the secretion vesicles membrane material is lost from the dictyosomes and added to the plasmalemma. This raises the question of how membrane material gets back to the dictyosomes; the scheme of Figure 7.2 gives various answers. First, the backflow of membrane material can occur in the form of vesicles which invaginate from the plasmalemma or evaginate from the ER and merge with the dictyosomes. Second, dictyosome membranes could be assembled from membrane-building components (lipid and protein molecules, small micelles) in the cytoplasm, not visible by electronmicroscopy. Such material could be given off from the plasmalemma and then be re-used for the formation of new ER or dictyosome membranes. Third, a possibility is provided by the interconnection and interdependence between the ER membrane system and the plasmalemma.

One certainly may assume that further similar vesicle transport processes occur in great number within the cell. However, they are not documented nearly as well as the polysaccharide transport in the Golgi vesicles, which therefore is a very important model case.

It is worth mentioning that in many glands, especially in salt-excreting glands of halophytes, numerous small vesicles are observed which may be involved in exocytosis (Osmond et al., 1969; Lüttge and Krapf, 1968; Lüttge, 1971a, 1975). Thomson and coworkers have investigated this in the salt glands of *Tamarix* (Thomson and Liu, 1967; Thomson et al., 1969). Rb^+ was used as an electron-dense tracer in electronmicroscopic studies of these glands. Only when salt-excreting plants have been supplied with Rb^+ can one observe an electron-dense precipitation in the vesicles which are found particularly in the vicinity of the cell wall. These experiments may suggest an Rb^+ extrusion by the vesicles, but they also leave many uncertainties. It is possible that Rb^+, as an ion which is toxic in higher concentrations, causes secondary effects during accumulation, which lead to particularly heavy density of the vesicles themselves. Even if the electronmicroscope correctly displays the distribution of Rb^+, a loading of the vesicles with Rb^+ could be an independent and parallel process, and not a direct pathway in salt excretion.

In nectary glands, which secrete large amounts of sugar, it must be assumed in most cases that secretion occurs by efflux of individual glucose, fructose, and sucrose molecules across the plasmalemma of the gland cells (eccrine secretion; review: Lüttge and Schnepf, 1976). However, some recent observations suggest that in particular cases dictyosomes and ER elements may also be involved in nectar secretion, so that vesicular extrusion of nectar (granulocrine secretion) may be a possibility in addition to eccrine secretion (Benner and Schnepf, 1975; Rachmilevitz and Fahn, 1975; review: Lüttge and Schnepf, 1976).

7.1.2 Endocytosis

Endocytosis, the reversal of exocytosis, is well known, in particular on a light-microscopic level in protozoa as the so-called phagocytosis. The cytoplasm flows around a particle, and the plasmalemma invaginates a bladder by which the particle is taken up into the cell interior. On the electronmicroscopic level of magnification, one also observes such plasmalemma invaginations with tiny vesicles entering the cytoplasm, i.e., pinocytosis. During this process an adsorption of particles at the plasmalemma surface is important for the invagination of vesicles. This binding may involve specific sites at the membrane surface and hence explain specificity of uptake processes. ATP energy may interact with the process at various stages. Hall (1970) detected ATPase activity cytochemically at the surface of vesicles near the plasmalemma of maize root cells; and Baker and Hall (1973) developed a hypothetical model along these lines, which explains selective Na^+ and K^+ uptake in the concentration ranges of systems 1 and 2 (cf. Chap. 6).

Although invaginations are very often observed at the plasmalemma of cells of higher plants, and vesicles in the vicinity of the plasmalemma, there is no unequivocal direct experimental evidence for the occurrence of pinocytosis in higher plants. In analogy with animal cells it may be justified to assume, however, that it does occur in plant cells as well. Especially transport of macromolecules across membranes can scarcely be envisaged other than by pinocytosis.

7.1.3 Transport in Vesicles Within the Cell

We have repeatedly mentioned transport of substances in vesicles within the cell. The scheme of MacRobbie (Sect. 6.4.2; Fig. 6.18b) postulates a pinocytotic uptake of ions, a merging of the invaginated pinocytosis bladders with the ER, a budding off of vesicles from the ER, and a merging with the vacuole. Jackman and van Steveninck (1967; van Steveninck, 1976b) found in disks of red beet tissue a correlation between the differentiation of the ER and ion uptake capacity, which, however, they did not describe in a statistically convincing way. Localization techniques, for instance precipitation of Cl^- ions with Ag^+ ions, leading to electron-dense AgCl precipitations visible in the electronmicroscope, suggest that ion transport across plant cells may occur in the cisternae of the ER (Stelzer et al., 1975; Läuchli, 1976a; see also Chap. 11). The scheme of Golgi-vesicle extrusion (Fig. 7.2) also is based on a quite extensive transport of vesicles with the cell.

It is worth mentioning here the ontogeny and the enzymic relations of the vacuolar system (= vacuome) of plant cells (review Matile and Wiemken, 1976). Electronmicroscopic investigations show that the large central vacuole during cell differentiation is formed by the merging of smaller vesicles budding off from the ER (Matile and Moor, 1968). Isolation of vacuoles and vesicles of various sizes by centrifugation in density gradients show that they contain the greatest activities of hydrolytic enzymes, i.e., that they have lysosome-like functions (Matile, 1966, 1968; Matile and Wiemken, 1976). Therefore in plants the vacuome must be the compartment where cytoplasm turnover is maintained; such hydrolytic digestion processes naturally have to be separated spatially from the synthesizing reactions in the cytoplasm. The hydrolytic enzymes in the vacuoles can act on the cytoplasm constituents, however, only when parts of the cytoplasm are expelled by vesicles into the vacuoles. Electronmicrographs give suggestions for the occurrence of tonoplast invaginations, and membrane-surrounded vesicles also have been observed within vacuoles.

If vesicles containing parts of the cytosol are expelled into the vacuole, then this is necessarily also associated with the transport of ions from the cytosol into the vacuole. Experimental evidence for this has been obtained by comparative investigations of ion uptake and ultra structure of young and old leaflets of the moss *Mnium* (Lüttge and Bauer, 1968; Lüttge and Krapf, 1968). The ion uptake isotherm in the concentration range of system 1 in old and young leaflets shows the expected saturation kinetics. In the concentration range of system 2, which according to the Torii-Laties hypothesis mediates transport into the vacuole, however, only old moss leaves have a hyperbolic isotherm. Young moss leaves here show a linear relationship of ion uptake to external ion concentration (Fig. 7.3). In young moss leaves particularly, numerous membrane invaginations are observed at the tonoplast and numerous vesicles occur within the vacuoles. This is quite clearly in agreement with a lysosomal function of the vacuole. The turnover of plasma material must be larger in young active cells than in old cells. Ion uptake associated with the formation of vesicles can easily explain the different transport kinetics of young and old moss leaves in the concentration range of system 2. Due to a prevailing transport of vesicles in young moss leaves, ion uptake proper at the tonoplast may not be rate-limiting and therefore shows no saturation isotherm, but rather a linear relationship obtained in the high concentration range. It should be noted, however, that in a recent investigation Butcher et al. (1977) question the lysosomal function of plant cell vacuoles. They suggest that hydrolases are compartmented in the cytoplasm in fingerlike

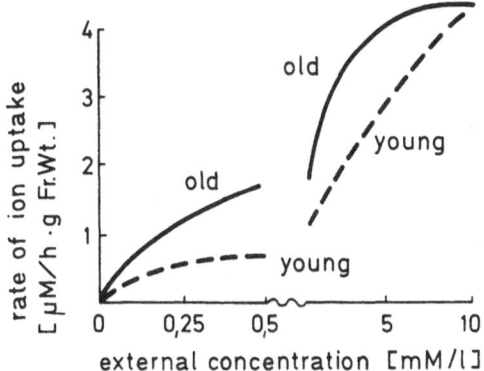

Fig. 7.3. Ion uptake isotherms of young and old leaves of the moss *Mnium*. (After Lüttge and Bauer. 1968.)

invaginations extending towards the vacuole. In the cytoplasmic phase the pH could thus be lowered locally by acids diffusing out of the vacuole across the tonoplast. This would allow action of hydrolases without vesicles budding off from the invaginations and being taken up into the interior of the vacuole.

7.2 Transport Functions of Cell Organelles

Just as the cell is protected from its surroundings, i.e., from the external medium by a membrane barrier (Sect. 1.1), so the organelles, such as the nucleus, the mitochondria, and the chloroplasts are separated from the cytosol by membranes and they attain a certain isolation or independence. This independence is particularly pronounced for mitochondria and chloroplasts and somewhat less so for nuclei.

The endosymbiosis hypothesis of the phylogeny of eukaryotic cells assumes that mitochondria and chloroplasts are microsymbionts once in evolution having been engulfed by a phagocytotic process. This speculation is supported by an array of facts which all stress the independence of these organelles as follows:

1. Mitochondria and chloroplasts resemble more primitive prokaryotic cells:
    a. their external membrane may be equivalent to the phagocytotic membrane of the host cytoplasm;
    b. their inner membrane which constitutes their major permeability barrier corresponds to the external membrane of prokaryotes; the inner membranes of mitochondria and chloroplasts, or rather the cristae and thylakoids derived from them, contain the electron

transport systems required for energy transfer reactions which in bacteria reside in the external membrane;

c. they multiply by division, and they divide asynchronically with their "host" cells;

d. they contain their own DNA and are genetically autonomous to a certain degree.

2. Symbiosis between eukaryotic cells in the animal and plant kingdom and prokaryotic (or also eukaryotic) green cells can be frequently observed. Most important in this context are natural and artificial endosymbioses of prokaryotic blue-green algae or even of chloroplasts within eukaryotic cells (Schnepf, 1964c; Taylor, 1967, 1968, 1970, 1971, 1973a,b; Taylor and Lee, 1971; Taylor et al., 1969, 1971).

This underlines the need for transport across the organelle "envelopes."

7.2.1 Nuclei

The nuclei certainly represent a unique case. Their external membrane is interrupted by relatively large pores. The regulatory function of the nuclei requires the extensive transport of macromolecules, particularly m-RNA, between the nucleus and the cytoplasm; RNA molecules may pass as a thread through the nuclear pores (Franke, 1970). Proteins synthesized in the cytoplasm may move into the nucleus. In addition there may be many other interactions between the nucleus and cytoplasm (review Brachet, 1976). Nevertheless, many observations suggest that the interior of the nuclei is not in unlimited contact with the cytoplasm by diffusion via the pores. Specific transport processes even of low molecular weight substances appear to be possible. On the giant chromosomes of animal cells, by analysis of puff formation, regulation of gene activity was followed and found to be correlated with specific changes of the ion content within the interior of the nuclei. Thus it is possible to imagine that ion transport processes participate in the regulation of metabolism (Lezzi, 1966; Kröger, 1967).

7.2.2 Mitochondria

The mitochondrion is surrounded by two membranes, and thus the following spaces can be distinguished: inner space or *matrix*, intermembrane space, and *external* space or cytosol (Fig. 7.4). The inner membrane forms cristae, tubuli, or sacculi by formation of infoldings towards the matrix space. This membrane is the site of the electron transfer systems of the respiratory chain. The matrix contains the soluble enzymes for the breakdown of substrates, i.e., of pyruvate in the tricarboxylic acid cycle and of fatty acids in β-oxidation. Pyruvate is formed in the extramitochondrial space via glycolysis, and fatty acids for β-oxidation also be-

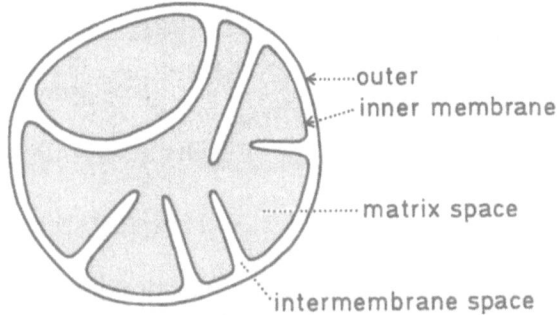

outer
inner membrane
matrix space
intermembrane space

Fig. 7.4. Schematic diagram of a mitochondrion from corn. (Malone et al., 1974; as modified by Heldt, 1976.)

come available outside the mitochondria. Energy in the form of ATP is generated inside the mitochondria, but must be available also for extramitochondrial endergonic reactions. This compartmentation requires an extensive transport of ions, metabolites, cofactors, and energy between the cytosol and the mitochondrial matrix.

For practical purposes we can neglect the outer membrane and the intermembrane space. The outer membrane of mitochondria is readily permeable for all solutes of low molecular weight, and the intermembrane space is easily accessible by diffusion. Conversely, the inner membrane is rather impermeable to most of the important solutes. Specific translocators or carriers residing in this membrane mediate the traffic of solutes; this is important for coordination and regulation of cytoplasmic and mitochondrial activities. The subject has recently been excellently reviewed, and we largely follow these reviews in depicting the more important features of this interesting system (Hanson and Koeppe, 1975; Heldt, 1976b; Wilson and Graesser, 1976). Most of this knowledge is based on work with animal mitochondria, but a substantial amount of evidence suggests that plant mitochondria behave rather similarly.

7.2.2.1 Electron Transport, Ion Transport, and Coupling of ATP-formation

Electrons are transferred along the chain of redox systems (flavoproteins, ubiquinone, cytochromes) residing in the inner mitochondrial membrane. Two conflicting models have been developed for coupling of ATP formation with this electron flow. First, a model of chemical coupling, in which a nonphosphorylated energy-rich covalent bond ($\sim$ or "squiggle") is formed during electron flow (i.e., $X \sim I$) the hydrolysis of which leads to ATP formation ($X \sim I + ADP + P_i \rightarrow X + I + ATP$). Second, a model of chemi-osmotic coupling after a charge separation across the inner membrane of the type of

$$
\begin{array}{c}
\text{H}_2\text{O} \\
\text{H}^+ \qquad \text{OH}^-
\end{array}
$$

in which the electrochemical gradient of H⁺ established can drive ATP formation (i.e., the Mitchell hypothesis).

The second hypothesis is depicted schematically in Figure 7.5. It has been much more widely accepted in recent years than the first one for various reasons. The nonphosphorylated high energy intermediate $X \sim I$ has never been identified. Conversely, several predictions of the Mitchell hypothesis have been verified experimentally, e.g., the H⁺-electrochemical gradient during electron flow and phosphorylation (see also Sect. 7.2.3.1); and this hypothesis also readily explains a number of secondary phenomena (e.g., Chap. 10). Figure 7.5 shows that uptake of mineral cations such as Mg^{2+} or Ca^{2+} into the mitochondria can be an alternative to ATP formation by dissipating the energy of the H⁺-

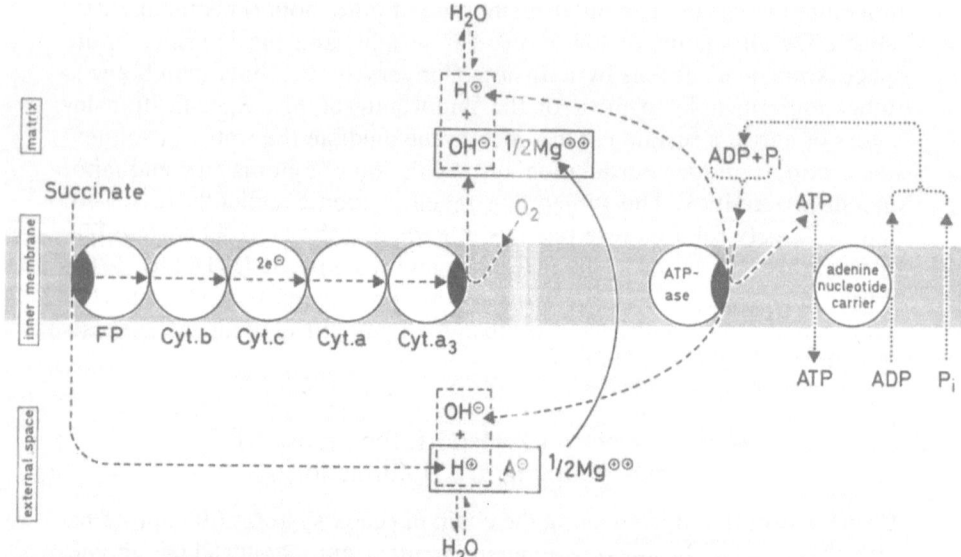

Fig. 7.5. Charge separation at the inner mitochondrial membrane by vectorial enzyme reactions during electron transfer along the respiratory chain and chemiosmotic coupling of oxidative phosphorylation after Mitchell (*broken arrows* and *broken rectangles*) or ion uptake as an alternative after R.N. Robertson (1968) (*solid arrows* and *solid rectangles*). The *dotted arrows* at the right hand of the scheme represent the adenine nucleotide carrier system after Klingenberg (1976). (Modified from Lehninger, 1964.) *FP*, flavoprotein; e^-, electron; A^-, anion; P_i, inorganic phosphate.

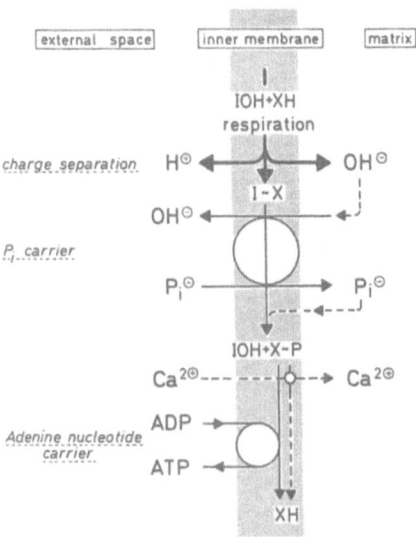

Fig. 7.6. Charge separation and chemical coupling of oxidative phosphorylation at the inner mitochondrial membrane after Hanson et al., 1972. (Modified from Hanson and Koeppe, 1975.) *IOH* and *XH*, metabolic intermediates; $I \sim X$, nonphosphorylated high energy intermediate; P_i, inorganic phosphate.

electrochemical gradient.

The hypothesis of chemical coupling was initially much promoted by the work in the laboratory of J.B. Hanson. These workers recently modified their model so that it combines aspects of the charge separation hypothesis and the old chemical hypothesis. This is explained in Figure 7.6. In this case Ca^{2+} uptake dissipates the energy of the squiggle of $X \sim P$ without formation of ATP, i.e., as in Figure 7.5, it is also an alternative to ATP formation.

7.2.2.2 Transport of Inorganic Phosphate and Adenine Nucleotide

Phosphorylation in the mitochondrial matrix and utilization of ATP energy in extra-mitochondrial compartments of the cell requires the transport of inorganic phosphate and of adenine nucleotide across the mitochondrial membrane. This transport is mediated by two carriers or translocators, as shown in the scheme of Figure 7.7. Inorganic phosphate (taken as $H_2PO_4^- = P_i^-$) can be transported independently of other solutes in exchange for OH^- (i.e., as an OH^-/P_i^- countertransport or antiport or an H^+/P_i cotransport or symport; the alternatives, OH^--countertransport and H^+-cotransport cannot be distinguished experimentally). Adenine nucleotides are transported by an adenine nucleotide carrier which exchanges ADP and ATP (Figs. 7.5, 7.6, and 7.7). Phosphate transport without

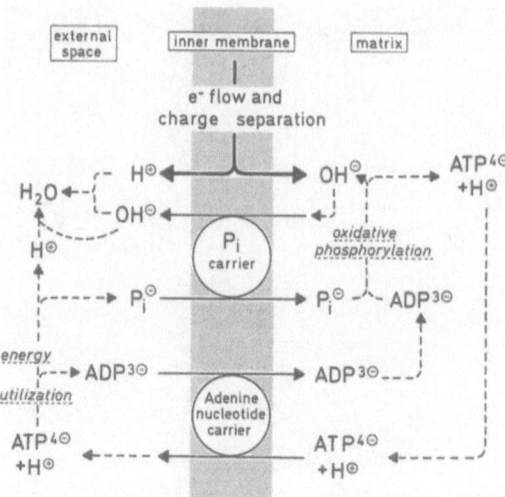

Fig. 7.7. Phosphate and adenine nucleotide carriers at the inner mitochondrial membrane according to the description of Heldt (1976b, p. 242).

concomitant oxidative phosphorylation and operation of the adenine nucleotide carrier would lead to an acidification of the mitochondrial matrix. This is regulated by utilization of the energy of charge separation (or of an active proton pump, which really means the same) for P_i accumulation within the mitochondria. During oxidative phosphorylation the acidification of the matrix is compensated by formation, consumption, and transport of H^+ and OH^- as shown in Figure 7.7.

7.2.2.3 Transport of Metabolic Intermediates: Transport Metabolites

A number of key intermediates of metabolism can be transported across the membrane envelopes of mitochondria (and also of chloroplasts). The term *transport metabolites* has been coined for these substances to highlight their important role in regulating and coordinating metabolic activities of various compartments (see also Chap. 10 Figs. 10.1, 10.16).

The more important mitochondrial carrier systems for metabolites are summarized in Figure 7.8. In the figure only those substrates are given for which these carriers have the highest affinities. In addition to the *phosphate carrier* already discussed above (Sect. 7.2.2.2), there are the following transport systems (Heldt, 1976):

1. a *dicarboxylate carrier,* which transports phosphate and the dicarboxylates malate and succinate (the latter with five times lower affinity, and perhaps also oxaloacetate);
2. a *tricarboxylate carrier,* which transports citrate, isocitrate, cis-aconitate and phosphoenolpyruvate (and, with lower affinity, the di-

carboxylates, malate and succinate); in corn mitochondria citrate can
be exchanged for P_i or SO_4^{2-} (Kimpel and Hanson, 1977);
3. an *α-ketoglutarate carrier*, which transports ketoglutarate and malate
   (and, with lower affinity, aspartate, glutamate, and oxaloacetate);
4. a specific *glutamate carrier;*
5. a *glutamate-aspartate exchange carrier;*
6. a specific *pyruvate carrier*, which may operate as an H^+-pyruvate symport mechanism or an OH^--pyruvate antiport mechanism (for maize
   mitochondria see Day and Hanson, 1977).

An important feature of these transport systems is that many of them
are linked indirectly with each other by common substrates of the various
carriers as indicated by the dashed lines in Figure 7.8. In this way the dicarboxylate, the tricarboxylate, and the α-ketoglutarate transport indirectly are all dependent on the transport of P_i, which in turn is driven by
charge separation or by the proton pump powered by the respiratory
chain. This makes respiratory electron flow the driving force of all these
transport processes. Charge separation (or the proton pump) constitutes
primary active transport in the sense defined in Section 2.5.5. The other
carrier mechanisms shown in Figure 7.8 constitute secondary active

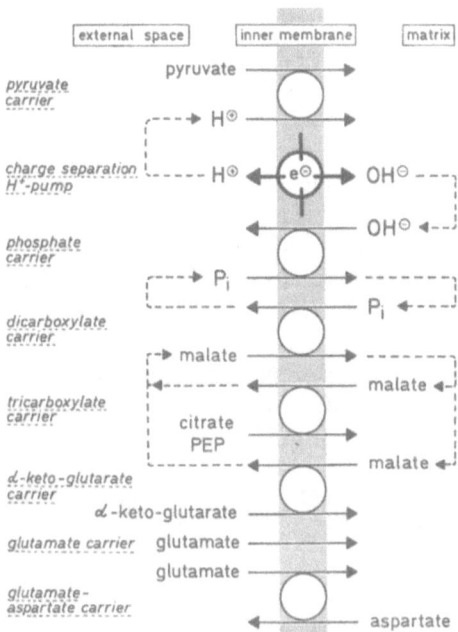

Fig. 7.8. Carriers of phosphate and various metabolites at the inner mitochondrial
membrane. (After Heldt, 1976b.) *Circles* and *solid arrows* represent direct links
between solute fluxes, *broken arrows* indicate indirect linkages of transport processes. *PEP*, phosphoenolpyruvate; P_i, inorganic phosphate.

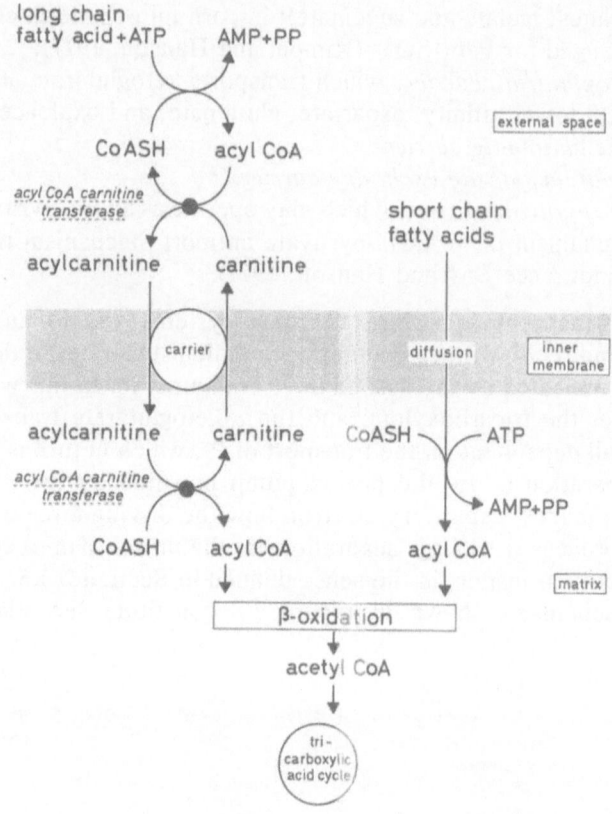

Fig. 7.9. Compartmentation of activation and oxidation of fatty acids with transport at the inner mitochondrial membrane. (Slightly modified from Heldt, 1976b.)

transport. This unifies an initially confusing multitude of individual metabolite movements and leads to a scheme which displays a high degree of conceptual unity (McGivan and Klingenberg, 1971; review: Heldt, 1976a).

Transport of fatty acids is depicted in the scheme of Figure 7.9. Short chain fatty acids probably diffuse into the mitochondria in their undissociated form. The transport of long chain fatty acids is mediated by carnitine $\left({}^{-}OOC-\overset{\overset{\displaystyle H}{|}}{\underset{\underset{\displaystyle H}{|}}{C}}-\overset{\overset{\displaystyle OH}{|}}{\underset{\underset{\displaystyle H}{|}}{C}}-\overset{\overset{\displaystyle H}{|}}{\underset{\underset{\displaystyle H}{|}}{C}}-\overset{+}{N}(CH_3)_3 \right)$ both in plant and in animal cells (but see Sect. 7.2.4.1).

7.2.2.4 Energy Transfer Between Mitochondria and Cytoplasm

Energy can be transported out of the mitochondria into the cytoplasm by the adenine-nucleotide carrier discussed above (Sect. 7.2.2.2, Figs. 7.5, 7.6, 7.7).

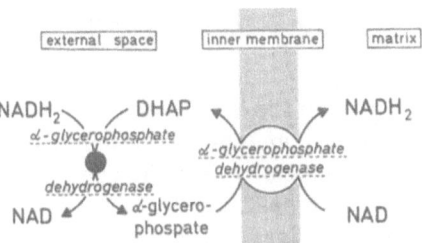

Fig. 7.10. Transport of reduction equivalents across the inner mitochondrial membrane by the α-glycerophosphate shuttle after Zebe et al., 1959 (according to the description of Heldt, 1976b, p. 248). *DHAP*, dihydroxyacetonephosphate.

Since reduction equivalents ($NADH_2$) generated by glycolysis in the cytoplasm can be used for oxidative phosphorylation in the mitochondria, transfer of energy must also be possible in the form of reduction equivalents. However, the inner mitochondrial membrane is impermeable for pyridine nucleotides, and cytoplasmic $NADH_2$ cannot be utilized directly within the mitochondria. The transfer of reduction equivalents across the mitochondrial membrane is mediated by shuttle systems of transport metabolites. Two possibilities can be discussed for mitochondria (see also Sect. 7.2.3.4 for chloroplasts): First, there is an α-glycerophosphate shuttle involving a soluble cytoplasmic α-glycerophosphate dehydrogenase and a membrane-bound α-glycerophosphate dehydrogenase as is shown in Figure 7.10. The shuttle can operate since the membrane-bound α-glycerophosphate dehydrogenase is accessible for α-glycerophosphate from the outside and to pyridine nucleotide from the inside. Second, another possibility is the malate-aspartate shuttle depicted in Figure 7.11. This shuttle combines the α-ketoglutarate and glutamate-aspartate carriers discussed in Section 7.2.2.3.

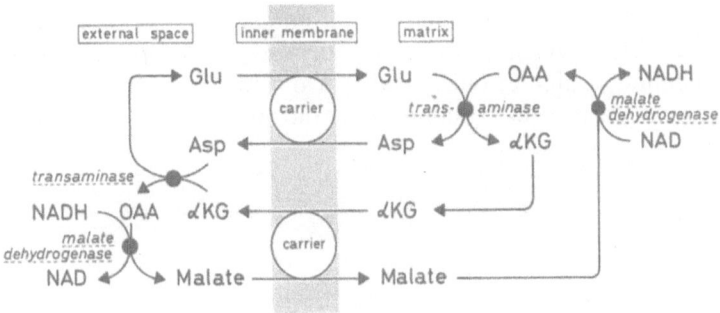

Fig. 7.11. Transport of reduction equivalents across the inner mitochondrial membrane by the malate-aspartate shuttle. (Slightly modified from Heldt, 1976b.) *Glu*, glutamate; *Asp*, aspartate; *OAA*, oxaloacetate; *αKG*, α-ketoglutarate.

7.2.3 Chloroplasts

Like mitochondria, chloroplasts are surrounded by two unit membranes. Infoldings and derivatives of the inner membrane are the thylakoid membranes forming an internal membrane system within the chloroplasts which is rather more complicated than the cristae, sacculi, or tubuli we observe in mitochondria. Thylakoid membranes can penetrate the chloroplast matrix or stroma as a single invagination but most frequently many invaginations are stacked together to form the grana. Thus the major spaces we can distinguish in chloroplasts are the cisternae enclosed by the thylakoid membranes, i.e., the thylakoids; the space outside the thylakoids but inside the inner membrane of the chloroplast envelope, i.e., the stroma, the intermembrane space and the *external space* (Figure 7.12).

As in the mitochondria, the outer membrane is quite permeable for many solutes and the traffic across the chloroplast envelope is controlled by the inner membrane. The thylakoid membranes are the site of photosynthetic pigments (mainly chlorophylls) and redox systems harvesting light energy and serving energy transfer to provide ATP and reduction equivalents for photosynthetic CO_2 reduction. Calvin cycle enzymes are localized in the stroma, but they must be in contact with the thylakoids. Starch synthesis also occurs within the chloroplasts. However, sucrose synthesis most likely resides in the cytoplasm. The chloroplast envelope is rather impermeable to sucrose and to hexoses. Starch synthesis in chloroplasts can proceed, however, in the dark from exogenous or cytoplasmic hexose. This immediately suggests that there must be rather specific transport processes across the chloroplast envelope; these will be discussed below. The utilization of photosynthetic energy by green cells for endergonic processes outside the chloroplasts also requires transfer of $\sim P$ and reducing potential between the chloroplasts and the cytoplasm (for light-dependent ion uptake across the plasmalemma and the tonoplast see Chaps. 8 and 10).

As in mitochondria, an extensive transport of solutes must occur across the chloroplast envelope. Electronmicrographs often reveal a tubular membrane system close to the surface of chloroplasts, the peripheral reticulum, which is quite distinct from the thylakoid system. Most frequently a peripheral reticulum is found in the bundle sheath chloro-

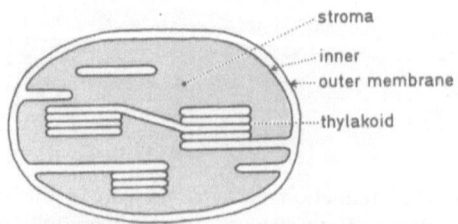

Fig. 7.12. Chloroplast structure.

plasts of C_4 plants (Laetsch, 1971; Chapman et al., 1975). Photosynthesis in C_4 plants requires a considerable transport of metabolites between the chloroplasts of mesophyll and bundle sheath cells (see Chaps. 10 and 11). It has therefore been speculated that the peripheral reticulum aids exchange of solutes between the cytosol and the chloroplasts. In hybrids of *Oenothera* Schötz and Diers (1975) have observed the formation of tubular structures towards the interior of the chloroplast and also towards the cytoplasm where the tubules form a network. These authors assume that the normal correlations between the plastids and the cytoplasm have been disturbed by hybridization, thus requiring adaptive features to maintain adequate regulation of transport.

In any event, transport between the cytosol and the chloroplasts must function in all green cells, and in the following we will deal largely with transport across the inner membrane of the chloroplast envelope and across the thylakoid membranes. This subject has been reviewed in several excellent publications which we can follow largely and suggest for further reading (Heldt, 1976a; McCarty, 1976; Nobel, 1975; Strotmann and Murakami, 1976; Simonis and Urbach, 1973; Walker, 1976).

7.2.3.1 Electron Transport, Charge Separation, and Coupling of ATP Formation

Electron transfer and coupling of ATP formation during photosynthesis occurs at the thylakoid membranes, and as in the case of the mitochondria a chemical and a chemiosmotic hypothesis of coupling can be distinguished (see Sect. 7.2.2.1). With chloroplasts, however, a number of elegant experiments resulted in significant evidence which supports the Mitchell hypothesis of charge separation and chemiosmotic coupling. During excitation of chlorophylls and transfer of electrons, protons are taken up into the thylakoid space, and this is accompanied by photophosphorylation and ATP formation (see also Chap. 8). Phosphorylation by isolated thylakoids is also possible in the dark with an artificial pH gradient established when the pH in the medium is rapidly raised from 4 to 8. This strongly suggests that the electrochemical H^+ gradient is responsible for phosphorylation and not excitation of chlorophyll and the associated redox reactions per se (Jagendorf and Uribe, 1966).

The electrical potential and the pH gradient across thylakoid membranes have been estimated indirectly mainly by studying the distribution of weak acids and bases (especially various amines) between the medium and the thylakoid space and by spectroscopic methods (cf. McCarty, 1976; Nobel, 1975; Junge, 1977). In addition direct measurements with microelectrodes have been made recently on unusually large chloroplasts within intact cells (Bulychev et al., 1971, 1972; Vredenberg et al., 1973; Vredenberg, 1974; Vredenberg and Tonk, 1975; Davis, 1974). The accuracy of this approach, especially of the application of pH microelectrodes, is still open to some criticism (see Findlay and Hope, 1976; Junge, 1977).

Distribution of amines and spectroscopic investigations of light-induced absorbance changes of pH-sensitive pigments suggest a thylakoid membrane potential of 50 to 250 mV in the steady state in the light (reviewed by Nobel, 1975; McCarty, 1976; Junge, 1977). Microelectrode work places the electrical potential across the envelope of giant chloroplasts of *Peperomia metallica* (Bulychev et al., 1972) and *Phaeoceros laevis* (Davis, 1974) either at $+10$ to $+15$ or -30 to -60 mV in the dark (measured with respect to the surrounding cytoplasm); the positive potentials are believed to represent measurements in which the electrode tip is in the thylakoid. In the light there is a transient change in the potential of $+10$ to $+30$ mV (reviewed by Findlay and Hope, 1976). The pH gradient across thylakoid membranes is at least 2 pH units, perhaps up to 3.5 units, in the steady state in the light. The maximum electrochemical gradient is 210 mV (reviewed by Nobel, 1975, and McCarty, 1976).

7.2.3.2 Ion Transport and Ion Accumulation

7.2.3.2.1 Mineral Cations and Chloride

Transport of mineral cations into and out of thylakoids and intact chloroplasts has received much interest in connection with the H^+ movements associated with photosynthetic electron transfer. A clear scheme is offered by Hind et al., 1974 (reviewed by McCarty, 1976), where at an external pH of 6.6 in the light H^+ uptake into the thylakoids is balanced by Cl^- uptake and Mg^{2+} efflux. There is little K^+ efflux which is entirely abolished at Mg^{2+} concentrations above 1 mM. In the dark the directions of these fluxes are reversed (Fig. 7.13). However, the stoichiometry of ionic relations of chloroplasts becomes more uncertain when in addition one considers the movements of these ions across the chloroplast envelope. There, in the light, H^+, Cl^-, and K^+ are released into the cytoplasm and Mg^{2+} is taken up (reviewed by Nobel, 1975). The exact magnitude of these processes is not known. Using an electron microprobe technique, de Filippis and Pallaghy (1973) have measured the light-dependent changes of ion levels in chloroplasts of intact *Elodea* cells. From their results they calculated rather substantial fluxes which are higher than most plasmalemma fluxes in Characean cells and closer to the tonoplast fluxes in such cells (e.g., Cl^-, 33 p eq cm^{-2} s^{-1}; K^+, 9 p eq cm^{-2} s^{-1}). However, the exact calibration is still a great problem in electron microprobe work, and there are still some inconsistencies in comparison with other methods (e.g., the unusually large sodium levels and fluxes obtained by de Filippis and Pallaghy, Table 7.1). The driving forces of mineral cation and chloride fluxes across the chloroplast envelope are also still in doubt. These ion fluxes may be active or, due to passive driving forces, H^+ transport may be the primary event. The changing pH of the stroma may affect membrane per-

Table 7.1. Ion concentrations in chloroplasts *in vivo*. (Largely from Nobel, 1975.)

Plant species	K⁺	Na⁺	Mg²⁺ m mol/l	Ca²⁺	Cl⁻	Reference
Nitella opaca	84	4				Saltman et al. (1963)
Nitella translucens	150				240	MacRobbie (1964)
Nitella flexilis	110	26			136	Kishimoto and Tazawa (1965a)
Tolypella intricata	340	36	14	21	340	Larkum (1968)
Pisum sativum	99	10	16	15	92	Nobel (1969)
Limonium vulgare	300	200			500	Larkum and Hill (1970)
Elodea densa:						
Light	200	170			320	de Filippis and Pallaghy (1973)
Dark	240	310			530	de Filippis and Pallaghy (1973)

meabilities (see Chap. 10) which together with the changing electric potentials (see Sect. 7.2.3.1) would contribute to passive driving forces (Sect. 2.2).

Therefore, it may be premature to summarize the major events in a scheme which is as simple as that given in Figure 7.13. Much work still needs to be done. The scheme does show that the mineral ion fluxes into and out of the chloroplasts are important not only in the function of chloroplasts per se but also in regulation of ionic contents of the cytoplasm and the cell as a whole.

Ion levels in chloroplasts of intact cells can be quite high. They have been estimated by using nonaqueous isolations of chloroplasts, by qualitative microautoradiographic observations, by electronmicroprobe analysis, and, indirectly, by kinetic measurements. Microautoradiographs of the leaf cells of the salt plant *Limonium* showed a considerable accumulation in the chloroplast of radioactive Cl⁻ which had been applied via the

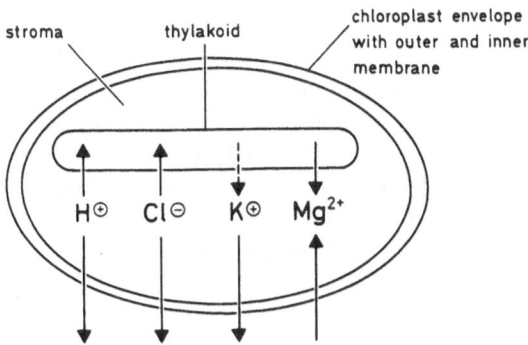

Fig. 7.13. Mineral ion fluxes in chloroplasts in the light. Combination of the results with isolated thylakoids reviewed by McCarty (1976) and the results on intact chloroplasts reviewed by Nobel (1975). In the dark the direction of these fluxes is reversed.

Table 7.2. Cl⁻-concentrations in various cell compartments of *Nitella flexilis* (Kishimoto and Tazawa, 1965a) and *Tolypella intricata* (Larkum, 1968; see also Table 6.3).

	Cl⁻-concentration m mol l⁻¹			
	External medium	Chloroplasts	Cytosol	Vacuole
Nitella	1.3	136	36	136
Tolypella	1.4	340	23–31	116–136

petiole (Ziegler and Lüttge, 1967). Kinetic compartmentation analysis in *Limonium* leaf cells suggests that Cl⁻ is distributed in four compartments, while for Na⁺ distribution only three compartments are apparent, namely the free space, the cytoplasm, and the vacuole. The fourth kinetic compartment which plays a role in Cl⁻ distribution could very well be the chloroplasts (Hill, 1970a,b; Larkum and Hill, 1970). Table 7.1 summarizes some values of mineral ion concentrations in chloroplasts of various plant species.

In a quite elegant way the ion content in chloroplasts can be analyzed in comparison to other cell compartments of the internodal cells of Characean algae. The chloroplasts here are densely packed in a layer of non-streaming cytoplasm adjacent to the walls. The content of the vacuole, the major part of the cytosol (streaming layer of cytoplasm), and the chloroplast layer can be pressed out one by one from cells which have been cut at the end, and so can be analyzed separately. For comparison, results obtained with chloroplasts prepared by nonaqueous processes of isolation can also be used. Among the mineral ions especially, Cl⁻ is highly accumulated in the chloroplast as compared to the ground plasm (see Tables 6.3 and 7.2).

7.2.3.2.2 CO_2 and HCO_3^-

A problem for some time has been whether CO_2 or HCO_3^- is the substrate for photosynthetic carbon reduction; now it appears to be quite generally accepted that RudP-carboxylase utilizes CO_2 rather than HCO_3^-. Intensive investigations in the laboratory of Heldt in Munich showed that CO_2 is also the species which moves across the chloroplast envelope (reviewed by Walker, 1976). HCO_3^- accumulation follows the pH gradient according to the following hydration equation:

$$HCO_3^- + H^+ \underset{\text{cytoplasm}}{\overset{\text{medium}}{\rightleftharpoons}} H_2O + CO_2 \;\Big|\Big|\rightarrow CO_2 + H_2O \overset{\text{chloroplast}}{\rightleftharpoons} HCO_3^- + H^+$$

chloroplast
envelope

In view of the normal rates of photosynthesis of about 200 μ mol CO_2 per h per mg chlorophyll (a similar value would be obtained on a fresh weight basis for most green tissues) it is clear that this transport must be extremely rapid.

7.2.3.2.3 Inorganic Phosphate

Inorganic phosphate is required as a substrate for photophosphorylation within the chloroplasts. To a fair extent transport into the chloroplasts appears to be mediated by a specific phosphate translocater or carrier mechanism which exchanges P_i for dihydroxyacetonephosphate (Heldt, 1976: Walker, 1976).

7.2.3.3 The Transport of Nucleotides

Adenylates cannot diffuse across the chloroplast envelope. As in mitochondria, there is a specific adenylate carrier which exchanges ATP and ADP with a tenfold higher affinity for ATP than for ADP. However, unlike mitochondria, quantitatively this mechanism plays no major role in energy transport across the chloroplast envelope. Pyridine nucleotides cannot permeate at all across the chloroplast envelope. Thus, the energy obtained in the form of ATP and reduced pyridine nucleotide ($NADPH_2$) by harvesting the light in the chloroplasts can be utilized for endergonic processes outside the chloroplasts only after indirect transport of $\sim P$ and reducing equivalents across the chloroplast envelope (Heldt, 1976; Walker, 1976). This is mediated by transport metabolites as described in the next section.

7.2.3.4 Transport Metabolite Shuttles

The term transport metabolite has already been defined (Sect. 7.2.2.3). There are three important transport metabolite shuttle systems which carry $\sim P$ and reducing equivalents concomitantly or reducing equivalents alone. These systems, together with the connecting reactions, are surveyed in Figure 7.14. For didactic reasons it is better to consider them first one by one before returning to the synopsis shown in the figure.

7.2.3.4.1 The Dihydroxyacetonephosphate-Phosphoglyceric Acid Shuttle of Stocking and Heber

The dihydroxyacetonephosphate (DHAP)-phosphoglyceric acid (PGA) shuttle is perhaps the earliest shuttle which was discovered. It was well established by the work of Stocking and Larson (1969) and of Heber and Santarius (1970). Figure 7.15 shows the metabolic reactions between 3-phosphoglyceric acid, 1,3-diphosphoglyceric acid, glyceraldehyde-3-phosphate, and dihydroxyacetonephosphate. The enzymes necessary for

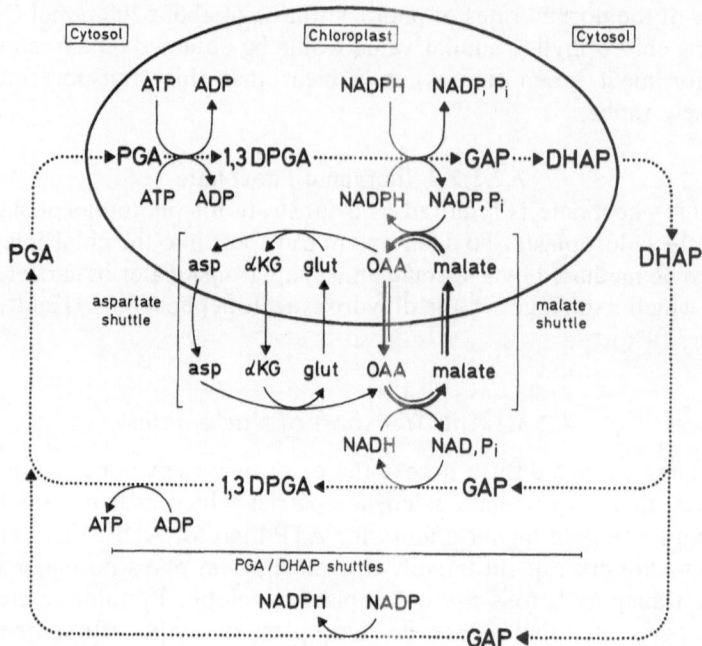

Fig. 7.14. Shuttle mechanisms mediating the transport of ~P and reduction equivalents across the chloroplast envelope. (Modified from Heber, 1975, and Strotmann and Murakami, 1976.) *DHAP*, dihydroxyacetonephosphate; *GAP*, glyceraldehyde-3-phosphate; *1,3-DPGA*, 1,3-diphosphoglyceric acid; *PGA*, 3-phosphoglyceric acid; *asp*, aspartate; *αKG*, α-ketoglutarate; *glut*, glutamate; *OAA*, oxaloacetate.

these reactions have been demonstrated to occur both in the interior of the chloroplast and in the cytoplasm. The shuttle thus transports stoichiometric amounts of ~P and reducing equivalents between the chloroplasts and the cytoplasm. The transport metabolites involved are DHAP (carrying the ~P and [H⁺] energy) and PGA. Glycerinealdehyde-3-phosphate most likely cannot pass the chloroplast envelope very readily. However, by the triosephosphate isomerase reaction it is in equilibrium with the transport metabolite dihydroxyacetonephosphate. This shuttle is reversible, DHAP can also move from the cytoplasm back into the chloroplast (Heber and Kirk, 1975). The operation of the shuttle carrying ~P and [H⁺] from the chloroplasts into the cytoplasm very much depends on the physiological state of the cell. In particular it needs a high $NAD/NADH_2$ ratio in the cytoplasm; at a ratio of 1 it stops entirely. It also needs cytoplasmic ADP for phosphorylation. Triosephosphate transport across the inner chloroplast membrane is obligatorily coupled with exchange of inorganic phosphate (see also 7.2.3.2.3) which also is very important for the regulation of chloroplastic and cytoplasmic metabolic activities (e.g., starch synthesis and mobilization; Heldt et al., 1977).

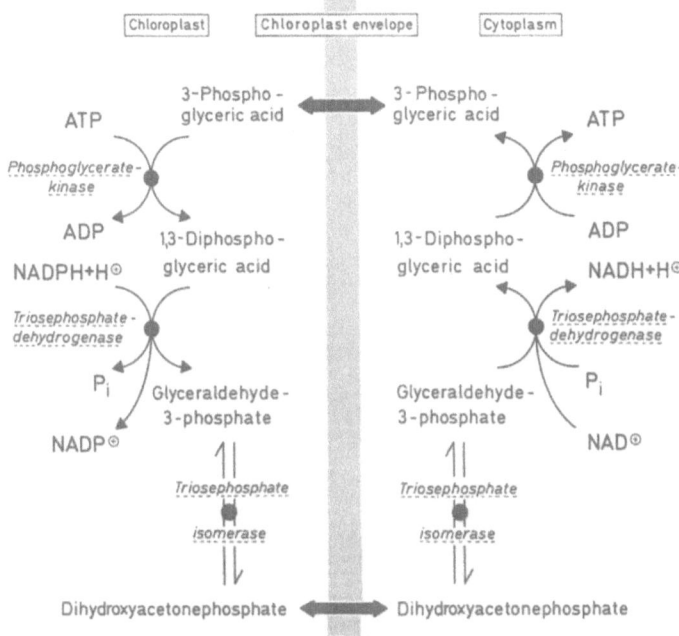

Fig. 7.15. Dihydroxyacetonephosphate-phosphoglyceric acid shuttle transporting ~ P and reduction equivalents across the chloroplast envelope. (According to the findings of Stocking and Larson, 1969, and Heber and Santarius, 1970.)

7.2.3.4.2 The Dihydroxyacetonephosphate-Phosphoglyceric Acid Shuttle of Kelly and Gibbs

More recently Kelly and Gibbs (1973) suggested a somewhat different nonreversible DHAP-PGA shuttle which is less dependent on the energy state of the cytoplasm, and which transports reduction equivalents out of the chloroplasts. This mechanism operates with the NADP-dependent nonphosphorylating glyceraldehydephosphate dehydrogenase in the cytoplasm, so that no ATP is regained in this compartment (Fig. 7.16).

7.2.3.4.3 The Malate-Oxaloacetate Shuttle

Like mitochondria, chloroplasts have a dicarboxylate carrier which transports malate and oxaloacetate as well as succinate, α-ketoglutarate, fumorate, aspartate, glutamate, asparagine, and glutamine (cf. Fig. 7.14; Gimmler et al., 1974; Strotmann and Murakami, 1976; Walker, 1976). The most important function of this carrier is that it can shuttle reduction equivalents across the chloroplast by exchanging malate and oxaloacetate (Fig. 7.17). This shuttle provides transfer of reduction equivalents not dependent on ~ P. In conjunction with the DHAP-PGA shuttle of Stocking and Heber (Sect. 7.2.3.4.1) this shuttle overcomes the problem incurred by the fact that ~ P can move across the chloroplast envelope

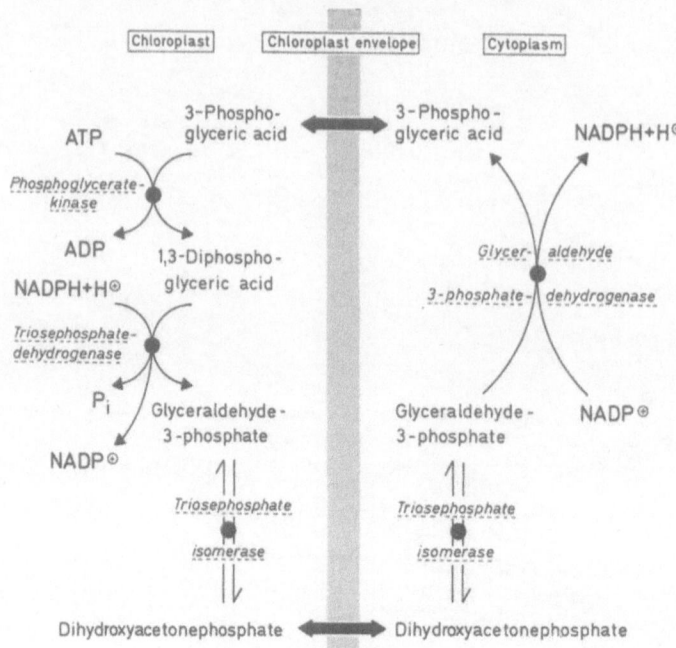

Fig. 7.16. Dihydroxyacetonephosphate-phosphoglyceric acid shuttle transporting reduction equivalents across the chloroplast envelope. (According to the findings of Kelly and Gibbs, 1973.)

only together with a stoichiometric amount of reduction equivalents. If, as suggested by Heber and coworkers, both shuttles are combined, the reversible malate-oxaloacetate shuttle will reintroduce reduction equivalents into the chloroplast and ~P transport across the envelope into the cytoplasm can proceed without a net loss of reduction equivalents from the chloroplast (Fig. 7.14).

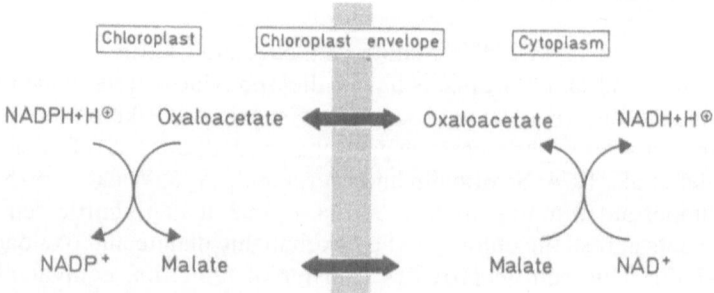

Fig. 7.17. Malate-oxaloacetate shuttle transporting reduction equivalents across the chloroplast envelope. (After Heber and Krause, 1971.)

7.2.3.5 Transport of Sugars and Amino Acids

Sugar transport into chloroplasts is mediated by a carrier transporting D-xylose, D-mannose, L-arabinose, D-glucose, and, somewhat more slowly, D-fructose and D-arabinose. The activity of the carrier at 1 mM D-glucose concentration is 3 μmol [mg chlorophyll]$^{-1}$h^{-1} as compared to an activity of the phosphate carrier of 200 μmol DHAP [mg chlorophyll]$^{-1}$h^{-1}. Although the activity of the carrier is rather low, it probably catalyzes a facilitated diffusion; sugar concentration reached within the chloroplasts is comparable to that in the external concentration (Schäfer et al., 1977).

McLaren and Barber (1977) described an amino acid carrier which transports leucine and isoleucine (but not glycine) against a concentration gradient into the chloroplasts. A conclusion by Nobel and Wang (1970) that isolated chloroplasts are freely permeable to amino acids is not confirmed (McLaren and Barber, 1977; Gimmler et al., 1974; for other amino acids see Sect. 7.2.3.4.3).

7.2.4 Microbodies

Microbodies form an interesting class of organelles which were first discovered and described in some detail in liver cells by De Duve and coworkers (Baudhin et al., 1965; De Duve and Baudhin, 1966). They are about 0.5–1.0 μm in diameter and surrounded by a single unit membrane. Liver microbodies contain catalase, uricase, and hydroxycarbonic acid oxidases forming H_2O_2, and thus largely function for detoxication. Plant microbodies have been studied mainly by H. Beevers and N.E. Tolbert and their associates (e.g., Breidenbach and Beevers, 1967; Breidenbach et al., 1968; Gerhardt and Beevers, 1970; reviews by Beevers et al., 1974; Tolbert, 1971). As in the liver, the microbodies of fat-storing plant organs, e.g., the glyoxysomes of castor bean endosperm extensively investigated in Beevers' laboratory, they contain catalase, uricase, and glycolic acid oxidase.

In plants we can distinguish two major classes of microbodies: (1) microbodies of fat-storing tissues, i.e., glyoxysomes, function together with mitochondrial and cytoplasmic enzyme systems in fat mobilization and gluconeogenesis (Beevers and coworkers); and (2) microbodies of leaves, i.e., leaf peroxisomes, which contain the enzymes of glycolate metabolism and function together with chloroplasts and mitochondria in photorespiration (Tolbert and coworkers). Both processes require considerable intracellular movements of transport metabolites. Schnarrenberger and Burkhard (1977) have reported a peculiar observation which may be important in this context. In vitro chloroplasts come in close contact with isolated peroxisomes via the membranes when P_i concentrations are added which are similar to P_i levels expected to be present in the cytoplasm. This effect occurs only when chloroplasts have an intact external membrane.

7.2.4.1 Gluconeogenesis From Fat:
Transport Between Glyoxysomes and Mitochondria

Somewhat different from the scheme shown in Figure 7.9, the mobilization of fat during germination of fat-storing seeds occurs in the glyoxysomes. Glyoxysomes are considerably increased in number during germination of fat storing seeds. Glyoxysomes contain the enzymes of β-oxidation of fats. It is as yet unclear how fatty acids are transported into the glyoxysomes. They are, however, degraded by β-oxidation within the glyoxysomes to give acetyl-CoA. Glyoxysomes also contain the enzymes of the glyoxylic acid cycle, and succinate is formed from acetyl-CoA. Succinate then moves as a transport metabolite from the glyoxysomes into the mitochondria, where oxaloacetate is formed which can serve as a substrate for gluconeogenesis by cytoplasmic enzyme systems.

Recent investigations in Beevers' laboratory suggest that glyoxysomes

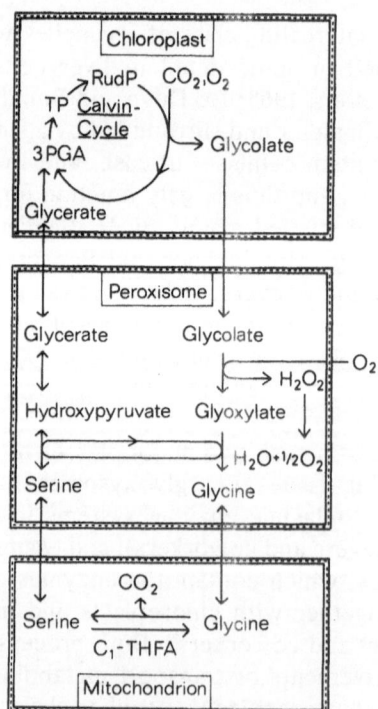

Fig. 7.18. Glycolate metabolism in leaves after Tolbert (1971). *3 PGA*, 3-phosphoglycerate; *RudP*, ribulose-1,5-diphosphate; *TP*, triosephosphate; *THFA*, tetrahydrofolic acid.

originate from the endoplasmic reticulum possibly by membrane flow (Gonzalez and Beevers, 1976).

7.2.4.2 Photorespiration: Transport Between Chloroplasts, Leaf Peroxisomes, and Mitochondria

In photorespiration, the activities of three kinds of organells are linked together: chloroplasts, peroxisomes, and mitochondria (Fig. 7.18). The substrate of photorespiration, glycolate, is formed in the light in the Calvin cycle within the chloroplasts, probably by ribulose-1,5-diphosphate carboxylase using O_2 as substrate (i.e., functioning as a ribulose-1,5-diphosphate oxygenase; Lorimer and Andrews, 1973). The oxidation of glycolate to glyoxylate and transamination to serine occurs in the peroxisomes. CO_2 is liberated from glycine in the mitochondrion. Serine is transported back to the peroxisome and, after transamination and reduction, is transformed into glycerate, which, after phosphorylation in the chloroplast, can flow back into the Calvin cycle. Thus we meet new transport metabolites, glycolate, glycine, and serine as well as glycerate.

7.3 Gland Cells as Particular Examples

Gland cells show a striking differentiation of their ultrastructure. The cytoplasm is very dense and rich in various organelles. Gland cells display unique transport properties, and thus they are an outstanding subject for the analysis of the relations between fine structure of the cytoplasm and transport functions.

7.3.1 Gland Functions

If the active secretory surface of salt glands of the mangrove *Aegialitis* has been correctly estimated, then those gland cells attain rates of transport of NaCl up to 90,000 nmol h^{-1} cm^{-2} (Atkinson et al., 1967). Compared with this rate, roots, the normal organs of salt transport of plants, at a KCl concentration of 1 mM in the external solution, transport the salt only with a rate of 35 nmol h^{-1} cm^{-2} into the exudate (Jarvis and House, 1970). Banana nectaries secrete in two to three days about 1–2 ml of an approximately 32% sugar solution, this means 25,000 to 50,000 nmol h^{-1} per banana flower. For comparison the glucose efflux, from the glucose-storing onion scale epidermal cells, into a sugar-free solution is only 10 nmol h^{-1} cm^{-2}. The active secretory surface of the nectaries of the banana flower with a comparable rate of transport of sugar should then amount to 2500–5000 cm^2 (Lüttge, 1971a).

This comparison underscores the extraordinary quantitative achievements in transport by the glands. Also in qualitative respects glands show outstanding capabilities. The composition of the secreted products usually is very specific. Chemically, secretion products consist largely of one particular substance or of only one class of substance, and other accompanying substances are observed only in very small amounts. A few examples provide a partial summary (cf. Lüttge, 1971a):

Secretion	Glands
Proteolytic enzyme (protein)	Digestive glands of carnivorous plants
Polysaccharides	Slime glands of carnivorous plants
Sugar	Floral and extrafloral nectaries
Wax	Wax glands on the surface of leaf veins of *Ficus* and Aracean species
Resins	Resin channel cells, shoot glands of *Viscaria*
Oils	Oil glands
Inorganic ions	Salt glands of halophytes and xerophytes

The biosynthesis of secretion products occurring within the gland cells may contribute considerably to the specificity of the secretion formed, especially during the secretion of secondary plant substances (e.g., wax, resins, oils). Since a large variety of substances is inside the cells, the compartments must possess specific transport properties. An exception, of course, is given by those glands in which secretions are found in lysigenous cavities formed by disintegration of cells (reviews: Lüttge, 1971a, 1975; Hill and Hill, 1976; Lüttge and Schnepf, 1976).

7.3.2 The Fine Structure of Gland Cytoplasm

A striking differentiation of the cytoplasm is associated with these remarkable achievements in function of the gland cells. Even at low magnification in the light microscope, the gland cytoplasm is surprisingly dense. This observation is primarily due to the fact that there are no large central vacuoles in most gland cells. Beyond this, however, in active gland cells the number, size, and activity of various organelles and membrane systems is considerably increased. The most important ones can be enumerated as follows:

Nuclei
Dictyosomes
Membrane-bound vesicles
The plasmalemma surface
Mitochondria

In addition, in glands there are particular cell wall depositions and numerous plasmodesmata connecting the gland cells with each other and the surrounding parenchyma cells. The importance of this is discussed elsewhere (Sect. 7.4).

Little is known as yet of the particular role of nuclei in the function of plant glands. There are a few suggestions of an increased activity of the nuclei of gland cells as compared to normal parenchyma cells. The nuclei in glands in relation to the cell volume are often strikingly large, which is not due only to an increased absolute size of the nuclei, but rather to the smallness of the gland cells. Special dictyosome activity during the secretion of polysaccharides has been discussed before (Sect. 7.1.1). It is a very characteristic property of gland cells that central vacuoles are missing, but instead there are numerous small vesicles in the cytoplasm.

We should still assess the importance of an enlarged plasmalemma surface and the role of the increased number of mitochondria. The former will be considered in a separate section (Sect. 7.4), since after this phenomenon had been observed first in gland cells it turned out to be the outstanding attribute of a more general class of cells mediating extensive short-distance transport.

The abundance of mitochondria in gland cells may simply be important for ample energy production during active transport of secretory materials. In nectaries some evidence for this has been obtained by Ziegler (1956; Fig. 7.19). In the nectaries of the calyx of *Abutilon* flowers, corre-

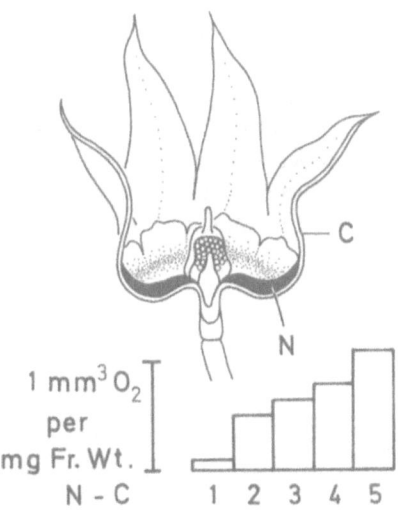

Fig. 7.19. Longitudinal section through an *Abutilon* flower after removal of the corolla and the difference between respiration rates of nectary tissue (*N*) and calyx tissue (*C*), i.e., N-C in $mm^3 O_2$ per mg dry weight, after Ziegler (1956). Different developmental stages of the flowers: *1,2,3,* buds of increasing size; *4,* flower open and fully developed; *5,* corolla wilted. In stages 3, 4 and 5 the glands secrete nectar.

lated with nectary secretion, there is a considerable increase in respiratory oxygen uptake as compared to nonsecretory neighboring calyx tissue.

A great density of mitochondria is found also in salt glands. By supplying ATP or redox carrier reactions (see Chap. 8) these mitochondria may affect active salt excretion. In a number of salt glands it was found, however, that light can increase salt excretion considerably by providing photosynthetic energy (Fig. 7.20, and Chap. 8). It is remarkable that in such systems (e.g., salt glands of *Atriplex spongiosa* and of *Limonium vulgare*) there is not also a striking increase in the number of chloroplasts comparable to that of mitochondria in the gland cytoplasm. By contrast the cytoplasm of the glandlike stalk cells of the epidermal bladders of *Atriplex spongiosa* and the gland cells of *Limonium* leaves are photosynthetically inactive (Fig. 7.20, cf. Lüttge, 1975). The problem of the coupling between spatially separated active ion transport processes and energy-providing processes will be discussed elsewhere in this volume (Chaps. 8 and 10). Here, this observation teaches us that although the energetic importance of the high density of mitochondria in gland cells appears to be self-evident, such a concentration of the energy-providing organelles in the gland cells themselves is not absolutely necessary.

Microautoradiographic and ultrastructural studies suggest that the abundance of mitochondria in the salt glands may also play a role which is different from that of energy provision for ion pumps at nonmitochondrial membranes. In order to develop this idea, knowledge of the theory of symplastic transport is required, which will be dealt with in Chaps. 10 and 11. Let us mention here only that symplastic transport can lead to a rapid equalization of concentration differences within the symplast, i.e., within the cytoplasm of all cells of a tissue which are connected by plasmodesmata. Let us remember further the capacity of the mitochondria mentioned in Section 7.2.2.1 to accumulate ions actively in their matrix. If one takes these three findings together, i.e., symplastic transport, active ion

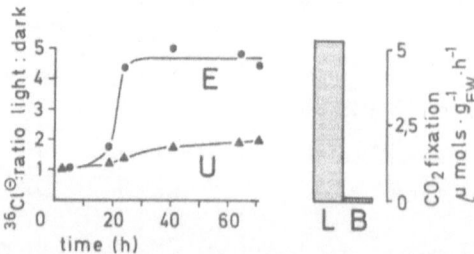

Fig. 7.20. Stimulation of chloride ($^{36}Cl^-$) elimination (E) into the bladder vacuoles of *Atriplex spongiosa* by light. U, uptake of $^{36}Cl^-$ by leaf lamina; L, photosynthesis of lamina; B, photosynthesis of bladders. Data are from experiments with 3 mm wide leaf slices floating on a solution of 0.5 mM $CaSO_4$, 5 mM $K^{36}Cl$, 0.25 mM $KH^{14}CO_3$. (Osmond et al., 1969; Lüttge, 1974, 1975.)

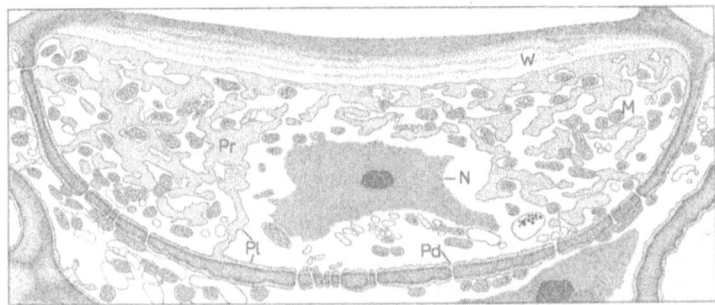

Fig. 7.21. Hydropote gland cell of the lower surface of a *Nymphaea* leaf. Schematic drawing after an unpublished original electron micrograph. *N*, nucleus; *M*, mitochondria; *Pd*, plasmodesmata; *Pl*, plasmalemma; *Pr*, cell wall protuberances; *W*, cell wall. About ×4,700.

accumulation within the mitochondria, and the particular richness of mitochondria in salt gland cells, then it becomes clear that by ion accumulation into the mitochondrial matrix, electrochemical gradients can be formed which eventually may lead to salt excretion.

The high ion content, which one can demonstrate by microautoradiography in salt-gland cells indeed may be due to a salt accumulation within mitochondria. The hydropotes on the lower surface of floating leaves of *Nymphaea* (water lily) are salt-transporting gland cells. Electronmicrographs of hydropote cells show that nearly all the total space not occupied by the cell wall protuberances and the large nuclei in the cells is occupied by mitochondria (Fig. 7.21). On the microautoradiographs one finds after uptake of radioactively labeled sulfate by the lower surface of *Nymphaea* an extraordinarily dense labeling in the hydropotes. Obviously, one can hardly understand this high accumulation of ions without the direct participation of ion uptake by the mitochondria themselves.

7.4 Transfer Cells

An extreme enlargement of the plasmalemma surface was observed long ago in gland cells and later in a multitude of other cell types (Gunning and Pate, 1969; Pate and Gunning, 1972). Common to all these cells is the location in tissue complexes where a particularly intensive short-distance transport is required. To describe these cells, Gunning and Pate have coined the very pertinent term transfer cells. The proliferation of the plasmalemma is due to an enlargement of the cell wall surface. The walls of transfer cells form protuberances, which extend into the cell interior where they branch and merge with each other. In this way a labyrinth is formed, which is lined by the plasmalemma and filled with cytoplasm (Figs. 7.21 and 7.22).

The widespread occurrence of such cell-wall protuberances in cells with important transport functions is enough reason for the assumption that these structures are of great importance for short-distance transport. New investigations continue to reveal further examples of transfer cells in various plant tissues, and the following enumeration is hardly complete. Transfer cells are found:

1. in glands,
2. at locations in bundles and veins of roots, shoots, leaves, and inflorescences, where substances must move between the long-distance transport pathways of the xylem and the phloem or must exchange with the adjacent parenchyma,
3. in companion cells of sieve tubes,
4. in ferns and in mosses, where the sporophyte, initially or continuously, has to be fed by the gametophyte via haustorial contacts,
5. in N_2-fixing symbiotic root nodules,
6. in haustoria of plant parasites.

(See Schnepf, 1969; Lüttge, 1971a; Pate and Gunning, 1972; Pate, 1976.)

This certainly is a very significant correlation at the structural level. The importance of the cell wall protuberances for transport becomes still more clear, however, by a functional correlation observed by Schnepf (1964c) some years ago. In the central nectaries of the gynoecium of the flowers of *Gasteria* he found the formation of the cell wall proliferations to be correlated in time with the maximum secretory activity. The cell wall protuberances are formed shortly before the onset of nectar secretion; and at the time of full secretory activity, a labyrinth was formed; in

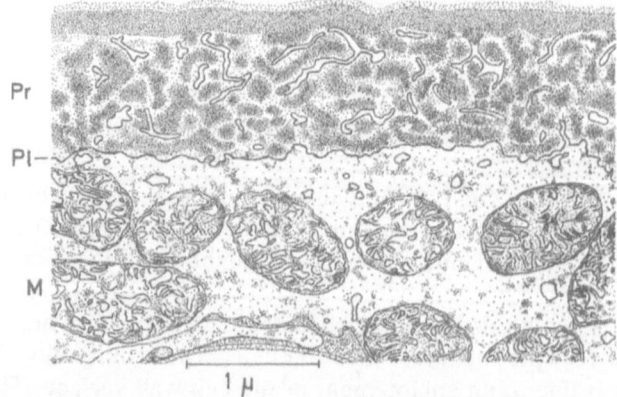

Fig. 7.22. External zone of a secretory gland hair cell of the scale-like leaves of *Lathraea clandestina*. *Pr*, external cell wall layer with a labyrinth of protuberances; *M*, mitochondria; *Pl*, plasmalemma. (Schematic drawing after Fig. 5 in Schnepf, 1964b.)

the post-secretory period, however, the protuberances degenerate, and the material is reabsorbed by the cells.

How do the cell wall protuberances function during secretion? A priori it is not clear whether the decisive function is an enlargement of the cell wall apoplast or of the plasmalemma surface. It has been pointed out that large apoplastic spaces are required in secretion as coupling spaces between the actively eliminated solutes and with the water following osmotically (reviews, Lüttge, 1975; Hill and Hill, 1976; Lüttge and Schnepf, 1976). Hill and Hill have even argued that the vesicles frequently observed at the surface of salt gland cells (see Sect. 7.1.1) are artifacts, and in fact represent sections of apoplastic spaces where the plasmalemma has moved away from the wall material of the protuberances, and where the secretion product is accumulating (Hill and Hill, 1973). It is nevertheless difficult to understand how passive transport in the cell wall should make the function of the transfer cells possible, and it appears likely that membrane transport processes are of greater importance in this connection. Plasmalemma proliferations similar to those found in transfer cells of plants have been observed also in secretory and excretory cells of animal tissues which do not have cell walls like the typical plant apoplast (see Schnepf, 1969). It is probably more sensible to ask the question: how can the increased plasmalemma surface effect the particularly intensive transport? There are different possibilities which, alone or jointly, can be realized.

Theoretically, the increased plasmalemma surface could assist secretion by the extrusion of vesicles (exocytosis). A larger surface energetically facilitates the merging of vesicles because surface energy = surface tension × area (Ziegler, 1968). Electronmicroscopic investigations, however, do not give any hints for the correlation of the formation of protuberances with a striking extrusion of vesicles (granulocrine secretion). It appears, therefore more likely that the cell wall labyrinth has a different function and affects the secretion of individual molecules (eccrine secretion) (Schnepf, 1969).

According to Eq. (2.3), the rate of passive diffusion through a limiting surface is directly proportional to the surface area. If the exit of substances from the gland cells occurs passively, then an enlargement of the plasmalemma surface at a given concentration difference $(c_o - c_i)$ would contribute considerably to the increase of the rate of exit. Conversely, the enormous flux rates (see above, Sect. 7.3.1) which have been calculated for glands might come down into the range normally found for plant transport processes if one uses the surface of the protuberances as the real size of the secretory surface instead of an estimation of the outer cell surface.

A third possibility is given when the exit of substances from the gland cells is active. No doubt the enlargement of surface is associated with considerable total increase of catalytic activity of the plasmalemma because it creates space for more membrane-bound pumps and carrier systems. In addition, a more intimate contact of the membrane is possible with organelles such as mitochondria and ER cisternae (Figs. 7.21 and 7.22). Direct hints for this are found in investigations of nectary glands. According to many observations a richness in mitochondria and a striking activity of acid phosphatase is very characteristic of nectary gland cells (Ziegler, 1956). Perhaps phosphorylations and dephosphorylations of sugars play a role in sugar transport (Ziegler, 1956; Lüttge, 1966c; Lüttge and Schnepf, 1976). A large number of cytochemical investigations show that phosphatases are particularly localized at the plasmalemma or in the vicinity of the plasmalemma. Very clear electronmicroscopic pictures of phosphatase localization have been obtained by Figier (1968) for the plasmalemma of extrafloral nectaries of *Vicia faba*. We have mentioned previously the importance energetically of the abundance of mitochondria in the gland cytoplasm. By a coordinated action of mitochondria (provision of ATP) and of the plasmalemma phosphatases, phosphorylation and dephosphorylation could drive energetically the transport of sugar in secretion by the nectaries.

Part III

Regulation and Control of Transport Processes by Cell Metabolism: Chapters 8 to 10

Processes of membrane transport of solutes are closely correlated with cell metabolism. One of the most important factors which link transport and metabolism is energy. The problems concern the kind of energy-providing processes discussed in Chapter 8, and the nature of the coupling between these processes and the transport mechanisms as discussed in Chapter 10. Metabolic regulation and control of transport processes, however, is not due only to the requirement of energy. Investigations of correlations between growth regulators such as phytochrome and phytohormones and transport mechanisms have become an important field of work in plant physiology in the last few years. In Chapter 9 it will become evident that membrane transport mechanisms also may be hormonally regulated.

Chapter 8

Sources of Metabolic Energy for Membrane Transport Mechanisms

This chapter concerns itself in particular with the energy sources for active transport. Any exergonic reaction of metabolism can drive active transport if the energy liberated in this reaction can be appropriately coupled with an active transport mechanism. Thus far this more general point of view has not been adequately appreciated in transport research. The overwhelming number of investigations on energetic coupling of active transport in plants more specifically deal with respiration and photosynthesis as energy-providing processes. The ATP which is formed in oxidative phosphorylation or in photophosphorylation, and, also the redox gradients formed by electron transport in cristae membranes of mitochondria or in thylakoid membranes of chloroplasts are discussed as possible sources of energy for active transport.

8.1 Respiration as the Energy Source for Active Transport

8.1.1 Electron Transfer Along the Respiratory Chain as a Direct Energy Source for Active Anion Transport

8.1.1.1 Salt Respiration and the Lundegårdh-Hypothesis

The first hypothesis on the coupling of energy-providing processes with active transport which was worked out in great detail was formulated in 1933 by Lundegårdh and Burström (Lundegårdh and Burström, 1933, 1935); it was then further developed by Lundegårdh. This hypothesis as-

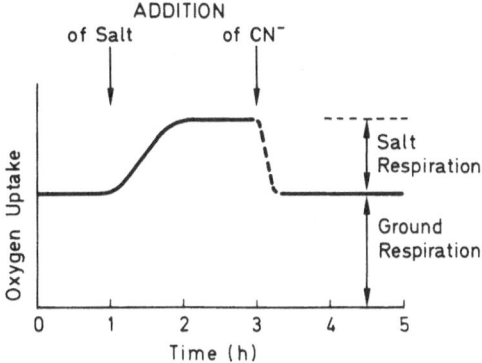

Fig. 8.1. Salt respiration. (Sutcliffe. 1962.)

sumed a direct dependence of active ion uptake upon electron transfer along the respiratory chain. The experimental basis for this concept was the finding that respiration of plant tissue is increased when the tissue is transferred into a salt solution from which it can take up and accumulate ions in the cell vacuoles (Fig. 8.1).

This phenomenon was termed anion or salt respiration and is defined as the increased O_2 uptake observed in the presence of ions in the external medium over that of the so-called ground respiration (respiration in distilled H_2O). Salt respiration is more cyanide-sensitive than ground respiration (Fig. 8.1). The experiments often suggest a stoichiometric correlation between the number of the transported anion equivalents and salt respiration. However, this stoichiometric correlation has not been confirmed in all investigations. For one O_2 molecule taken up four electrons can be transferred along the respiratory chain, therefore, according to the Lundegårdh hypothesis, maximally four anion equivalents can be taken up actively for each O_2 molecule. Considerably larger values have been found, however, and in some plant tissues during salt uptake no salt respiration at all is observed. Nevertheless, in many plant tissues, there is a clear correlation between salt respiration and ion uptake.

The coupling of ion uptake with electron transport along a cytochrome chain as suggested by Lundegårdh is schematically drawn in Figure 8.2. Each cytochrome molecule in the oxidized state can bind one anion equivalent more than in the reduced state. In this way, the ions can migrate along the electron ladder in a direction opposite to the flow of electrons.

When Lundegårdh was formulating his hypothesis it was not yet known that cytochromes are bound within the cristae membranes of the mitochondria and that the respiratory chain is strictly localized there. Lundegårdh initially assumed that in accord with the scheme of Figure 8.2, the cytochromes are arranged polarly in the cytoplasm in order to ex-

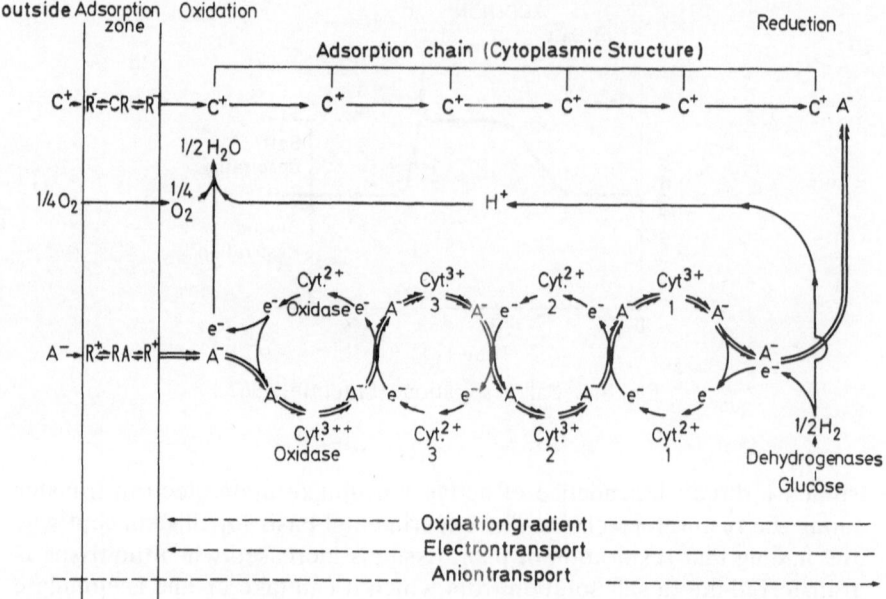

Fig. 8.2. Correlation between electron transfer along the cytochrome chain and anion plus cation uptake according to Lundegårdh (1950), (Lüttge, 1969). C^+, cation; A^-, anion; R^- and R^+, cation and anion carriers respectively; e^-, electron; Cyt, cytochrome. In the cytochrome chain *double arrows* refer to anion transport, *single arrows* to electron transport.

plain ion transport from outside to inside. Later, Lundegårdh tried to adapt his hypothesis to more modern cytological knowledge. He postulated positively and negatively charged carriers which mediate the introduction into the cells of anions and cations via the plasmalemma membrane. The positively charged anion carriers were assumed to deliver anions directly to the cytochrome systems while the cations migrated passively in the cytoplasm along the gradient built up by anion transport. In this process the activity of metabolism would nevertheless be necessary for the maintenance of the appropriate absorption structures (Fig. 8.2). In this form also the hypothesis is not very acceptable cytologically. It does not consider the plasmalemma alone but in addition the cytoplasm itself or, at least a part of it, as the external barrier across which the accumulation of ions occurs. To bring his hypothesis into agreement further with modern knowledge on cell compartmentation Lundegårdh considered active ion transport as a statistical result of the movement of little particles carrying the electron ladders in a redox gradient. This is based on the idea that mitochondria, so to speak, act as vehicles for ion transport (Lundegårdh, 1950, 1955, 1958a,b,c).

8.1.1.2 Model of a Redox Pump After Robertson and Conway

The basis for the hypothesis of the direct energetic coupling between ion transport and electron transfer along the cytochrome chain has been further evaluated and confirmed by Robertson and his coworkers by a detailed experimental characterization of the phenomenon of salt respiration (Robertson, 1960, 1968). The model developed by this school is simpler than the Lundegårdh hypothesis itself although it is similar in principle. A carrier localized within the membrane at the interior face takes up electrons from the electron transfer cofactors and is thereby reduced; at the exterior face of the membrane the carrier is oxidized so that a redox gradient exists within the membrane. A model much like this has been suggested by Conway (Conway, 1955). The redox carrier system can, according to these models, transport cations from the membrane face having carrier oxidation to the membrane face having carrier reduction and move anions in the opposite direction.

The various theoretical possibilities of such a redox carrier are schematically summarized in Figure 8.3. The stoichiometry is based on the

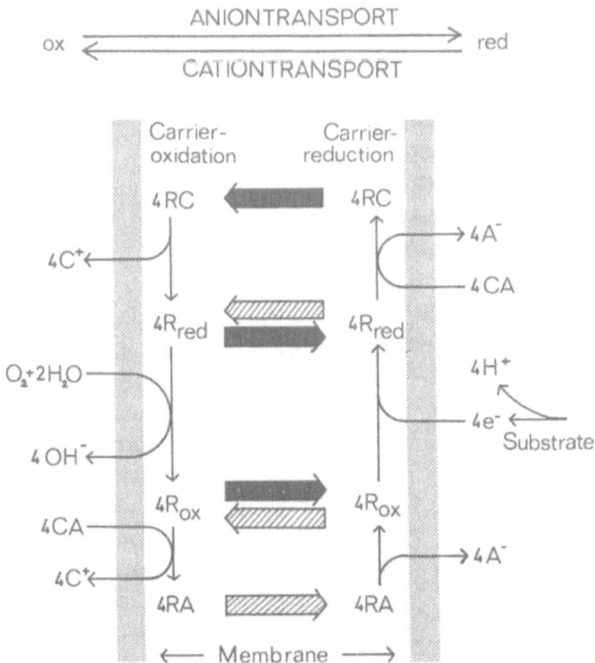

Fig. 8.3. Scheme of the redox-carrier model for cations and anions according to Conway (1955) and Robertson (1960, 1968). C^+, cation equivalent; A^-, anion equivalent; R, carrier; *red*, reduced; *ox*, oxidized; e^-, electron; *thin arrows*, reactions of carriers and substrates; *thick arrows*, transport; *shaded*, anion transport; *bold*, cation transport.

assumption that each carrier molecule can take up one electron. According to this, then, one O_2 molecule can oxidize four carrier molecules which then can transport four anion equivalents and so on. Let us briefly consider various theoretically possible cycles of this carrier system:

1. *Anion Transport—Anion Transport Without Redox Reactions on the Carrier.* The oxidized carrier, R_{ox}, takes up one anion equivalent, the carrier-anion complex, RA, is split at the opposite face of the membrane; the oxidized carrier can diffuse back and can re-enter the cycle. This carrier mechanism is electrogenic; cations and anions are separated, going to opposite faces of the membrane. An electropotential gradient is created and cations can migrate along this gradient passively following the anions. Energy is required to maintain the oxidized state of the carrier; but there is no immediate and compelling stoichiometric correlation between respiration and anion transport.
   The Carrier Works as Redox System. Four reduced carrier molecules are transferred into their oxidized form at the oxidizing face of the membrane by an O_2 molecule. Now four anion equivalents can be bound. On the membrane face of carrier oxidation four OH^- ions remain plus four cation equivalents. Electroneutrality is maintained. The carrier anion-complex diffuses to the reducing face of the membrane. There the four anions are set free, the four carrier molecules are reduced by uptake of electrons from the electron transfer system, and the reduced carrier diffuses to the oxidizing membrane face and enters the cycle anew. On the membrane face of carrier reduction four protons came into existence which compensate for the electrical charge of the four anions. Thus, here as well, electroneutrality is maintained, but a pH gradient has built up across the membrane.

2. *Cation Transport.* Cation transport occurs in the direction opposite to anion transport. If there are no redox reactions on the carrier, the mechanism is electrogenic. To allow cation transport the carrier must be maintained by metabolism in the reduced state. If the carrier works as a redox system, the consequences for ion distribution at the membrane can be discussed in a way similar to that for anion transport.

3. *Coupled Anion and Cation Transport.* This combined mechanism is also electrogenic. Let us begin with the oxidized carrier on the membrane face of carrier oxidation; it takes up four anions and leaves behind four cation equivalents. On the membrane face of carrier reduction then four anion equivalents are liberated. Furthermore, by carrier reduction on this membrane face four protons are formed and when the reduced carrier now takes up four cation equivalents then four anion equivalents remain behind. On the membrane face of carrier reduction the net result thus is $4 A^- + 4 H^+ + 4 A^-$. On the opposite membrane face the four cations are liberated. During carrier oxidation, on the

other hand, four OH^- ions are formed. Together with the four cation equivalents which remained behind during anion transport this gives a net result of $4 K^+ + 4 OH^- + 4 K^+$. Thus at the membrane a pH gradient and an electrical gradient is built up.

8.1.1.3 Criticisms of the Model of Respiratory Electron Flow as a Direct Energy Source of Active Ion Transport

A number of arguments against the model of respiratory electron flow as a direct energy source of active ion transport have been mentioned previously (Sect. 8.1.1.1). The stoichiometric requirements for the relation between salt respiration and ion transport are not fulfilled (e.g., Sutcliffe, 1962; Sutcliffe and Hackett, 1957). The model also cannot sufficiently explain the specificity of ion uptake. We know from Section 6.1.2.1 and Table 6.1, that one must distinguish quite a number of specific ion transport mechanisms. Redox processes alone fail to provide this specificity; in addition one would have to assume that the redox reactions are associated with specific carrier molecules.

The greatest difficulties for the understanding of the coupling of electron flow and membrane transport result from cytomorphological considerations. How can electron transport in the cristae membranes of mitochondria be coupled with ion transport in membranes spatially separated? This principal question is also pertinent in the discussion of coupling of electron flow in photosynthesis with ion transport. The Mitchell hypothesis of charge separation at cristae and thylakoid membranes and modern ideas on the mode of coupling between energy metabolism and membrane transport, however, let the old Lundegårdh hypotheses appear in a new light (see Chap. 10). Redox pumps in the original sense of Lundegårdh, Robertson, and Conway (Figs. 8.2 and 8.3) can mediate only the active transport of ions. However, further developments of the redox pump hypothesis may also allow an explanation of active transport of nonelectrolytes. One possibility is an activation of the carrier by sterical changes as a consequence of redox reactions, rather than by a change in charge. Other possibilities are cotransport and countertransport models (Chap. 10).

8.1.2 ATP as the Energy Source for Active Transport

8.1.2.1 ATP, the General Energy Currency of the Cell

It is widely accepted that ATP is the general energy currency of the cell. ATP powers many energy-consuming processes. ATP formed in the mitochondria, however, can penetrate the mitochondrial envelope only by mediation of specific transport mechanisms. The same is true also for the

transport of ATP across the chloroplast envelope (Chap. 7.2). Nevertheless, despite this compartmentation ATP is available at all the sites and locations within the cell where it is needed. With ATP as a source of energy
various possible transport mechanisms could be driven; ion pumps, carrier systems (see Sect. 6.1.2.1; Fig. 6.1) and processes for transfer of nonpolar compounds.

8.1.2.2 Experiments With Inhibitors

As mentioned previously (Sect. 2.5.4) the use of metabolic inhibitors is
important for characterization of active or metabolic transport in biological systems. We distinguish between inhibitors of electron transport, and
uncouplers of phosphorylation. Uncouplers block the formation of ATP
which is normally coupled with electron transport in the cristae and thylakoid membranes, or they inhibit formation of an energy-rich phosphorylated intermediate ($X \sim P$) on the way to ATP. At lower, i.e., reasonably
specific concentrations, uncouplers do not inhibit electron transport; in
fact electron flow is often increased. Some of the most widely used inhibitors in connection with transport studies are cyanide (CN^-) and azide
(N_3^-) which block electron flow in mitochondria and chloroplasts, and
dichlorophenyl dimethyl urea (DCMU) specifically as an inhibitor of noncyclic electron flow in photosynthesis (see Sect. 8.2.3.3); 2,4 di-
nitrophenol, DNP, or the derivatives of carbonylcyanide phenylhydra-
zones are used as uncouplers of phosphorylation. Regrettably, however,
the use of inhibitors usually does not lead to unambiguous results. The
specificity of the reactions often is not known with the necessary accuracy. Thus, in complex intact cells, in addition to affecting important
primary processes of electron flow and ATP formation, respectively,
they can have numerous less specific effects and secondary actions.

CN^-, for instance, blocks ion uptake and salt respiration by plant cells.
This was considered as a proof of the Lundegårdh hypothesis (Fig. 8.1).
Ion transport, however, is also inhibited by uncouplers, which suggests
the participation of ATP or energy-rich phosphate ($\sim P$). AsO_4^{3-}, which
can compete with PO_4^{3-} in phosphorylation reactions (competitive inhibition), also blocks ion uptake. In maize-root tissue AsO_4^{3-} prevents formation of energy-rich phosphate compounds to the same degree as the uptake of Cl^-, SO_4^{2-} and PO_4^{3-}; at the same time O_2-consumption (electron
flow) is increased. This is an important indirect evidence for the dependence of these ion transport processes on energy-rich phosphate (Fig. 8.4;
Weigl, 1964). A combination of inhibitors can also be applied to show that
ATP is the source of energy for ion transport in roots. Under anaerobic
conditions some energy-dependent Cl^- uptake in maize roots can be
maintained by the addition of the artificial electron acceptor ferricyanide.
The respiratory chain in this treatment is considerably shortened but ATP
is formed. Uncouplers and arsenate inhibit the ferricyanide-sustained Cl^-

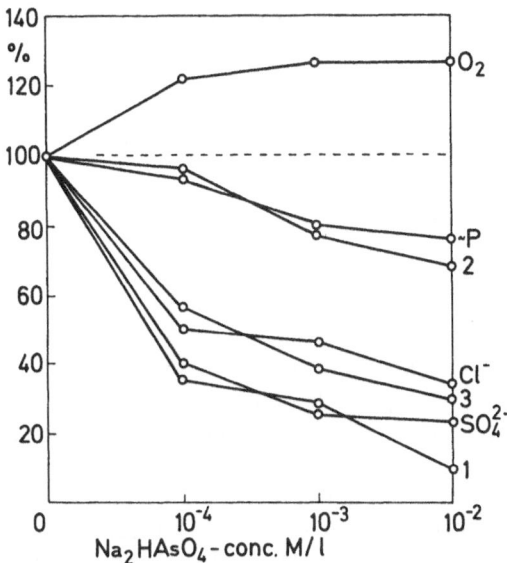

Fig. 8.4. The effect of AsO_4^{3-} on the uptake of O_2, Cl^-, and SO_4^{2-}, and the formation of energy-rich phosphate ($\sim P$) and some other organic phosphate compounds in maize roots. *1*, unknown; *2*, UDPG, ATP and UTP plus diphosphates; *3*, hexose phosphate, UMP and AMP. (Weigl, 1964.)

transport (Budd and Laties, 1964). This experiment suggests that for the functioning of active Cl^- uptake in root tissue ATP formation is more important than electron flow along the complete respiratory chain.

On the other hand, Ginsburg and Ginzburg (1970), found that CN^- inhibits Cl^- uptake in maize roots whereas the uncoupler DNP increases Cl^- influx threefold. From this one should conclude that Cl^- uptake is

Table 8.1. Different energy sources for various ion fluxes in storage tissues.

Ion flux	Source of energy	Reference
Cl^- uptake at the plasmalemma of carrot root cells	e-flow	Cram (1969b)
Anion and cation uptake from 0.5 mM solutions by beet root cells, i.e., probably plasmalemma influx	e-flow	Polya and Atkinson (1969)
Cl^- uptake at the tonoplast of carrot root cells	$\sim P$	Cram (1969b)
Na^+, K^+ uptake at the tonoplast of carrot root cells	ATP or $\sim P$	Lüttge, Cram et al. (1971)
Anion and cation uptake from 40 mM solutions by cells of carrot root steles, i.e., probably tonoplast influx	e-flow	Atkinson and Polya (1968)

ATP-independent and more directly coupled to electron flow. This discrepancy perhaps is explained by the fact that Budd and Laties, and Ginsburg and Ginzburg, respectively, were working under somewhat different conditions and hence dealing with different Cl^- fluxes in the maize root (see Sect. 8.1.3, and Table 8.1).

8.1.2.3 Salt Respiration and ATP-Driven Ion Transport

The phenomenon of salt respiration has long been the most important basis for the hypothesis of electron transfer reactions as the direct energy source for ion transport. This poses, however, a fundamental question: In which way can the utilization of respiratory energy by ion transport lead to increased O_2 consumption by the tissue? Is there an answer which offers an alternative to the hypothesis of direct coupling of electron transport and ion transport? It is possible to assume that respiratory electron flow is limited by the consumption of ATP, or an energy-rich intermediate $(X \sim P)$ which is formed during oxidative phosphorylation prior to ATP synthesis. An increased consumption of ATP or $X \sim P$ provides an increased amount of acceptors (e.g., ADP) for the formation of energy-rich phosphate coupled with electron flow. Thus, a more rapid turnover due to increased ATP consumption would lead to greater O_2 consumption. Therefore, an increased utilization of ATP, or $X \sim P$, by ion transport, after addition of salt to the medium, could explain salt respiration just as well as a direct utilization of electron flow as an energy source.

Lüttge, Cram et al. (1971a) have investigated this alternative using carrot-root tissue. With uncoupling of electron flow and phosphorylation in carrot-root tissue by Cl-CCP (m-chloro-carbonylcyanidephenylhydrazone), a maximum O_2 uptake is obtained at 10^{-6} M. The maximum salt respiration is observed in 60 to 80 mM KCl (Fig. 8.5). If salt (60–80 mM KCl) is added to the tissue, when there is already maximal uncoupling by previous addition of 10^{-6} M Cl-CCP, no further increase in respiration is obtained beyond the rate already reached by the uncoupling. O_2 uptake in 10^{-6} M Cl-CCP and in 10^{-6} M Cl-CCP plus 60–80 mM KCl is considerably lower than O_2 uptake in 80 mM KCl without uncoupler. In a somewhat different experiment the maximum salt respiration was measured initially in 60 mM salt and then 10^{-6} M Cl-CCP was added. In this case the uncoupler still caused an increase in the rate of respiration. From these experiments, first, one can conclude that neither in 10^{-6} M Cl-CCP alone nor in 60–80 mM salt alone is a rate of respiratory electron flow obtained which is limited by the structure of the respiratory system, and which would give an absolute upper ceiling possible for the amount of respiratory units in the cristae membranes. This is a prerequisite for the second conclusion which is decisive in our context: Although in principle electron transfer could proceed at faster rates and still more O_2 could be taken up, salt does not increase respiration when it is uncoupled from phosphorylation. From this we can finally conclude that salt acts on respiration by an increased consumption of ATP or ~P and not by direct coupling of ion transport with electron flow, because the uncoupler blocks only phosphorylation, not electron transfer. In summary, these experiments show that the phenomenon of salt respiration is well in agreement with a model of an ATP-driven mechanism of ion transport.

The increased electron flow in salt respiration must be correlated with an increased consumption of substrate. Pitman, Mowat et al. (1971) have investigated ion and sugar uptake by barley roots and compared it with ion and sugar content. Low salt levels are correlated with high sugar contents of the root tissue and vice versa. Salt in the external medium inhibits sugar uptake (glucose, fructose), possibly by inhibition of sugar transport into the root cells at the tonoplast. At the same time because of sugar diffusion from the vacuole into the cytoplasm, cytoplasmic sugar level is raised temporarily. As a consequence after transfer of a tissue of low salt content into a salt solution, respiration is increased (salt respiration). In time, respiration and ion uptake decrease due to the depletion of the sugar substrate in the cells.

8.1.2.4 Membrane ATPases and Active Transport

It has been mentioned earlier that ATPases may be involved in membrane transport in plants (Sect. 5.3.3). This, of course, would provide the best evidence for a utilization of ATP in active transport. A correlation of ATPase activity with membrane transport is suggested by three types of evidence:

1. ATPase activities have been detected in the fractions of homogenates of higher plant cells which were enriched in plasmalemma and, in some cases, in tonoplast fragments (review: Hodges, 1976; Rungie and Wiskich, 1973; Wilkins and Thompson, 1973; Leigh and Wyn Jones, 1975; Leonard and van der Woude, 1976). However, the identification of the membrane preparations as plasmalemma and tonoplast respectively is a great problem (Sect. 5.3.3). Thus the early optimism has been dampened somewhat. Histochemical studies have revealed ATPase activity at almost all membranes within higher plant cells (Läuchli, Kramer et al., 1977).

2. ATPase activities have been correlated with ionic relations of plants. Examples come from comparative studies with related plant species or varieties. Fischer et al. (1970) and Leonard and Hodges (1973) have studied K^+ uptake and K^+ activation of ATPases in roots of four grass species. As shown in Figure 5.14 they obtained very similar isotherms for both processes at varied KCl or RbCl concentrations. Although this correlation is most striking, it must be born in mind that it may be coincidental; a direct relation has not yet been established (Leigh and Wyn Jones, 1975). Kylin and coworkers (Karlsson and Kylin, 1974;

Hansson, 1975) worked with four varieties of sugar beet of different salinity resistance. A good correlation between the presence and the
properties of ATPases and Na^+ resistance of the four varieties can be
constructed from their data (Table 6.6).
3. Substrate (i.e., chloride) induction of a Cl^--dependent ATPase in salt
glands of *Limonium* is correlated with induction of active Cl^- excretion by the glands (A. E. Hill and Hill, 1973a,b; cf. Lüttge, 1975; see
Sect. 5.3.2; see Fig. 5.11).

8.1.3 Various Ion Fluxes in Cells of Nongreen Storage Tissues Driven by Different Sources of Energy

The preceding section has shown that in certain situations a very strong
case can be made in favor of ATP as the energy source of active membrane transport. In no way, however, can this be a generalization that
ATP is considered as the only ubiquitous energy source of transport. Individual fluxes must be specified by their localization and direction within
the compartmented cell, by the kind of particles transported, and by the
plant organ and species or variety. Fluxes of various solutes may be
driven by different energy sources. We will give one example here, of energy relations with some ion fluxes in nongreen storage tissues. More examples will be described when particular transport processes in green
cells (Sect. 8.2.4) and molecular modes of energetic coupling (Chap. 10)
are discussed.

Let us first return to the experiments on salt respiration in carrot-root
tissue explained in Section 8.1.2.3. Ideally to find out which particular
flux is causing increased O_2 uptake, it would be desirable to vary the flux
of each ion alone in a complex compartmented system and to investigate
its effect on salt respiration; but this is not possible. In the discussion of
various transport models of the plant cells we have seen that many fluxes
are mutually interdependent (Sect. 2.3). One can, however, by planning
the experiments, arrange conditions so that one flux changes considerably
but that the other fluxes vary only slightly.

In the experiments with carrot-root tissue described in Section 8.1.2.3
there is good reason for the assumption that active ion influx into the
vacuole, across the tonoplast, is the flux affected by the addition of 60–
80 mM KCl. In Section 6.2.1 the fact that this flux is considerably altered
over the range of high external salt concentrations (larger than 1 mM) was
considered in detail. The salt respiration isotherm and the isotherm for ion
uptake into the vacuoles of carrot-root tissue are reasonably identical in
the concentration range of 1–80 mM (Fig. 8.5). In the low-concentration
range (0–0.5 mM), where the tonoplast influx is not the factor which decisively affects total ion uptake, there is often no salt respiration at all.

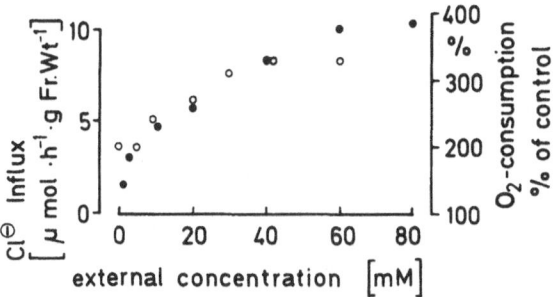

Fig. 8.5. Salt respiration isotherm (*open symbols*) and Cl⁻-influx isotherm (*closed symbols*) of carrot root tissue. (From the data of Figs. 3 and 4 in Lüttge, Cram et al., 1971.)

Therefore the flux, which in Section 8.1.2.3 has been characterized as ATP- or ~P-dependent, must be an influx at the tonoplast.

The driving force for other energy-dependent fluxes in carrot cells is not necessarily dependent on the same source of energy. Atkinson and Polya (1968), in contrast to Lüttge, Cram et al., (1971a) came to the opinion that, in carrot-root tissue, anion and cation uptake from a 40 mM KCl-solution is not driven by ATP or ~P but is more closely coupled to electron transport. In many experiments ion uptake appeared to be inhibited by anaerobiosis and uncouplers. The inhibition of ion uptake by anaerobiosis commenced very rapidly, whereas the ATP level in the tissue decreased more slowly. Conversely, with ethionine, the ATP level in the cells could be lowered drastically without a particular impairment of ion uptake. Under the conditions of the experiments of Atkinson and Polya, therefore, electron flow along the respiratory chain, with O_2 as a final oxidant, appears to be the most important supply of energy for ion uptake. The absence of a correlation of ion uptake with ATP level, however, is no unequivocal proof for the independence of ion uptake from ATP. Because of the ATP compartmentation it would be important in addition to know the turnover of cytoplasmic ATP pools which could supply energy to ion uptake mechanisms at the plasmalemma and tonoplast.

In another series of experiments with beet tissue the ATP level in the first 2½ h after the transition to anaerobiosis almost did not decrease at all, whereas ion uptake from 0.5 mM KCl or NaCl solution was inhibited drastically (Polya and Atkinson, 1969). In this case, also, the decrease of ATP level by means of ethionine had scarcely any effect on ion uptake. In these experiments since ion uptake was at low external concentration (see Sect. 6.2.1) it may be assumed with a certain degree of probability that, in the cells of red beet, influx at the plasmalemma is not dependent on ATP or ~P, but is more directly driven by electron flow.

According to Cram (1969b), the same is true for the plasmalemma influx in carrot-root tissue. Cram investigated the effect of various inhibitors on Cl^- influx at the plasmalemma and at the tonoplast. The plasma-lemma influx is unaffected by the uncoupler Cl-CCP and by oligomycin. Both inhibitors uncouple ATP formation from electron flow, Cl-CCP probably by discharge of the energy-rich $X \sim P$ which is formed in a se-quence prior to synthesis of ATP, and oligomycin by preventing the reac-tion leading to $X \sim P$. Electron flow in intact carrot-root cells is not im-paired by oligomycin. Anaerobiosis (pure nitrogen atmosphere) inhibits plasmalemma influx. This means that only redox reactions of the respira-tory chain can be responsible for driving Cl^- influx at the plasmalemma of carrot root cells; influx at the tonoplast of carrot-root cells is inhibited by anaerobiosis, Cl-CCP, and oligomycin. This confirms the conclusion reached before that this flux very likely is supplied with energy by ATP or the energy-rich phosphorylated intermediate $(X \sim P)$.

The most important result to be stressed in this discussion is that in nongreen cells different sources of energy may be responsible for driving the various ion fluxes across individual membranes (Table 8.1).

8.2 Photosynthesis as an Energy Source for Active Transport

8.2.1 The First Proofs for the Dependence of Transport Processes on Energy of Photosynthesis

Of basic importance in the history of the research on the utilization of en-ergy of absorbed light for active transport is a paper published in 1957 by van Lookeren-Campagne. This investigation shows that the action spec-trum of light-induced Cl^- uptake by *Vallisneria* leaves is essentially iden-tical with the action spectrum of photosynthesis of these leaves. Both spectra are also identical with the absorption spectrum of the leaves de-termined by measuring light transmission by the leaf. Other investigations from the same laboratory (Arisz and Sol, 1956) somewhat earlier had made clear that light-induced Cl^- uptake by *Vallisneria* leaves is inde-pendent of the formation of carbohydrates in photosynthesis. The photo-synthetic formation of substrate for respiration therefore could not be responsible for the energetic relations between photosynthesis and ion uptake. It is evident that energy equivalents formed in the primary reac-tions of photosynthesis can be utilized directly for light-dependent Cl^- transport.

Three years earlier inhibitor experiments performed using *Chlorella* cells led Kandler (1954, 1955) to the conclusion that in this organism the light-driven glucose uptake is correlated with ATP formation in photo-synthesis. These earlier experiments, however, do not make clear whether the stimulation of glucose uptake by photosynthesis occurs by way of an energetic coupling with a membrane transport mechanism or is due to an increased metabolism of glucose (photoassimilation of glucose) in the interior of the cells creating a sink.

The analysis of light-dependent Cl⁻ uptake of *Vallisneria* leaves by van Lookeren-Campagne, and Kandler's experiments on the light-stimulated glucose uptake in *Chlorella* cells laid the basis for further investigations of the energetic coupling between photosynthesis and energy-dependent transport mechanisms. In the time following, light-dependent ion uptake by green cells has been investigated quite intensively with a large variety of organisms, i.e., algae and higher plants, in a number of laboratories (Sect. 8.2.4). Before we can understand these experiments we must briefly recall the more important energy transfer reactions of photosynthesis.

8.2.2 Simplified Scheme of Photosynthetic Energy Transfer Reactions

The scheme of the photosynthetic energy transfer reactions shown in Figure 8.6 has been deliberately simplified. Investigations on the energetic coupling of active transport processes with primary photosynthetic reactions have not proceeded to an extent which would require all the details of these reactions to understand the following discussion. (Some of these details, for instance arrangement of some cofactors participating in energy transfer reactions, are also still controversial.)

In the light reactions of photosynthesis there is energy transfer; chemical energy and reduction equivalents in the form of ATP and NADPH are produced. In noncyclic electron flow two photosystems act together. By

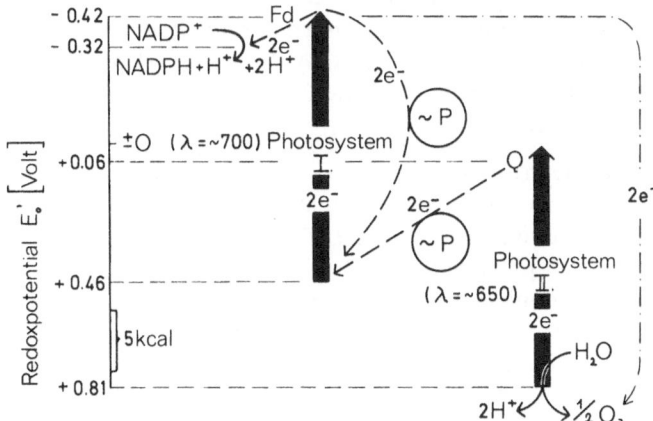

Fig. 8.6. Simplified scheme of the primary reactions of photosynthesis. *NADP⁺*, nicotinamide adenine dinucleotide phosphate; *Fd*, ferredoxin, *~P*, phosphorylation; *e⁻*, electron; *Q*, plastoquinone.

the light reaction of photosystem II, electrons are excited so that they can be transferred from water to the plastoquinone-plastohydroquinone system (Q). The electrons then fall back via various redox systems to a more positive redox potential and eventually are raised in the light reaction of photosystem I to a very negative redox potential level. Via various intermediates they eventually reduce pyridine nucleotide (nicotinamide adenine dinucleotide phosphate, $NADP^+$) to $NADPH + H^+$, which in this form can supply the reduction equivalents necessary for CO_2 assimilation. During noncyclic electron transport, at least at one site, ATP can be formed ($\sim P$ in the scheme of Fig. 8.6). In the total balance for each $\frac{1}{2} O_2$ released two electrons can be transferred via the noncyclic redox chain, during which 1 ATP and 1 $NADPH + H^+$ are formed. The question of whether in noncyclic electron flow phosphorylation can occur at a second site, i.e., whether the ratio ATP/2 electrons is possibly larger than unity, does not need to occupy us here because stoichiometry between the energetic coupling of transport mechanisms and photosynthetic energy is not established well enough experimentally.

Independently of photosystem II, ATP can be formed also in cyclic photophosphorylation. In this process electrons are not transferred from the excited photosystem I to ferredoxin (Fd) and then to the pyridine nucleotide. Instead they flow in cyclic electron transport via various cofactors from the highly reducing negative redox potential back to the positive redox potential of the nonexcited pigment system I.

Another possibility when electrons cannot flow via ferredoxin to the pyridine nucleotide is pseudocyclic electron flow or the Mehler-reaction. This involves both photosystems but, instead of $NADP^+$, O_2 serves as an electron acceptor from pigment system I. As in noncyclic electron flow ATP can be formed also during pseudocyclic electron flow. Possibly during such photoreduction of O_2 in the absence of CO_2 reducing power is also generated and consumed at a normal rate (Radmer and Kok, 1976).

In various investigations of photosynthesis as related to active transport mechanisms one or more of the following energy sources have been considered (Fig. 8.6):

1. Noncyclic electron flow:
   a. electron transfer as such (in analogy to the Lundegårdh hypothesis or to the redox carrier model, Figs. 8.2 and 8.3)
   b. energy-rich phosphate (ATP or $\sim P$)
   c. reduction equivalents ($NADPH + H^+$)
   d. intermediary metabolites as they are formed during photosynthetic CO_2 assimilation with the consumption of ATP or $NADPH + H^+$ or both.
2. Cyclic electron flow:
   a. electron transfer as such
   b. energy-rich phosphate.

3. Pseudocyclic electron flow:
   a. electron transfer as such
   b. energy-rich phosphate
   c. perhaps reduction equivalents.

8.2.3 Experimental Alteration of Energy-Transfer Reactions in Photosynthesis and Correlation With Energy-Dependent Transport Processes

Similar to the investigation of respiration as an energy source for active transport are the attempts to alter experimentally the energy-transfer reactions of photosynthesis in order to find correlations of particular parts of the scheme shown in Figure 8.6 with the energy-dependent transport processes. There are principally four possibilities:

1. Variation of wave-length of the irradiated light.
2. Variation of the gas content of the atmosphere, in particular of CO_2 supply.
3. The use of inhibitors and artificial electron donors and acceptors.
4. The use of mutants.

The results which photosynthesis research has obtained with these methods in experiments with isolated thylakoids, isolated chloroplasts, and intact cells are a prerequisite for each investigation of photo-synthesis-dependent transport mechanisms. Naturally it is impossible to describe all details here. However, it is important to understand a few basic types of experiments of particular significance in research on photo-synthesis-dependent ion uptake processes.

8.2.3.1 Varying the Wavelength of Irradiated Light

By the choice of the appropriate wavelengths in the red region of the spectrum one can either excite photosystem I alone or both photosystems together. Up to a wavelength of 705 nm photosystems I and II are active, but above 705 nm largely photosystem I alone is excited, and above 730 nm there is no photosynthetic electron flow at all. Experimentally it is simple to use various cut-off filters, which eliminate the part of the spectrum as indicated above (Fig. 8.7). In this way experimental evidence can be obtained showing whether a transport process is dependent on cyclic electron flow, on noncyclic electron flow, or on the photophosphorylation associated with these electron flows (MacRobbie, 1965). A certain difficulty in working with these wavelengths results from the fact that a clear enough separation cannot be obtained between an effect on photosystems and on the phytochrome system. The elimination of light under 730 nm prevents photosynthesis, but at the same time active phytochrome (P-730)

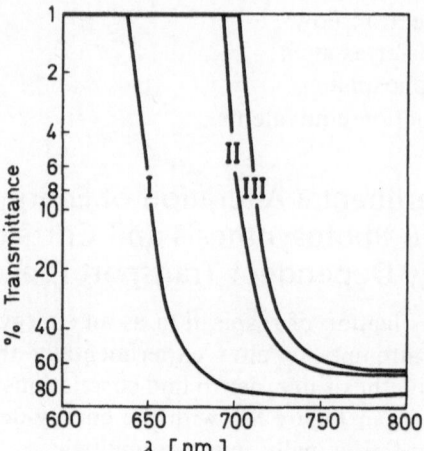

Fig. 8.7. Transmittance of gelatine cut-off filters of the type used by MacRobbie (1965). (After an original spectrophotometer scan obtained by Lüttge and Pallaghy, 1969.) Filter *I* transmits approximately equal numbers of quanta between 705 and 730 nm and below 705 nm, this filter excites both photosystems; filter *II* cuts out most of the light at 705 nm and therefore blocks photosystem II, but leaves photosystem I in operation; filter *III* prevents excitation of both photosystems.

is converted to the nonactive configuration (P-660; Sect. 9.1.1). When red light of wavelengths greater than 705 nm is used (photosystem I active) part of the phytochrome is in its active form. On using light which excites both photosystems, the equilibrium between P-660 and P-730 is shifted further towards P-730 (Sect. 9.1.1). Nevertheless the participation of phytochrome systems is very unlikely when light acts on the transport process via the *energy supply*. (However, see Sect. 9.7 for some limitation of this conclusion.) Experimentally, one can analyze this by combination of light qualities of defined wavelengths.

In order to understand this, we must know the Emerson effect, which was the first important evidence for the joint action of two photosystems in photosynthesis. By the aid of Figure 8.6 this effect can be easily understood. If simultaneously light bands of the wavelength 650–660 nm and 700–710 nm are provided, photosystems I and II are excited at the same time and the rate of photosynthesis observed (CO_2 fixation or O_2 evolution) is larger than the sum of the rates of photosynthesis measured when each light band is applied alone. One can also use this Emerson-enhancement effect to show whether a photosynthesis-dependent process (e.g., an active transport process) needs both photosystems or only one of the two photosystems alone. Since, at 700–710 nm there is less active phytochrome present than at 650–660 nm, one could make a mistake in such experiments in case an Emerson-enhancement effect is superimposed by a phytochrome inhibition quenching this enhancement. Active phytochrome, however, is also changed to the inactive configuration by red light of 740 nm, which is not effective in photosynthesis. Furthermore, one can make use of the characteristic reversibility of the phytochrome system. As the scheme on p. 235 shows, an irradiation with $\lambda = 660$ nm quenches the effect of a preceding irradiation with $\lambda = 730$ nm and vice versa. By such experiments Raven (1969a,b) was able to show that the phytochrome system has no effect on light-dependent ion transport in *Hydrodictyon africanum* (see Sect. 8.2.4.1.3).

A good understanding of the correlation between transport and photosynthesis can be obtained from a comparison of action spectra, especially in the far-red part of the spectrum. It was mentioned in Section 8.2.1 that action spectra gave the first good evidence for an energetic linkage between photosynthesis and Cl^- or phosphate uptake in *Vallisneria*. Action spectra for Rb^+ and Cl^- uptake by *Vallisneria* leaves, obtained many years after the work of van Lookeren-Campagne in the same laboratory at Groningen by Prins (1973), are very similar but more detailed in the far-red region (Fig. 8.8a,b). Since photosystem II is excited up to 705 nm only, it can be observed that in the red part of the spectrum (in fact just above 680 nm) a sharp drop occurs in the action spectrum of all processes requiring photosystem II. This so-called red drop naturally is observed for O_2 evolution and CO_2 fixation of photosynthesis itself which requires noncyclic electron flow and hence both photosystems (Figs. 8.8, 8.9). Processes which share this red drop such as Rb^+ uptake by *Vallisneria* (Fig. 8.8) and Cl^- uptake by *Hydrodictyon* (Fig. 8.9) must also depend on photosystem II. Since light is still absorbed at higher wavelengths by photo-

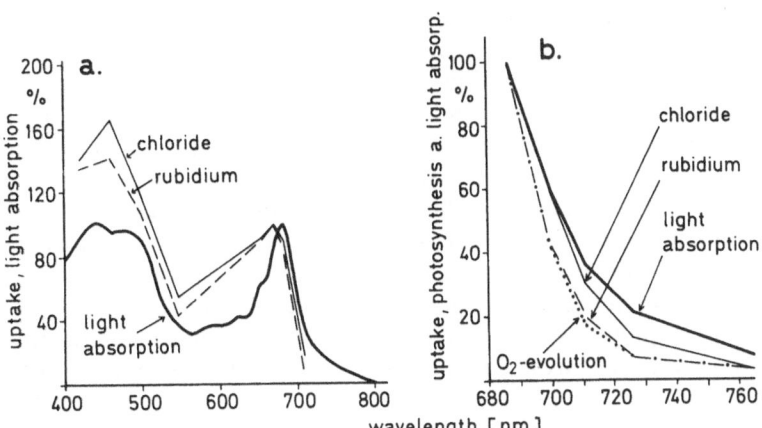

Fig. 8.8a and **b.** Light absorption spectrum by *Vallisneria* leaves and action spectra of Cl^- and Rb^+ uptake and photosynthesis. **a** covers the whole range between 400 and 800 nm. **b** gives details in the far red region showing a sharp "red drop" of photosynthetic O_2 evolution and Rb^+ uptake but not of Cl^- uptake. (Prins, 1973.)

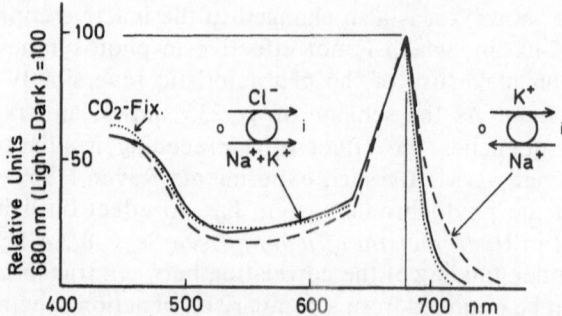

Fig. 8.9. Action spectra of CO_2-fixation and of two different ion pumps in *Hydrodictyon africanum*. (After Raven, 1969a.)

system I, the light absorption spectra and action spectra of photosystem I, cyclic electron flow, and associated processes, do not show the sharp red drop. This is the case for instance with Cl^- uptake by *Vallisneria* and with K^+ uptake by *Hydrodictyon,* so that we may conclude that these ion transport processes can be driven by cyclic electron flow or cyclic photophosphorylation (Figs. 8.8, 8.9).

A good action spectrum for uptake of Na^+ and Cl^- by cells of *Nitella flexilis* showing a close correlation with photosynthetic CO_2 fixation has been obtained by Takeda and Senda (1974).

8.2.3.2 Effect of CO_2 Concentration in the Atmosphere

Elimination of CO_2 from the atmosphere has been useful in transport investigations since it effects a transition from noncyclic to cyclic electron flow. Photosynthetic CO_2 reduction is the main sink for reduction equivalents from the reduced pyridine nucleotide formed in noncyclic electron transport. If by elimination of CO_2 the continuous reoxidation of $NADPH + H^+$ to $NADP^+$ is prevented, the electron acceptor at the end of the chain of noncyclic electron transport is absent. An electron transfer then must occur mainly via the cyclic pathway (Jeschke, 1967; Jeschke and Simonis, 1967, 1969). The usefulness of this method, however, may be limited. Reduction processes other than CO_2-fixation (e.g., nitrate reduction) may utilize the reduction equivalents of $NADPH + H^+$ and therefore supply $NADP^+$ as an electron acceptor, thus maintaining a certain noncyclic electron flow (Ullrich, 1971, 1972; see Lüttge, 1973a,b; Eisele and Ullrich, 1975; Raven and Glidewell, 1975b). Noncyclic photophosphorylation coupled to nitrate photoreduction in the alga *Hydrodictyon africanum* may provide ATP for growth (Raven, 1977a). When all $NADP^+$ of the cell is fully reduced, electrons may flow also to O_2 as an acceptor, and this "pseudo-cyclic" electron flow (Fig. 8.6) can be coupled with ATP-formation (e.g., Egneus et al., 1975; Glidewell and Raven, 1975).

8.2.3.3 Experiments with Inhibitors

Some of the inhibitors useful in investigations of photosynthesis of isolated thylakoids and chloroplasts have been used successfully also in experiments with intact plant cells. With some other inhibitors, the results gained with isolated membranes could not be confirmed in intact systems or there are still not enough data available. In complex systems there are greater difficulties in the use of inhibitors than in isolated thylakoids and chloroplasts. The inhibitors must be taken up through the plasmalemma into the cells. There inhibitors not only affect thylakoid membranes, but rather they may also act on other membranes in the system. Therefore, the disappointing result of a comparative investigation of the effects of four inhibitors, phlorizin, Dio-9, imidazol, and Cl-CCP on intact cells and on isolated chloroplasts of *Chara corallina* is not surprising. With one inhibitor only, Cl-CCP, was the effect obtained *in vitro* (isolated chloroplasts) correlated with the observations made in vivo (intact cells; Smith and West, 1969).

CCP (Cl-CCP or FCCP, i.e., m-chloro- or p-trifluoro-methoxy carbonyl cyanide phenylhydrazone) is an uncoupler widely used in transport studies. CCP inhibits photosynthetic ATP formation. Since ATP formed in noncyclic electron transport is necessary for CO_2 fixation, CCP also inhibits photosynthesis. It has been possible to characterize systems with ion transport processes not affected by CCP (or even being enhanced) while photosynthetic CO_2 fixation was inhibited. This discrepancy between the effect of the uncoupler on the transport process and on photosynthetic CO_2 fixation, for reasons mentioned above, could serve as a criterion for an ATP-independent supply of photosynthetic energy to a transport process. Although in various algae (Smith and West, 1969) and leaf cells of higher plants (Lüttge et al., 1971b) CCP is a potent inhibitor of photosynthetic CO_2 fixation, in *Chlorella* cells it seems to block only cyclic photophosphorylation but not photosynthesis (Tanner et al., 1969). Usually, however, it is possible by the use of CCP to obtain evidence for the coupling of transport processes with ATP, or $\sim P$, from noncyclic and cyclic photophosphorylation. CCP inhibits not only CO_2 fixation but also O_2 evolution in the light. As an uncoupler, used in not too high concentration, it does not block electron flow directly; electron transfer may even be increased (Teichler-Zallen and Hoch, 1967). When CCP inhibits CO_2 reduction by preventing ATP formation, however, noncyclic electron transport is suppressed indirectly in a way similar to that with the elimination of CO_2 (Sect. 8.2.3.2). In this process O_2 evolution also is reduced and electrons must now be transferred largely via the cyclic pathway.

Another widely used inhibitor in transport investigations is DCMU (N'-(3,4-dichlorophenyl)-N,N-dimethylurea). DCMU is supposed to prevent noncyclic electron transfer. The exact site of action is still disputed; however, it appears to block photosynthesis (O_2 evolution and CO_2 fixation) at concentrations between $5 \cdot 10^{-7}$ and $2 \cdot 10^{-6}$ rather specifically via

Table 8.2. Conditions required for the isolation and inhibition of various mechanisms of ATP synthesis in green cells. (Raven and Glidewell, 1975a.)

Mechanism of ATP synthesis	Conditions for isolation	Conditions for inhibition
Oxidative phosphorylation	Dark plus O_2	Presence of cyanide or antimycin A
Cyclic photophosphorylation	Light, gas phase N_2, plus DCMU	Presence of antimycin A
Non-cyclic photophosphorylation, CO_2 as electron acceptor	Light, gas phase N_2, plus CO_2, plus antimycin A	Presence of cyanide or DCMU; absence of CO_2
Non-cyclic photophosphorylation, NO_3^- as electron acceptor	Light, gas phase N_2, plus nitrate and antimycin A	Presence of cyanide or DCMU; absence of NO_3^-
Non-cyclic photophosphorylation, O_2 as electron acceptor (= Mehler reaction)	Light, gas phase N_2 plus O_2, plus antimycin A and cyanide	Presence of DCMU; absence of O_2

the inhibition of noncyclic electron flow. Very high concentrations apparently impair cyclic electron flow also and a still larger concentration has other nonspecific effects.

In the investigations of the role of respiration as an energy source for light-independent transport mechanisms and in studies of photosynthesis-dependent transport processes various combinations of uncouplers and inhibitors of electron transport have been used to gain evidence on the nature of energetic coupling. Besides the inhibitors described here, quite a number of other chemicals are used. Valinomycin, nigericin, dinactin, mycostatin, and other substances which influence ion permeability of isolated thylakoids can also be of interest in transport studies with intact cells. The experience with intact systems, however, is still very limited.

Table 8.2 taken from Raven and Glidewell (1975a) lists some conditions under which in green cells various mechanisms of ATP synthesis can be isolated (i.e., made the only ATP source available) and inhibited respectively by combining inhibitor methods with varying the gas composition of the atmosphere (Sect. 8.2.3.2).

8.2.3.4 The Use of Mutants

Both in algae and to a limited extent in higher plants, mutants have been isolated with partially defective photosynthetic electron transfer chains and pigment systems. Some have been used successfully in photosynthesis research. Of particular interest in context with ion uptake studies are mutants in which only one of the two photosystems is active (e.g., *Scenedesmus;* Bishop, 1964; *Chlamydomonas;* Levine, 1969; *Oenothera;* Fork and Heber, 1968). Also, from normal leaves of certain higher plants, cells can be obtained with modified photosynthetic apparatus. In C_4-plants (see Sects. 8.2.4.1.5, 10.2.1 and 11.3.2.3 mesophyll cells have both active photosystems I and II, but the green bundle-sheath cells in some C_4-plants have no active photosystem II (Woo et al., 1970; Downton et al., 1970; Polya and Osmond, 1972; for a different view see Andersen et al., 1972; Bishop et al., 1972; Smillie et al., 1972). Ion transport experiments with such systems are rare though (Lüttge and Ball, 1971; Hope et al., 1974).

8.2.4 Special Energy-Dependent Transport Mechanisms in Green Cells and Tissues

8.2.4.1 Ion Transport

8.2.4.1.1 Blue-Green Algae

Not until recently have blue-green algae—especially *Anacystis nidulans* —become objects for ion transport studies. Energetically these pro- karyotic green cells display some interesting and unique properties. In eukaryotic green cells light often has little effect on steady state ATP levels under aerobic conditions (see Fig. 8.11; Sect. 8.2.4.1.5). In *Anacystis nidulans* respiration in the dark is very low and thus, as in isolated chloroplasts, ATP levels are very low in the dark and highly increased in the light (Bornefeld and Simonis, 1974). Dewar and Barber (1974) suggest that Cl^- uptake is active against a gradient of electrochemical potential. These authors make an interesting comment in this context: Cl^- uptake appears to be active in all photosynthetic microorganisms, but not in het- erotrophic cells such as *Escherichia coli* or yeasts. K^+ influx and efflux in *A. nidulans* appear to be passive while Na^+ efflux may be active (Dewar and Barber, 1973). Phosphate is complex; the metabolic uptake of PO_4 is highly dependent on P-metabolism (Falkner et al., 1974; Bornefeld et al., 1974; Simonis et al., 1974).

8.2.4.1.2 Eukaryotic Microalgae

Eukaryotic microalgae, especially Chlorococcales such as *Chlorella, Scenedesmus* and *Ankistrodesmus* are outstanding objects of photo- synthesis research. Just remember that initially it was studies with *Chlorella* and *Scenedesmus* cells which led to the understanding of the path of carbon in photosynthesis and to establishment of the reductive pen- tosephosphate or Calvin cycle (Bassham and Calvin, 1957). Hence trans- port research with these algae has often involved the metabolized inorganic anions, phosphate, P_i, and HCO_3^-. A few examples may show that some of these transport processes display quite interesting properties.

The pH-dependence of phosphate uptake by *Ankistrodesmus braunii* suggests that a metabolically dependent uptake mechanism transports $H_2PO_4^-$ but not HPO_4^{2-}. Na^+ stimulates $H_2PO_4^-$ uptake at the plasmalemma (Ullrich-Eberius and Simonis, 1970).

HCO_3^- uptake by *Scenedesmus obliquus* depends on adaptation of these algal cells either to air, or to an increased CO_2 supply. The latter treatment reduces the activity of carbonic anhydrase in the cells to about $1/20$ of that observed in air (Findenegg, 1974). This enzyme obviously is important in facilitating the availability of OH^- ions from intracellular HCO_3^- and maintaining an OH^-/HCO_3^- exchange mechanism. Thus air-grown cells take up large amounts of HCO_3^- even at a pH below 6, whereas cells grown in air plus 1.5% CO_2 take up little HCO_3^- (Findenegg, 1977a,b). Cl^- uptake can compete with HCO_3^- uptake by air-adapted cells under certain conditions (Findenegg, 1974).

A most interesting example of regulation and interactions between ion transport and metabolism is provided by the stoichiometry between photosynthetic nitrate reduction and alkalinization of the medium by *Ankistrodesmus braunii* cells. In general terms this stoichiometry is

$$OH_{efflux}^- = NO_{3\,influx}^- - NO_{2\,efflux}^- + NH_{4\,efflux}^+ \qquad (8.1)$$

$NO_{2\,efflux}^-$ under most conditions is very small. In the presence of an adequate carbon source (CO_2 or glucose) NH_3 produced by NO_3^- reduction within the cells is largely utilized for amino acid synthesis, thus OH_{efflux}^- equals $NO_{3\,efflux}^-$. In the absence of a carbon source the cells still take up and reduce NO_3^-. Now relatively large amounts of NH_4^+ are released from the cells because NH_3 can no longer be fixed in amino acids. Under this condition the cells appear to take up NO_3^- and to waste reductive power, since the reduced nitrogen is released again. This, however, allows the cells to maintain noncyclic photosynthetic electron flow and photophosphorylation when the reductive pentosephosphate cycle, as the usual consumer of $NADPH + H^+$, is inoperative. NO_2^- is also taken up and reduced by *A. braunii* cells and then a similar stoichiometry can be worked out:

$$OH_{efflux}^- = NO_{2\,influx}^- + NH_{4\,efflux}^+ \qquad (8.2)$$

(Eisele, 1976; Eisele and Ullrich, 1975; Ullrich and Eisele, 1977).

Nonmetabolized ions have also been investigated. For instance, *Chlorella* has a K^+/Na^+ exchange mechanism which is independent of light under aerobic conditions. Under anaerobic conditions it appears to be driven by cyclic photophosphorylation because then K^+ uptake and Na^+ release are highly stimulated by light, and are insensitive to DCMU and severely inhibited by uncouplers (Barber and Shieh, 1973).

The filamentous green alga *Mougeotia* seems to have a light-dependent Cl^- uptake mechanism which, like that of *Nitella* or *Atriplex spongiosa* (see Sects. 8.2.4.1.3 and 8.2.4.1.5 respectively), depends on noncyclic electron flow. Cl$^-$ uptake is much more sensitive to DCMU than K$^+$ uptake (Wagner, 1974).

8.2.4.1.3 Giant Algal Coenocytes

In 1965 MacRobbie showed that light-dependent chloride uptake by giant internodal cells of the Characean alga *Nitella translucens* is strongly inhibited when the excitation of photosystem II is prevented by appropriate means (cut-off filters of 705 nm, or use of DCMU). K$^+$ uptake is, however, not affected by a transition from noncyclic to cyclic electron flow. Conversely, uncouplers do not inhibit Cl$^-$ uptake or even stimulate it, whereas K$^+$ uptake is inhibited (MacRobbie, 1966). MacRobbie concluded from these experiments that light-dependent K$^+$ uptake and Cl$^-$ uptake are driven by different photosynthetic energy sources, namely K$^+$ uptake by ATP formed in cyclic photophosphorylation and Cl$^-$ uptake by noncyclic electron transport but ATP-independent.

These publications have been received with great interest and have triggered many further investigations. A certain difficulty in the argument was seen initially in the fact that the K$^+$ and Cl$^-$ concentrations used in MacRobbie's experiments were different. Therefore, it was not clear whether she may have measured different transport mechanisms dominating in the respective ranges of concentration.

It has also been suggested that the correlation of the inhibition of noncyclic electron flow and Cl$^-$ uptake (by DCMU or by cut-off filters) alone is no proof for the dependence of Cl$^-$ uptake on electron transport by photosystem II. Investigations with *Elodea* leaves showed that DCMU inhibits light-dependent Cl$^-$ uptake in the presence of CO_2 more strongly than in CO_2-free solution. These results led Jeschke and Simonis (1967, 1969) to the following argument: we have seen before that CO_2 fixation is necessary for the supply of NADP$^+$ as an electron acceptor at the end of the noncyclic electron transport chain (Sect. 8.2.3.2). If photosystem II is inhibited by DCMU or by light of 705 nm, very likely the lowered electron flow is used entirely for a residual CO_2 fixation and Cl$^-$ uptake is strongly inhibited. In the absence of CO_2 the final acceptor for the electrons is missing, since all pyridine nucleotide is rapidly accumulated in its reduced form. The electrons then are transferred via the cyclic pathway with concomitant ATP formation by cyclic photophosphorylation. This ATP could drive Cl$^-$ uptake. In this way, however, the effect of a weaker DCMU inhibition of light-dependent Cl$^-$ uptake in the absence of CO_2 can be readily explained by an ATP-driven Cl$^-$ uptake mechanism. A still different argument which is possible would be that Cl$^-$ uptake itself can use up redox equivalents. The lower inhibition by DCMU in the absence of CO_2 then could simply be explained as the absence of CO_2 reduction as a competitive inhibitor.

Nevertheless, the lesser sensitivity to uncouplers of Cl$^-$ uptake in *Ni-*

tella as compared to *Elodea* suggests that in these plants different sources of energy are utilized for Cl^- uptake; and that in *Nitella*, unlike *Elodea*, a photophosphorylation-independent mechanism may be present.

In *Hydrodictyon africanum*, like *Nitella translucens*, an ATP-dependent cation pump and an ATP-independent anion pump can be distinguished. The ion transport systems of *Hydrodictyon africanum* have been very carefully investigated by Raven in numerous publications (Raven, 1967 to 1971; Raven et al., 1969; Raven and Glidewell, 1975a,b; Glidewell and Raven, 1975).

The cation transport mechanism is a K^+–Na^+ exchange pump, which is uncoupler-sensitive and also is inhibited by ouabain. Since the better-known animal ATPase systems are recognized by ouabain-sensitivity, one may assume that in *Hydrodictyon* we also deal with an ATPase-like mechanism (see Sect. 5.3.3). For light stimulation of this K^+–Na^+ exchange mechanism it is sufficient to excite photosystem I alone (see below). It can also be powered by ATP from pseudocyclic photophosphorylation.

The second ATP-independent mechanism largely serves for Cl^- transport into the cells. This is coupled with a certain K^+–Na^+ influx (the light-dependent Cl^- uptake rate is 1.5 pmol cm^{-2} s^{-1}; the Na^+ + K^+ uptake, 0.3 pmol $cm^{-2} s^{-1}$). This mechanism depends on the excitation of photosystem II (elimination of light under 705 nm or use of DCMU), and its sensitivity to the uncoupler Cl-CCP is lower than that of photosynthetic CO_2 fixation.

Particularly elegant are Raven's experiments with various light qualities. We have already seen how by this procedure participation of the phytochrome system can be excluded as a factor in ion regulation in *Hydrodictyon africanum* cells (Sect. 8.2.3.1). The action spectrum of the

o $\overset{\overset{\textstyle Cl^-}{\longrightarrow}}{\underset{\underset{\textstyle K^+, Na^+}{}}{\bigcirc}}$ i pump is identical with the action spectrum of photosynthesis.

The action spectrum of the o $\overset{\overset{\textstyle K^+}{\longrightarrow}}{\underset{\underset{\textstyle Na^+}{}}{\bigcirc}}$ i exchange pump, however, does

not have the characteristic red drop above 680 nm which is typical for photosynthesis (see Sect. 8.2.3.1); however, below 680 nm it corresponds quite well to the action spectrum of photosynthesis (Fig. 8.9). This suggests that the activity of this K^+–Na^+ exchange system needs only light reaction I of photosynthesis.

In view of the Emerson-enhancement effect (see Sect. 8.2.3.1) it becomes likely that the Cl^- and K^+–Na^+ uptake pump requires only the excitation of photosystem II. An Emerson effect has been found only for photosynthesis itself, but not for the Cl^- uptake pump. Hence this ion transport mechanism needs the activity of only one photosystem and this must be photosystem II because of its sensitivity to DCMU and light of 705 nm.

If this is correct, then here we have an important concept. The observation of a reduced Cl⁻ uptake during inhibition of photosystem II by DCMU or 705 nm does not allow us to decide whether photosystem II alone, or photosystem II plus I (i.e., noncyclic electron transport), is required for chloride uptake (see Fig. 8.6). In other investigations, where ATP-dependence could be excluded by use of an uncoupler, the latter possibility leads to the assumption of reduced pyridine nucleotides (NADPH + H⁺) as an energy source for active ion uptake. The absence of an Emerson effect on the light-dependent Cl⁻ uptake pump of *Hydrodictyon africanum* therefore is the first, and to the knowledge of the authors, so far, the only experimental finding suggesting a direct coupling of the photosynthesis-dependent ion transport with the electron excitation of photosystem II. (It may be noted though that a limited amount of photophosphorylation appears to be possible with photosystem II alone in a cyclic electron flow at photosystem II: Yocum 1977a,b; Yocum and Guikema, 1977).

Thus by a combination of the methods discussed in Sections 8.2.3.1 and 8.2.3.3 a considerable amount of evidence has been gained for *Hydrodictyon africanum* showing that photosynthesis can drive certain ion fluxes by the provision of ATP, whereas other ion fluxes are directly coupled to photosynthetic electron transfer. A light-dependent, ATP-independent Cl⁻ uptake mechanism has been found also in *Tolypella intricata* (Raven et al., 1969) but apparently does not occur in *Chara corallina* (Smith and West, 1969).

8.2.4.1.4 Angiosperm Water-Plant Leaves

Light-dependent Cl⁻ uptake by the leaf cells of angiosperm water plants apparently is not directly coupled to photosynthetic electron transport as is the case for Cl⁻ uptake by *Nitella* and still more clearly by *Hydrodictyon*. The action spectra of Cl⁻ and K⁺ uptake by *Hydrodictyon* and *Vallisneria* differ in the far-red region as discussed in Sections 8.2.3.1 and 8.2.4.1.3. In *Vallisneria* K⁺ but not Cl⁻ uptake shows a "red drop"; conversely in *Hydrodictyon* the Cl⁻ pump, but not the alkali cation exchange pump, obligatorily depends on excitation of photosystem II. In the preceding section we have seen that DCMU-inhibition of light-dependent Cl⁻ uptake by *Elodea* cells depends on the CO_2 content of the medium and suggests that here the ATP formed in cyclic photophosphorylation plays an important role as a source of energy. The results of experiments with uncouplers, and the action spectrum also do not indicate clearly that there is an ATP-independent ion uptake mechanism in *Elodea* in the light; rather they may be explained easily with the assumption of ATP or ~P as a source of energy (Jeschke, 1967, 1970a, 1971, 1972b,c). The situation in another well-investigated higher water plant, *Limnophila*, is similar (Penth and Weigl, 1969, 1971).

Naturally this does not mean that a light-dependent but ATP- or ~P-independent anion pump in higher water plants or even in higher plants, as a rule, does not occur. The number of species investigated thus far is much too small to allow such a generalization. As one can see in the algae, the nature of the uptake mechanism can be different between species.

8.2.4.1.5 Green Cells of Aerial Leaves of Higher Plants

Water plants under natural conditions are surrounded by an ion solution and they absorb with their entire surface. Although it is not so clearly evident at first glance, this is not entirely different from the cells of aerial leaves of higher plants. With the transpiration stream the leaves continuously obtain an ionic solution which can be distributed via the fine branches of the veins in the apoplast of the leaves. (In Sect. 12.1 we will describe this situation in more detail and discuss various models.) Thus, the conditions are quite similar to those of the cells of water plant leaves. Nevertheless it is extremely difficult to describe ion uptake by the cells of aerial leaves experimentally. While the external solution of water plants as a natural source of ions can easily be manipulated in its composition, the composition of the solution distributed in the aerial leaves by transpiration in the free space is dependent on too many factors to be varied experimentally without difficulty.

For this reason experiments with intact leaves are of limited value only. With leaf pieces which are floating on top of the solution little progress can be made because the diffusion pathways are too long. In addition, there are anatomical obstacles so that an equilibration of the apoplastic space with an external solution chosen by the experimenter is difficult or impossible. To a large extent, this problem can be overcome by preparing small leaf slices. With such leaf slices ion uptake and also photosynthesis can be investigated in an experimental medium just as with algae or water plants. The equilibration of apoplastic space depends on the geometry of the tissue slices. The diffusion problems become smaller with decreasing size of the tissue slices (Smith and Epstein, 1964; Osmond, 1968). If ion uptake is investigated by such a preparation a sharp maximum of the rate of ion uptake is obtained at a width of the slices of about 0.3 to 0.5 mm. The equilibration of the apoplastic space here is optimal and rates of ion uptake are obtained which correspond to those observed with water plant leaves or with root tissues. With increasing thinness of leaf slices too many cells are destroyed and one arrives very quickly at a width at which the percentage of intact cells is very low or no intact cells are present at all (Fig. 8.10).

The introduction of this method of preparation of leaf material has permitted a good deal of work investigating the nature of energetic coupling of ion uptake by cells of aerial leaves just as with algal cells or water plant leaves. Nevertheless, recent critical evaluations of the method raise a note of caution. When metabolic reactions in leaf slices are correlated with ion uptake by measuring photosynthetic and respiratory gas ex-

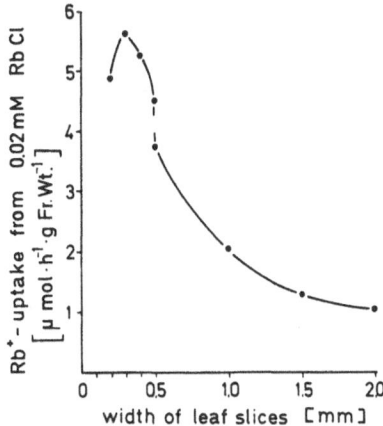

Fig. 8.10. Rates of ion uptake by leaf slices of varied widths. (Smith and Epstein, 1964.)

change, problems of gas diffusion may occur (MacDonald, 1975; MacDonald and Macklon, 1975; Ullrich-Eberius et al., 1976a; Macnicol, 1976).

Experiments with leaf slices using various wavelengths of light, gas compositions of the atmosphere, and inhibitors have usually shown a great versatility. Ion uptake mechanisms of leaf cells are able to utilize any possible energy source. It is generally found that any given energy source, if isolated in the appropriate way (e.g., Table 8.2), can drive ion uptake by leaf slices, and that cells can easily switch from one energy-providing pathway to another as experimental conditions change. Thus, the result of a large body of work in many laboratories is that uptake of at least the major inorganic ions (e.g., K^+, Na^+, Cl^-, phosphate) can be powered by a general ATP pool or "energy state" of the leaf cells (e.g., Johansen and Lüttge, 1974, 1975; MacDonald et al., 1975; Ullrich-Eberius et al., 1976b; Lüttge and Ball, 1976; reviews by Jeschke, 1976; Lüttge and Pitman, 1976).

This can be illustrated by experiments on K^+ and Cl^- uptake by slices of young (6-day-old) barley leaves in which the rates of respiratory and photosynthetic ATP production (as estimated from gas exchange measurements assuming P/O ratios of 3 for respiration and 1 for noncyclic photosynthetic electron transport) have been varied considerably (Fig. 8.11; Lüttge and Ball, 1973, 1976; Lüttge, Kramer et al., 1974). It is obvious that ion uptake is rather independent of the particular pathway and the rate of ATP production as long as experimental conditions allow any appreciable energy transfer at all. Limitations occur only in the dark $+N_2$ when the low ATP production by glycolysis is the sole energy source possible or in the light $+N_2$ when etiolated tissue incapable of photosynthesis is used (0 h greening time in Fig. 8.11). Thus, light stimulates ion uptake

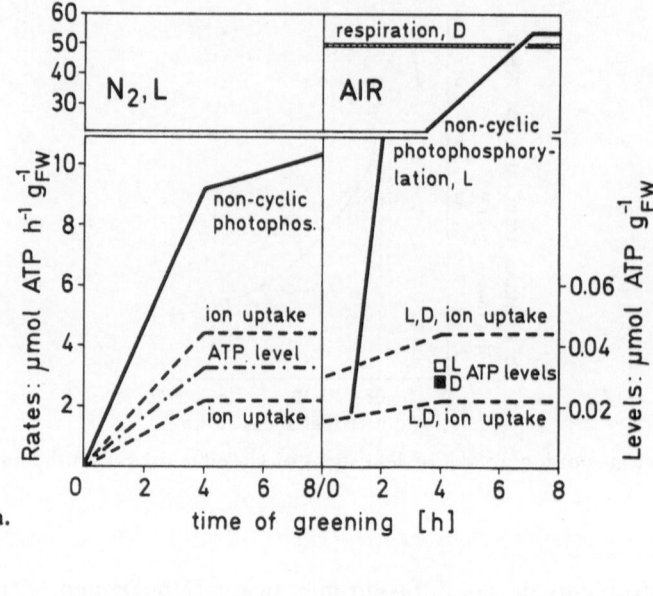

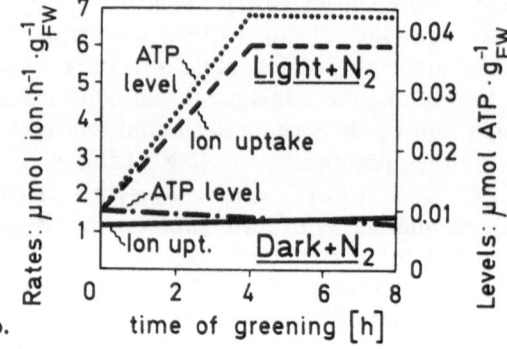

Fig. 8.11a and **b.** Rates of energy transfer, ion uptake and ATP levels in slices of 6-day-old barley leaves at various times during greening. ATP production rates (a) were estimated from measurements of O_2 exchange; two alternatives of ATP consumption by ion uptake are given in a assuming ½ and 1 ATP per univalent ion respectively. ATP levels and ion uptake in Dark $+N_2$ have been subtracted in a but are shown in b, where ion transport rates are not given in terms of μmol ATP h^{-1} g_{FW}^{-1} but in μeq ion h^{-1} g_{FW}^{-1}. L, light; D, dark. (Lüttge and Ball, 1976.)

only under anaerobic conditions. It can be seen, that ATP levels respond to the varied conditions in a way very similar to ion uptake or, conversely, that ion uptake is closely correlated with ATP levels of the cell. This becomes particularly obvious when ion uptake and ATP levels in the light $+N_2$ are compared in de-etiolating barley leaves, i.e., in dark grown leaves greening when illuminated ("time of greening" in Fig. 8.11). The kinetics of the increase of ATP levels and of the rate of ion uptake by leaf slices are very similar in relation to greening time of the barley plants from which the leaf slices are obtained.

This versatility appears to be fairly general in aerial leaves. Only in a few cases of plants with the C_4-metabolism of photosynthesis is a portion of the Cl^- uptake genuinely light-dependent. This Cl^- uptake appears not to be controlled in general by the supply of ATP or $\sim P$, but is more directly linked with noncyclic photosynthetic electron flow. Light-dependent Cl^- uptake is less uncoupler-sensitive than CO_2 fixation. In the C_4-plant *Atriplex spongiosa* detailed experiments have separated a light-

dependent $\text{o} \underset{K^+}{\overset{Cl^-}{\bigcirc}} \text{i}$ uptake mechanism from a $\text{o} \underset{Na^+}{\overset{K^+(H^+,Cl^-)}{\bigcirc}} \text{i}$ ex-

change mechanism. This closely resembles the situation in *Hydrodictyon africanum* (Sect. 8.2.4.1.3). The latter transport system is independent of light, i.e., it operates at the same rate in light and darkness, and it is inhibited by uncouplers. Probably this system is a K^+-Na^+ exchange mechanism in which K^+ uptake is accompanied by a Cl^- uptake and a small proton uptake, and the electrical balance is equalized by a sodium release (Lüttge, Pallaghy et al., 1970).

Ion uptake by chloroplast-containing stomatal guard cells (see Sect. 12.3.5) may have energetic relations different from those of the bulk of the green leaf cells. Jacoby et al. (1973; Jacoby, 1975) note that light stimulation of K^+ uptake by *Phaseolus* leaf preparations may be due to stomatal K^+ accumulation. Na^+ uptake is not affected by light in these leaves.

8.2.4.2 The Hexose Uptake System of *Chlorella* Cells

The energy-dependent sugar uptake system of *Chlorella* which has been investigated by Kandler, Komor, Tanner, and their coworkers has a number of remarkable properties (Tanner, 1969; Tanner and Kandler, 1967; Tanner et al., 1970; Komor and Tanner, 1971; Decker and Tanner, 1972; Komor, 1973; Komor et al., 1972, 1973; Komor and Tanner, 1974a,b; Haass and Tanner, 1974; Tanner et al., 1974).

This transport system is inducible (Sect. 5.3.2). It has a general specificity for hexoses. Sugar derivatives which cannot be metabolized, for instance 3-O-methylglucose, are also transported. This is important since during energy-dependent uptake of metabolizable substances it is usually extraordinarily difficult to find out whether the driving force is an energy-coupled transport mechanism or the use of the particular substrate in metabolism and therefore the continuous maintenance of a sink for passive uptake of that particular substance. Since metabolism cannot change the internal concentration of 3-O-methylglucose once taken up, it can be shown easily that this sugar derivative is accumulated against a large concentration gradient, i.e., it is actively transported in the strict thermodynamic sense.

In our consideration of photosynthesis as an energy source for transport mechanisms it is interesting to note that hexose uptake by *Chlorella* is enhanced in the light. In the absence of O_2 and CO_2 (pure N_2 atmosphere), i.e., during entire, or at least drastic, inhibition of respiration and of noncyclic electron transfer of photosynthesis, hexose uptake is highly stimulated by light. Hexose transport is far less sensitive to DCMU than photosynthetic O_2 evolution. Hence, the active hexose uptake system of *Chlorella* can be powered by energy from cyclic photosynthetic electron flow. However, ATP or $\sim$P does not appear to be the direct energy source of active hexose transport. It is very likely that the sugar transported is not phosphorylated during uptake. Komor and Tanner (1971) used a large number of hexose derivatives (such as 3-O-methylglucose and others) in which the phosphorylation by substitution at various C-atoms was prevented. By this procedure they showed that no specific carbon atom was required for phosphorylation of the sugars transported. Phosphorylation of a different C-atom in the transport of each derivative would imply a most unspecific phosphorylating system, and this is unlikely. Also ATP is not used for a carrier phosphorylation or "activation". Metabolic energy probably serves to maintain a pH gradient at the plasmalemma and hexose transport is coupled to H^+ transport in a cotransport (or symport) mechanism. This contributes to the evaluation of coupling mechanisms between energy transfer reactions and membrane transport which is discussed in Chapter 10.

Chapter 9

Phytochrome and Phytohormones Affecting Membrane Transport Mechanisms

Phytochrome and phytohormones are "signaling systems," which regulate many events in living plants. Most well known are their effects on morphogenesis; but also there are many interactions with physiological and biochemical processes, which have been extensively investigated. Many transport processes are associated with phytochrome- or phytohormone-regulated events. A few selected examples are discussed in this chapter.

9.1 Phytochrome

9.1.1 The Phytochrome System

Phytochrome is a biliprotein which occurs in two configurations, as P-660 (= P_R) and as P-730 (= P_{FR}). The P-660 form has a light absorption maximum at a wave length of 660 nm (red = R) and by the quanta absorbed is changed to the P-730 form. This has an absorption maximum at 730 nm (far-red = FR) and by the absorption of appropriate quanta is changed

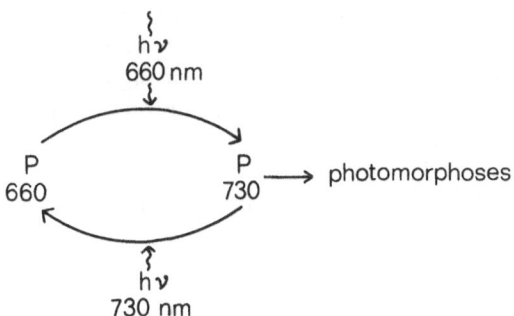

back to the P-660 form. P-730 is the active form of phytochrome which is responsible for triggering a large number of reactions, so-called photomorphoses (Mohr, 1969).

9.1.2 Phytochrome Association with Membranes

For many reasons, recently reviewed by Marmé (1977), it can be assumed that phytochrome in plant cells is localized within various membranes, or at least associated with them. Marmé classifies the approaches leading to this conclusion as follows:

> physiological localization
> spectrophotometric localization
> immunochemical localization, and
> cell fractionation

Some of the best evidence for an association of phytochrome in sterically ordered arrays with the plasmalemma comes from physiological analysis of phytochrome-dependent chloroplast movement in the cells of the filamentous green alga *Mougeotia*. Experiments with microbeams of red and far-red light show that the sensitivity resides in peripheral regions of the cylindrical cells of this alga. Experiments with polarized light of 730 and 660 nm show that P-730 molecules absorb light vibrating perpendicularly to the membrane surface and P-660 molecules absorb light oscillating in parallel with the membrane surface. The change in orientation of the molecules is perhaps brought about by sterical changes in the tetrapyrrole system of phytochrome (Fig. 9.1). Thus there is a close correlation between phytochrome and the plasmalemma membrane (Haupt, 1968, 1970a,b; Haupt et al., 1969).

More recently a series of papers on cell fractionations have shown that in homogenates phytochrome is bound to a particulate fraction which consists of small membrane-bound vesicles (microsomes) derived from the plasmalemma or rough ER or both. Irradiation with red light (R) increases binding when the intact tissue is irradiated but not when irradiation is given after extraction (Grombein et al., 1975; Marmé and Schäfer, 1972; Marmé et al., 1973; Quail et al., 1973; Boisard et al., 1974; Marmé et al., 1974; Quail, 1974a,b; Quail and Schäfer, 1974; Williamson et al., 1975; Yu, 1975a,b). Some care is needed, however, in interpreting these results on particle-bound phytochrome in terms of pigment-membrane interactions. Phytochrome can bind to ribonucleoprotein, and this binding may be an artifact occurring during homogenization when phytochrome and ribonucleoproteins are brought into contact. P_{FR} binds to ribonucleoprotein more readily than P_R. Since ribosomes preferentially absorb basic proteins, this may reflect changed surface properties of the phytochrome protein due to photoconversion (Quail, 1975a,b).

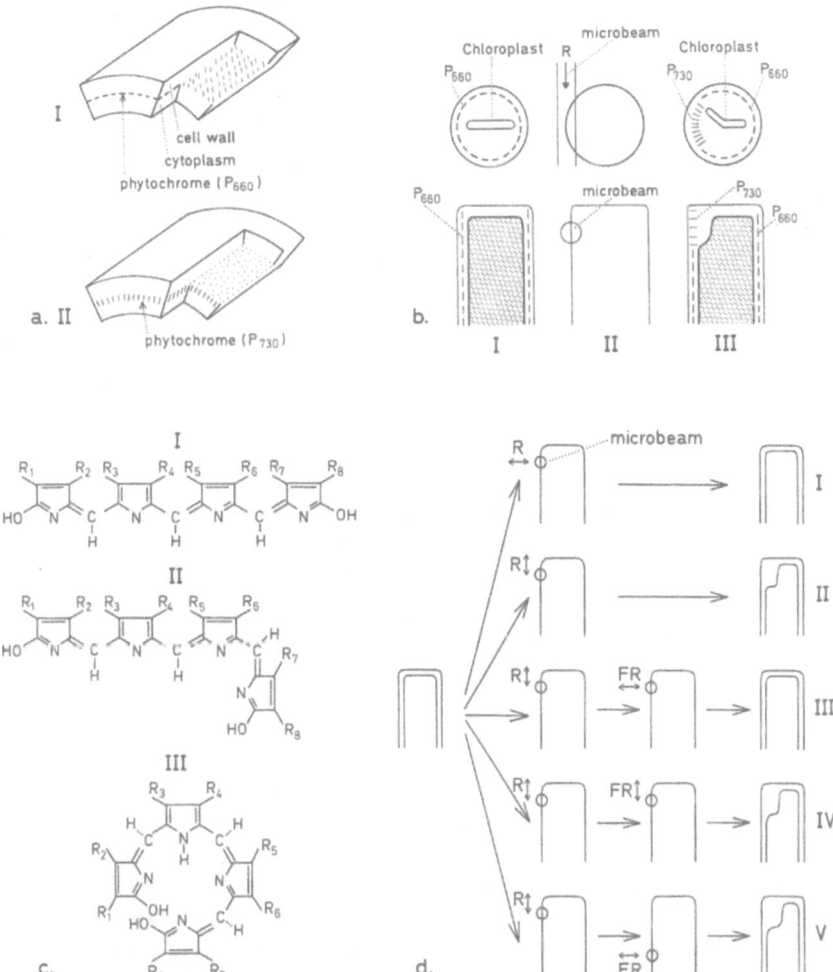

Fig. 9.1a–d. a Part of a cylindrical *Mougeotia* cell with phytochrome P-660 and P-730 molecules absorbing polarized light oscillating in parallel and normal to the surface, respectively. (Haupt, 1970b.) **b** schemes of cross sections and surface views of *Mougeotia* cells I before, II during and III after irradiation with a microbeam of red light (*R*). The localized chloroplast movement and the changes in dichroic orientation of phytochrome molecules are shown in III as compared with I. (Haupt et al., 1969.) **c** possible configurations of the 4 pyrrole rings in phytochrome. (Haupt, 1968.) **d** experiments with microbeams of polarized red (*R*) and far red (*FR*) light, where double arrows indicate the electrical vector. *I*, wrong direction of electrical vector of R: no reaction; *II*, right direction of electrical vector of R: reaction; *III*, right direction of R followed by right direction of FR: no reaction; *IV*, right direction of R followed by wrong direction of FR: reaction; *V*, right direction of R followed by right direction of FR but at wrong location: reaction. (Haupt et al., 1969.)

9.1.3 Physiological Effects of Phytochrome Related to Membranes

As pointed out by Marmé (1977) one must distinguish between physiological observations suggesting localization of phytochrome at or within membranes, and physiological effects demonstrating the action of phytochrome on membranes. The latter may or may not be due to association of phytochrome with membranes. Physiological effects of phytochrome related to membranes are of great interest to transport physiology. Some examples are discussed.

9.1.3.1 The Phytochrome-Dependent Photoelectric Membrane Effect

Light possibly can affect membrane transport in different ways:

1. by affecting the membrane directly; i.e., by interaction with some membrane compound(s) regulating membrane parameters such as permeability,
2. by affecting the photosynthetic apparatus; i.e., by changing the availability of metabolic energy for transport or by changing proton electrochemical gradients, or both.

In this section we will consider the first possibility. The second point is discussed in Section 8.2 and Section 10.3 and will be taken up again.

In 1937–38, L. and M. Brauner described changes of electrical properties of parchment-paper model membranes brought about by light. Results of a large number of physiological transport investigations also suggest that light can directly affect membrane properties. Such an action of light becomes particularly clear when a sudden illumination, or a sudden darkening, causes the membrane potential to move away from its resting level. A photoelectric effect whose duration is in the order of a few minutes up to about one hour is well known in green cells and is dependent on an action of light via the photosynthetic apparatus (Sect. 10.3.1.2). In addition a very rapid photoelectric effect was found in *Acetabularia* whose duration is on the order of milliseconds to a second, and which is not dependent on absorption by chlorophyll (Schilde, 1968; Fig. 9.2). Weisenseel and Ruppert (1977) observed that phytochrome and Ca^{2+} are involved in light-induced depolarization of membrane potential in *Nitella*. Red light (R) depolarizes, red light plus far-red light (R + FR) shows a reduced depolarization, and far-red light after red light (R $\rightarrow$ FR) enhances the repolarization. These authors, however, could not entirely exclude possible effects of photosynthesis. Löppert et al. (1978) inhibited photosynthesis of *Lemna* fronds (duckweeds) with 10^{-6} M DCMU (see Sect. 8.2.3.3) which prevented the membrane potential transients triggered by white and red light (R). In the presence of DCMU there was a small but

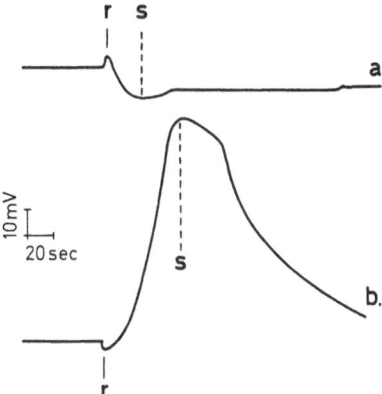

Fig. 9.2. Rapid (*r*) and slow (*s*) photoelectric effect after illumination (**a**) and darkening (**b**) of *Acetabularia* cells. (Schilde, 1968.) The action spectrum of the slow effect corresponds to that of photosynthesis.

clear hyperpolarization with far-red light (FR); red light (R) was only effective when irradiated after far-red light (FR) and led to a depolarization to the original level of the potential.

We know very few details about the molecular mechanisms by which light can change the properties of the membranes important for transport of particular solutes. Light must affect the molecular fine structure of the membrane; for this perhaps, photosynthesis can serve as a model in which molecular mechanisms of such changes have been intensively investigated. With the primary reactions of photosynthesis in the thylakoid membranes by absorption of light quanta, electrons are activated and transfers of electrons are triggered; redox reactions occur in the membrane which eventually lead to utilization of the absorbed-light energy for the formation of ATP and reduction equivalents. A spatially strictly determined arrangement of the light-absorbing pigment molecules, chlorophylls and accessory pigments, and the various redox systems in the thylakoid membranes is a prerequisite for the transfer of the energy from the light quanta to other forms of energy which can be utilized by the cells. With light absorption and excitation of electrons, and transfer of electrons, ion movements across thylakoid membranes are directly connected. These light-dependent ion fluxes underline the model character of the thylakoid system for the more general problem of light-affected membrane activities.

Although the molecular interactions of phytochrome at the membrane level are much less clear (Marmé, 1977) phytochrome-dependent bioelectrical effects indicate the possibility of a direct correlation between membrane transport and the phytochrome system. A phytochrome-dependent bioelectrical reaction (named the Tanada effect after its discoverer), supports this conclusion. The polarity of the electrical field at the tips of sec-

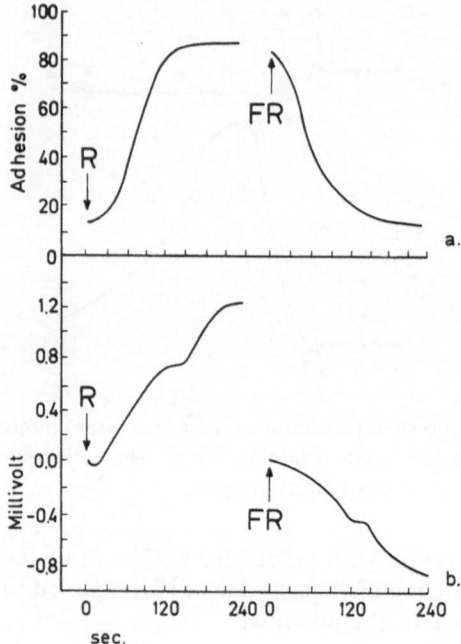

Fig. 9.3. Tanada effect; **a** phytochrome-controlled adhesion of tips of secondary roots of *Phaseolus aureus* to a negatively charged glass surface; **b** change of polarity of the electric field around the root tips after P-730 to P-660 transitions. *R*, irradiation with red light leading to P-730; *FR*, irradiation with far-red light leading to P-660. (Jaffe, 1968.)

ondary roots of *Phaseolus aureus* changes during P-660 → P-730 transitions in a reversible way. This can be directly seen by adhesion and release of root tips, respectively, from a negatively charged glass surface (Tanada, 1968; Jaffe, 1970; Junghans and Jaffe, 1972; Racusen and Miller, 1972; see Fig. 9.3).

Racusen and Etherton (1975) investigated the movement of single root cap cells of mungbeans in an electrical field. Changes in electrical charges of the cell surface were governed not only by the phytochrome system but also by manipulation of Ca^{2+} and H^+ concentrations in the medium. With plasmolyzed root cap cells these authors showed that the changes in surface charge are associated with the plasmalemma and not with the cell wall. When cell wall and plasmalemma were separated by plasmolysis in the electrical field protoplasts moved within their cell wall "envelopes." Racusen and Etherton concluded that protein conformational changes in the plasma membrane of root cap cells were the molecular mechanism of fixed charge alterations leading to the Tanada effect.

It had been suggested that acetylcholine, the membrane-active agent well known from animal systems, plays a role in the phytochrome-mediated Tanada effect (Jaffe, 1970; Jaffe and Thoma, 1973). However, an interference of acetylcholine with phytochrome reactions and membrane processes in plants has not been confirmed (Kasemir and Mohr, 1972; Satter et al., 1972; Marmé, 1977).

9.1.3.2 Phytochrome and Ion Fluxes: Nyctinastic Movements

The preceding sections have already shown that "photomorphoses" triggered by active phytochrome (P-730) may not be growth reactions only (formative reactions) but also purely physiological events such as the movements of the chloroplasts of *Mougeotia* cells and membrane electrical phenomena. Most interesting in the context of light-dependent membrane transport processes are rapid plant movements which are based on the regulation of turgor (turgor movements). The phytochrome-dependent nyctinastic movements of leaves (sleep movements of leguminous plants) and the photosynthesis-dependent opening and closing of stomata are among the more outstanding examples.

The phytochrome regulation of nyctinastic leaflet movements in *Mimosa* (Fondeville et al., 1967) and *Albizzia* (see references below, and in the figures) has been investigated intensively. These movements are caused by rapid changes of turgor in the cells of certain motor tissues (pulvini), in which rapid and drastic changes of membrane permeability play an important role. The movement is depicted in Figure 9.4. Darkness after irradiation with red light (phytochrome active) leads to a closure of leaflets, i.e., to sleep movement.

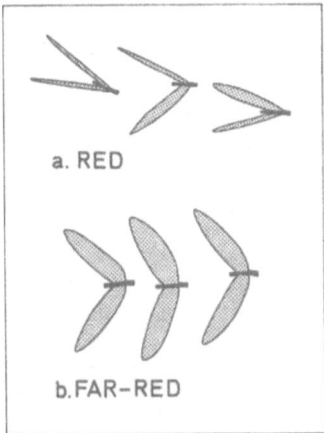

Fig. 9.4. Phytochrome-controlled leaflet movements of *Albizzia julibrissin*. Position of pairs of leaflets 20 min after a short irradiation of red light **a** and far-red light **b**, followed by darkness. After irradiation with red light phytochrome is present in its active form (P-730) the pinnule pairs perform a sleep movement. By irradiation with far-red light phytochrome is converted to its inactive form (P-660) and the pinnules in subsequent darkness remain in the open fully unfolded position. (Drawn after Fig. 23 in Hillman and Koukkari, 1967.)

Table 9.1. Sleep movements of pinnules of *Albizzia julibrissin* (see Fig. 9.4) and electrolyte efflux from the cut end of the rachis of excised pinnae (from Jaffe and Galston, 1977).

	Change in 30 min	
Light regime	Pinnule movement (degrees)[a]	Electrolyte efflux (micromhos)
Continuous white light	+ 3	2.61
Continuous darkness	−115	4.44
10 min FR light followed by continuous darkness	− 38	2.85
10 min FR light followed by 10 min R and then continuous darkness	−116	4.86

[a] Positive numbers refer to an opening, negative numbers to a closing (i.e., sleep movement of pinnules). FR, far red; R, red light.

An important criterion for the presence of a phytochrome mechanism is the reversibility of the action. The light presented last (see scheme on p. 235) determines the reaction occurring. Table 9.1 shows an example. Furthermore, this table shows that electrolyte efflux is associated with the nyctinastic closure movement of the leaflets.

The motor tissue of *Albizzia* is depicted in Figure 9.5. From this figure it becomes clear that the upper and the lower face of the pulvinus must react to the same irradiation in a different way. The upper face becomes shorter during the closing movement, the lower face becomes longer; during an opening movement the opposite occurs. Indeed, one can observe in the microscope that the cells of one side become turgescent and the cells on the opposite side shrink (Satter, Schrempf et al., 1977). By aid of an X-ray microanalysis technique it was shown that potassium transport within the pulvinus decisively takes part in this turgor regulation. Table 9.2 shows that the potassium transport during the closure move-

Table 9.2. K^+ transport between ventral and dorsal parts of pulvini during the sleep movements of *Albizzia julibrissin* pinnules (after Satter, Marinoff et al., 1970).

	Pinnule closing movement (degrees)	K^+-content rel. units		
Light regime		Ventral part of pulvini	Dorsal part of pulvini	Ventral minus dorsal
Initial position in white light	—	156 ± 29	135 ± 54	+ 21
After 100 min in white light	0	140 ± 31	134 ± 36	+ 6
4 min R, 80 min D	160	104 ± 40	218 ± 48	−114

Key: R, red light; D, darkness.

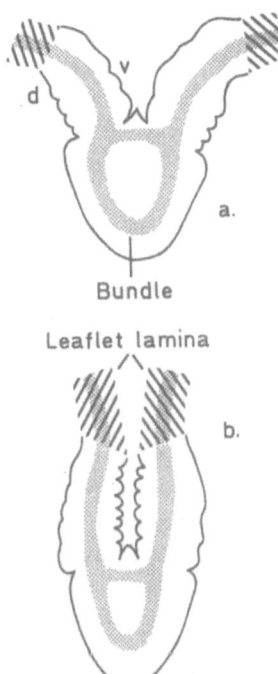

Bundle

Leaflet lamina

Fig. 9.5. Pulvini of *Albizzia julibrissin* in section; **a** pinnule in the open position (day position). Cells of the upper, adaxial or ventral face of the pulvinus (*v*) are turgescent, cells of the lower, abaxial or dorsal face (*d*) are shrunken; **b** pinnule in the closed position ("sleep" position); *v*, shrunken; *d*, turgescent. (After Satter, Sabnis et al., 1970b) ×36.

ment occurs from the upper side of the pulvinus to its lower side. However, K^+ transport between the two sides is not sufficient and an extra supply from adjacent tissues, functioning as a K^+ reservoir, must be assumed. Chloride ions nearly balance the charge of moving K^+ ions but some other ions may be involved in addition (Schrempf et al., 1976; Satter, Schrempf et al., 1977). Perhaps H^+ ions also are transported in the system. In *Avena* coleoptiles Pike and Richardson (1977) have observed phytochrome-controlled H^+ excretion. Also it is interesting in this context to note that in the nonmotor tissue of bean epicotyls red light inhibits the uptake of phosphate and K^+ and that far-red light releases this inhibition (Tezuka and Yamamoto, 1975).

The seismonastic and nyctinastic movement reactions of the leaflets of *Mimosa pudica* involve shrinkable vacuoles of the pulvinus cells. The "contraction" of the vacuole during nyctinasty is under phytochrome control. During stimulation of movement Ca^{2+} ions move into the central vacuole and they displace K^+ flowing out, probably acting as an osmotic agent, and water thus is lost by the cells of the motor tissue into the adjacent tissue. This brings about the very rapid abaxial loss of turgor and

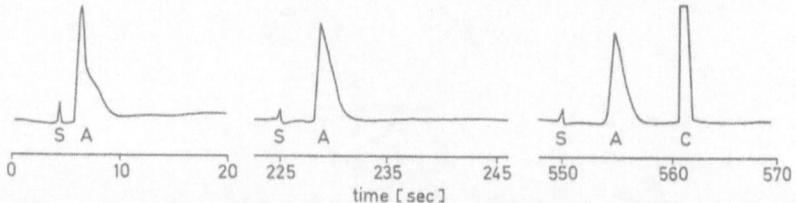

Fig. 9.6. Action potentials (*A*) of *Mimosa pudica* after electrical stimuli (*S*). The potential was measured externally on the primary leaf petiole. *C* is a calibration mark of 0.1 volt. (Experiment by K. Umrath drawn after Fig. 99 of Bünning, 1939.)

therefore the leaf movement (Toriyama and Jaffe, 1972; Setty and Jaffe, 1972).

Racusen and Satter (1975) recently observed rhythmic phytochrome-regulated changes in the transmembrane potentials of *Samanea* pulvini. It is well known since the measurements of Umrath in the early 1930's that the pulvini movements in Leguminosae species, e.g. in *Mimosa*, are associated with action potentials (Fig. 9.6). A more detailed analysis of the sequence of events in action potentials in plant cells is available for the giant internodal coenocytes of *Chara* (Sect. 2.2.2.5; Fig. 2.10). In the present context it is interesting to note that turgor pressure changes seem to accompany the changes in ion movements during action potentials, and that concomitantly volume changes occur even in the *Chara* cells (Oda and Linstead, 1975).

Leaf positions of Leguminosae species are well known to oscillate also in the form of circadian endogenous rhythms (Bünning, 1973). Phytochrome interacts with the rhythmic oscillator of these endogenous

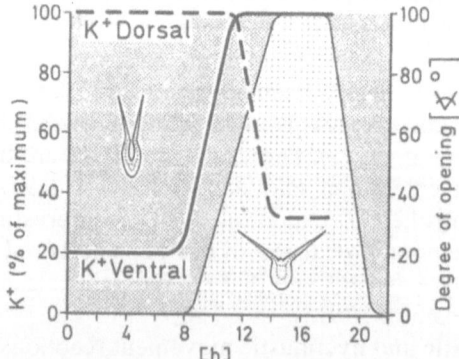

Fig. 9.7. Rhythmic movement and K^+ flux between dorsal (=abaxial) and ventral (=adaxial) motor cells of darkened leaflets exposed to a brief FR-illumination at 0 or at 8 h. *Dashed curve*, K^+ in dorsal; *solid curve*, K^+ in ventral motor tissue. *Heavily shaded area*, closed position; *lightly shaded area*, open position of leaflets. (After Fig. 1 of Satter, Applewhite et al., 1974.)

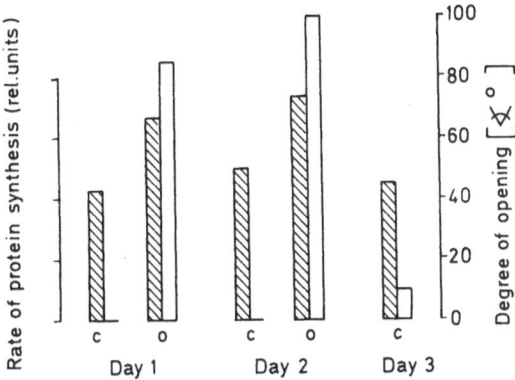

Fig. 9.8. Rhythm of the rate of protein synthesis (*left shaded bars*) and leaflet opening (*right open bars*) of *Albizzia julibrissin*. Protein synthesis was measured by following rates of ^{14}C-leucine incorporation. *c*, closed; *o*, open position of leaflets. (After Fig. 1 of Applewhite et al., 1973.)

movements in *Albizzia julibrissin* and also in *Samanea saman*. In a number of investigations of these phenomena Galston and Satter and coworkers have gained further insight into the role of K$^+$ transport during the swelling and shrinking of motor tissue cells (Satter and Galston, 1973; Applewhite et al., 1973; Satter, Applewhite et al., 1973; Satter, Applewhite, and Galston, 1974; Satter, Geballe et al., 1974a,b). Opening of the leaflets appears to be the active part of the rhythm. Opening results from an energy-dependent K$^+$ pumping into the ventral motor cells and out of the dorsal motor cells. Closing is passive and is due to K$^+$ leakage in the opposite direction. The interaction between the two processes results in the observed rhythmic oscillation (Fig. 9.7). The rhythm of active K$^+$ pumping is accompanied by a rhythm of cycloheximide-sensitive protein synthesis (Fig. 9.8). This suggests that a protein in the membranes of ventral motor cells is part of a molecular entity providing the basis for a rhythmically active K$^+$-pump. Jose (1977) observed phytochrome modulation of ATPase activity in a membrane fraction from *Phaseolus*. Gradmann and Mayer (1977; Mayer, 1977; Gradmann, 1977), however, do not believe in a dominating role of K$^+$-pumping in turgor movements. Due to the generally high K$^+$-permeability of plant cell membranes they consider K$^+$-pumping as much too inefficient and prefer Cl$^-$-pumping, which is supported by experiments with pulvini of *Phaseolus coccineus*.

9.1.3.3 Water Permeability

Investigations of the effects of phytochrome on the kinetics of plasmolysis and deplasmolysis of *Mougeotia* cells indicate that P-730 increases the hydraulic conductivity of the cell membranes (L$_p$, see Sect. 2.2.3.5). After irradiation with red light (R) the lag phase between the onset of plas-

Table 9.3. Effects of R and FR irradiation on the lag phase (s) between the change of osmotic pressure and the onset of plasmolysis or deplasmolysis of *Mougeotia* cells (after Weisenseel and Smeibidl, 1973). Errors are standard errors.

| | Plasmolysis (s) | | Deplasmolysis (s) |
| | 0.5 M | 1.0 M | |
Irradiation	mannitol		
5 min R	13.7 ± 0.64	3.5 ± 0.14	19.7 ± 0.79
5 min FR	33.9 ± 1.72		30.0 ± 0.36
5 min R + 10 min FR	27.7 ± 1.16	7.3 ± 0.22	28.7 ± 0.87

molysis or deplasmolysis and the change of the osmotic pressure of the medium is much reduced, and this effect is reversed by irradiation with far-red light (FR); (Table 9.3; Weisenseel and Smeibidl, 1973). Conversely, Pike (1976) found that phytochrome has no effect on permeability to tritiated water in the etiolated tissues of several species. Whether this may mean that the permeability change is restricted to green cells is not clear; however, etiolated seedling tissues certainly have phytochrome.

9.2 Indole Acetic Acid (IAA)

9.2.1 The IAA-Dependent H^+–K^+ Exchange Pump of *Avena* Coleoptiles

The growth regulator β-indole acetic acid (IAA) affects the structure and viscosity of plant cell membranes (Morré and Bracker, 1976; Helgerson et al., 1976) and interacts with ion transport in plant tissues. Under certain circumstances, cation and anion uptake are selectively affected (Higinbotham, Latimer et al., 1953; Higinbotham, Pratt et al., 1962; Lüttge, Higinbotham, and Pallaghy, 1972; Haschke and Lüttge, 1973; Bentrup et al., 1973; Wanless et al., 1973; James et al., 1977).

Obviously proton concentrations in the cell wall play a decisive role in regulation of cell wall plasticity which is important for the extension growth of cells. Based on intensive investigations, Hager et al. (1971) arrived at the assumption of an IAA-regulated ATP-dependent proton extrusion mechanism. IAA, according to this hypothesis, acts as an activator of a proton pump localized in the cell membrane (Fig. 9.9). The pump entity presumably can be separated from the membrane by osmotic shock. Recent investigations by Rubinstein and coworkers (Rubinstein, 1977; Rubinstein et al., 1977) show that osmotic shock very selectively affects the plasmalemma of *Avena* coleoptile cells. Osmotic shock eliminates growth and auxin-stimulated H^+ efflux, it inhibits uptake of α-amino-isobutyric-acid, 3-O-methylglucose, and leucine by 74%–90%

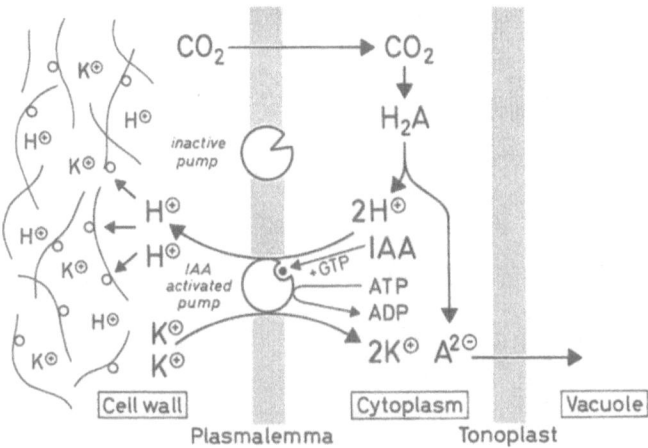

Fig. 9.9. Regulation of an H^+-K^+ exchange pump by IAA as the primary action of the growth regulator on extension growth. (After Hager et al., 1971; incorporating the findings of Haschke and Lüttge, 1973, 1975a,b, 1977a,b; see also Figs. 6.6 and 10.4.)

and Cl^- uptake by 30%; however, protein synthesis is inhibited by only 11% and respiration and polar auxin transport in the coleoptiles are not affected.

The hypothesis of Hager et al. raises some questions, regarding the stoichiometry of the pump mechanism:

1. which is the counterion of the H^+ pumped out of the cells?
2. how is intracellular pH regulated during continuous H^+ extrusion?

The answers to these questions are provided by studies of ionic charge balance during IAA-stimulated extension growth (Haschke and Lüttge, 1973, 1975a,b). It turns out that K^+ is taken up during H^+ release and that malic acid is synthesized by CO_2 dark fixation. Extension growth is stimulated by K^+ in the outer medium of *Avena* coleoptile segments and it is accompanied by K^+ and malate accumulation in the tissue (Fig. 9.10). The stoichiometry of the IAA dependent H^+-K^+ exchange mechanism (Fig. 9.9) is

½ molecule of malic acid synthesized;
1 H^+ pumped out;
1 K^+ taken up;
½ dicarboxylic malate anion accumulated together with the K^+ taken up.

It should be mentioned in passing that a similar mechanism with K^+ and malate accumulation may be involved in elongation growth of cotton fibers (Dhinsda et al., 1975).

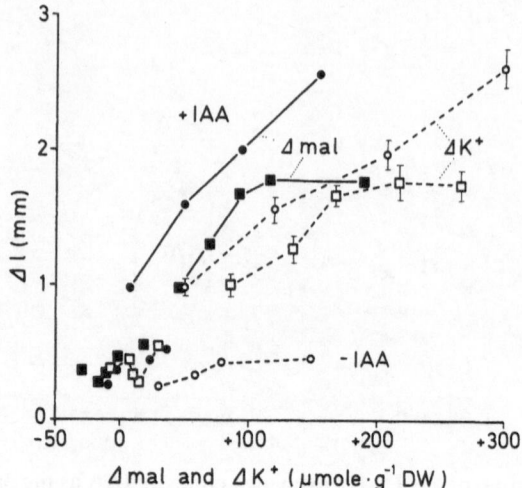

Fig. 9.10. Changes in malate and K⁺ levels (Δ mal and Δ K⁺) in IAA-induced extension growth (changes in length = Δl) of 10-mm long segments of *Avena* coleoptiles. (After two experiments, i.e., *squares* and *circles* respectively, of Haschke and Lüttge, 1975a and b.) *Open symbols,* Δ K⁺; *closed symbols,* Δ mal.

The enzyme most important in malate synthesis by CO_2 dark fixation is phosphoenol pyruvate carboxylase. This enzyme is active in oat coleoptiles. IAA has no direct effect on the enzyme in vitro. In vivo IAA-stimulated H^+ efflux could indirectly lead to increased malic acid synthesis due to the alkaline pH optimum (pH 8) of phosphoenol pyruvate carboxylase (Dymock et al., 1977). Thus enhanced acid synthesis during IAA-stimulated H^+–K^+ exchange provides both protons to be extruded and organic acid anions to balance the charge of K^+ taken up without the intracellular H^+ concentration being affected. In this way malic acid synthesis functions as in a more general mechanism of homeostatic pH control proposed by Raven and Smith (1974) and Smith and Raven (1976).

The original model of Hager et al. (1971) implies that ATP, or other nucleoside triphosphates, provides the energy for active H^+ extrusion. The uncoupler DNP, the senescence hormone ABA (see Sect. 9.4), the potassium ionophore valinomycin, and the inhibitor of protein synthesis cycloheximide (see Sect. 12.2.5) inhibit IAA-induced growth (Rayle, 1973). Kasamo and Yamaki (1973) have shown recently that IAA stimulates the activity of an Mg^{2+}-dependent ATPase in a mungbean hypocotyl fraction enriched with plasmalemma fragments. This stimulation results from an IAA effect on enzymes already present and is not due to new ATPase synthesis.

A conflicting result was reported by Penny et al. (1975) who introduced pH microelectrodes into xylem vessels of *Avena* coleoptiles and *Lupinus* hypocotyls and observed no pH change in this apoplastic space before the onset of IAA-induced growth. This work has been criticized by Cleland (1975), and indeed rapid apoplastic acidification after application of IAA has been observed in various laboratories with normal glass electrodes and with micro-pH electrodes (Cleland, 1976; Jacobs and Ray, 1976). Vanderhoef et al. (1977a,b) also have reservations about the acid growth hypothesis. They suggest there is a distinction between two biochemically different phases of growth. Only the first one, an initial transient response to auxin, shows the characteristics of acid growth. Pope (1978) observed synergistic effects of IAA and acid pH and concludes that IAA-induced acidification "is not involved in regulating the rate of growth, but is the result of a separate direct or indirect action of auxin." Ray (1977) found that the auxin-binding sites of maize coleoptiles are localized on the membranes of the rough endoplasmic reticulum. His modified acid growth hypothesis suggests an IAA-stimulated H^+ uptake into the cisternae of the endoplasmic reticulum and a granulocrine extrusion via Golgi vesicles (cf. Sect. 7.1.1).

The whole field at present is in rapid development and very difficult to review briefly. Nevertheless, it is anticipated that the basic features of the acid growth hypothesis and the IAA-stimulated H^+-K^+ exchange mechanism as depicted in Figure 9.9 will prove useful in the future and survive forthcoming developments.

9.2.2 Bioelectric Effects of IAA

IAA also plays a role in tropistic reactions of plant roots or shoots induced by light or by gravity stimuli. On one side of an organ an increased growth is triggered leading to a movement and to a specific arrangement of the organ with respect to the gradient of the stimulus. During this process, asymmetric distributions of growth regulator can be observed.

Tropistic reactions are associated with electrical phenomena. Electric potential differences across entire plant organs are generally observed following gravity stimuli (Grahm and Hertz, 1962; Brauner and Diemer, 1967). For an entire organ like a root or a coleoptile this geoelectric effect may range from 30–70 mV; for individual cells 2 mV have been recorded (Fig. 9.11; Etherton and Dedolph, 1972). Conversely, the growth reaction can be triggered by an external electric field (Fig. 9.11; Brauner and Bünning, 1930). IAA affects electric fields along plant organs, such as roots and coleoptiles, and electric currents passing through these organs (review: Scott, 1967). Lateral electrical potentials are induced in maize coleoptiles by asymmetric auxin application (Morath and Hertel, 1978). These electrical phenomena cannot be explained without assuming a contribution of ion regulation at membranes.

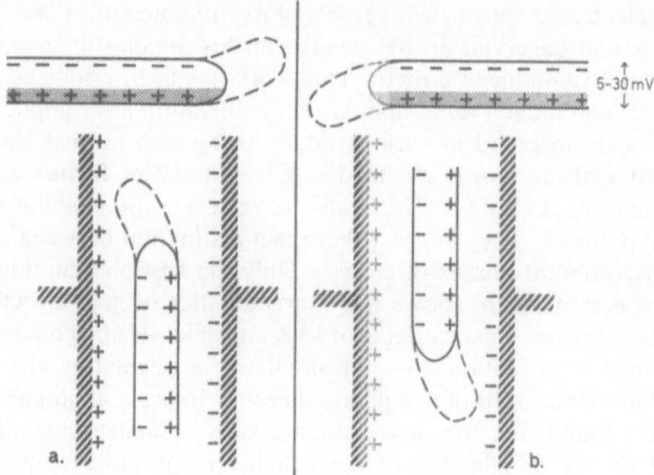

Fig. 9.11. Geoelectrical effect. In the earth's field of gravity the lower side of the organs becomes positive as compared with the upper side. **a** coleoptiles; **b** roots; the two different organs show different reactions in the field of gravity and also in an electrical field. IAA enrichment is indicated by *heavier shading*. (In part after Brauner and Bünning, 1930.)

There is a voluminous literature on binding of IAA to the plasmalemma and other membrane fractions of plant cells (e.g., Batt and Wilkins, 1976; Batt and Venis, 1976). Until very recently cell membrane electric potentials measured with intracellular glass microelectrodes did not appear to be affected by IAA unless excessive concentrations were used (Lüttge, Higinbotham, and Pallaghy, 1972; Nelles, 1975a). Only occasionally has a hyperpolarizing effect been reported (Marrè, Lado et al., 1974a,b,c); effects observed in 1970 by Etherton were quite small. However, in corn coleoptiles IAA affects relative membrane permeabilities of Na^+ and K^+. In long-term experiments (8–24 h) with *Avena* coleoptile, IAA was found to cause depolarization of the cells, while K^+ exchange and accumulation were enhanced (James et al., 1977). IAA interacts with Ca^{2+}-dependent K^+-induced depolarization of membrane potential. It also prevents small depolarizations caused by sucrose (Nelles, 1975a,b; Nelles and Müller, 1975a,b). Differences in the behavior of membrane potentials in coleoptiles of dwarf and normal mutants of maize are eliminated by treatment of coleoptile segments obtained from dwarf mutants with IAA (and gibberellic acid; Nelles, 1976a,b,c; Neumann and Jánossy, 1977).

Most recently Cleland et al. (1977) have reported a rapid hyperpolarization of the oat coleoptile transmembrane potential by IAA and fusicoccin in short term experiments (~ 1h; see Sect. 9.3) due to stimulated H^+ extrusion.

9.3 Fusicoccin (FC)

Recently the effects on transport of a fungal toxin, fusicoccin (FC), isolated from *Fusicoccum amygdali,* has received a great deal of study (for
reviews see Marrè and Ciferri, 1977). Fusicoccin induces K^+-dependent
H^+ excretion with concomitant malate accumulation and H^+-dependent
elongation growth just as observed with IAA. However, FC acts not only
in coleoptiles and young shoot axes but also in many other plant tissues,
most interestingly including roots (Lado et al., 1974, 1976a,b; Marrè, Colombo et al., 1974; Marrè, Lado et al., 1974a,b; Marrè, Lado et al.,
1975; Pitman et al., 1975a,b; review: van Steveninck, 1976a). FC significantly hyperpolarizes the membrane potential, and probably the FC-
induced H^+ excretion is electrogenic. This has led to an interesting argument about the nature of the coupling between K^+ uptake and H^+ release
in the H^+-K^+ exchange mechanism and the mode of action of FC as compared with IAA. While Pitman et al. (1975a) prefer the concept of an electrogenic coupling mechanism between active H^+ excretion and passive
K^+ uptake, both Marrè's group and Cleland assume a chemical coupling
at a membrane carrier system or ATPase. Cocucci et al. (1976) suggest
that such a mechanism may slightly favor H^+ release over K^+ uptake, thus
explaining the hyperpolarization of the membrane potential. Lado et al.
(1976a,b) feel that FC acts directly on plasmalemma receptor sites, while
IAA acts via protein synthesis having primary receptor sites at the ER.
Conversely, Cleland (1976) speculates that auxin activates an electrogenic
membrane ATPase and that FC stimulates one or more steps in malic acid
synthesis of the pH-stat mechanism. Like IAA, fusicoccin does not seem
to act on enzymes directly, but affects metabolism indirectly via its stimulation of H^+ transport (Marrè, 1977; Stout et al., 1978; Stout and Cleland,
1978).

Thus at the moment the use of FC proves to be an important new tool
for the study of hormone effects on membrane transport mechanisms; at
the same time it injects new uncertainties.

9.4 Abscisic Acid (ABA)

The "senescence" hormone of plants, abscisic acid (ABA), which leads to
leaf fall of trees, has manifold effects on membrane transport processes. It
increases water permeability of root cells (Glinka and Reinhold, 1971;
Collins, 1974). It affects the K^+/Na^+ selectivity of beet root tissue (van
Steveninck, 1972). Most widely known are its effects of antagonizing the
stimulating hormones such as cytokinins, IAA, and gibberellins. This also
reflects on transport. For example in *Helianthus* epicotyls IAA or gibberellic acid causes an increase, ABA a decrease, of K^+ and phosphate

levels. Given together with the stimulating hormones ABA exerts an antagonistic effect on ion levels (Dörffling et al., 1973a,b). In the explanation of such phenomena, however, there always remains an unresolved ambivalence. One cannot show unequivocally where the causal chain begins, and whether the membrane effects indeed are primary or only secondary consequences of growth. In this context hormonally regulated transport processes which are not associated with growth processes are of particular interest.

Such a mechanism is represented by the H^+-K^+ exchange mechanism of stomatal guard cells, which regulate the width of stomata by aid of a turgor mechanism. K^+ influx and efflux from the guard cells appears to be required for the regulation of the osmotic pressure (π) in the cells which leads to osmotic water flows and the turgor pressure changes for this movement [see Eqs. (2.7) and (2.10); see also Sect. 12.3.5]. Various hormones affect the stomatal opening and closing (e.g., Tal and Imber, 1970; Tal et al., 1970). ABA inhibits K^+ uptake into the guard cells which is required for the opening movement (Jones and Mansfield, 1970, 1972; Cummins et al., 1971; Horton and Moran, 1972; Cooper et al., 1972; Kriedemann et al., 1972).

With its effects on stomatal regulation and a selective effect on ion transport across the roots ABA is probably a key messenger in regulation of transport in the plant as a whole (see Chap. 13).

9.5 Other Phytohormones and Membrane Active Agents

The effects of the phytohormones IAA and ABA and of the phytotoxin FC have been discussed in detail to exemplify the action of such agents on membrane transport. Most of the other phytohormones known, e.g., a gibberellins, kinins (see also Sect. 12.2.5), ethylene, also affect membrane transport and permeabilities (review: van Steveninck, 1976a). In addition a fair number of membrane active compounds can be listed which also interact with membranes and transport, such as sterols, polylysines and histones, ionophores, various phytotoxins, etc. (review: van Steveninck, 1976a).

9.6 "Aging"

Isolated storage tissue, and also root tissue, directly after preparation takes up ions from an external medium slowly and only after 10 to 20 h are higher ion uptake rates attained (Fig. 9.12). The Na^+-K^+ selectivity of tissue can also change after slicing (Rains and Floyd, 1970; Floyd and Rains, 1971; Poole, 1971a,b). This phenomenon has been poorly named as

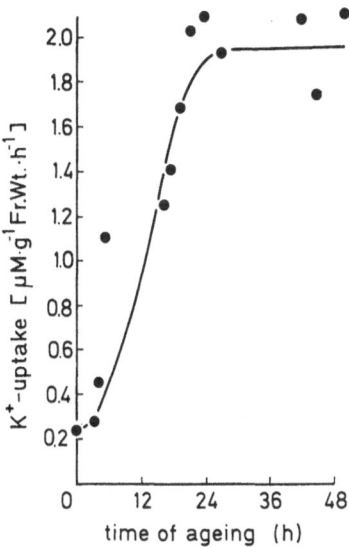

Fig. 9.12. Changes in the rate of K$^+$ uptake from 0.1 mM KCl by isolated maize root steles during "adaptive aging." (Lüttge and Laties, 1967).

"aging." During "aging" not only are ion transport processes altered, but also a vast number of physiological and biochemical functions are subject to dramatic changes (review: van Steveninck, 1976b). This has led to the delicate question of whether freshly prepared or "aged" tissue preparations more closely represent the situation in situ, i.e., in the intact plant. From his early investigations on the biochemistry of potato tuber slices, Laties (review: Laties, 1975) found good reasons to assume that freshly cut tissue has more of the properties of the tissue in situ than aged tissue. The same was assumed for isolated root stele (Lüttge and Laties, 1967; see Fig. 9.12). Thin leaf slices of aerial leaves of higher plants (Sect. 8.2.4.1.5) also increase in ion uptake rate during "aging," although not as pronouncedly so as storage tissue slices and root steles. A thorough analysis has confirmed that freshly cut leaf slices regarding both ion transport (Pitman, Lüttge et al., 1974a,b) and energy metabolism (Ullrich-Eberius et al., 1976a,b) more closely resemble the tissue in situ than aged slices.

Van Steveninck (1976b) has considerably improved the terminology by coining the term "adaptive aging" for the kind of processes discussed above in contrast to "senescent aging" describing true senescence. In both cases, aging is accompanied by considerable changes of metabolism and of membrane transport. Phytohormones interact at both levels.

In a discussion of membrane transport it is interesting to pick out three examples, from the many phenomena reviewed by van Steveninck (1976b), in order to show the interactions between adaptive aging and membranes: (1) Jackman and van Steveninck (1967) showed that practi-

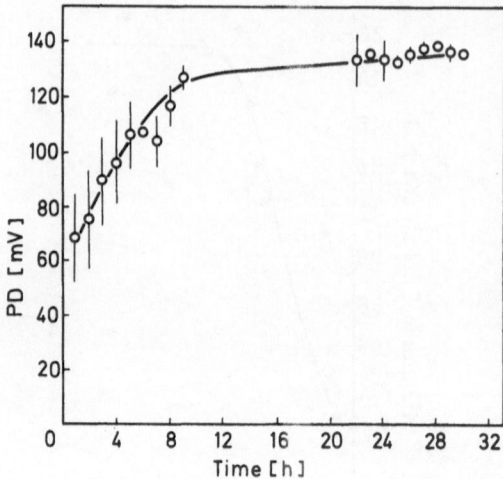

Fig. 9.13. The increase in cell electropotential following excision of 1-cm long pea epicotyl segments. Errors are the standard deviation. (Higinbotham and Pierce, 1974.)

cally all ER had disappeared from beetroot cells after slicing of the tissue and an ER membrane system was developed concomitantly with the appearance of salt uptake capacity during aging. (2) Hanson et al. (1973) found, although not consistently, that the tonoplast showed increased electron density in electron micrographs after aging. (3) The electrogenic membrane potential of root and pea epicotyl segments seems to increase during aging (Macklon and Higinbotham, 1968; Pitman, Mertz et al., 1971; Lin and Hanson, 1974; Fig. 9.13).

9.7 A General Problem: Membrane Effects as Primary Actions of Phytochrome and Phytohormones?

Phytochrome and phytohormones could exert an effect on membrane transport in three different ways by:

1. regulating enzymes;
2. acting directly on membranes; and
3. affecting energy metabolism and the availability of metabolic energy for transport.

The first possibility has long dominated as an explanation of the molecular level of phytochrome and hormone effects in general. Various ways in which phytochrome may control enzymes are excellently reviewed by

Schopfer (1977). These include de novo synthesis of enzymes via operation of the genetic code (transcription, translation), enzyme degradation and various types of enzyme activation. Membrane and transport effects could be explained by alterations of membrane constituents through enzyme activity or as secondary consequences of growth requiring uptake and redistribution of materials.

Considerable evidence of the kind described in this chapter with a few selected examples (for many more examples see van Steveninck, 1976a; Marmé, 1977) strongly suggests that a direct effect on membranes is possible [see (2) above] or that this may even be the principle molecular mechanism of phytochrome and phytohormone action. Transport processes then may lead to changes in the nucleocytoplasmic environment and hence secondarily affect enzyme and gene activity (e.g., Göring and Mardanov, 1976).

We may assume that a certain amount of time, on the order of many minutes up to one or two hours, is required for a complicated regulation via the genetic code. Therefore there has been a search for fast responses to support the idea of a primary action on the level of membranes (e.g., Nissl and Zenk, 1969; Evans, 1974). Indeed some of the effects of phytohormones on transport are very fast. In particular the phytochrome-dependent movements occur very rapidly after the appropriate change of illumination (see Table 9.1). They are often clear all-or-none reactions. The phytochrome system functions here as a typical trigger mechanism. (Light clearly acts on membrane transport in these cases and not by affecting the provision of energy for active transport.) Furthermore, some of the effects described can also be observed during inhibition of RNA and protein synthesis.

Whether hormone actions generally are due primarily to membrane effects, which then indirectly affect the realization of genetic information, or whether the primary effect is a gene and enzyme regulation, however, still remains an unsolved and controversial problem. There is no convincing reason why both actions should not occur independently or in parallel (cf. reviews of Marmé, 1977, and Schopfer, 1977). In our context it is decisive that hormones and light absorbed by phytochrome, in principle, can rapidly interact with transport processes.

Findings highly suggestive of a direct interaction between hormones and membranes lead to the question of what is the molecular basis of such interactions. Many authors think of a binding or an incorporation of the hormone molecules into the membrane and thus of a change of the molecular fine structure or allosteric effects on catalytically active membrane proteins (e.g., Thimann, 1963; Muir et al., 1967; Weigl, 1969a,b,c; Hertel et al., 1972; Marmé, 1977). When phytochrome molecules are associated with membranes, a direct effect on membrane transport by light quanta absorbed by phytochrome appears possible (Sect. 9.1.2).

The third possibility mentioned above, i.e., action of plant growth regulators via effects on energy metabolism so far has received little attention. A series of recent papers appears to urge, however, that in future research the interaction of phytochrome with energy turnover reactions be considered more seriously (Thibault and Michel, 1971; Frosch and Wagner, 1973a,b; Frosch et al., 1973; Michel and Thibault, 1973; White and Pike, 1974; Haupt and Trump, 1975).

Chapter 10

Coupling Between Energy-Transfer Processes in Mitochondria and Chloroplasts and Transport Mechanisms at Spatially Separated Membranes

10.1 Introduction: Coupling at the Anatomical, at the Cellular and at the Membrane or Molecular Level of Organization

During the discussion of sources of metabolic energy for membrane transport in Chapter 8 we had to pose the question of how energy-transfer reactions at the cristae membranes of mitochondria and at the thylakoid membranes of chloroplasts can be coupled to active transport processes at distant membranes, such as the plasmalemma and the tonoplast. This question is particularly pertinent for those transport processes at the plasmalemma or tonoplast which apparently are directly linked to electron transfer at the cristae or thylakoid membranes, but the question also had to be raised for ∼P dependent mechanisms (Sects. 8.1.1.3 and 8.1.2.1). These considerations deal with coupling at the *cellular level of organization*. Of course, this is not only important for transport mechanisms but rather extends to all processes outside mitochondria and chloroplasts which utilize respiratory or photosynthetic energy. (For a monograph with numerous contributions on this general problem of intracellular communication see Stocking and Heber, 1976.)

Also there is the problem of coupling at the *anatomical level of organization*, i.e., in situations in which metabolic energy is involved in the cooperation of differentiated tissues or organs in a plant. This is a greater problem than that already realized at the cellular level due to the spatial separation of energy-providing reactions and energy-dependent membrane transport processes. We can illustrate this, for example, by describing some findings on the mechanism of salt excretion in *Atriplex spongiosa* leaves. The leaves of *A. spongiosa* and of other *Atriplex* and *Chenopodium* species have epidermal hairs which consist of large bladderlike, highly vacuolated cells and glandlike stalk cells. The active Cl^- excretion into the large bladder vacuole is a highly light-stimulated and photosynthesis-dependent but ATP- or ~P-independent process (see Sect. 7.3.2, Fig. 7.20; Sect. 11.3.2.2.1, Fig. 11.6; Sect. 12.3.1, Fig. 12.1a). The cytoplasm of the bladder and stalk cells, however, contains no chloroplasts and the cells are photosynthetically inactive. The energy for active chloride excretion in the light must be derived from the chloroplasts in the distant leaf mesophyll.

A third organizational level, at which the question of coupling between transport and energy becomes pertinent is the *membrane level*. Here, of course, the question merges into the problem of molecular interactions between various forms of energy and active transport mechanisms residing in the membrane across which transport occurs. This has also been referred to elsewhere in this book (Sect. 5.2.3.2 and 5.3.3).

In a complex system of different organs showing coordinated activities, the coupling of energy with transport suggests that there must be messenger signals or mediating substances. The usage of most of these terms is fraught with difficulties of definition arising from attempts to develop logically consistent models, particularly cybernetic feedback systems (e.g., Cram, 1976). Here, we will try to avoid getting involved in semantic quarrels and approach the question rather pragmatically.

Mediators bringing about co-operation at the anatomical level are *hormonal messengers, transport metabolites* carrying energy in the form of energy-rich phosphate or redox potential, and *physical signals* which are mainly represented by changing potential gradients especially electrical or electrochemical ones. We will discuss hormonal signals in some detail in a later chapter (Chap. 13). When *physical signals* are involved at the cellular and at the anatomical level of organization we may talk of a *biophysical mode of coupling,* and when *transport metabolites* are involved we may talk of a *biochemical mode of coupling* (Lüttge, 1971b, 1973a). These terms strictly apply only to the overcoming of compartmentation at the cellular and anatomical level. They supplement but do not interfere with the distinction between chemical and chemi-osmotic coupling of energy and transport at the membrane level (i.e., the distinction between primary and secondary active transport as defined in Chap. 2; for another treatment of some of the points made in this introduction see Lüttge and Pitman, 1976):

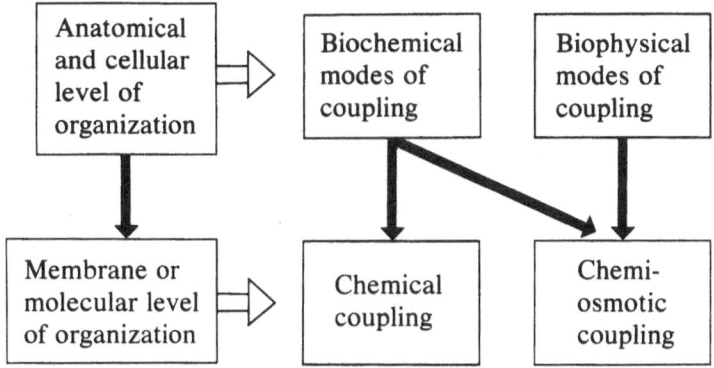

10.2 Biochemical Modes of Coupling

In section 7.2 we have described transport processes across mitochondrial and chloroplast envelopes. We have seen that transport metabolites can move across mitochondrial and chloroplast membranes and carry energy with them, either as reduction equivalents or as ~P, or both. Thus biochemical modes of coupling by the aid of transport metabolites can explain two possibilities outlined in Chapter 8. They explain coupling of transport at the plasmalemma or tonoplast to redox energy from electron flow within mitochondria and chloroplasts. They also explain utilization of ~P from phosphorylation coupled to electron transfer within these organelles at the remote membranes of the plasmalemma and the tonoplast.

The 3-phosphoglyceric acid-dihydroxyacetonephosphate shuttle which carries both ~P and a stoichiometric amount of reduction equivalents across the chloroplast envelope (Figs. 7.14 and 7.15) introduces an interesting problem in this context. As a net result for 1 mol of transport metabolite 1 mol of ATP and 1 mol of reduction equivalents are transported from the chloroplast into the cytoplasm without a direct migration of adenylate or pyridinenucleotide for which the chloroplast envelope is not readily permeable. This stoichiometry is unavoidable if the transport of ~P is mediated by transport metabolites. If ATP diffusion across the chloroplast envelope is impossible, and if a specific adenylate transport system is absent or very ineffective (Heber and Santarius, 1970), then ~P obviously can be transported only across the chloroplast envelope in the company of reduction equivalents. Under these conditions, it would not be possible for the ATP formed in cyclic photophosphorylation, independently of noncyclic electron flow, to supply energy for processes occurring outside the chloroplasts, because during cyclic photophosphorylation no reduction equivalents are formed (Fig. 8.6) and ATP could not be transported. We have, however, discussed in Section 8.2.4 membrane transport processes which are driven by ATP formed in cyclic photophosphorylation. In addition there are quite a number of other cell functions which can be supplied with energy in this way. This would mean that either our ideas on the role of cyclic and noncyclic photophosphorylation in such processes are not correct, or that a considerable transport of $\sim$P in the form of adenylate is possible across the chloroplast envelope.

A way out of this dilemma is offered by the additional assumption of a shuttle carrying reduction equivalents from cytoplasm back to the chloroplasts (see Sects. 7.2.3.4 and 8.2.2). In this way, with cyclic photophosphorylation only, reduction equivalents transported from the cytoplasm into the chloroplast maintain transport of $\sim$P mediated by transport metabolites (Heber and Krause, 1971; Krause, 1971).

10.2.1 Coupling of Membrane Transport Processes With Energy From Reduction Equivalents Carried by Transport Metabolites

Experiments of Kaback and coworkers (Barnes and Kaback, 1970; Kaback and Milner, 1970) with membrane preparations (ghosts) of *Escherichia coli* have shown that products of intermediary metabolism can indeed fulfill such a role. The oxidation of D–(–)–lactate to pyruvate catalyzed by isolated membranes is a prerequisite for amino acid transport across these membranes. By aid of a very similar mechanism uncharged particles also (e.g., β-galactosides) are transported across *E. coli* membranes. During these processes no phosphorylations are required; ATP and other energy-rich phosphate compounds have no effect on amino acid and β-galactoside transport. One has to assume that redox reactions in the membranes catalyze the transport observed across the membranes. The oxidation of lactate by a membrane-bound D–(–)–lactate dehydrogenase, which is coupled with a flavoprotein could be a first step, and a redox reaction on a carrier a last step in an electron transfer chain localized within the membrane. Membrane vesicles prepared from a mutant of *E. coli* deficient in D–(–)–lactate dehydrogenase are inactive in transport. When the enzyme is added as a homogeneous preparation, it binds to these vesicles and they become able to oxidize D–(–)–lactate and to perform D–(–)–lactate-dependent transport (Short et al., 1974).

In those higher plant leaves in which Cl$^-$ uptake appears to be rather directly linked to photosynthetic redox reactions (Sect. 8.2.4.1.5) it has been speculated that malate functions as a transport metabolite carrying redox equivalents (Sect. 7.2.3.4.3). All the plants in which such a Cl$^-$ uptake mechanism appears to be active are C$_4$ plants (*Atriplex spongiosa, Zea mays,* and *Amararanthus caudatus;* Lüttge et al., 1971a,b). In C$_4$-plants transport metabolites play a special role (see also Chap. 11). The

CO_2 assimilatory tissue is differentiated into the so-called mesophyll and bundle-sheath parenchyma. Corresponding to this anatomical differentiation there is a biochemical differentiation (Fig. 10.1). The net synthesis of carbohydrates and the accumulation of starch occurs here largely in the chloroplasts of the bundle-sheath parenchyma. The CO_2 from the air is fixed in the mesophyll cytoplasm by phosphoenolpyruvate-(PEP-) carboxylase which carboxylates phosphoenolpyruvate. From the resulting oxaloacetate malate is formed in the mesophyll chlorolasts by consumption of redox equivalents (NADPH + H$^+$) formed in light-dependent reactions. Malate then is transported from the mesophyll chloroplasts to the bundle-sheath chloroplasts. During this process, membrane transport is necessary twice (i.e., transport across the envelopes of mesophyll chloroplasts and of bundle-sheath chloroplasts). In addition a symplastic transport in the cytoplasmic phase is required (see Sect. 11.3.2.3.2). In the bundle-sheath cells malate is decarboxylated. The pyruvate formed is transported back to mesophyll cells. The CO_2 and the redox equivalents liberated from the malate can be utilized for synthesis of carbohydrates by

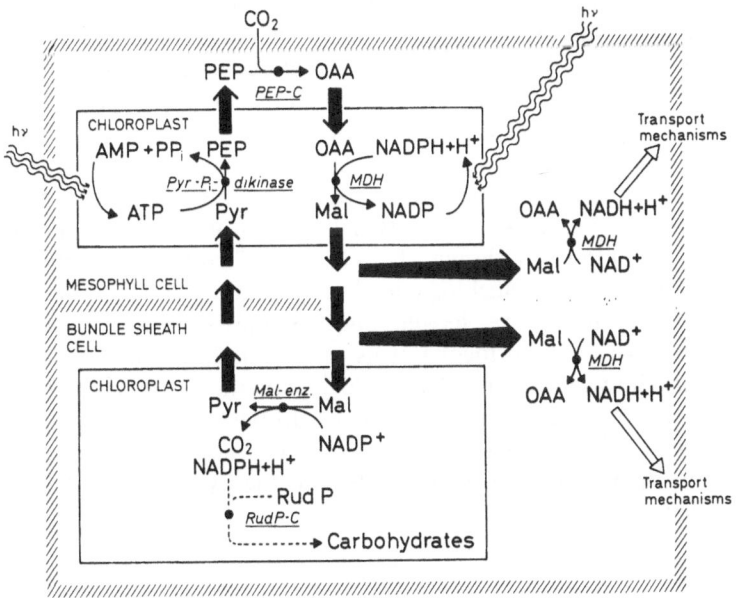

Fig. 10.1. Strongly simplified scheme of the C_4 pathway of photosynthesis with malate and pyruvate as transport metabolites and with the possible utilization of reduction equivalents from malate for membrane transport mechanisms. (After Hatch and Slack. 1970: Hatch et al.. 1971: Osmond and Smith. 1976.) *PEP*, phosphoenolpyruvate; *PEP-C*, phosphoenolpyruvate carboxylase; *OAA*, oxaloacetate; *Mal*, malate; *MDH*, malate dehydrogenases; *Pyr*, pyruvate; *Mal-enz.*, decarboxylating malic enzyme; *RudP*, ribulose-1.5-diphosphate: *RudP-C*, ribulose-1,5-diphosphate carboxylase.

CO_2-fixation via ribulose-1,5-diphosphate-(RudP-) carboxylase in the Calvin cycle.

Malate transport from the mesophyll to the bundle-sheath thus mediates a CO_2 transport and a transport of reduction equivalents. The latter are particularly important for CO_2 assimilation in those bundle-sheath cells where the chloroplasts do not have an active photosystem II, in other words, where chloroplasts cannot perform noncyclic electron flow. This is the case in several species. The CO_2 transport can be envisaged as a mechanism for the concentration of CO_2 in the bundle-sheath cells. The advantage of this is an increase of the reaction rate of RudP-carboxylase. This enzyme has a much lower affinity to CO_2 ($K_M = 450 \times 10^{-6}$ M) than the PEP-carboxylase ($K_M = 7.1 \times 10^{-6}$ M; cf. Hatch et al., 1971: p. 149).

In photosynthesis of C_4-plants thus a considerable transport of intermediary metabolites is necessary (see Figs. 11.8, 11.9), i.e., according to our simplified scheme of Figure 10.1 especially of malate and pyruvate which migrate in opposite directions.

The energy of the photosynthetically formed reduction equivalents carried by the malate is not necessarily transported entirely to the bundle-sheath chloroplasts. It is also possible that a part of the malate is subject to reactions within the cytoplasm through which it is passing during symplastic transport. By cytoplasmic NAD-dependent malate dehydrogenase, redox equivalents could be regained in the form of NADH + H^+, and it appears that this redox potential could be utilized by active transport processes.

Apart from the considerable transport of intermediary products in C_4-plants, and from the known enzyme distributions, which make possible the reactions in the way shown, the scheme in Figure 10.1 is hypothetical. There is, of course, the striking correlation mentioned above, i.e., that the higher plants, in whose leaves photosynthesis-dependent but ATP-independent Cl^- uptake mechanisms were found, are all C_4-plants. The hypothesis shown in Figure 10.1 would explain this correlation. However, this hypothesis still has some drawbacks. In the system discussed for the maintenance of the shuttle traffic ATP is also necessary, namely at the site of the reintroduction of pyruvate into the cycle where it has to be phosphorylated to phosphoenolpyruvate.

10.2.2 Coupling of Membrane Transport With Energy from $\sim$ P Carried by Transport Metabolites

In the preceding section we have described in some detail two cases in which redox energy appears to be used for energy-dependent membrane transport. This is important since until quite recently, for cytomorphological reasons, there has been a considerable reluctance to accept membrane transport at the plasmalemma or tonoplast driven by redox-potential energy (Sect. 8.1.1.3). As pointed out in Section 8.1.2.1, ATP is the general energy currency of the cell. Thus, in contrast to transport driven by reduction potential, the problems of coupling at the cellular and at the anatomical level did not seem serious when ATP or ~P appeared to be the energy source involved. Transport metabolites mediating the distribution of ~P within the cell have been described in Section 7.2.

According to this situation modern models explaining molecular mechanisms of coupling between energy and transport at the membrane level have largely been developed for ATP or ~P as the energy source.

10.2.3 Coupling at the Membrane or Molecular Level of Organization

Carrier models presented in Figures 2.21 and 5.7 for ATP-driven membrane transport and in Figures 8.2 and 8.3 for redox potential-driven membrane transport are essentially dealing with molecular mechanisms of coupling between transport and energy-yielding reactions. More modern ideas are the membrane ATPase type mechanisms of chemical coupling and the charge separation (Mitchell hypothesis!) mechanisms of chemiosmotic coupling.

10.2.3.1 Chemical Coupling

Chemical coupling occurs in all cases in which an ATPase protein both catalyzes hydrolysis of ATP and, by reversible binding of a particle mediates its transport across a membrane. These chemically coupled transport processes constitute primary active transport. They may mediate one-way transfer of a given solute species such as the widely distributed 1 H^+ or 2 H^+ pumps (Fig. 10.2a,b). Such mechanisms then cause secondary active transport because chemi-osmotic coupling (see Sect. 10.2.3.2 below) must lead to electrical charge balance. They also may mediate two-way exchange reactions when two particle species moving in opposite directions are reversibly bound to the same entity. Many such exchange systems have been described, in particular the Na^+-K^+ exchange by the ouabain-sensitive animal ATPase (Fig. 10.2c) and the H^+-K^+ exchange of plant and fungal cells (Fig. 10.2d). It is often a matter of conjecture though, whether chemical or chemi-osmotic coupling occurs in such cases (e.g., Sect. 9.3).

Chemical coupling between ~P and transport can occur also independently of ATPases. A well-established example is the phosphoenolpyruvate-glucose-phosphotransferase system of bacteria (Kaback, 1970a,b). This is essentially a vectorial-group transfer reaction. First, ~P energy from glycolysis is used. A phosphate group from phosphoenolpyruvate (PEP) generated in glycolysis is bound to a heat-stable, low molecular-

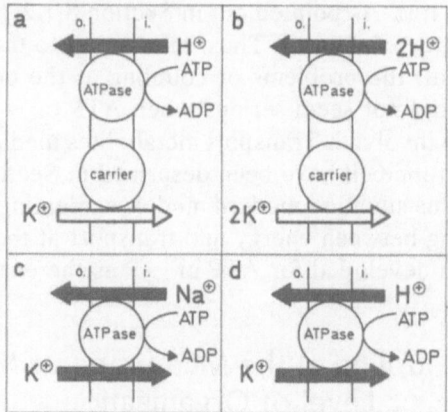

Fig. 10.2. Membrane ATPases mediating exchange mechanisms by chemical coupling (**c, d**) and by chemi-osmotic coupling of charged particle fluxes (**a, b**). *Solid arrows*, chemically coupled (primary active) transport; *open arrows*, chemi-osmotically coupled (secondary active) transport. *o*, outside; *i*, inside the cell.

weight protein (HPr), which is part of a carrier system in the membrane:

$$1. \quad PEP + HPr \xrightleftharpoons[Mg^{2+}]{Enzyme\ I} pyruvate + P\text{-}HPr.$$

Second P-HPr can react with a sugar molecule outside the cell, and the sugar-phosphate formed by conformational changes of the carrier system is moved inside the cell

$$2. \quad P\text{-}HPr + sugar_{out} \xrightleftharpoons[Mg^{2+}]{Enzyme\ II} sugar\text{-}P_{in} + HPr.$$

(See Fig. 10.3.)

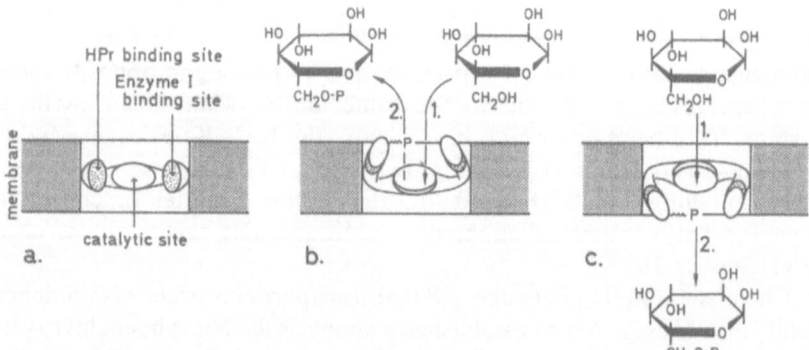

Fig. 10.3. Schematic model for the mechanism of vectorial (i.e., transport: **c**) and nonvectorial phosphorylation (**b**) of glucose by isolated bacterial membrane preparations. (From Kaback, 1970.)

10.2.3.2 Chemi-osmotic Coupling

We have already mentioned the occurrence of chemi-osmotic coupling as a consequence of an ATPase transport mechanism which moves only one ionic species in one direction across the membrane (Fig. 10.2a,b). Electrical charge balance does not require specificity of such a transport. However, specific transport, e.g. of K^+ and Na^+, is often observed and thus the involvement of specific carriers is assumed (Fig. 10.2a,b).

A more general scheme explaining the major, energy-dependent ion fluxes of *Chara* cells by charge separation at the plasmalemma has been presented by Smith (1970, Fig. 10.4). This concept can be generalized further and then a large number of energy-dependent transport processes can be explained. To understand this generalization it is necessary to remember the electrogenic pump described in Section 2.2.2.4, and the co-transport and counter-transport carriers discussed in Section 2.5.2.3. Remember that by the electrogenic pump an electrochemical potential is built up at the membrane. This is not necessarily an H^+-potential, but could be an electrochemical potential of any charged particle ($=$ion) species transported by the electrogenic pump. However, it appears that H^+ transport is the most widely distributed if not the generally prevailing mechanism of electrogenicity in bacterial, fungal, and plant cells (e.g., Slayman et al., 1973; Saito and Senda, 1973; Spanswick, 1974a,b; Slayman and Gradmann, 1975; Walker and Smith, 1975). The electropotential gradient generated by an electrogenic pump (see Fig. 2.7) for H^+ could drive fluxes of ions of the same charge in the opposite direction (as in Fig. 10.2a,b; and in Fig. 10.4 upper part, where an H^+–K^+, Na^+ exchange transport is assumed) or fluxes of ions of the opposite charge in the same direction (as in Fig. 10.4 bottom where an OH^-–Cl^- exchange transport is assumed). If we now assume that the transport of uncharged particles of a substance S is coupled to these ion fluxes in the fashion of counter-transport or co-

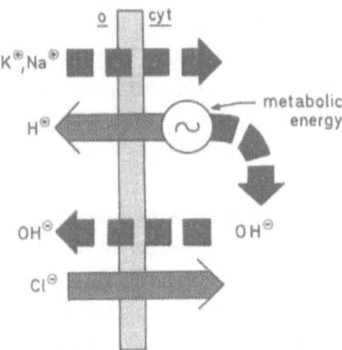

Fig. 10.4. Metabolic energy dependent charge separation at the plasmalemma of a *Chara* cell and coupled anion and cation uptake processes. (After Smith, 1970.) *o*, outside the cell; *cyt*, cytoplasm.

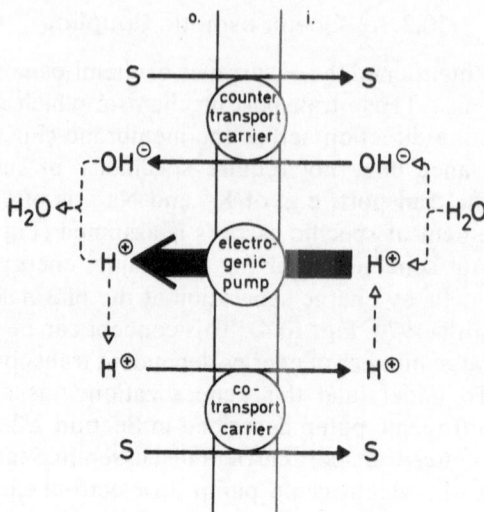

Fig. 10.5. Chemi-osmotic coupling of transport of solutes (S) by counter-transport (-antiport) or co-transport (-symport) utilizing the energy of an electrochemical potential gradient established by an electrogenic H^+ pump. o, outside; i, inside of cell. (Modified from Slayman and Gradmann, 1975.)

transport as depicted in Figure 10.5, indeed we can explain all sorts of energy-dependent fluxes. Bentrup (unpublished) recently claims that if we have an outwardly directed proton pump at the plasmalemma and an inwardly directed proton pump at the tonoplast most energy-dependent transport processes in higher plants can be explained. These proton pumps appear to be likely when H^+ electrochemical gradients are considered (Raven and Smith, 1974).

Systems like those depicted in Figure 10.5 have indeed been found experimentally, especially for sugar transport in lower organisms such as bacteria (West, 1970; West and Mitchell, 1972, 1973; Kashket and Wilson, 1973), the fungus *Neurospora crassa* (Slayman and Slayman, 1974), and the unicellular green alga *Chlorella vulgaris* (Komor, 1973; Komor and Tanner, 1974a,b; also see Sect. 8.2.4.2). A 1:1 (approximately) hexose/H^+ co-transport or hexose/OH^- counter-transport seems to be involved in all these cases.

Whether such a mechanism of sugar transport may operate also in higher plants has been considered recently in a number of laboratories. A prediction of the model of Figure 10.5 is that when transport of S sets in and the utilization of energy of the electrochemical gradient at the membrane is started, the gradient must be lowered. With an increased supply of metabolic energy (ATP) the pump may start to operate at an increased rate and thus reestablish the electrochemical gradient and maintain it at its original level. Nevertheless, a transient decrease of the electrochemical

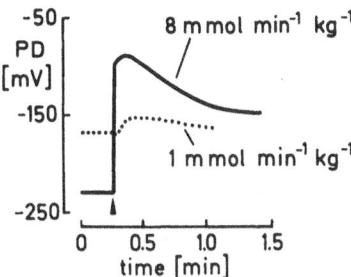

Fig. 10.6. Depolarization of membrane potential of *Neurospora crassa* hyphal cells, having a derepressed (*solid line*) and repressed glucose uptake system (*dotted line*) respectively, after the addition of 1 mM glucose to the medium (*arrow*). Rates of glucose uptake are also given, where kg refers to cell water. (Modified from Slayman and Slayman, 1974.)

gradient is expected immediately after the addition of the transportable particle species S when there is a small lag between the onset of transport and an increase of the rate of pumping.

Since an electrochemical gradient always has a component of chemical concentration and a component of electrical potential difference [see Chap. 2, Eqs. (2.17), (2.19)], a transient change of membrane potential is expected also at the onset of transport of S. Indeed, with *Neurospora* a depolarization of the membrane potential by about 120 mV has been observed at the onset of glucose transport at a saturating concentration (Fig. 10.6; Slayman and Slayman, 1974). The earliest similar observations in higher plants were reported for *Impatiens balsamina* root cells by Jones et al. (1975; Fig. 10.7). In fronds of duckweed (*Lemna gibba*) a number of substrates and ions (sugars, amino acids, inorganic phosphate, nitrate)

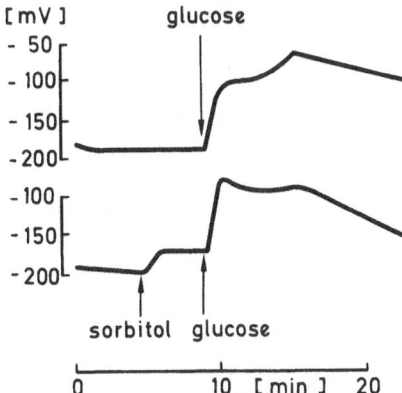

Fig. 10.7. Depolarization of the membrane potential of *Impatiens balsamina* root cells by the addition of 50 mM glucose but not by 50 mM sorbitol. (After M.G.K. Jones et al., 1975, from Lüttge, 1977.)

cause depolarizations of membrane potential at the onset of their transport (Ullrich-Eberius et al., 1978; Novacky et al., 1978a,b; Fig. 10.8). The
maximum depolarization of the membrane potential in all these cases depends on the concentration of the transported solute and on the rate of
transport (see also Fig. 10.6).

Repolarization, i.e., recovery of the potential difference, occurs after
removal of the transported solute from the medium, but also spontaneously when the solute is still present (Figs. 10.7–10.10). The latter indicates increased effectiveness of the active pump reestablishing the electrochemical gradient at the membrane as suggested above. This is seen
also when, after spontaneous repolarization, the transported solute is removed from the external solution. We observe a small transient hyperpolarization which can be explained by the pump still operating at the faster
rate immediately after withdrawal of solute particles from the co-
transport mechanism (Fig. 10.9). In the green cells of *Lemna* the energy
dependence of repolarization also becomes evident by effects of light.
The maximum depolarization is smaller and recovery is faster in the light
than in the dark, indicating an effect of photosynthetic energy (Figs. 10.8
and 10.10).

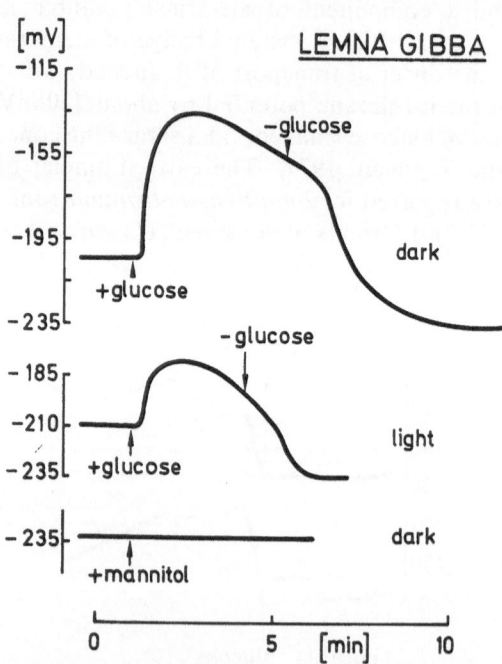

Fig. 10.8. Depolarization of the membrane potential of dark grown sugar starved
Lemna gibba (strain G 1) cells by 20 mM glucose but not 20 mM mannitol. Spontaneous recovery of membrane potential and recovery after glucose removal are
more pronounced in the light than in the dark, suggesting an additional photosynthetic energy supply. (From Lüttge, 1977.)

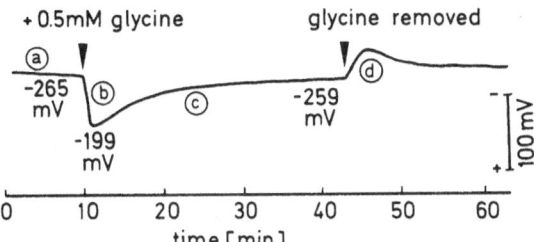

Fig. 10.9. Membrane potential of *Lemna gibba* G 1 and glycine transport. *a*, resting potential; *b*, transient depolarization due to the onset of glycine transport; *c*, energy dependent repolarization; *d*, transient hyperpolarization after removal of glycine. (Fischer, 1978.)

The involvement of protons has been shown, or at least strongly suggested, by the pH-dependence of the depolarization due to solute transport and the subsequent repolarization, and also by pH changes in the medium during transport of solutes (especially of sugars: hexoses, sucrose) in various systems, e.g., *Ricinus* cotyledons and leaves (Komor et al., 1977; Komor, 1977; Malek and Baker, 1977; Hutchings, 1978a,b), beet leaves (Giaquinta, 1977), *Samanea* pulvini (Racusen and Galston, 1977), *Lemna* fronds (Novacky et al., 1978a,b). Nelles (1975) has shown that in corn coleoptiles sucrose depolarizes the membrane potential by about 13 mV. Indoleacetic acid (IAA) which stimulates H^+ excretion in coleoptiles

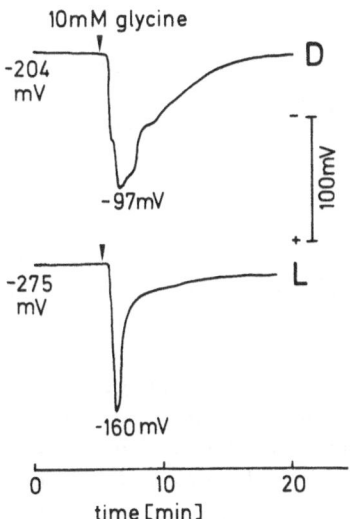

Fig. 10.10. Transient depolarization of the membrane potential of *Lemna gibba* G 1 due to the onset of glycine transport (10 mM added at *arrow*) in the dark (*D*) and light (*L*). Note that the depolarization in the light is much shorter. (Fischer, 1978.)

(Sect. 9.2) prevents this effect. Colombo et al. (1978) observed an enhancement of 3-O-methylglucose uptake by fusicoccin (FC) which is known to increase H^+ pumping out of the cells (Sect. 9.3).

In conclusion H^+-sugar co-transport or OH^--sugar counter-transport mechanisms (Fig. 10.5) appear to be ubiquitous in bacteria, fungi, eukaryotic algae and higher plants.

10.3 Biophysical Modes of Coupling

In Section 9.1.3 we described cases in which light affects electrical phenomena in membranes independently of the photosynthetic apparatus. In this section we will deal with light effects involving photosynthetic energy transfer.

We have seen in Chapter 7 that ion transport (especially H^+ and K^+) across the thylakoid membranes is associated with light-dependent electron flow within the membranes (Sect. 7.2.3). A direct pumping of protons due to light absorption has been discovered recently in the halophilic microorganism *Halobacterium halobium*. The outer membranes of these cells contain a purple pigment, bacteriorhodopsin, which closely resembles the visual pigments of animals. Absorption of light by this system creates an electrochemical proton gradient which is not linked to any electron flow; and it may be assumed that the light-generated proton gradient arises from a vectorial release and uptake of protons by bacteriorhodopsin without involvement of biochemical reactions. Thus this pigment acts directly as a light-driven proton pump, a purely biophysical mode of coupling at the membrane or molecular level of organization. The H^+ electrochemical gradient established in this way can be utilized by the cells for ATP synthesis, ion transport (K^+, Na^+), uptake of amino acids, just as in the thylakoids of chloroplasts or with co-transport mechanisms of plasmamembranes of other cells. Hence, the cells of *H. halobium* may represent a very interesting model system (Oesterhelt and Stoeckenius, 1973; Danon and Stoeckenius, 1974; Wagner and Hope, 1976; Oesterhelt et al., 1977).

In the chloroplasts other ion fluxes across the thylakoid membranes can be coupled with movements of H^+ or K^+. These migrations of ions change the ion concentrations outside and inside the thylakoids. From the topography of the cell it would follow that also the electrochemical potential difference between the chloroplast stroma and the cytoplasm and as a further consequence potential differences at the plasmalemma and the tonoplast should be affected. Eventually ion uptake and ion release by the cell as a whole could be determined and would be a clear example in this way, of biophysical coupling as outlined in Section 10.1.

10.3.1 Light Effects on Membrane Potentials

10.3.1.1 The Resting Potential

Reports on differences between resting potentials in continuous light and in continuous darkness are conflicting. Often light does not have any effect at all on the resting potential, but both depolarizing and hyperpolarizing effects of photosynthetically active light have also been reported (Bentrup, 1974b; Felle and Bentrup, 1974a,b; Jeschke, 1970a,c). It appears that in these cases light affects the membrane potential via the energy state of the cell which is influenced by photosynthetic energy-transfer reactions. Thus, light-dependent electrogenic pumps have been assumed to be involved (Gradmann, 1970; Saddler, 1970a,b; review: Bentrup, 1974b). But when this is so, we are not dealing with a purely biophysical mechanism but rather with a biochemical mode of coupling as defined above. The situation resembles somewhat that of ion uptake by spinach leaves, in which light can stimulate or inhibit ion uptake depending on the physiological condition of the tissue (Lüttge et al., 1971b).

Caution is needed also because of the light-triggered membrane potential oscillations described below. Since these transient phenomena which occur after light–dark or dark–light changes are often longer than 1 h, it is sometimes not clear whether the experiments establishing changes of resting potential have been pursued for a long enough time.

10.3.1.2 The Photosynthesis-Dependent Photoelectric Membrane Effect

Photosynthesis-dependent changes in membrane potentials often occur as transitory phenomena. The membrane potential which is altered from its resting value by a light–dark transition or by a dark–light transition, after a number of oscillations in many cases returns to the level of the initial resting potential. Figure 10.11 shows the typical course of such a transitory change of membrane potential of a green leaf cell in light–dark–light transitions. In this experiment the measuring electrode was in the vacuole of the cell and the reference electrode in an external medium of 5 mM KCl and 0.1 mM $CaSO_4$. The potential measured, therefore, is composed of the tonoplast, the plasmalemma, and the cell wall potentials, where, however, the plasmalemma potential makes by far the largest contribution (Sect. 2.2.3.2, Table 2.2). After a transition from dark to light, the membrane potential reacts initially with a depolarization. A lag phase has not been found with the equipment used, i.e., 3 N KCl-filled glass electrode, agar bridges, calomel half cells and a line recorder attached to an electrometer. Shortly after the beginning of the depolarization, the potential swings back once more and then the depolarization continues; within

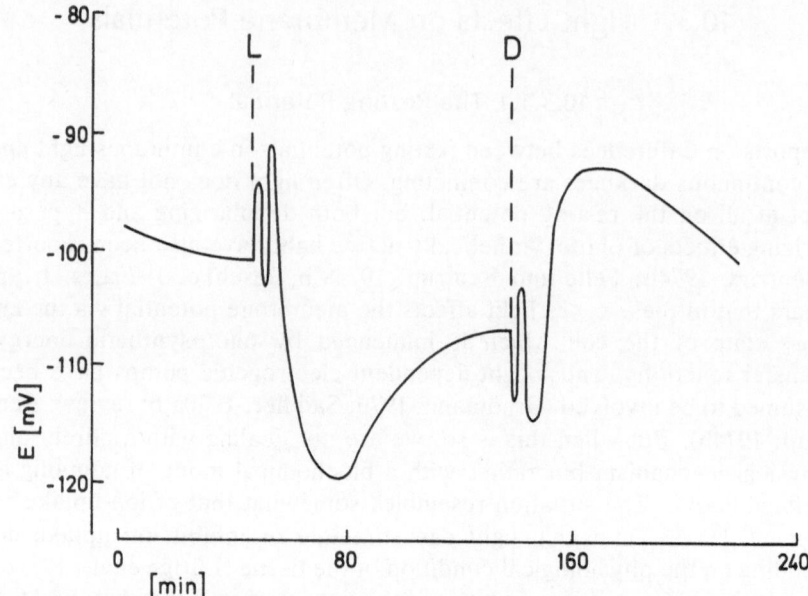

Fig. 10.11. Transient membrane potential changes of green mesophyll cells of *Atriplex spongiosa* after dark–light and light–dark changes. (After Pallaghy and Lüttge, 1970.)

2 min it reaches a maximum, whereupon the potential returns slowly to the level of the resting potential in damped oscillations of increasing length and decreasing amplitude. After a light–dark transition we observe a similar effect; the initial change after a light–dark transition, however, is a mirror image of the dark–light transition. The first change here is a hyperpolarization.

A transient change of the potential away from the resting level by changes in illumination has been shown to occur in a large number of green plant cells. Experiments with inhibitors, with artificial electron donors and acceptors, with mutants and with light of defined wave length (cf. Sect. 8.2.3) clearly show that it is a phenomenon depending on noncyclic electron flow in photosynthesis (see Pallaghy and Lüttge, 1970, with further references; Throm, 1970, 1971a,b, 1973; Brinckmann and Lüttge, 1974; Fig. 11.7; review: Bentrup, 1974). *Acetabularia* appears to be the only case in which it may be triggered by photosystem I alone (Gradmann, 1970).

The phenomenon of a photosynthesis-dependent photoelectric membrane effect leads to the question of how the potential is changed in different directions during light–dark and dark–light transitions. Some hints are obtained by the observation of pH changes of the medium which are correlated in time with the membrane potential changes. Let us evaluate whether H^+ ion movements can make an important contribution to the photosynthesis-dependent potential changes. If light is suddenly switched on, immediately photosynthetic electron transfer reactions commence and H^+ ions are taken up into the interior of the thylakoids. In this way the chloroplast stroma becomes more alkaline and eventually the cytoplasm could lose H^+ ions. A decrease of H^+ ion concentration in the cytoplasm according to the Nernst [Eq. (2.19)], or the Goldman equation [Eq. (2.28)], must be accompanied by a depolarization of the membrane potential which we indeed observe. At the same time by the lowering of the cytoplasmic H^+ concentration an increased amount of H^+ should get into the cell from the outside across the plasmalemma. During a light–dark change these events should occur in the opposite direction. In fact, the direction of the pH changes in the external solution after light–dark and dark–light transitions is altered as predicted (Fig. 10.15). With intracellular pH-microelectrodes Davis (1974) also observed cytoplasmic pH transients in the green cells of the liverwort *Phaeoceros laevis* which are in agreement with such a sequence of events. As shown in Figure 10.12, an initial depolarization of the electric membrane potential after light is switched on is paralleled by an alkalinization of cytoplasmic pH; an initial hyperpolarization of membrane potential after light is switched off is paralleled by an acidification of the cytoplasm, etc. (Fig. 10.12 also shows transients of vacuolar pH.)

There are, nevertheless, a few difficulties in the interpretation given above. First of all there are two different possibilities for explaining the apparent proton net fluxes of intact cells. In the sense described above they could mirror the physical coupling of the electron transport and H^+ flux at the thylakoid membranes. On the other hand apparent proton

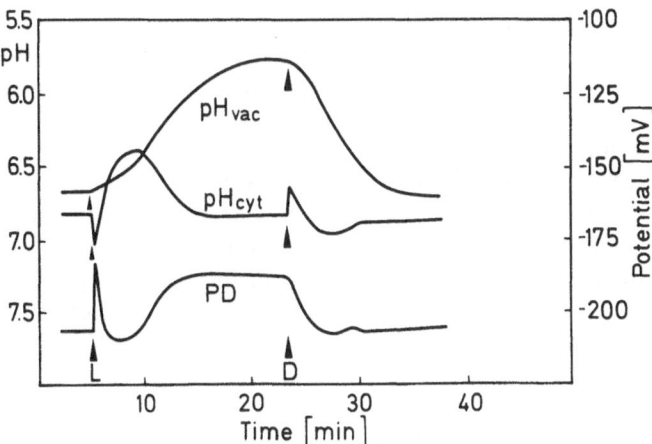

Fig. 10.12. Typical time course curves for the effect of light on the membrane potential and the pH of the cytoplasm and vacuole of *Phaeoceros laevis*. (Davis, 1974.)

fluxes could simply be a consequence of photosynthetic CO_2 assimilation according to the following scheme:

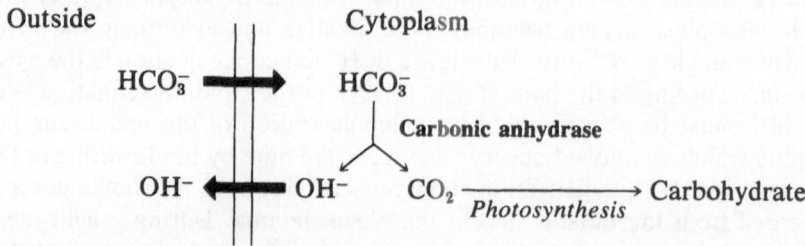

If, in this scheme HCO_3^- and OH^- are transported by two pumps operating with different kinetics, then oscillations of membrane potentials could occur (Denny and Weeks, 1970).

Most important for this question are experiments in which apparent proton net fluxes can be observed in the absence of CO_2 fixation. According to Neumann and Levine (1971) this is not possible. These authors worked with a *Chlamydomonas* mutant whose secondary photosynthetic reactions were impaired so that no CO_2 could be fixed, although the primary reactions of photosynthetic energy transfer were normal. In intact cells no net proton flux could be observed. Isolated thylakoids, however, showed a normal proton flux. This suggests that H^+ movements at the thylakoids are not reflected at the chloroplast envelope or the plasmalemma.

On the other hand, Hope et al. (1972) found that the uncoupler F-CCP and the artificial electron acceptor p-benzoquinone (pBQ) dramatically inhibit CO_2 fixation of *Elodea* leaves, whereas an appreciable apparent proton flux is maintained. According to this experiment, at least part of the proton flux of intact cells could be coupled directly to electron flow of photosynthesis. F-CCP and pBQ, however, also increase membrane permeability. Therefore, it remains unclear as to whether these treatments are not affecting plasmalemma permeability and thus allowing recording of the H^+ flux at the thylakoids with a pH electrode in the external medium. Experiments with etiolated barley leaves, which have been allowed to de-etiolate and turn green to such an extent that noncyclic electron flow of photosynthesis is operating, but CO_2 fixation does not yet occur, clearly show that part of the proton flux and membrane potential oscillations are directly related to electron flow (O_2 evolution) and independent of concomitant CO_2 fixation (Figs. 10.13 and 10.14). This can be taken as a good example of a physical mode of coupling.

There still remain, however, a number of problems. In view of the behavior of isolated chloroplasts with intact envelopes the finding of a CO_2 fixation-independent but O_2 evolution-correlated H^+ uptake by intact cells in itself is a puzzle. The envelope of chloroplasts is not very permeable for H^+ (Heber and Krause, 1971). Those apparent H^+ movements

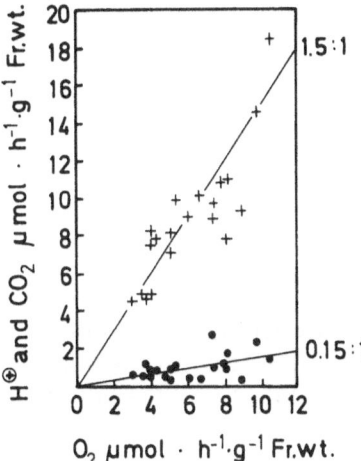

Fig. 10.13. Correlation between net H^+ influx and O_2 evolution (*crosses*) and between CO_2 fixation and O_2 evolution (*circles*) in de-etiolating barley leaves which have been greening in the light for various periods between 30 and 120 min. (Lüttge, 1973.)

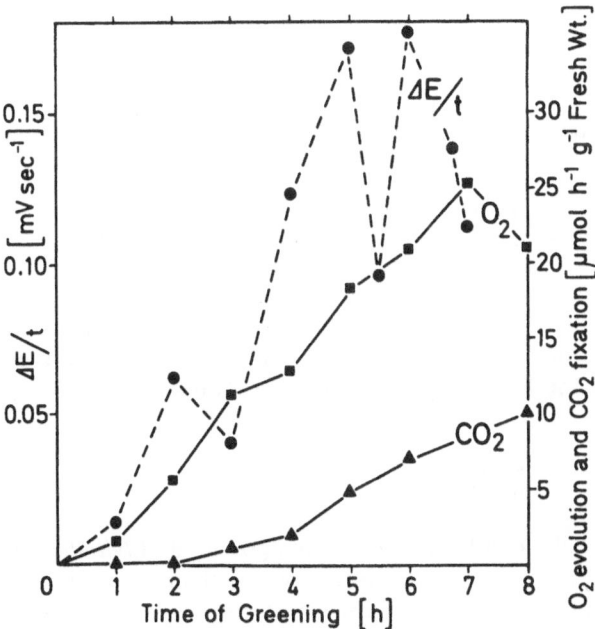

Fig. 10.14. Initial rates of light triggered membrane potential changes ($\Delta E/t$), photosynthetic O_2 evolution and CO_2 fixation by barley leaf cells in relation to time of greening (de-etiolating). (Data from Lüttge, 1973b, and Lüttge, Kramer et al., 1974.)

which are observed result in an acidification of the external medium of intact isolated chloroplasts rather than in an alkalinization. This discrepancy between observations with isolated chloroplasts and intact cells perhaps might be due to a different behavior of chloroplasts when they are embedded in their natural medium, i.e., the cytoplasm of living cells, as compared to artificial isolation media.

This point apparently has received surprisingly little attention by investigators working with isolated chloroplasts, but there are some pertinent observations of plant physiologists studying intact cells. During intracellular pH measurements using antimony covered and plastic insulated solid pH microelectrodes in the liverwort *Phaeoceros laevis* Davis (1974) has found transient pH changes in the cytoplasm after light–dark–light changes (Fig. 10.12) but little difference of cytoplasmic pH in continuous light and continuous darkness; at external pH values between 4.7 and 6.7 cytoplasmic pH was between 6.5 and 6.9 in the light and 6.6–6.8 in the dark. There are still a number of reservations about the technique of using pH microelectrodes. Measuring the distribution of the weak acid 5,5-dimethyl-2,4-oxazolidinedione (DMO), Walker and Smith (1975) observed somewhat higher pH values in the cytoplasm of *Chara corallina*. At an external pH between 5 and 6 cytoplasmic pH was 7.4 in the dark and 7.7 in the light, i.e., a little more alkaline in the light. Spanswick and Miller (1977) compared various methods of cytoplasmic pH measurements in *Nitella translucens*. They obtained similar results with DMO and glass electrodes with a rather large diameter with sealed tips, but lower pH values with the antimony electrode probably due to damage of the plastic insulation as the electrode passed the cell wall:

DMO	7.42 ± 0.07	(SE)
glass electrode	7.54 ± 0.15	(SE)
antimony electrode	6.74 ± 0.15	(SE).

A possible role of pH values of media in this context is stressed by the finding that membrane permeabilities appear to be sensitive to pH. For *Chara* cells the following relations were found:

	P_K	P_{Na}	P_{Cl}
pH 6.5	$4.84 \cdot 10^{-7}$	$3.46 \cdot 10^{-7}$	$2.06 \cdot 10^{-10}$ (cm s^{-1})
pH 4.5	$0.49 \cdot 10^{-7}$	$14.0 \cdot 10^{-7}$	$5.11 \cdot 10^{-10}$ (cm s^{-1})

(Lannoye et al., 1970). The membrane potential of *Nitella* cells in the pH range of 8 to 6 is relatively insensitive to changes of H^+ concentration; however, in the pH range 6–4 the membrane reacts to changes of H^+ concentrations like an H^+ electrode (i.e., membrane potential changes of 56 mV per pH unit; Kitasato, 1968).

Another obstacle for causally relating the photosynthesis-dependent membrane potential oscillations with changes in the direction of H^+ movements during light–dark–light transitions is that the H^+ fluxes at the thylakoids appear not to be large enough. Furthermore, if the hypothesis is correct, then according to Eq. (2.28), an experimental change of H^+ concentration in the external medium of intact cells must affect the membrane potential. An effect of the pH of the external solution on membrane potential has not been observed in all investigations (Jeschke, 1970a,c) but they have been seen frequently enough (Kitasato, 1968) so that at least qualitatively this cannot be used as an argument against the hypothesis.

Vredenberg and Tonk (1973) appear to agree that the oscillations of photosynthesis-dependent membrane potential are associated with transport processes at the chloroplast envelope. In *Nitella translucens* they observed a fast light-triggered reaction which causes a decrease in the membrane resistance. They think that this is due to chemical or ionic changes in the cytoplasm "caused by a relatively fast light-induced translocation of a reaction product, an intermediate, or ions across the tonoplast envelope." A slower light-dependent reaction then causes a stimulation of the electrogenic ion pump at the plasmalemma of the *Nitella* cells. (See also Vredenberg, 1976, for a review.) The potential oscillations may be due to the interplay of the faster and slower reactions. This combines elements of what we called physical coupling (fast reaction) and biochemical coupling (slow reaction).

Many of the qualitative and quantitative problems of explaining the photosynthesis-dependent membrane potential oscillations can be avoided if we follow a suggestion of Bentrup (1974). He assumes that small light-dependent pH changes in the cytoplasm (Davis, 1974; Walker and Smith, 1975) lead to significant changes of protonation and electric charge of the plasmalemma (and tonoplast) membrane structures. In this case potential oscillations would be triggered by a typically physical mode of coupling as defined above (Sect. 10.1). Nevertheless, we must conclude that despite much work the exact mechanism of coupling between photosynthetic electron flow in the thylakoid membranes and the photosynthesis-dependent photoelectric membrane effect remains unresolved.

10.4 Kinetic Correlation of Transient Phenomena and Regulation in the Steady State

In Figure 10.15 light-dependent potential changes and apparent net proton fluxes discussed in Section 10.3.1.2 are compared with a number of other light-on or light-off signals which can be observed in intact green cells. Of particular interest are the changes of light absorption and transmission respectively of intact *Ulva* and *Porphyra* cells, which occur with very similar kinetics. These effects could also be correlated directly with ion

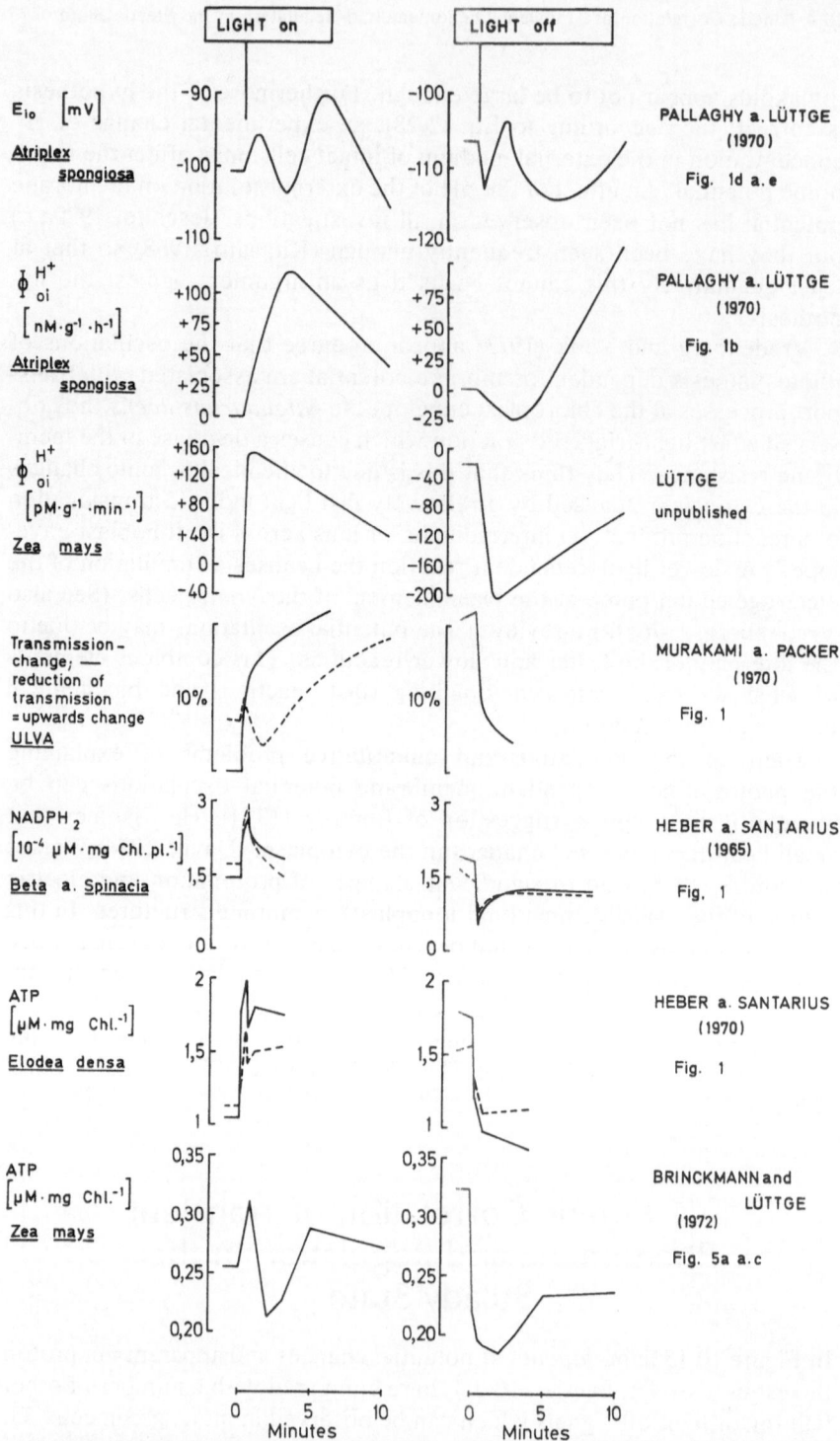

Fig. 10.15. Comparison of various signals obtained with intact green plant cells when the light is switched on or off. (μM, nM, pM refer to $-mol$.)

transport processes in the thylakoid membrane leading to shrinking and swelling of thylakoids. When light is switched on proton uptake by the thylakoids leads to protonization of the membranes, whereby the gaps between the individual thylakoids become smaller and the thickness of the thylakoid membranes decreases. The pH value in the interior of the thylakoids decreases the dissociation equilibrium of weak organic acids, thus increasing the undissociated acid molecules which are electrically neutral, and can flow out of the thylakoids; this loss of osmotically effective substances results in a decrease in the volume of the thylakoids. At the onset of darkening, all these processes occur in the opposite direction (Packer et al., 1970; Murakami and Packer, 1970). Furthermore it is interesting that the reduction level of pyridinenucleotide, which is formed in noncyclic electron flow, and largely utilized in CO_2 reduction, also shows transient changes of a duration in the range of minutes after light–dark–light transitions (Heber and Santarius, 1965). The ATP level in the cells is also subject to light-triggered transient changes (Strotmann and Heldt, 1969; Holm-Hansen, 1970; Heber and Santarius, 1970; Urbach and Kaiser, 1971; Brinckmann and Lüttge, 1972; Simonis and Urbach, 1973) which kinetically are similar to the other transient phenomena. In *Anacystis nidulans*, a blue-green alga in which ATP levels in the light are about three times as high as in the dark under aerobic conditions, ATP levels oscillate in the light after transitions from N_2 (anaerobiosis) to N_2 + CO_2 (photosynthesis under anaerobic conditions; Bornefeld and Simonis, 1974). From a simple schematic formulation of ATP hydrolysis or synthesis it becomes clear that changes of ATP level must also be correlated with changes of pH in the cell.

$$\begin{array}{ll}
ATP^{4-} + & H_2O \rightleftharpoons ADP^{3-} + \quad P_i^{3-} + 2\,H^+ \\
ADP^{3-} + & H_2O \rightleftharpoons AMP^{2-} + \quad P_i^{3-} + 2\,H^+ \\
\hline
ATP^{4-} + & 2\,H_2O \rightleftharpoons AMP^{2-} + 2\,P_i^{3-} + 4\,H^+
\end{array}$$

The transient increase of cytoplasmic ATP level when light is switched on, according to this, would also result in a lowering of the proton concentration. Here again the quantitative discussion is made difficult by the fact that numerous unknown parameters would have to be used in an appropriate equation; these include pH values in the critical compartments and the dissociation of the particular ionic bonds of ATP, ADP, AMP, and P_i (Hofmann and Zundel, 1974a,b). Walker and Smith (1975) assume that the small changes in cytoplasmic pH between light and dark, i.e., the small alkalinization which they observed in *Chara corallina* in the light, arises from changes in cytoplasmic ATP/ADP ratios (see Sect. 10.3.1.2).

The remarkable fact that all these signals as shown in Figure 10.15 have kinetics which are in the range of minutes makes it likely that they all have a similar cause, namely switching photosynthetic electron flow on and off.

In the steady state, however, many of these parameters also remain constant and often are similar in continuous light and darkness respectively. Often the membrane potential, after its oscillations, returns to the same resting potential in continuous light and in continuous darkness (see Sect. 10.3.1.1). The ATP level in many green cells is similar also in continuous light and in continuous darkness (Lüttge et al., 1971b). Finally one may not assume that the cytoplasmic pH value in the steady state is much different in the dark and in the light. Cytoplasmic reaction systems are too sensitive to pH for the cell in the long run to withstand such changes.

Some regulation must be involved here. The highly damped oscillatory character of the transitory phenomena summarized in Figure 10.15 may indicate such regulation. Oscillations in metabolic pathways (e.g., glycolysis: Betz, 1966, 1968; Betz and Hinrichs, 1968) or membrane-bound processes (e.g., oscillations of an electrogenic pump in *Neurospora:* Gradmann and Slayman, 1975) indicate the existence of oscillating metabolic feedback systems (review: Hess, 1977; Fig. 10.16).

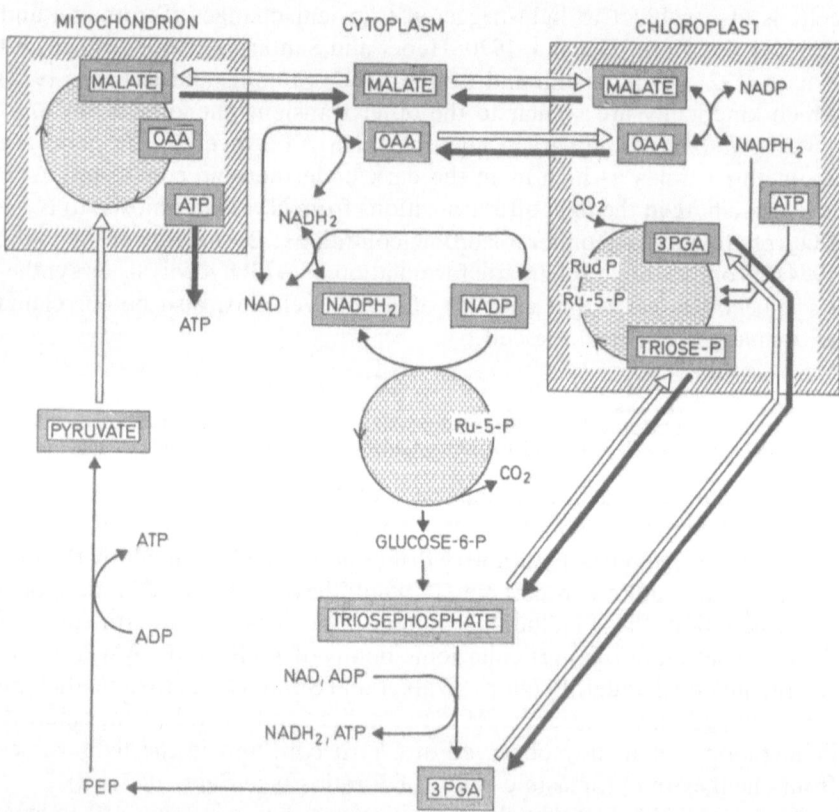

Fig. 10.16. Interactions between cofactor systems and transport metabolites in regulation of metabolic co-operation between cytoplasm, mitochondria and chloroplasts. *PEP*, phosphoenolpyruvate; *OAA*, oxaloacetate; *PGA*, phosphoglyceric acid. The other abbreviations have their usual meaning.

10.5 Correlations of Electrical Fields and Growth

It has been shown that growing cells or organs generally drive ion currents through themselves and external fields occur as well (L.F. Jaffe and Nuccitelli, 1977). In developing *Fucus* or *Pelvetia* eggs, each day several pulses of positive current, believed to represent cation flow, enter the rhizoidal end (L.F. Jaffe et al., 1974); within the cell the current may reach 1 nA cm^{-2}. The Ca^{2+} current seems particularly important. A 10- or 30- fold increase in external K$^+$ concentration may result in bursting of cells. Lowering external osmotic strength of the artificial sea water results in bursts of K$^+$ plus Cl$^-$ efflux accompanied by current pulses (Nuccitelli and L.F. Jaffe, 1976); this ion efflux restored the original turgor, approximately at least, and that shows the process is osmoregulatory.

It has also been shown that external stimuli, e.g., light or an electropotential gradient, may determine the polarity of developing Fucoid eggs; the effects appear to be dependent on external K$^+$ concentration. Bentrup (1974) has found that the photoreceptor, which is presumed to have highly ordered dipolar molecules, responds to plane-polarized but not naturally polarized light. The model developed by Bentrup to explain polarity induction in *Fucus* eggs involves an H$^+$ export system.

It should be mentioned that in the absence of external stimuli of a suitable strength polarity appears to be established by forces internal to the eggs, presumably an electrical field generated by ion currents.

Associated with the polarity of developing Fucoid eggs is the evidence for localized channels allowing selectivity in ion currents at the poles; Jaffe and Nuccitelli (1977), in their review, consider the hypothesis of, and evidence for, lateral or perimembrane electrophoresis. Presumably this may apply to the membrane components constituting the selective ionic channels.

External electrical fields are also associated with higher plant organs (Scott, 1967) and with pollen tubes (Weisenseel and L.F. Jaffe, 1976). It may be concluded that there are electrical circuits in both the symplast and apoplast perhaps basically driven by transport across the plasmalemma; however streaming potentials could arise where solute flow is appreciable, as in the xylem. The significance of these phenomena remains to be elucidated although Newman and Sullivan (1976) have proposed a simple model to explain the phytochrome-mediated electrical pulse generated by red or far-red light in coleoptiles.

Part IV

Inter-Cellular and Inter-Organ Transport: Chapters 11 to 13

The term "short-distance transport" usually is intended to exclude the long-distance transport in the phloem and xylem. Short-distance transport may deal with movement across a membrane, within a cell, or even between different cells of a tissue which are close to each other. In contrast to the term long-distance transport, short-distance transport is therefore often not defined strictly in the literature and the term is used by various authors in different senses. If, however, one defines short-distance transport more strictly, meaning only membrane transport and movement within a cell (i.e., intracellular transport), and at the same time introduces the new term "medium-distance transport" for movement occurring between various cells of a complex system independently of the xylem and phloem (i.e., the pathways for long-distance transport), then we arrive at three analogous terms. These terms appear useful because they are quite descriptive. It is a particular advantage of these three expressions that it is immediately clear which basic transport processes are necessary in the complex system of the intact higher plant in order to make it functional as a living unit.

After having dealt with short-distance transport in the preceding chapters, we aim in Chapter 11 to discuss the importance of the various pathways of transport for medium-distance and long-distance transport. In Chapter 12, on the basis of a few particularly well-known models of highly differentiated organs of higher plants, the interactions between short-distance, medium-distance, and long-distance transport will be assessed. Finally, the regulation of transport in the plant as a whole will be discussed in Chapter 13. In so doing we close the circle in an attempt to describe highly integrated systems, and come back to the claim we made initially, i.e., that the final aim of our discussion in general of research in transport will have to be the understanding of all the individual transport processes which are necessary for the function of an entire higher plant.

Chapter 11

Medium-Distance Transport and Long-Distance Transport

11.1 The Importance of Particular Pathways for Medium-Distance and Long-Distance Transport

According to Münch (1930) we can distinguish in each plant apoplastic and symplastic spaces. Symplastic spaces are delineated by the plasmalemma constituting the external surface of the living protoplasm. When the protoplasts of different cells are interconnected by plasmodesmata, they form a symplastic system. Apoplastic spaces conversely are situated outside the plasmalemma barrier. Among the apoplastic spaces are the intermicellar and the interfibrillar spaces of the cell wall, but also the intercellular spaces, usually air-filled. Similarly, vessels of the conducting elements serving long-distance transport of water and salts in the xylem have to be considered as belonging to the apoplastic system because their protoplasts have died. Since the structure, and, especially, the activity of the living cytoplasm provide conditions for transport which are very different from those present in the apoplast it is very useful to distinguish between symplastic and apoplastic transport pathways.

11.2 Apoplastic Transport Pathways

11.2.1 Cell Wall Transport

We have already described the conditions for transport of material in the apoplastic free space of cell walls (Sect. 4.3). Chemical and electro-chemical potential gradients are the driving forces for the movement of particles in this space. For water transport capillary forces are also important. In addition, we have to consider water potential gradients; for instance, by the evaporation of water from the surface of leaves (transpiration) a stream of water (volume flow) in the leaf apoplast is maintained in which solute particles may be dragged along. At the places of maximum water evaporation these solute particles may accumulate.

The cell wall phase must always be crossed by solutes moving through the plasmalemma in short-distance transport between the symplast and the external medium. Conversely, the cell wall apoplast also provides a pathway for medium-distance transport. This function is often of importance. When the apoplastic free space of roots, for instance, is equilibrated with the external medium, not only the root epidermal cells can take up ions actively from the medium, but also all cortex cells are connected to the medium via the apoplastic space and can participate in active ion uptake. This leads to a considerable increase in the effective surface involved in ion uptake by roots (Lüttge and Laties, 1966, 1967). Similarly, as we have seen, this is true for ion uptake by leaf cells from the apoplast (Sects. 8.2.4.1.5 and 12.1).

Certainly, however, transport between individual cells of a tissue will not occur exclusively in the wall space. Under certain conditions the apoplastic route can be blocked for transport of exogenous solutes and transport across the cytoplasm can be enforced (Sect. 4.2), so that a metabolic control of transport is possible. For a similar reason the transport of substances of endogenous origin usually synthesized in the cytoplasm will not occur in the apoplast; in this case the metabolic control on the substances in question would be lost. We will see below that the transport in the symplast, as compared to transport in the apoplast, has not only the advantage of the metabolic control of transport, but also it can be more rapid (Sect. 11.3). Notwithstanding these limitations medium-distance transport in the free space is important and irreplaceable.

11.2.2 The Transpiration Stream

Considerable volumes of water with dissolved nutrients reach their terminal point passing through the vessels of veins into the apoplastic space of leaves. Outside an aquatic environment only plants of minimal size and mass can afford to transport water and dissolved nutrients exclusively in the apoplastic wall space without having specifically differentiated elements for long-distance transport. Even in the larger mosses particular transport tissues have evolved from lengthy prosenchyma cells in the midribs of leaves and in stems (Eschrich and Steiner, 1967, 1968a,b; see Frey, 1977, for a phylogenetic interpretation). Paleobotanical studies show that conductive tissues occurred in the first land plants which evolved on earth. Knowledge of the anatomy of conductive tissues of extant higher plants is basic but cannot be discussed in detail here. In brief, the conducting elements of the xylem consist of vessels or tracheids, or both, and, at maturity lack protoplasts. Tracheids, which predominate in the vascular plants below the angiosperms, are elongate with pitted walls, but the walls are not perforated. Vessels, characteristic of angiosperms, develop from linear series of cells but at the time of maturation the cross walls become perforated; thus they form a pipelike system, generally having less resistance to flow than tracheids. The xylem in either case forms a continuous apoplastic pathway for solute flow from the root to the leaf veins.

The driving force for the flow of water in these conduction elements is composed of two components. First, transpiration, and, second, uphill accumulation of salts brought about by the activity of living cells in the root. To describe the water flow from the root medium (soil or culture solution) through the plant into the atmosphere it is useful to separate the total water potential gradient between the soil-root system and the shoot-atmosphere system into single components (e.g., Table 11.1):

Table 11.1. Representative values for the various components of the water potential in the soil-plant-atmosphere system. $\Psi = \rho_w gh$ in the liquid phases [Eq. (2.8)], and $(RT/\bar{V}_w)$ ln (% relative humidity/100) $+ \rho_w gh$ in the gas phases, all at 25°. (Nobel, 1974.)

Location	P (bars)	$-\pi$ (bars)	$\rho_w gh$ (bars)	$\dfrac{RT}{\bar{V}_w} \ln \left(\dfrac{\text{\% relative humidity}}{100}\right)$ (bars)	Ψ (bars)
Soil 0.5 cm below ground and 1 cm from root	−2	−1	0		−3
Soil adjacent to root	−4	−1	0		−5
Xylem of root near ground surface	−5	−1	0		−6
Xylem in leaf at 10 m above ground	−8	−1	1		−8
Vacuole of leaf mesophyll cell at 10 m	2	−11	1		−8
Cell wall of leaf mesophyll cell at 10 m	−4	−5	1		−8
Air in cell wall pores at 10 m (water vapor assumed to be in equilibrium with water in cell wall)			1	−9	−8
Air just inside stomata at 95% relative humidity			1	−70	−69
Air just outside stomata at 60% relative humidity			1	−702	−701
Air just across unstirred layer at 50% relative humidity			1	−951	−950

$$\Delta\Psi_{\text{root medium-atmosphere}} = \Delta\Psi_{\text{leaf tissue-atmosphere}}$$
$$+ \Delta\Psi_{\text{conducting system-leaf tissue}}$$
$$+ \Delta\Psi_{\text{root tissue-conducting system}}$$
$$+ \Delta\Psi_{\text{soil-root tissue}} \tag{11.1}$$

A water deficit of the atmosphere as compared to tissue leads to a continued evaporative water loss from the aerial parts of the plant, i.e., transpiration. Such a water deficit is maintained mainly by solar energy radiation, which thus is the source of energy for transpiration. Mainly as a consequence of transpiration the other potential gradients described in the equation above are built up and volume flow through the plant is maintained. This constitutes the transpiration stream. Under particular environmental conditions, for instance in forests in a tropical fog belt, or in montane regions in the spray of torrential streams, the water potential gradient may have an opposite direction and plants can have a water deficit relative to air and thus may take up water from the atmosphere.

For maintenance of water flow in the vessels usually the plant does not have to use its own metabolic energy. We cannot neglect, however, the contribution of the work done in concentrating solutes which is accomplished by living cells. Active ion uptake from the root medium via the root tissue leads to an osmotic driving force, and thus contributes to the water potential gradient $\Delta\Psi_{\text{soil-root tissue}}$. An active transport of ions from the living root tissue into the dead conducting elements (although still controversial, see Sect. 12.2) would add to the water potential gradient $\Delta\Psi_{\text{root tissue-conducting}}$ system. In any case, active ion transport in the root establishes osmotic driving forces which result in a positive hydrostatic pressure in the root xylem and, with excised roots, to root-pressure exudation. An active ion uptake by the leaf cells from the leaf apoplast would also add to the total water potential gradients, $\Delta\Psi_{\text{conducting system-leaf tissue}}$ (Sect. 12.3).

A consequence of the water movement caused by active transport of solutes becomes visible especially under conditions in which transpiration is eliminated as a force maintaining the water potential gradient; for instance this is the case when the atmosphere has a high relative humidity and the gradient $\Delta\Psi_{\text{leaf tissue-atmosphere}}$ therefore disappears. In many plants under these conditions an intensive water transport in the direction root-shoot system to external surface may still occur. This water transport becomes visible in guttation, which is an exit of liquid droplets from small water pores which may be nonfunctional stomata or from glands. The phenomenon often can be observed on leaves. A particularly descriptive example is the lady's mantle, *Alchemilla*, in which the leaves have hydathodes on every little dent of the leaf which can yield droplets of guttation. *Alchemilla* grows on moist meadows in the mountains and very frequently in the early hours of the morning when the solar radiation is not

Fig. 11.1. Guttation at the leaf dents of Lady's mantle (*Alchemilla*).

very significant, one can observe droplets on the leaf margin as shown in Figure 11.1.

Transpiration is completely eliminated when the entire shoot system is removed; then, when glass tubing is connected to the root, the xylem sap rises in the tubing, a clear demonstration of root pressure. This phenomenon occurs in the angiosperms but apparently not in conifers, which have tracheids only (Kramer, 1969).

The guttation fluids or the exudates are never pure water but normally a solution of various inorganic ions and organic acids such as amino acids and Krebs cycle acids (Gračanin, 1964; Anderson and Reilly, 1968; Anderson and Collins, 1969; Perrin, 1972). This is not an active water transport as such but rather an active transport of solute particles with a passive (osmotic) movement of water. The unlikelihood of a true active transport of water was mentioned in Section 2.5.2.2.

With a few exceptions the positive hydrostatic pressures developed in the root xylem have not been considered seriously as an important driving force in the ascent of sap. As early as 1727, however, a "root-pressure" process, dependent on "vital" tissue, was hypothesized by Stephen Hales. Since the development of the cohesion theory of the ascent of sap (Dixon and Joly, 1895) root pressure phenomena have received little attention. Nevertheless in 1938 White, using a manometric technique requiring minimal volume flow, found that tomato root exudation continued undiminished at a pressure of 6 atm, the upper limit of the manometer he was using. White believed that the pressure could have reached 10 atm. At 6 atm sap could be raised to more than 200 feet (60 m). It seems quite clear that with smaller plants in conditions in which $\Delta\Psi_{\text{root medium–atmosphere}} = 0$, or nearly so, ascent of xylem sap could continue.

Except for unusual conditions, e.g., an aquatic habitat, Ψ_w shows a large negative gradient from the root medium through the plant to the atmosphere; instead of a positive hydrostatic pressure in the xylem, water is under tension, i.e., P is negative. This can occur because of the cohesive and adhesive properties of water which can retain its continuity under a tension (experimentally) of about 300 bars (Nobel, 1974). The cohesive forces (tensile strength) of water are quite adequate to maintain the continuity of the water filaments occupying the vessels, or tracheids; the "pulling" force arises from transpiration in the leaves. Adhesive forces attracting water to the wall surfaces of the conducting elements are commensurate with cohesive strength. Since one atmosphere corresponds to approximately 10 m of water filament, a cohesive force of 50 atm would allow the accent of sap to 500 m. If we take into account a loss by friction of about 0.1–0.2 atm/m there is still enough left for a supply of water to canopies of trees which are 120–130 m high (cf. Mohr, 1969). Indeed we can observe in nature that this is just about the highest limit that trees can reach. Redwoods (*Sequoia sempervirens*) in the coast range of Southern Oregon and California can attain heights up to 350 feet (about 110 m; Berry, 1964), the highest redwoods found in Humboldt County north of San Francisco were 367.8 feet (i.e., 114 m; Boerner, 1969). The highest giant sequoia (*Sequoia gigantea*) in Yosemite National Park is 300 feet (about 90 m; Cole, 1963). Sequoias are matched by some Australian gum trees, i.e., giant *Eucalyptus*, which attains similar heights.

Resistances play an important role. An analogy to Ohm's Law can be made:

$$\text{Potential} = \text{current} \times \text{resistance}$$
$$V = I\,R \tag{11.2}$$

and one can formulate for transpiration (T)

$$\Delta \Psi = T\,R. \tag{11.3}$$

In Eq. (11.3) R is a complex term corresponding to the various resistances which limit water flow through the various parts of the environment–plant system as depicted in Figure 1.1.

$$R = R_{soil} + R_{root} + R_{shoot} + R_{leaf} + R_{leaf\ surface}. \tag{11.4}$$

These single resistances again can be composed of various components. For instance, the resistance for water exit via the cuticle of the leaves (cuticular transpiration) will certainly be much larger than the resistance for the loss of water across open stomata (stomatal transpiration). In this example we consider two parallel resistances. Also the structures in sequence, for instance intercellular systems, cell walls, membranes, cytoplasm, pathways for long-distance transport, etc., pose quite different resistances for water flow. Here the principles of the cooperative action of short-distance, medium-distance, and long-distance transport become clear, and one can easily imagine that within the plant the resistance of the pathways of long-distance transport is the lowest and that the resistance for the transport across membranes is the highest.

For details of structural and other properties including a physical treatment of the problem of water transport see Slatyer (1967), Kramer (1969), and Larcher (textbook treatment, 1973, 1975). In summing up, we conclude that mass flow in the vessels is important for regulation of the water supply of plants, and that at the same time it is necessary for the supply of all parts of the shoot system with nutrients which have been taken up by the roots. The simplest, and probably largely correct consideration of long-distance transport of mineral nutrients is that they are just dragged along by the mass flow of the transpiration stream. More recent investigations show, however, that there is also an important exchange of ionized particles on fixed charges at the walls of the vessels (Läuchli, 1967, 1972).

11.3 Symplastic Transport

The conception of the term symplast is credited to Münch. The phenomenon of a particularly good communication between cells whose protoplasts form a symplast is regularly observed in plants. The velocity of transport between cells of plant parenchyma is $1-6$ cm h^{-1} and considerably larger than that of simple diffusion (Arisz and Wiersema, 1966; Webb and Gorham, 1965).

11.3.1 Plasmodesmata, the Structural Basis for Symplastic Transport

An important prerequisite for the rapid communication in tissue obviously is the occurrence of cytoplasmic bridges, plasmodesmata, between the adjacent cells (Fig. 11.2). When particularly good communication occurs in animal cells, there is often a very close association of the plasma membranes of neighboring cells (e.g., tight junctions). In plant cells "tight junctions" are prevented by the cell wall, and thus plasmodesmata connections become essential. Plasmodesmata were first discovered by Tangl in 1879, who thus provided the basis for the symplast theory and for the field of intercellular communication in plants which recently began to develop rapidly (reviews in Gunning and Robards, 1976).

In the primitive orders of the green algae congenital colonies are formed in which the cells remain associated after cell division. Thus some algae form colonies (e.g., Volvocales) with increasing differentiation and division of labor in the different parts. In such colonies cells have plasmodesmata between them. Conversely, cell colonies may originate from secondary aggregation of cells, which had separated after division; these do

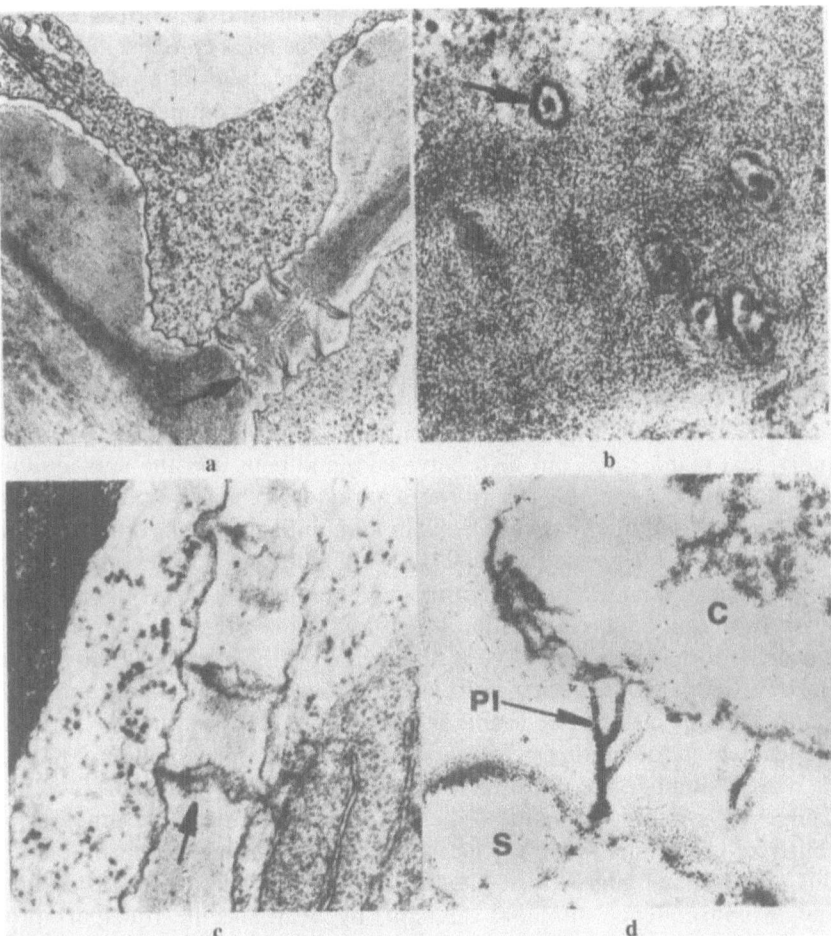

Fig. 11.2. a Longitudinal sections of plasmodesmata (*arrow*) between xylem parenchyma cells of *Phaseolus coccineus* roots. ×30,600. (Original electron micrograph by courtesy of Prof. Dr. A. Läuchli.); **b** cross sections of plasmodesmata between xylem parenchyma cells of *Phaseolus coccineus* roots. ×100,900. (Original electron micrograph by courtesy of Prof. Dr. A. Läuchli.); **c** longitudinal sections of plasmodesmata between mesophyll cells of barley leaves. ×20,600. (Original electron micrograph.); **d** plasmodesmata (*Pl*) between a sieve tube (*S*) and a companion cell (*C*) of a *Limonium* leaf. The leaf was kept with its petiole in a 550 mM NaCl solution and thus was transporting much Cl⁻. The medium of fixation for electronmicroscopy included Ag⁺ ions precipitating Cl⁻ as AgCl in the tissue and thereby localizing Cl⁻ as an electron dense precipitate. The strong correlation of precipitates with the plasmodesmata suggests that Cl⁻ can move in these structures. ×13,500. (Ziegler and Lüttge, 1967.)

not have plasmodesmata between individual cells and do not reach as high a level of organization. This comparison of true algal colonies and simple algal aggregates stresses the role of plasmodesmata for intercellular communication which probably allowed evolution of complex higher organisms. In fact, of course, evolution in general was much more complex than this simple example might imply. Plasmodesmata like structures are found between cells in filaments of prokaryotic blue-green algae in which there is a division of labor between heterocysts and ordinary cells (Marchant, 1976). In many higher plants there is not only a primary formation of plasmodesmata during cell division, but also secondary formation after establishment of a nonperforated cell wall occurs (Jones, 1976).

In tissues of higher plants we find plasmodesmata in increased numbers, usually at locations where particularly good communication appears necessary for a special function (e.g., in gland tissues between the individual gland cells, but also between gland cells and the surrounding parenchyma; Lüttge, 1971a). In roots particularly large numbers of plasmodesmata are found between cells in longitudinal files, i.e., on the anticlinal walls, while periclinal walls are interrupted by fewer plasmodesmata. Cells of longitudinal files obviously function in a highly coordinated way. In the root tip, cells within longitudinal files divide synchronously, but cells of adjacent files have a different dividing rhythm (Juniper and Barlow, 1960; tables giving sizes and frequencies of plasmodesmata in many plant species can be found in Robards, 1976). The function of plasmodesmata in facilitating symplastic transport becomes obvious by their increased numbers at locations where considerable transport occurs. Conversely, plasmodesmata are scarce or absent where the function of cells requires their isolation from the surrounding tissue. An example for this are stomatal guard cells. Guard cells regulate their turgor pressure, and in consequence stomatal opening, by very specific transport processes. Guard cells behave differently from surrounding epidermal cells, and between the guard cells and the neighboring cells plasmodesmata are absent, or at least are not very frequent (see also Sect. 12.3.5.1).

The movement of solutes in plasmodesmata has been demonstrated by various cytochemical localization techniques. The now widely used Ag-precipitation technique developed by Komnick (1962) for Cl-localization in animal cells was first applied to plant material by Ziegler and Lüttge (1967). They used leaves of the halophyte *Limonium vulgare* fed with NaCl via the cut petioles. Fixation of leaf samples for electron microscopy was performed with Ag-acetate present in the fixation medium. In the electron microscope an electron dense precipitation of AgCl was particularly localized in the plasmodesmata (Fig. 11.2). This has been confirmed many times now with other plant tissues (van Steveninck et al., 1973; Stelzer et al., 1975) and with other precipitation techniques (van Iren and van der Spiegel, 1975; review: van Steveninck, 1976c). Thus, although these cytochemical localization techniques are still subject to some criticism because of possible artifacts, they show at least that plasmodesmata are accessible for solute particles.

Some of the best evidence for the function of plasmodesmata in symplastic transport is indirect and comes from observations of physiological functions which would appear to be impossible in the absence of a symplastic continuity. Some of these will be described in Section 11.3.2.

It now seems clear that plasmodesmata are living protoplasmic bridges between individual cells. The details of their fine structure have long been a matter of dispute among electronmicroscopists. Also there is some variability as shown in Figure 11.3. Some plasmodesmata are simple pores in the cell wall, lined by the plasmalemma. In others, the endoplasmic reticulum (ER) passes from one cell to the other. This provides two separated and compartmented pathways for transport in the plasmodesm, firstly, in the cytoplasmic annulus around the strand of ER, and secondly, within the ER cisternae or desmotubule. The latter is of particular interest for transport physiology. Often it has been speculated that the channels and cisternal systems of the endoplasmic reticulum serve a transport function within the cytoplasm of single cells (see Sects. 6.4 and 7.1). If the ER-system is continuous via the plasmodesmata from one cell to the next, these membrane channels may also serve symplastic transport through the parenchyma. Investigations with barley roots using the Ag-precipitation technique for Cl⁻ localization provide experimental evidence for such transport (Stelzer et al., 1975). The desmotubule may also be tightly bound and then a constriction appears to block the pathway via the cytoplasmic annules. Median nodules and cavities in the mid-line of the wall may occur and desmotubules may also anastomose or branch (Figs. 11.2 and 11.3; review Robards, 1976). Most recently it has been suggested that membrane-associated particles occluding the cytoplasmic

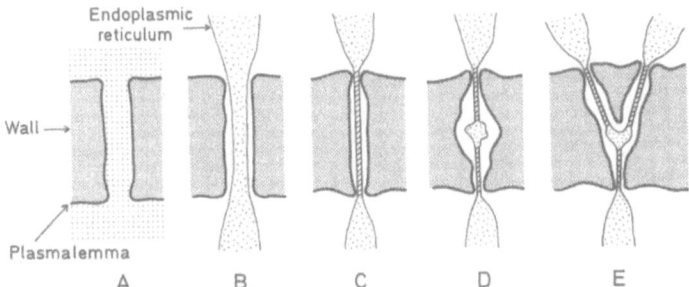

Fig. 11.3. Plasmodesmatal variation. **A** Plasmalemma-lined pore without ER (desmotubule) as found mainly in algae, occasionally in higher plants; **B** pore with strand of ER (desmotubule) loosely bound, having a cytoplasmic annulus; **C** tightly bound desmotubule; transport in the annulus, i.e., between the plasmalemma and the desmotubule appears to be blocked; **D** plasmodesm with median nodule; **E** anastomosing desmotubules. (Robards, 1976.)

annulus may function as sphincters controlling symplastic transport by functioning as plasmodesmatal valves (Willison, 1976; Evert et al., 1977).

11.3.2 Physiological Functions Depending on Symplastic Transport

11.3.2.1 Arisz' Experiments on Symplastic Transport

Although the term symplast was coined by Münch, Arisz was the authoritative investigator of symplast transport. For his experiments, which extended over many years, he used the bandlike submerged leaves of the higher water plant *Vallisneria* (Arisz, 1954, 1960, 1969; Arisz and Wiersema, 1966; reviews: Helder, 1967; Lüttge, 1969; Spanswick, 1976). This plant easily provides strips of leaves of identical length that can be used for experiments. Conductive tissues with xylem and phloem are present in these leaves, but in particular the xylem does not appear to be as conductive as in aerial leaves of land plants. There is no doubt that the presence of conductive tissues is a certain drawback of this material. An artificial interrupting of the central bundle, which is the only particularly well-developed one in *Vallisneria* leaves, shows, however, that the transport observed between two zones of the leaves is independent of whether this bundle is intact or whether there are only parenchyma bridges between the two parts of the leaves.

The basic scheme of the experimental set-up in most of the experiments of Arisz is always the same (Fig. 11.4). The strips of leaves are mounted between two or three chambers and are sealed so that the solu-

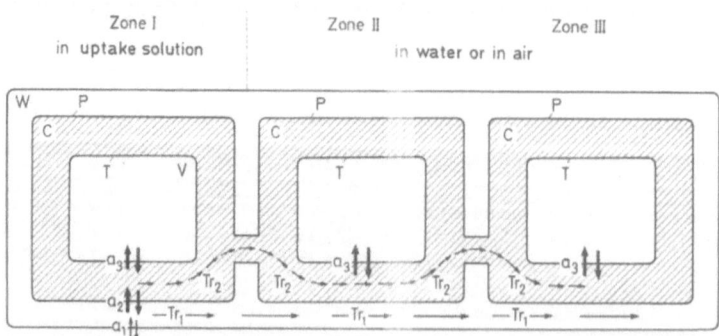

Fig. 11.4. Simplified scheme of the division of a *Vallisneria* leaf into various zones in Arisz' experiments on symplastic transport. *Zone I*, uptake region; *Zones II and III*, regions into which transport occurs; a_1, a_2 and a_3, influx and efflux of the compartments apoplastic free space, cytoplasm, and vacuole. *Thick arrows*, membrane-controlled transport. Tr_1, apoplastic transport in the free space; Tr_2, symplastic transport; W, cell wall; P, plasmalemma; C, cytoplasm; T, tonoplast; V, vacuole. (Lüttge, 1969.)

tions of the individual chambers are connected only via the leaf parenchyma. If the solutions in the individual chambers are varied, experiments can be done measuring transport through the leaf parenchyma. A few typical experiments will be described:

1. If a radioactively labeled substance (for instance, ^{36}Cl a tracer not metabolized) is applied to zone I and if zones II and III are in water, velocity of transport within the parenchyma can be measured from I to II and III. Velocities between 2 and 4.4 cm/h are found. At the same time it can be shown that an apoplastic transport (Tr_1 in Fig. 11.4) does not make an appreciable contribution to the transport rate observed, because an efflux of radioactivity does not occur into the medium of zones II and III, which should be the case if there were appreciable tracer transport in the apoplastic free space.

2. Application of an inhibitor to zone II does not block the transport of tracer from zone I to zone III. This shows that the symplastic transport as such is not hampered by the inhibitor taken up in zone II. That the inhibitor nevertheless is effective is demonstrated by the fact that ion accumulation into the vacuoles of zone II is inhibited (see also experiment 6).

3. Application of the inhibitor in zone I blocks the uptake of tracer and transport within the parenchyma. The driving force for the symplastic transport, therefore, must be the concentration gradient built up by active uptake in zone I, thus developing a gradient between zones I and II.

4. Another test is the application in zone I of radioactively labeled medium and in zone II of a nonlabeled solution of otherwise equal composition. Rapid tracer transport from I to II is no longer observed. The isotopic exchange evidently is much slower than symplastic transport.

5. With Arisz' set-up simultaneous transport of different substances in opposite directions can also be shown. If to zone I a radioactively labeled KCl solution is applied, and to zone II a ^{14}C-labeled asparagine solution, KCl transport takes place in the direction from I to II and an asparagine transport in the opposite direction from II to I; this transport must depend on the concentration gradient for each species of particle. For a more detailed interpretation, however, one would also have to take into account the electropotential gradient, because the relevant driving force for the transport of charged particles is the electrochemical potential.

Apart from these conclusions on symplastic transport, the system of Arisz also makes possible elegant experiments on membrane transport across plasmalemma and tonoplast, respectively. Therefore, we want to describe two further experiments here, although, their results would be more relevant in the context of the compartmentation models discussed in Chapter 6.

6. Using the set-up of experiment 3, a tracer and an inhibitor were applied to zone I; uptake by the tissue is a function of inhibitory effects on uptake across the plasmalemma and the tonoplast. If the inhibitor is added not to the medium of zone I but rather to the medium of zone II, but the tracer as before to zone I, then tracer accumulation in zone II can be measured. In this case only the inhibitory effect on the transport across the tonoplast into the vacuole is seen because the tracer which arrives in zone II from zone I has already been taken up into the cytoplasm within zone I. In this way Arisz found that chloride transport at the plasmalemma is more sensitive to cyanide, arsenate, and uranyl ions than transport across the tonoplast. On the other hand transport across the tonoplast is more sensitive to azide.

7. A selective light effect on membrane transport processes was shown by Arisz in the following way. If the leaf zone absorbing Cl^- (zone I) is in the light but the zone not exposed to Cl^-, zone II, is in the dark, then very little Cl^- is transported to zone II. The bulk of the ions taken up are accumulated in the illuminated zone I. If conversely zone I is in the dark and zone II is in the light, then the bulk of the ions taken up in the darkened zone I will be transported into the vacuoles of the illuminated zone II. By alternating darkness and light, respectively, of zones I and II, the ions already absorbed can be retranslocated via the symplast between the vacuoles of zone I and zone II. Additional experiments show that the effect of light lies in utilizing photosynthetic energy, thus an important effect of photosynthesis on chloride transport into vacuoles is demonstrated.

11.3.2.2 Electrical Coupling Demonstrating Symplastic Continuity

11.3.2.2.1 Passage of Electrical Current

Spanswick and coworkers (Spanswick and Costerton, 1967; Spanswick, 1972b; reviews Spanswick, 1976; Goodwin, 1976) studied the passage of electrical current from cell to cell in higher plant tissues and in Characean thalli to demonstrate electrical coupling via plasmodesmata. Their experimental set-up in measurements using the leaves of the angiosperm water plant *Elodea* is explained in Figure 11.5. With the glass microelectrode I a current is injected into the vacuole of cell A, and the current can pass to the electrode I_0 in the external solution. E_1 and E_2 record the membrane potentials. (According to Ohm's law resistance can be calculated from current and potential.) In small higher plant cells glass microelectrodes normally can only be inserted into the vacuoles; insertion into the cytoplasm is not possible unless further experimental procedures are used, such as the centrifugation technique mentioned in Figure 2.9. Thus in Spanswick's experiment of Figure 11.5a the current injected via I passes

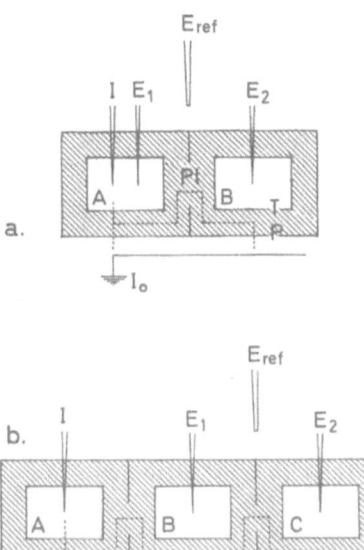

Fig. 11.5. Experimental setup for injection of current and determination of tonoplast (T), plasmalemma (P) and plasmodesmatal (Pl) resistances in *Elodea* leaves. I, E_1, E_2 are glass microelectrodes inserted into cells A, B, and C; I_o and E_{ref} are electrodes in the external solution. I and I_o serve passing a small electrical current through the system; E_1 and E_2 measure intracellular membrane potentials in relation to E_{ref}. (After Spanswick, 1972a, 1976.)

from cell A across the tonoplast and the plasmalemma of cell A to the external electrode, I_o. E_1 measures in series the membrane potentials of tonoplast and plasmalemma, which are both affected by the current. The current can also pass from I across the tonoplast of cell A, via the plasmodesmata between cells A and B and across the plasmalemma of cell B to the external electrode, I_o. E_2 in cell B again measures the tonoplast and plasmalemma potentials (of cell B) in series, but in this case only the plasmalemma potential is affected by the current; i.e., the tonoplast and the plasmalemma resistances contribute to the potential measured by E_1 (E_A), whereas the potential recorded by E_2 (E_B) is determined by the resistances of the plasmodesmata and the plasmalemma. Therefore, a comparison between E_A and E_B cannot yield the true electrical coupling via the plasmodesmata between cells A and B. (The "coupling ratio" E_B/E_A is 0.29 for *Elodea* cells.) This problem is overcome by the setup of Figure 11.5b. The pathways of current between cells A and B are as before. However, as can be seen in the figure, E_1 and E_2 record potentials (E_B and E_C respectively), which both depend on plasmalemma and plasmodesmata resistances in the same way. Thus E_C/E_B gives the true coupling ratio via the plasmodesmata, which is 0.72 in the case of *Elodea*.

By a combination of measurements as depicted in Figures 11.5a and b, Spanswick calculated the various specific resistances as follows:

plasmalemma	0.31	Ωm^2
tonoplast	0.1	Ωm^2
plasmodesmata	0.0051	Ωm^2

(Similar experiments were performed with other plant tissues, e.g., cortical cells of maize roots, *Avena* coleoptiles, and Characean thalli.)

These results prove that an intercellular electrical coupling via the plasmodesmata is much more efficient, i.e., of much lower resistance, than a coupling via the two plasmalemmas and the cell wall space between adjacent cells.

On the other hand the resistance of the plasmodesmata in *Elodea* is about 60 times higher than would be expected for unlimited diffusion in an aqueous phase. In *Nitella* this "impediment factor" is 330 (Spanswick and Costerton, 1967) and in *Chara* values between 7 and 700 have been considered (Bostrom and Walker, 1975; Tyree et al., 1974). Thus, these electrophysiological studies confirm results of ultrastructural investigations (Sect. 11.3.1) showing that plasmodesmata are not simply open aqueous channels but are occluded or constricted to a certain degree. It might be noted though, that Fischer and MacAlister (1975) come to the conclusion that occluding of plasmodesmata in electron microscopic preparations is an artifact due to sudden reduction of pressure. This may also occur during impalement of electrodes, so that electrophysiological work would overestimate plasmodesmatal resistances. Plasmodesmata of *Chara* cells in vivo might well be occupied only by viscous cytoplasm (Tyree et al., 1974).

The symplastic pathway in plants most likely is also the path for the propagation of action potentials (Williams and Spanswick, 1976). Intercellular electrical coupling can be controlled by growth regulating systems such as phytochrome (Racusen, 1976).

Experiments like those of Spanswick described in Figure 11.5 are technically very difficult. It is painstaking to insert more than one microelectrode (3 in Fig. 11.5) into small cells of a higher plant tissue. This perhaps is the reason why in some cases electrical coupling in higher plants was not observed (e.g., Goldsmith et al., 1972). Therefore, it is useful to apply another approach to demonstrate electrical coupling which will be explained in the next section.

11.3.2.2.2 Propagation of Light-Dependent Membrane Potential Oscillations

In Section 10.3.1.2 we have learned about membrane potential oscillations in green plant cells. These electrical signals are generated only in photosynthetically active cells when light is switched on or off. Using tissues which comprise both green and nongreen cells it has been possible

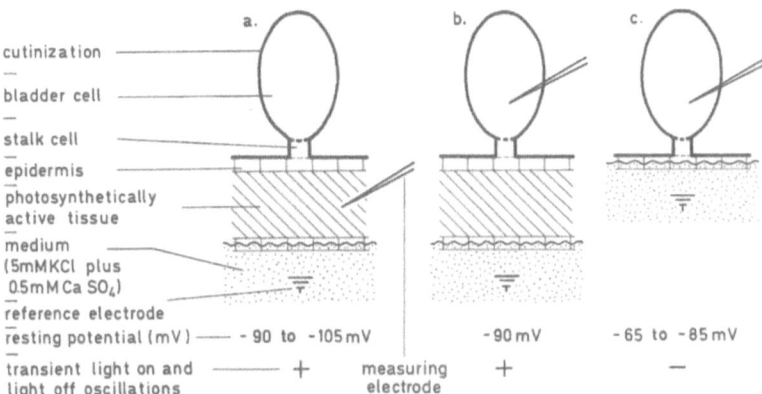

Fig. 11.6. Electrical coupling between vacuoles of salt bladders and mesophyll cells of *Chenopodium album*. Transient light-on and light-off oscillations of membrane potential generated in the mesophyll cells **a** are also picked up in the bladder cells attached to leaf mesophyll **b**, but no oscillations are observed in bladder cells of isolated epidermis **c**. (After Osmond et al., 1969; Pallaghy and Lüttge, 1970; Lüttge and Pallaghy, 1969.)

to demonstrate electrical coupling by triggering the signal in the green cells and monitoring it with a microelectrode in a nongreen cell, i.e., by inserting one microelectrode only.

Very suitable systems are the leaves and large epidermal bladder cells of Chenopodiaceae such as *Atriplex spongiosa* and *Chenopodium album*. The bladder cells have large central vacuoles in which salt accumulates (see Sect. 7.3.2), and only a thin layer of cytoplasm along the cell wall. They are connected by a glandlike stalk cell to the leaf surface. Plasmodesmata in the bladder and stalk cell walls appear to allow symplastic coupling with the green mesophyll cells. The bladder and stalk cell cytoplasm does not contain photosynthetically active chloroplasts, these cells are photosynthetically inactive (Sect. 7.3.2). The photosynthesis-dependent membrane potential oscillations, however, are not observed only with an electrode in a green mesophyll cell, but also when the electrode is inserted in the nonphotosynthetic bladder cells of *Atriplex* or *Chenopodium* (Fig. 11.6). In *Chenopodium album* the epidermis with the bladders can easily be stripped off the green mesophyll tissue. In this case light-triggered membrane potential oscillations cannot be observed in the bladders any more, although the high resting potential assures intactness of the cells. Thus the signal observed in the situation of Figure 11.6b, i.e., with the tip of the electrode in the bladder cell of an intact piece of leaf, must be transmitted to the bladder from the green mesophyll. These observations strongly suggest electrical coupling, but they do not provide a rigorous proof because of the structural complexity of the system. The cutinization and suberization of the surface of the leaf epidermis and

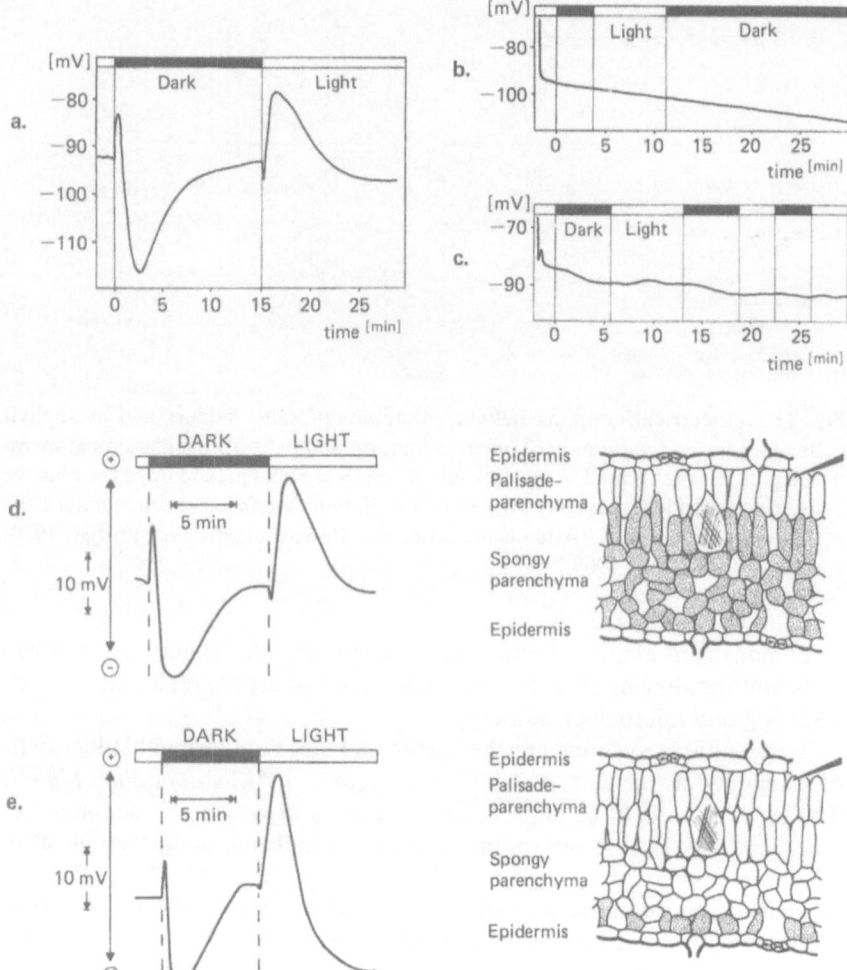

Fig. 11.7. a–c Membrane potentials of *Oenothera* leaf samples comprising only nonmutated or mutated cells respectively, where light-triggered membrane potential oscillations can only be obtained with nonmutated cells. **a** Light-triggered membrane potential oscillations of a nonmutated green palisade parenchyma cell of *Oenothera* (wild type); **b** membrane potential of a palisade parenchyma cell from a mutated leaf area of *Oenothera albicans* × *hookeri* IV/Vγ where photosystem II is inhibited; **c** mutated palisade parenchyma cell from *Oenothera albicans* × *hookeri* IV/IIα where photosystem I is inhibited. **d–e** Membrane potentials of *Oenothera* leaf samples comprising both mutated and nonmutated cells, where membrane potential oscillations can be obtained with an electrode in a mutated palisade parenchyma cell (*upper right corner* in diagrams of leaf sections) of the pale leaf areas. **d** Only the uppermost cell layer of the palisade parenchyma is mutated, all other mesophyll cells (*dotted cells*) are normally green; **e** only the lowest cell layer of the spongy parenchyma is normally green (*dotted cells*), all other mesophyll cells are mutated. (Brinckmann and Lüttge, 1974.)

bladders and of the wall between bladder and stalk cells makes an apoplastic transmission of the signal unlikely. Spanswick (1976) has pointed out, however, that for the same reason coupling between the electrode in the bladder and the reference electrode in the medium may be a problem. If the pathway of least resistance between the two electrodes included mesophyll cell membranes, a signal could be observed with the electrode in the bladder even in the absence of symplastic connections.

Experiments with variegated leaves avoid these problems. Of special interest are mutants of *Oenothera*. As shown in Figure 11.7 light-triggered membrane potential oscillations are observed only with the electrode in green nonmutated leaf (Fig. 11.7a). No signal is observed, if leaf pieces are used comprising only mutated pale cells with inactive photosystem II (Fig. 11.7b) or photosystem I (Fig. 11.7c). If, however, the mutated tissue is left in contact with nonmutated green tissue a signal can readily be detected with the electrode in a pale mutated cell (Fig. 11.7d,e). This occurs over a fair distance without apparent lag. In one experiment the distance between the electrode in a mutated cell and the closest green cells was 0.9 mm. In these experiments, the leaf apoplast clearly provides a low resistance pathway between the microelectrode inserted into the leaf cells and the reference electrode in the external solution. Thus they unequivocally prove symplastic electrical coupling. Propagation of the signal in the symplast is too fast to be due to transport of solutes, e.g., of ions or of some product of photosynthesis. This fast propagation of the signal suggests kind of an "electrotonic" mechanism of coupling, whatever this means. In this context it also appears worth mentioning experiments by Penel and Greppin (1975) showing that in spinach leaves a phytochrome effect can be communicated very rapidly from illuminated parts to nonilluminated parts suggesting a fast electrochemical communication.

11.3.2.3 Metabolic Co-Operation Between Parenchyma Cells Over a Medium Distance

11.3.2.3.1 The Nature of Metabolites Transported

In Chapter 7 we introduced the useful term transport metabolites. This refers to intermediary products of metabolism which can pass intracellular compartmental barriers, especially the envelopes of mitochondria and chloroplasts. Transport-metabolite shuttles are essential for intracellular regulation of metabolic pathways and for energy-dependent processes outside the organelles (Chap. 10). Beyond intracellular regulation transport metabolites also provide for intercellular metabolic co-operation and regulation between parenchyma cells over a medium distance.

A particular example is given by C_4-photosynthesis. A simplified scheme of C_4-photosynthesis was introduced in Section 10.2.1 (Fig. 10.1). We have seen there that malate and pyruvate are shuttled between mesophyll and bundle-sheath, i.e., transport metabolites move not only

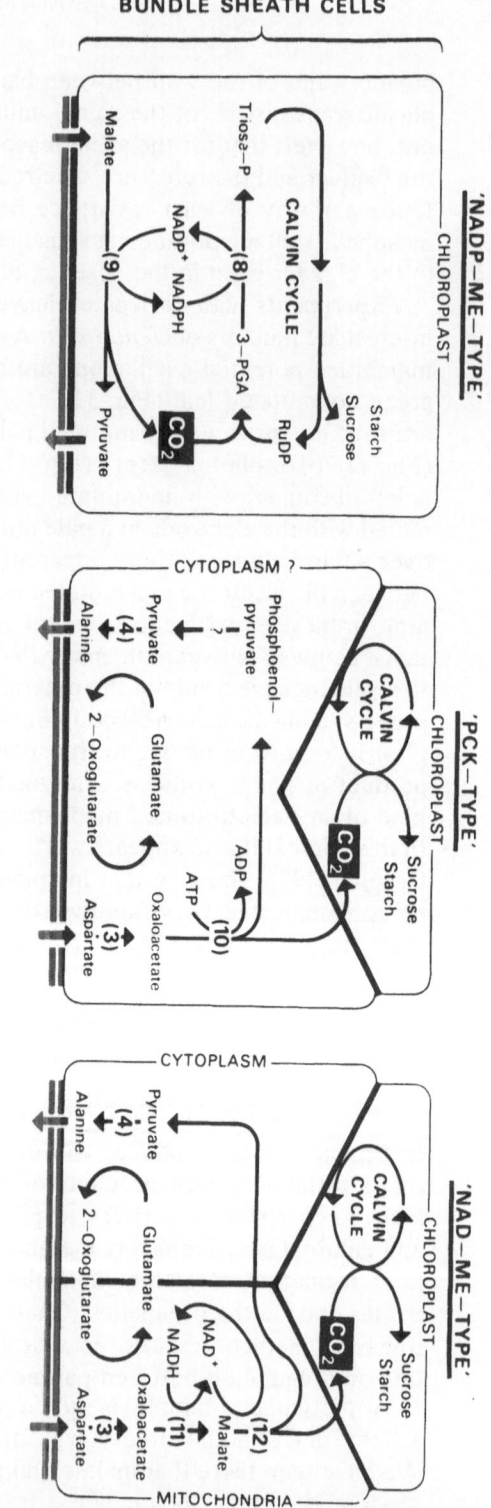

Fig. 11.8.

between organelles within given cells, but also between organelles of different, specifically differentiated cell layers. C_4-photosynthesis provides an exciting, natural model for symplastic transport.

In recent years investigations of enzyme and substrate compartmentation, photochemical reactions, and energy metabolism in C_4-photosynthesis have supplied a wealth of new information. In contrast to the original German version of this book, written only a few years ago (Lüttge, 1973), it is now far beyond the scope of our text to review the salient features of C_4-photosynthesis and comprehensively discuss the transport processes involved. This has been done recently by Hatch and Osmond (1976), and Figure 11.8, taken from this review, may give an idea of the complexity of intra-cellular and inter-cellular compartmentation which plays a role in C_4-photosynthesis and which requires communication mediated by transport metabolites.

Photochemical and energy-producing events in chloroplasts of mesophyll and bundle-sheath cells, C_4-acids synthesized in the mesophyll cells, and decarboxylation mechanisms of C_4-acids in the bundle-sheath cells all contribute to diversification in C_4-photosynthesis (Gutierrez et al., 1974; Hatch et al., 1975; Mayne et al., 1974). On the basis of the decarboxylation mechanism of C_4-acids in the bundle-sheath three major groups of C_4-plants can be distinguished as shown in Figure 11.8 (Hatch et al., 1975; Hatch and Osmond, 1976; Gutierrez et al., 1974). The figure shows many intracellular transport metabolite shuttles in the mesophyll cells and bundle-sheath cells, respectively. These are consistent with the mechanisms discussed in Chapter 7. Events in the mesophyll cells are similar in all three groups of C_4-plants (scheme in bottom of Fig. 11.8). Differences in the bundle-sheath cells of the three groups (three schemes at top of Fig. 11.8) show that at least two intercellular shuttle systems occur:

1. a malate-pyruvate shuttle in the NADP-ME-TYPE (Fig. 11.8 left and Fig. 10.1);

Fig. 11.8. Intracellular metabolite shuttles in C_4-photosynthesis as given by schemes outlining the reactions of C_4-pathway photosynthesis in mesophyll (*lower*) and bundle sheath (*upper*) cells and the intracellular location of these reactions. Separate schemes for bundle sheath cells show the different C_4 acid decarboxylation mechanisms for NADP-malic enzyme-type (NADP-ME-TYPE), PEP carboxykinase-type (PCK-TYPE) and NAD-malic enzyme-type (NAD-ME-TYPE) species. The malate-pyruvate shuttle from mesophyll cells applies to NADP-type species and the aspartate-alanine shuttle to PCK-type and NAD-ME-type species. The enzymes involved are: *1*, PEP carboxylase; *2*, NADP malate dehydrogenase; *3*, aspartate aminotransferase; *4*, alanine aminotransferase; *5*, pyruvate, P_i dikinase; *6*, adenylate kinase; *7*, pyrophosphatase; *8*, 3-PGA kinase, NADP glyceraldehyde-3-P dehydrogenase and triose-P isomerase; *9*, NADP malic enzyme; *10*, PEP carboxykinase; *11*, NAD malate dehydrogenase; *12*, NAD malic enzyme. (Hatch and Osmond, 1976.) For abbreviations see usage in Chapter 7.

2. an aspartate-alanine shuttle in the PCK-TYPE and the NAD-ME-TYPE (Fig. 11.8 middle and right).
3. a 3-PGA-DHAP shuttle appears to be important in addition. This requirement results from the fact that reduction equivalents regained from C_4-acid decarboxylation in the bundle-sheath cells only suffice to reduce one molecule of 3-phosphoglyceric acid (3-PGA). In CO_2 fixation via RudP-carboxylase, however, two molecules of 3-PGA are formed. Quite a number of C_4-plants have no active photosystem II in their bundle-sheath chloroplasts, which means that they can form ATP only by cyclic photophosphorylation but not produce their own photosynthetic reduction equivalents. This results in a deficit in reduction equivalents in these bundle-sheaths which can be taken care of by the 3-PGA-DHAP shuttle between the mesophyll and the bundle-sheath as shown in Figure 11.9.

It is very tempting here to add the interesting example of metabolic co-operation in the brown alga *Laminaria hyperborea* worked out by Willenbrink and co-workers (Schmitz et al., 1972; Willenbrink et al., 1975). In this case co-operation occurs over a much larger distance than in C_4-

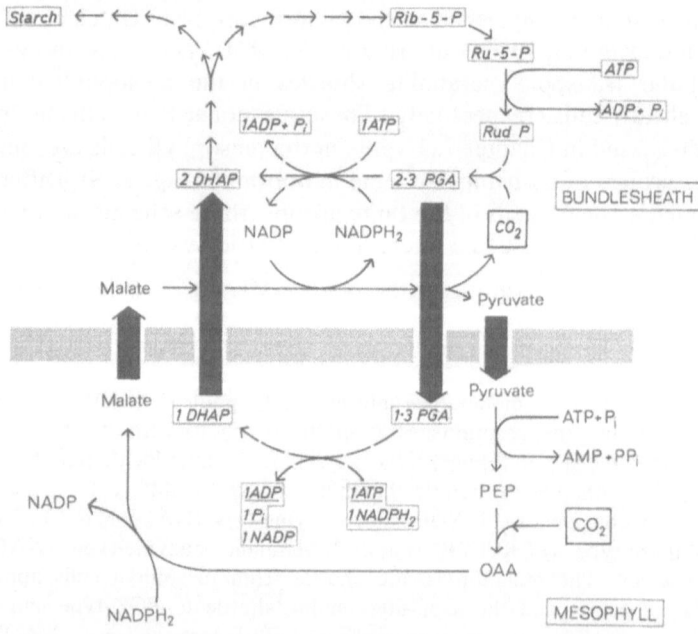

Fig. 11.9. The need of a DHAP-3-PGA shuttle in some cases of C_4-photosynthesis. Metabolic reactions (*thin arrows*) and intercellular metabolite transport (*thick bold arrows*). C_4-acid pathway = *solid arrows, roman symbols;* Calvin cycle = *dotted arrows, italic and shaded symbols* (see Hatch et al., 1971, Fig. 4, p. 146). For abbreviations see usage in Chapter 7.

photosynthesis. Also it is not clear whether possibly a special translocation tissue akin to pathways of long-distance transport in higher plants is
involved. Metabolites predominantly move within certain groups of cells
in the medulla of the phylloid (Schmitz et al., 1972). On the other hand
this example fits well here as it demonstrates co-operation of different
tissues in CO_2-assimilation. Photosynthetic characteristics of *Laminaria
hyperborea* are that in all parts of the fronds the Calvin cycle is the dominant pathway and the only light-dependent one for CO_2 fixation. There is
rapid incorporation of label form $^{14}CO_2$ into amino acids and appreciable
amounts of mannitol are also formed. Mannitol synthesis is enhanced
with increasing age of the phylloid. The special problem of *Laminaria hyperborea* is that in its sublittoral habitat in the early growth season light
intensities are so small that its compensation point of net photosynthesis
is exceeded only for a few hours per day. Nevertheless the young fronds
grow rapidly. There is a very substantial CO_2 dark fixation mediated by
PEP-carboxykinase in the young growing fronds which amounts to 25% of
the CO_2 fixation and which continues at unchanged rates in the light. In
the old mature fronds the proportion of dark fixation is only 5%. Carbon
balance sheets suggest that in the young fronds an appreciable part of
carbon required for growth is acquired by CO_2 dark fixation. However, a
net gain of carbon by dark fixation is not possible without the input of energy. This is achieved in *Laminaria hyperborea* by the transport of mannitol from the old fronds to the young growing fronds. Thus reducing

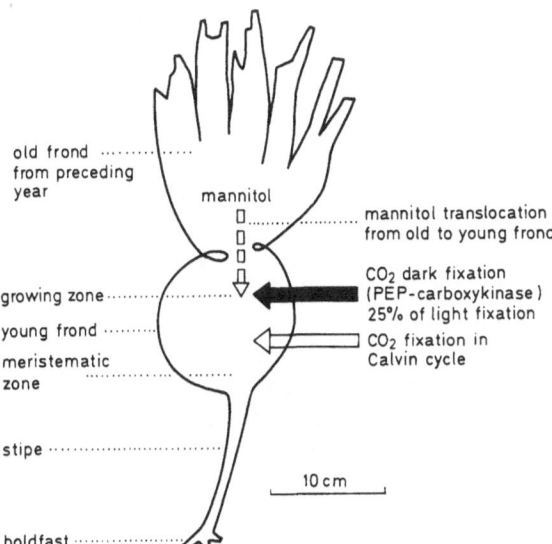

Fig. 11.10. Metabolic co-operation over medium distances during growth in the
brown alga *Laminaria hyperborea*. (After Willenbrink et al., 1975.)

equivalents stored in the old fronds in the preceding growth season can be utilized by the young fronds for growth in the early spring (Fig. 11.10).

11.3.2.3.2 Pathways of Symplastic Transport and Transport Efficiency During C_4-Photosynthesis

The symplastic transport of metabolites between the mesophyll and the bundle-sheath cell layers of C_4-plants is facilitated by a particularly high number of plasmodesmata between the two cell types (O'Brien and Carr, 1970; Laetsch, 1971, 1974; Carolin et al., 1973; Olesen, 1975; Osmond and Smith, 1976). In monocotyledonous C_4-plants (but not in dicotyledonous ones) the apoplastic transport pathway is blocked and hindered by suberization of the cell walls between the tissues (O'Brien and Carr, 1970; Laetsch, 1971; Osmond and Smith, 1976; Evert et al., 1977). It appears that in maize the suberin lamella is arranged between mesophyll and bundle-sheath cells in such a way that plasmodesmata become the only possible pathway for the exchange of metabolites between the two cell types, but that transpirational water loss from the veins still can occur via the apoplastic route (Evert et al., 1977). As stressed by Osmond and Smith (1976), it is important that plasmodesmata facilitate the transport of small metabolite molecules, but at the same time prevent movement of large enzyme molecules and thus maintain the metabolic compartmentation in C_4-leaves. According to Evert et al. (1977) the endoplasmic reticulum provides a direct connection between chloroplasts of mesophyll and bundle-sheath cells via the desmotubules of the plasmodesmata between the two cell types, and transport of metabolites might occur within the ER cisternae.

Experiments to measure the efficiency of symplastic transport in C_4-photosynthesis have been designed by Osmond (1971). The uptake and the transport of most metabolites from outside across the plasmalemma into the symplast occurs at rates similar to those observed for uptake of inorganic ions, i.e., in the range of $1-2 \mu$ mol/h/g fresh weight. The uptake of dicarboxylic acids, such as malic acid from an external medium, may be more sluggish (Lüttge and Ball, 1977). Conversely, photosynthesis is 100-fold more rapid, i.e., $100-200 \mu M CO_2$/h/g fresh weight. If, in its transport from the mesophyll to the bundle-sheath layers, malate would have to cross cell barriers (cell walls and plasmalemmas of mesophyll and bundle-sheath cells) this transport would be slow enough to reduce photosynthesis greatly. This is not the case, however; obviously metabolites transported in C_4-photosynthesis have to cross only organelle membranes but not cell membranes. The latter may be facilitated by chloroplast peripheral reticulum (Osmond and Smith, 1976) and ER-connections (Evert et al., 1977).

In conclusion, C_4-photosynthesis closely depends on efficient symplastic transport (Osmond and Smith, 1976).

11.3.3 The Mechanism of Symplastic Transport

The historical experiments by Arisz make clear that nonequilibrium states within the symplast are the driving forces for symplastic transport. Regrettably, these experiments had been largely forgotten for some time, although the chambers introduced by Weigl and by Pitman for investigations of transport in roots (see Fig. 12.3) closely mimic Arisz' experimental setup. Recently Bräutigam and Müller (1975a,b,c) have returned to Arisz' material, *Vallisneria*, with a study of α-aminoisobutyric acid transport. Using kinetin, which creates a sink or which increases the strength of a sink already present, they obtained new evidence for Arisz' conclusion that nonequilibria within the symplast are the major driving forces for symplastic transport and that there is no intrinsic polarity in the *Vallisneria* leaves.

The causes for the build-up of such nonequilibria can be of a complex nature. Membrane transport processes as well as synthesis and degradation reactions can establish gradients within the symplast. Symplastic transport has a tendency to equal out these gradients so that, as such, it is of a passive nature as shown for instance in experiment 2 in Section 11.3.2.1.

Using disks of tissue from the pitcher of the carnivorous plant *Nepenthes* it has been possible to demonstrate this leveling out of the gradient within the symplast by using a microautoradiographic technique (Lüttge, 1966b). Disks were allowed to take up labeled Cl^- on the face which corresponds to the exterior wall of the pitcher and to secrete the Cl^- by the glands on the opposite side, corresponding to the interior wall of the pitcher. Quantitative microautoradiography showed that the cytoplasm of all cells, i.e., both in nonspecialized mesophyll cells and in the specialized gland cells was equally labeled.

Tyree (1970) has attempted to calculate how large the concentration gradient must be for symplastic movement across the onion root. Using experimental data from the literature on KCl transport in roots, the amounts, the cross-sectional area, and the length of plasmodesmata, he arrived at the conclusion that the concentration difference per cell needs to be only 0.1 mN to allow symplastic transport. Since the distance between epidermis and stele is about ten cells, it turns out that for symplastic transport over the whole distance a concentration gradient of about 1 mN would be required.

The fact that passive symplastic transport is more rapid than simple diffusion poses a problem for the explanation of its mechanism. The most simple explanation is based on cytoplasmic streaming. If cytoplasmic streaming leads to very rapid equilibration of all substances within the flowing cytoplasm, or within the cytosol in each single cell, then only the relatively short plasma bridges between the cells would have to be negotiated by diffusion. A modern theory of symplastic transport on the basis of thermodynamics of irreversible processes was made by Tyree (1970) using data from a considerable amount of literature. The result is largely identical with what we have concluded from our more qualitative consideration above: cyclosis of cytoplasm and the occurrence of plasmodesmata play key roles in symplastic transport. For distances up to about 50 μm simple diffusion, which occurs very rapidly over short distances, is superior to the velocity of cyclosis (5 cm h^{-1}). However, due to the presence of organelles, ER-membranes, etc., the path of diffusion is tortuous and it may turn out that cyclosis makes the major contribution to the equalization of gradients within the cytosol of a given cell. Intercellular transport across the nodes between the lengthy internodal cells of Characeae is normally limited by the rate at which cytoplasmic streaming brings solutes to the plasmodesmata, rather than by movement through plasmodesmata (Tyree et al., 1974; Bostrom and Walker, 1976), but in the shorter onion root cells and in staminal hairs of *Tradescantia* cyclosis does not appear to be the rate-limiting step (Tyree et al., 1974; Tyree and Tammes, 1975). The rate-determining step in symplastic transport then is the movement from one cell to the next with other processes such as diffusion and protoplasmic streaming in the lumen of the cells leading to perfect mixing in the cytoplasm (Tyree and Tammes, 1975). The least resistance for movement of low molecular substances from one cell to the next is provided by the plasmodesmata. Perhaps water is exempt from this because membrane permeability for water is high (Sect. 2.5.2.2), and it can easily travel via both plasmalemmas and the wall of adjacent cells. For larger molecules most likely plasmodesmata constitute the only possible pathway for intercellular transport.

As mechanisms for transport in the plasmodesmata, in principle, the following possibilities can be discussed:

1. a pump
2. carriers
3. a volume flow (convection)
4. diffusion

A pump mechanism, assumed by some authors (Williams and Fensom, 1975) denied by others (Bostrom and Walker, 1975) is difficult to envisage, because it would have to operate over a comparatively enormous distance if one compares the length of the plasmodesmata (about 500 nm) with the thickness of lipoprotein membranes (about 7–8 nm) for which such pump mechanisms are usually considered.

Carrier mechanisms would not be very effective since the carrier complex would be most likely less mobile within the pores because it is certainly larger than the transported particles alone.

The relative contribution of convection and diffusion depends on the length of the pores, the streaming rate, and the diffusion coefficient within the pores. Under the assumption of a maximal streaming rate of 5 cm/h, it turns out that most likely diffusion plays a major role in transport of low molecular nonelectrolytes, and of electrolytes through plasmodesmata (Tyree, 1970).

Observations reported by Arisz that symplastic transport in *Vallisneria* leaves occurs unimpaired in the absence of any visible cytoplasmic streaming appear to rule out a particular importance of cyclosis in this process. This would certainly remove much of the basis for the above discussion. One has to remember, however, that even in the absence of a movement of organelles which can be seen in the light microscope the cytosol can stream. It is possible that Arisz would not have observed this. In *Nitella translucens* Williams and Fensom (1975) did not see any clear correlation between visible cyclosis and transport of carbonate, urea, acetate, Cl^- and Na^+; but laser light scattering data suggest that there is a more rapid submicroscopic streaming of the cytoplasm. In 1971 Booij, in interpreting the theory of Arisz on symplastic transport, suggested electrical currents associated with active ion translocation between apoplast, cytoplasm, and vacuole continuously keep the cytosol moving. By different rates of active ion transport in different cells of a tissue, or in different parts of an organ or a whole plant, nonequilibria are supposed to be built up so that they cause cyclosis. In *Elodea* Ca^{2+} inhibits and Mg^{2+} stimulates cytoplasmic streaming while Cl^- increases the differences between the Ca^{2+} and Mg^{2+} treatments (Forde and Steer, 1976). According to the hypothesis of Booij the driving force for symplastic transport would indirectly be active ion transport.

We note that discussion of the mechanism of symplastic transport merges with the evaluation of the function and the mechanism of cytoplasmic streaming. In the present context we cannot deal with this interesting problem in detail. Although the latest extensive review is not recent, it is still highly recommended (Kamiya, 1959).

11.3.4 Two Important Advantages of Symplastic Transport: Efficiency and Metabolic Control During Transport of Metabolites

Symplastic transport has two obviously important advantages. First, over medium distances it is much more efficient than diffusional transport in the apoplast. Second, it allows close metabolic control over the solutes transported. These points are illustrated by examples of metabolic cooperation of parenchyma cells over a medium distance as described in Section 11.3.2.3. Especially the consideration of C_4-photosynthesis has shown that only symplastic transport is efficient enough to allow the observed rates of photosynthesis; "the fluxes of metabolites between the mesophyll and bundle-sheath cells . . . . are evidently too rapid to involve transport across cell membranes and the cell wall'' (Osmond and Smith, 1976).

Apart from the standpoint of efficiency, another important advantage of symplastic metabolite transport becomes clear. During translocation within the symplast the plant does not relinquish metabolic control of the transported metabolites into which the plant invests much energy. This would be the case in apoplastic movement.

The combination of the two principles becomes still clearer when we consider long-distance transport of metabolites, in particular of carbohydrates. With the exception of a few special cases, carbohydrates are not transported in the apoplastic pathways of xylem vessels over larger distances. One of these exceptions is the mobilization of carbohydrates in the stems of trees at the beginning of the growth season in the spring, when much sugar moves in the xylem of the stems upward to the opening buds. This sugary sap can be collected from wounds cut into the stems, as is done with sugar maple, the source of the delicacy maple syrup. With the phloem a pathway for long-distance transport has evolved which has an efficiency similar to that of the xylem and especially carries on sugar translocation. The important difference is that in the phloem transport occurs behind the plasmalemma barrier. In this sense phloem transport can be considered as a particularly efficient example of symplastic transport.

11.4 Transport in Sieve Tubes

11.4.1 Transport of Assimilates as a Special Case of Symplastic Transport

There are a number of reasons to consider the long-distance transport of assimilates as a particular case of symplastic transport.

If in symplastic transport within parenchyma the transport of particles from one cell to the other, i.e., through the plasmodesmata, is the limiting step, the following conditions should considerably facilitate such transport:

1. files of long cells;
2. files of cells having increased numbers and increased diameters of plasmodesmata in their cross walls (or increased plasmodesmatal area relative to the rest of the cell wall).

Transitions from transport in parenchyma with isodiametric cells to development of such systems appearing more efficient in translocation should have occurred during evolution of the elements conducting photosynthetic products. Indeed, such developments are observed in comparative investigations of extant plants having different levels of organization. In the higher mosses, Musci, which do not have highly differentiated conductive tissues as in higher plants, there are special prosenchymatic cells in the stems serving assimilate transport (Eschrich and Steiner, 1967). In the brown algae there is a whole sequence of such developments. The conductive elements of the cauloids (i.e., the stem-like parts of the thallus) of *Laminaria* have cross walls with a greatly increased number of perforations. The further development of highly effective pathways for long-distance transport in the brown algae, however, is not an increase in number but an increase in cross-sectional area of perforations, which then are called pores; the number decreases (Ziegler, 1968) (Table 11.2). Wide pores are also found in the cross walls (sieve plates) of the sieve tube members in the phloem of angiosperms. The migration of substances which is facilitated between single cells leads in an increasing degree to a loss of cell individuality. To a high extent the individual parts form a continuum, i.e., the sieve tube. That the sieve tubes belong to the plant symplast is clearly indicated by the numerous plasmodesmata via which they are connected to the neighboring parenchyma and, in particular, to the neighboring companion cells. The sieve tube members have a semipermeable plasmalemma and can be plasmolyzed. The lumen of the sieve tubes, however, is not occupied by vacuoles, since at maturity there is no tonoplast. The transition from the sieve tube cytoplasm along cell walls of the sieve tube to the interior of the sieve tube lumen is continuous and not blocked by a membrane barrier. Therefore, the whole interior of the sieve tube belongs to the symplast. Mature and transporting sieve tubes do not have a nucleus and are not as rich in organelles as normal cells. Nevertheless, by no means are sieve tubes metabolically inert. There are numerous enzymes and cofactors within the sieve tubes (reviews Ziegler, 1975; Eschrich and Heyser, 1975) so that another important aspect of symplastic transport is realized here: the transported metabolites are not released from metabolic control.

The investigations of the fine structure of sieve tube content and of the sieve plates with their pores (the sieve pores) have led to recognition of numerous interesting phenomena. Besides the sieve plates fibrillar strands of protein (P-protein) are the most striking structural peculiarities of sieve tubes (Fig. 11.11). Do these protein fibrils and do ultrastructural elements of the sieve pores play a role in driving long-distance transport?

Table 11.2. Number, density and diameter of pores in the cross walls of assimilate conducting elements in the order of *Laminariales* (brown algae) (from Ziegler, 1968).

	Laminaria	*Pelagophycus*	*Macrocystis*
Number of pores	20 000–30 000	1000–2000	100–200
Density of pores (number per μm^2)	50–60	4–6	0.1
Diameter of pores, μm	0.06	0.3–0.8	2–3

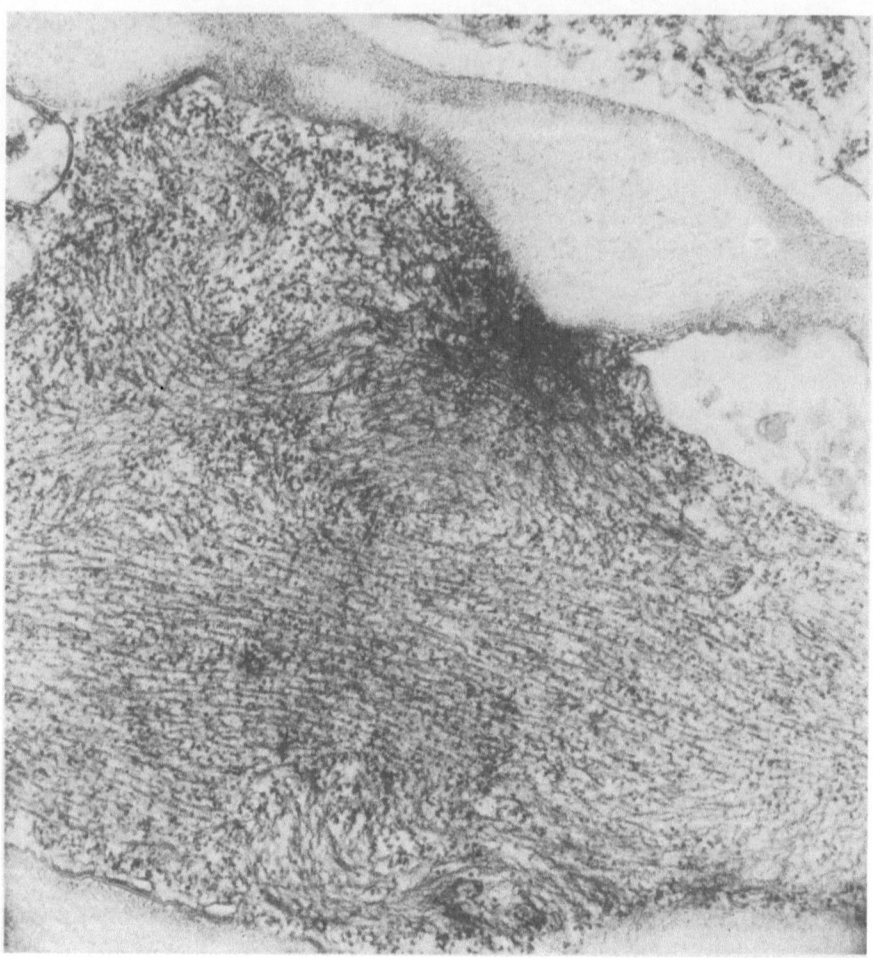

Fig. 11.11. P-protein in the sieve tube of *Ficus bengalensis*. Fix. OsO_4. Prim. magn. 17,800, sec. magn. 28,500. Original H. Ziegler.

We will mention below how little agreement there is among investigators of sieve tube transport on the function of those structures. There is only clarity about the fact that sieve tubes are indeed the anatomically differentiated pathways for long-distance transport of organic substances. This is shown by ringing experiments among others. In these experiments transport of photosynthate in trees is interrupted by cutting the bark in which the sieve tubes are located. The first ringing experiment, as far as it is known, was performed in 1675 by Marcello Malpighi. In the present day elegant microautoradiography clearly shows that long-distance transport of assimilate occurs in the sieve tubes (e.g., Fritz and Eschrich, 1970; Eschrich and Fritz, 1972; early microautoradiography of ion transport in the phloem out of the leaves: Biddulph, 1956).

11.4.2 The Problem of the Mechanism of Sieve Tube Transport

The most important difference between symplastic transport in a strict sense and sieve tube transport is a quantitative one. The maximum velocity of symplastic transport is on the order of about 6 cm/h; the transport in the sieve tubes, however, can reach velocities of 50–100 cm/h or more. This considerable quantitative difference of 1–2 orders of magnitude almost suggests a qualitative difference; the differences are gradual though. According to Canny (1971) the velocities of phloem transport measured are in the range between 1 and 200 cm/h, and 50–100 cm/h appears to be a generally accepted medium value (Canny, 1975). Of course, the speed alone is not all that important; it is important to consider the capacity of sieve tube transport, i.e., the amount of substance transported in an amount of time over a unit of area. MacRobbie (1971a) has recalculated the rates of phloem transport reported in the literature into the units of pmol s^{-1} cm^{-2} (which are the usual units considered in discussion of flux rates) and she has compared the rates of phloem transport with typical rates of ion fluxes across membranes (in pmol s^{-1} cm^{-2}):

Transport of sucrose in phloem	$2.5–20 \times 10^6$
Most ion fluxes across membranes	1–10
Cl$^-$ fluxes across membranes in *Acetabularia*	500–700
Salt excretion by the glands of the mangrove *Aegialitis*	5×10^3

Thus it becomes clear that phloem transport is more efficient by some orders of magnitude than membrane transport even in the most outstanding case of ion excretion by *Aegialitis* (see also Sect. 7.3.1). The mechanism of phloem transport, therefore, cannot be based on mechanisms of membrane transport.

The extreme difficulty in explaining sieve tube transport as a particular case of symplastic transport is also evident from the fact that in symplastic transport itself there are many details which are still not well understood (Sect. 11.3). An alternative would be the search for a qualitatively quite different mechanism. The dispute, now extended over many years, regarding the different hypotheses on the mechanism of translocation in the sieve tubes is not yet resolved. The most important hypotheses will be very briefly described here. Among other possibilities a classification of these hypotheses can be made according to the role which is assumed to be played by the fine structures of the sieve tube, its cytoplasm, and the sieve pores. (Reviews: Eschrich, 1970; MacRobbie, 1971a; Crafts and Crisp, 1971; Kollmann, 1975; encyclopedic treatment in Zimmermann and Milburn, 1975.)

11.4.2.1 Münch's Pressure Flow Hypothesis
and Its Variations

The hypothesis of pressure flow, convincing by its clarity and simplicity, was first formulated by Münch, 1926 (see Münch, 1930). According to this hypothesis the photosynthetic cells of the leaf create a source of carbohydrates in the symplast; the assimilate-utilizing or storing (e.g., as starch) tissues create a sink. Long-distance transport from source-to-sink will be driven by the osmotic and pressure gradient built up, for example, between leaf and root. The pipe systems of the phloem serve as pathways for long-distance transport in which a stream of sap moves carrying assimilate along with it. By aid of a simple model (Fig. 11.12) it is easy to demonstrate such a streaming of water with dissolved particles [volume flow, see also Eqs. (2.37) and (2.42)].

Naturally, in the course of time this hypothesis had to be modified. Presently, nobody assumes a direct connection of the assimilate-forming and the assimilate-utilizing parts of the plant. There are too many membrane barriers, for instance, between a carbohydrate formed in a chloroplast and the amyloplast of a starch-accumulating root or tuber cell. In the companion cells of the sieve tubes one finds that the cytoplasm is particularly full of organelles (nucleus, mitochondria) and that these cells are particularly active metabolically almost akin to gland cells. It is assumed that these cells are effective by active loading (vein loading) and unloading to set up the gradient within the sieve tubes which leads to volume flow

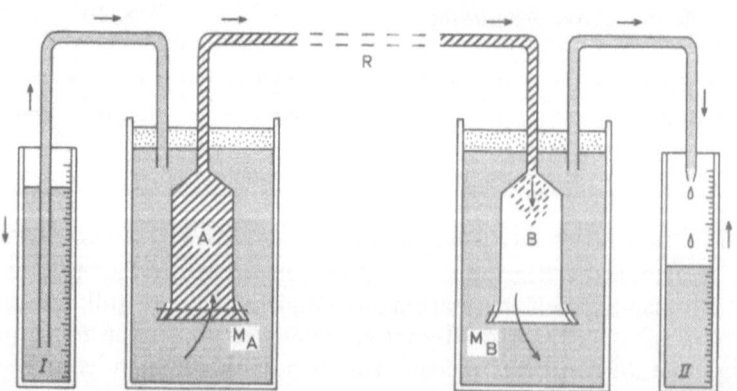

Fig. 11.12. Model experiment of Münch demonstrating a pressure flow driven by an osmotic gradient. (After Münch, 1930; from Ziegler, 1963.) *Cell A*, 10% sucrose solution stained with Congo red; *cell B*, water; *R*, joining glass tubing; *M*, semipermeable membranes. Cell A takes up water across the semipermeable membrane M_A, thus a hydrostatic pressure is created by which water is pressed out of the system from cell B across the semipermeable membrane M_B. Sugar and Congo red are transported from cell A to cell B. Water flows from the cylinder *I* to the cylinder *II* until the osmotic gradient A–B has disappeared.

(e.g., Ziegler, 1956). The loading and unloading, the uptake and release of sugar in a sieve tube, therefore become decisive factors in translocation. Eschrich and coworkers (Eschrich et al., 1972; Young et al., 1973) have demonstrated this further by very instructive model experiments.

The pressure flow hypothesis explains a large number of observations in a simple way; for instance, the fact that a large variety of solute particles is transported with the same velocity or that a strictly localized small heat impulse is rapidly translocated polarly in one direction (Ziegler, 1963; Ziegler and Vieweg, 1961). It also explains the observation that the sieve tube content is under pressure. One can demonstrate the latter by simple wounding of the phloem (for instance by cutting the bark of a tree with a sharp knife), on which the sieve tube content flows out of the cut surface. Aphids which feed from the phloem insert their mouth parts into individual sieve tubes; if one separates the body of the insect from the mouth part one can observe for quite a while a flow of sieve tube sap out of the cut end of the mouth parts. These techniques also permit the collection of sieve tube sap for a chemical analysis of its composition (see Ziegler, 1963). For the same purpose phloem exudation from cut flower stalks of certain monocotyledons, e.g., *Yucca*, has also proved very useful in giving particularly copious volumes (van Die and Tammes, 1975).

The cytoplasmic structures within the sieve tubes are not considered particularly important by the pressure flow hypothesis. It is only important that there is a sieve cell plasmalemma as a diffusional barrier toward the outside, so that the transported metabolites and substances remain under metabolic control and also cannot flood the apoplast and then serve as substrate for intruding microorganisms. The sieve tube plasmalemma as a semipermeable membrane is the basis of the functioning of the pressure flow mechanism [see Eqs. (2.37) and (2.42)]. Further, it appears important that the sieve tube cytoplasm has a large amount of enzymes and co-factors which may serve the metabolic control on the transported substances. Beyond this, the structures in the interior of the sieve tubes have only negative qualities. They must not be of a kind that blocks streaming in the lumen of the tubes or the sieve pores.

The hypothesis of mass pressure flow has repeatedly been questioned. Questions are posed as to whether the source-sink gradient is large enough to trigger a considerable mass flow, and other driving forces are sought. It has been reported that particles can move simultaneously in opposite directions. Canny (1971) records a continuous retreat of the protagonists of the mass flow theory. First, bidirectional transport was attributed to different bundles in the pathway; then to different sieve tubes within the phloem of a given bundle. Since it has been reported now that within one and the same sieve tube transport may occur in opposite directions, a border has been reached by this retreat. This indeed would be true, but on the other hand, it is not certain that the latter report is valid.

An aphid having its mouth parts in the phloem sap of a given sieve tube which is under pressure of course creates a new effective sink. When the exudate flow from the cut end of the mouth parts, after removal of the insect's body, is considered it is not at all surprising that the solution streams to this place from both ends of the sieve tube. The best evidence for the occurrence of bidirectional transport in one and the same sieve tube perhaps is supplied by microautoradiography (Trip and Gorham, 1967, 1968). MacRobbie (1971a) discusses this and other suggestions for bidirectional transport. She concludes that bidirectional translocation in a given sieve tube is possible only when there are weak gradients in the sieve tubes and that it is not possible to envisage a bidirectional transport when there is a strong source-to-sink gradient, which leads to the mass transport of a rate of 20×10^6 pmol/s/cm^2. In this case, the sieve tube must be unidirectional. In general, the evidence for true bidirectional transport within one and the same sieve tube does not appear very convincing. (Review on bidirectional transport: Eschrich, 1975.)

The strongest objections to the pressure flow hypothesis come from ultrastructural research. The controversy in this case regards the meaning of the fine structure which is observed. The most important point is to what extent in electronmicrographs one really observes the structures of the functioning sieve tubes; and, conversely, to what extent are there fixation artifacts. This, of course, is a general problem in electronmicroscopy which in the best of cases shows fixation equivalent to reality, but in fixing the content of the pipe system of the phloem, this problem is amplified. It is not possible here to give the details of this controversy. One of the major difficulties is, that in view of the sieve tube fine structure with the beautiful filamentous, fibrillar, net-type protein structures shown in electronmicrographs, it appears hard to believe that the translocation mechanism is as simple as assumed by the pressure flow hypothesis. (For detailed discussions see also in Ziegler, 1968; articles in Zimmermann and Milburn, 1975.)

11.4.2.2 Hypotheses in Which Functional Contributions of P-Protein and Sieve Pore Structure Are Important

A number of mechanisms have been discussed which are based on more or less important functional contributions of the protein strands observed in sieve tubes and sieve pores (P-protein) and on other details of sieve pore fine structure. According to the disposition in Zimmermann and Milburn (1975) these mechanisms can be classified as follows:

1. protoplasmic streaming;
2. electroosmotic flow;
3. other possible mechanisms: peristalsis of cell walls, microelectrokinesis, surface active movement mechanisms, reciprocating flow hypothesis, contractile proteins.

The evidence and arguments in favor and against these various mechanisms are thoroughly evaluated in the various articles in the volume of
Zimmermann and Milburn (1975). Here only a few brief remarks can be
made.

In general, all these hypotheses appear to be based on feasible ideas;
but, excepting electroosmotic flow, none of them has reached the convincing logical clearness of the pressure flow hypothesis. Canny (1975)
concludes in his survey that "protoplasmic streaming of the kind observ-
able in other plant cells does not occur in sieve elements." Fensom (1975)
suggests that a single mechanism is not necessarily responsible for
long-distance transport in the phloem but that a multi-modal action might
be involved; and he lists a number of open questions that have to be
answered experimentally and analytically before the involvement of a
variety of mechanisms can be assessed. We conclude that the evidence
for all these possibilities is too disputable at the moment to deserve further
discussion in our brief survey.

The hypothesis of an electroosmotic mechanism, on which much early
work has been done (e.g., Fensom, 1957), has been developed more recently by Spanner (Spanner and Jones, 1970; Spanner, 1975) into a quite
clear model, although still based on a number of hypothetical assumptions. This interesting and informative model will be briefly described
here. Sugar transport across the sieve pores from one sieve cell to the
next is assumed to be driven by the migration of K^+ ions in an electroosmotic process. Decisive in this context is the way in which the electrical
fields are built up at each sieve plate and how K^+ ions can stream back,
i.e., how a re-translocation of K^+ functions, without which electroosmosis cannot be continuously effective. The model by Spanner and Jones
is shown in Figure 11.13. To understand this model we have first to mention two experimental findings:

> Membrane aggregates of sieve tubes seem to be associated with
> ATPase activity (Yapa and Spanner, 1974). These membranes
> might be the cytological basis for an effective K^+ pump.
> Sieve tubes contain large amounts of ATP (0.07–0.059 μM; Kluge and
> Ziegler, 1964).

Spanner and Jones (1970) assume that K^+ is taken up actively into the
sieve tube lumen on the side of the sieve plates at which the stream of sap
arrives. The ATP of the sieve tube is enriched at these faces of the sieve
plates by the assimilation stream and is assumed to serve as a source of
energy for the K^+ pump which acts here. This accumulation of ATP is an
important prerequisite because in this way the polarity of the mechanism
is explained. On the other side of the sieve plates the K^+ is supposed to
leak passively from the sieve tubes into the apoplast and to diffuse back to
the sites of active uptake. Companion cells may also participate in potassium retranslocation. The K^+ gradient at the sieve plates may not be due

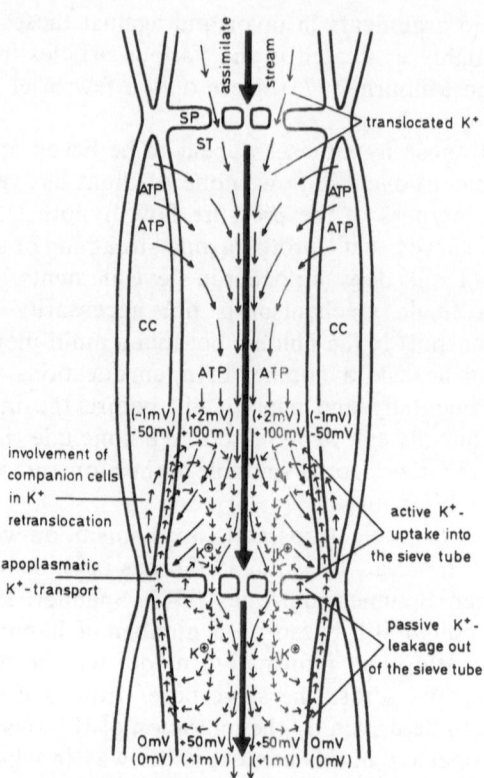

Fig. 11.13. Electroosmotically driven transport in sieve tubes drawn after the ideas of Spanner. *Thick bold arrow,* direction of assimilate stream; *longer thin arrows,* ATP distribution; *shorter thin arrows,* K⁺-movements (K⁺-recirculation and K⁺-translocation). *SP,* sieve plate with sieve pores; *ST,* sieve tube; *CC,* companion cell. The electrical potential gradients [mV] are values assumed by Spanner and Jones (1970) and Spanner (1975; mV values in *brackets*) respectively.

alone to K⁺ circulation but also to a considerable degree to the K⁺ transported in the assimilation stream. The slimy material (slime plugs) or the strands of P-protein observed in electronmicrographs of the sieve pores (Fig. 11.11) are assumed to occlude the pores to an extent making the ultimate channels minute enough to allow generation of electroosmotic forces (see Sect. 2.5.2.2). There may remain only fine pores for the movement of K⁺ of a diameter of 3–10 nm at the most. This means that hydrated sucrose molecules of 0.9 nm diameter can be transported electroosmotically. The high sucrose concentration in the sieve tubes could amplify this selective movement of sucrose. A very low structure-dependent selectivity effect in favor of sucrose, as compared to water, would already lead to an osmotic gradient which would considerably amplify the electroosmotic gradient built up by active potassium transport.

The material in the sieve pores also is postulated to provide a sufficiently high density of fixed negative charges at the alkaline pH of sieve tubes, which is necessary for formation of electrical double layers in the minute channels extending through the sieve pores. As indicated in Section 2.5.2.2 this is a prerequisite for electroosmosis.

The electroosmotic model shows clearly how an active ion pump could drive transport within the sieve tubes. Questions left open are largely of a quantitative nature. What is the relation of the consumption of energy to the amount of transport? Is ATP used not only in vein loading but also in long-distance transport itself? Are the sizes of the required K^+ membrane flow and K^+ translocation flows quantitatively feasible? (For a critical review of this hypothesis see MacRobbie, 1971a.)

11.4.3 Phloem Loading

As we have seen above (Sect. 11.4.2.1) vein loading and unloading of phloem (review Geiger, 1975) is essential in the modern version of the Münch pressure-flow mechanism. Active membrane transport processes during loading or unloading, or both, create pressure gradients within the sieve tubes and thus provide the driving force for translocation. Active loading is also important in many of the other hypotheses on translocation mechanisms. According to our nomenclature (p. 283) it is essentially a problem of medium- and short-distance transport. In fact, the problem of how photosynthetic products move from the source, chloroplasts, to far-removed sinks of consumption, or storage, constitutes one of the most outstanding examples of coupling between short-distance, medium-distance and long-distance transport and the interconnections between the different pathways. (This is the theme of the following chapter.) Photosynthetic products move out of the chloroplasts across the membranes of the chloroplast envelope, then they move to the minor leaf veins via symplastic or apoplastic routes, or both, and are loaded into the sieve tubes by specific membrane transport mechanisms. After long-distance transport in the sieve tubes again membranes and apoplastic and symplastic routes must be passed to reach the sites of utilization.

We have stressed before that one of the virtues of symplastic transport is the continuing metabolic control over the solutes transported. This appeared to make unlikely the occurrence of considerable apoplastic movement of metabolic substrates. There is, however, an increasing amount of evidence that sieve tube loading occurs from the apoplastic space. It can be envisaged that parenchyma transport from the chloroplasts to the veins mostly occurs in the symplast, but the apoplast very likely is passed before eventual sieve tube loading. This loading is a specific carrier-mediated process, requiring metabolic energy as shown by inhibitor studies. Modification of SH-groups of membranes by water soluble SH-agents, which are not taken up into the interior of cells, leads to an inhibition of phloem loading although photosynthesis and respiration are not affected (Giaquinta, 1976).

The small veins provide a very large surface area in the leaf for the active loading mechanism, and distances which have to be negotiated by symplastic and apoplastic transport are minimized. Presumably the companion cells play an important role in the loading mechanisms. Along the small veins they acquire the structural characteristics of transfer cells (Sect. 7.4) having invaginations of the cell wall with a greatly increased area of plasmalemma and with an associated ATPase activity (Bentwood and Cronshaw, 1978).

Companion cells and sieve tubes are linked by special plasmodesmata. These features of transport to the veins and sieve tube loading have been ably reviewed by Geiger (1975).

Even more recently than this review evidence from four laboratories independently suggests that sugar loading from the leaf apoplast into the sieve tubes may be mediated by an H^+-sugar cotransport mechanism (see Sect. 10.2.3.2). The pH-dependence of sucrose transport, interaction with electrical membrane potential, and apparent H^+-movements in association with sucrose transport have been shown in cotyledons of *Ricinus communis* seedlings (Hutchings, 1976; Komor et al., 1977), in the hollow petiole of mature *Ricinus communis* leaves (Malek and Baker, 1977), and in beet leaves (Giaquinta, 1977), where these transport processes presumably lead to phloem loading.

11.5 Concluding Remarks

An assessment of possible routes for transport shows that all plant organisms are provided with two principally different pathways for translocation of solutes: an apoplastic and a symplastic system. The special requirement of effective long-distance transport in more highly developed land plants led to a selective pressure, which resulted in the evolution of very specialized apoplastic and symplastic pathways: tracheids and vessels in the xylem and sieve tubes in the phloem, respectively.

Chapter 12

Coupling of Short-Distance, Medium-Distance and Long-Distance Transport in Special Plant Organs and Interconnections Between Different Pathways of Transport

12.1 General Models of Roots and Leaves

The highly complex system of the higher plant, composed of cells, tissues, and organs, could not function if short-distance, medium-distance, and long-distance transport were independent of each other. Chapter 11 has clearly shown these transport systems are interdependent; there are couplings and interconnections throughout the plant. Specific mechanisms localized at strategic points in the system must play an important role in such coupling. It seems self-evident that these must be metabolically controlled membrane-transport processes; but not all membrane-transport processes present in the system have this controlling function. We will discuss two particular models, namely ion transport across roots and transport of substances within leaves, and try to identify specific mechanisms involved in coupling of short-distance, medium-distance and long-distance transport. Figure 12.1 shows that the problem in both cases is basically the same.

Roots take up ions from the soil or from an external medium into their apoplastic free space; influx and efflux at the plasmalemma of the root hairs, and of epidermal and cortex cells control ion uptake and ion release of the cytoplasmic phase. Influx and efflux at the tonoplast determine the amount of accumulation or mobilization of ion reserves in the vacuole. Symplastic transport regulates the distribution of ions in the whole organ and also across a barrier in the apoplastic space which is marked by the Casparian strip. The efflux from the stelar parenchyma cells loads the pathways for long-distance transport.

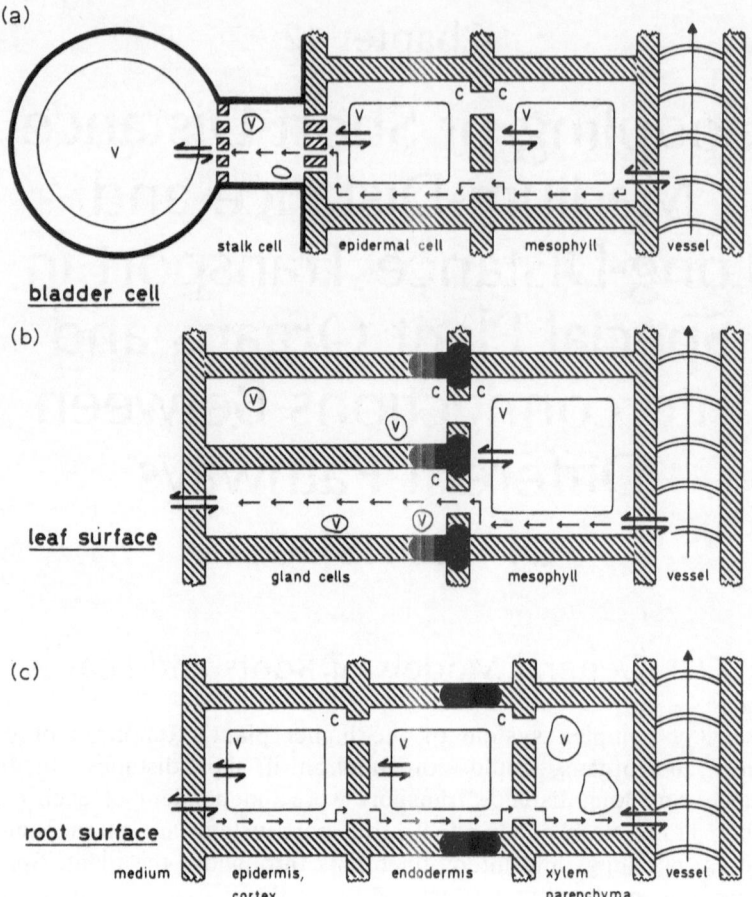

Fig. 12.1. Simplified structural schemes of a leaf with salt bladders **a**, of a leaf with salt glands **b**, and of a root **c**. Normal cell walls are *shaded*, blocks of apoplastic transport by cell wall lignification, cutinization and suberization are indicated by *heavy black lines*. The schemes show 1. short-distance transport at membranes ($\rightleftharpoons$), i.e., uptake into the symplast from the ambient apoplast; release into the bladder vacuoles, to the leaf surface or into the vessels of the xylem; fluxes into and out of vacuoles; 2. symplastic transport between co-operating cells of leaf and root tissues respectively; 3. long-distance transport in the xylem vessels. *V*, vacuole; *C*, cytoplasm. (Lüttge 1971a, 1974, 1975.)

It is made clear in Figure 12.1 which processes of ion transport in leaves are in analogy to this. The transpiration stream in the xylem of the leaves with its finely branched tracheal elements in the minor veins and the free space of the walls provides an external solution. Plasmalemma and tonoplast fluxes and symplastic transport correspond to similar processes in the root. Cell wall depositions (for instance in glands, Fig. 4.10) have a function analogous to that of the Casparian strip. Elimination of salt by glands at the external leaf surface can be compared with the release of ions into the apoplastic pathways in the vessels. The scheme of

Table 12.1. Ion content in the leaf free space.

Method of measurement:	Perfusion	Centrifugation	Determination of equilibrium concentration
Authors:	Bernstein (1971)	Jacobson (1971)	Pitman et al. (1974b)
Material:	Cabbage	*Dionaea muscipula* (Venus' fly trap)	Barley seedlings
Ion concentration in free space:	2–10 mEq/l	cations 27.9 mEq/l anions 16.5 mEq/l	
Concentration of certain ions:		K$^+$: 6.4 mEq/l Na$^+$: 8.9 mEq/l Cl$^-$: 13.8 mEq/l	K$^+$: 5 mEq/l Cl$^-$: 5 mEq/l

Figure 12.1 in addition shows the ion accumulation in the particularly large vacuoles of epidermal bladder cells (see Sect. 7.3.2., Fig. 7.20). Somewhat more problematic in leaves as compared to the roots is the question of the composition and the concentration of the ion solution in the external space. With roots one can analyze the soil or the soil solution. The solutions which are present in the free space of the leaf under natural conditions depend on transpiration and on long-distance transport into and out of the leaf. Table 12.1 lists results of analyses which have been obtained by aid of a perfusion technique (Bernstein, 1971), by aid of a centrifugation method (Jacobson, 1971), or by determination of the external concentration with which leaf cells are in flux equilibrium (Pitman et al., 1974b). Free space characteristics of barley leaves with water free space and Donnan free space were analyzed by Pitman et al. (1974a,b) and *Citrus* leaves by Smith and Fox (1975). In *in vitro* experiments one can use leaf slices of 0.5 mm width and can allow them to equilibrate with an experimental external solution. Thus transport processes in leaf cells can be investigated in a way similar to methods used for roots, algae and water plants (Fig. 8.10, Sect. 8.2.4.1.5).

12.2 The Model of the Root: Various Hypotheses on the Mechanism of Ion Transport From an External Medium Across the Root Into the Xylem Vessels

Transport from an external medium into the root xylem is usually measured by an analysis of the exudate, i.e., a solution which flows out from the cut surface of an excised root, or by analysis of the ion content of the shoot of intact plants, which reflects the input by the transpiration stream. In such experiments the use of radioisotopes has proven to be very useful.

Particular applications of the exudate method are those experiments in which the cut surfaces of isolated roots are dipped into a solution in which the exuded ions can be detected (Falk et al., 1966; Weigl, 1969d, 1970, 1971; Pitman, 1971, 1972a). Surprisingly it is seldom that the analysis of guttation fluid, e.g., from the tips of leaves and coleoptiles of cereal seedlings (Perrin, 1972), has been used to measure root-pressure exudation (see Sect. 11.2.2). This would provide an excellent nondestructive method.

For an active movement of ions from an external medium across the root into the xylem vessels it is *sufficient* on the basis of thermodynamic principles that active membrane transport occurs at only one of the possible sites shown in Figure 12.1. However, thermodynamic principles *do not rule out* the possibility that active membrane transport occurs at more than one site. In the course of more than half a century of extensive research on the problems of ion transport from an external solution, or from the soil, across the root into the xylem pathway of long-distance transport, the several possibilities given by the model of Figure 12.1 and allowed by thermodynamic principles have led repeatedly to controversies. For a better survey we shall try to distinguish the five more important hypotheses in the following way:

1. The hypothesis of the endodermis pump.
2. The hypothesis of cortex plasmalemma transport and symplastic movement across the root.
3. The hypothesis of xylem element differentiation.
4. The hypothesis of the stelar parenchyma pump.
5. The two-pump hypothesis, combining hypotheses 2 and 4.

These hypotheses will be treated separately in the following sections.

12.2.1. The Hypothesis of the Endodermis Pump

The hypothesis placing the site of active transport in ion movement across the root within the endodermis goes back to Ursprung and Blum (1921). These authors observed an abrupt change of the so-called *Saugkraft (suction pressure)*, showing a sudden increase of the negative water potential in the cells of the stele behind the endodermis. This observation of a discontinuous jump at the endodermis (*Endodermissprung*) later on received little attention. Today we can be sure, that apart from the Casparian strip in their radial walls (Sect. 4.2.1) the cells of the endodermis represent nothing unique; the endodermal cells are highly important only in providing a blockage of apoplastic transport. Ions moving inwards to the stele must be taken up into the symplast before reaching the endodermis or, at the latest, at the endodermis. According to Glinka (1977) in an electrical analog we can regard symplastic and apoplastic pathways in the cortex as two resistances in parallel with the symplastic route across the endodermis arranged in series. Only mistakenly has this been interpreted as implying that the endodermis functions as a site pumping ions inwardly. Modern measurements of gradients across the roots (e.g., of ion activities and membrane potentials; see Sect. 12.2.2) do not show discontinuities at the endodermis. Endodermal cells do not have the cytological characteristics of gland cells and of transfer cells; i.e., of cell types involved in solute pumping (see Sect. 7.3 and 7.4). Anatomically the endodermis represents the innermost cell layer of the cortex, and, in some special cases such as the genus *Erica* it is the only cortical cell layer. Transport experiments and utrastructural investigations prove that the cytoplasm of the endodermis cells is part of the root symplast extending from the cortex across the endodermis into the stelar cylinder (e.g., Robards and Robb, 1974; Haas and Carothers, 1975; Stelzer et al., 1975; Roboards and Clarkson, 1976). Thus the postulated mechanism of an endodermis pump can be excluded on the basis of many experiments and observations.

12.2.2 The Hypothesis of Cortex Plasmalemma Transport and Symplastic Movement Across the Root

This hypothesis assumes that during ion transport across the roots there is only one pumping site, which is localized in the plasmalemma of root epidermis, root hair, and cortical cells. This hypothesis was originally conceived by Crafts and Broyer (1938). It found support in the 1950's by Arisz (1956) and in the 1960's by Laties and his group (cf. Laties, 1967, 1969; Lüttge, 1969). The cortex cells outside the endodermis play an important role in this hypothesis. The cortex apoplast can be penetrated by water and ions from the external medium as an apparent free space, and thus the plasmalemma of all cortex cells provides a highly enlarged surface for active ion uptake into the root symplast. This uptake then is assumed to be followed by symplastic transport to the stele and passive leakage of ions into the xylem vessels. Malone et al. (1977), investigating the properties of an ATPase in the epidermis and outer cortex of maize roots, conclude that these tissues must have a primary energy-linked role in ion absorption by the root.

That symplastic transport occurs across the root is now very well established by data from physiological experiments, structural observations, and ion localization techniques; the latter also support the concept that plasmodesmata and also the endoplasmic reticulum extending through them are important pathways (see Sect. 11.3.1; Stelzer et al., 1975; van Iren and van der Spiegel, 1975; review Robards and Clarkson, 1976). Particularly elegant are experiments by Clarkson and co-workers on selective ion transport across the secondary and tertiary stages of endodermis (Clarkson et al., 1971; Harrison-Murray and Clarkson, 1973; Clarkson and Robards, 1975; Ferguson and Clarkson, 1976a). In these heavily suberized stages of endodermal cell differentiation (Sect. 4.2.2., Fig. 4.9) the endodermal plasmalemma becomes inaccessible for water and solutes from the cortex apoplast. This has severe effects on radial movement of Ca^{2+} and Mg^{2+}, which are not transported symplastically. The radial movement of these ions is greatly inhibited and virtually blocked eventually by endodermis suberization. However, at the same time, ions like potassium and phosphate appear to be unhindered. These ions can be transported symplastically via plasmodesmata extending through the secondary and tertiary endodermis.

Symplastic transport across the root is also essential in the remaining hypotheses to be discussed here (i.e., 3, 4, and 5 as listed in Sect. 12.2). In contrast to the general agreement on symplastic transport, controversy has arisen over the second requirement of the Crafts-Broyer hypothesis, i.e., passive leakage of ions into the vessels. Originally this appeared to be supported by the contention that O_2 partial pressure is lowered in the interior of the roots and that therefore due to inhibited respiration stelar parenchyma cells cannot retain ions. Bowling (1973b) measured the O_2 gradient across sunflower roots using a polarographic micro-O_2-electrode and found a small gradient (Table 12.2) which is probably not sufficient to explain leakiness of stelar cells. Leonard and Hotchkiss (1978) found cation ATPases in cells of both cortex and stele at similar levels, which does not support the hypothesis that stelar parenchyma cells are deficient in ion pumps.

Freshly isolated steles of roots do not have a capacity for ion accumu-

Table 12.2. O_2 gradient (Bowling, 1973b) and pH gradient (Bowling, 1973c) across *Helianthus annuus* roots.

Cell layer	O_2 partial pressure [mm Hg]	Vacuolar pH[a]
External solution	159 ± 1	6.38
Epidermis	151 ± 3	5.70 ± 0.12
Cortex 1	147 ± 5	5.85 ± 0.13
2	140 ± 8	5.98 ± 0.12
3	137 ± 9	6.08 ± 0.12
4	135 ± 8	6.21 ± 0.20
5	not measured	6.41 ± 0.33
6	not measured	not recorded
Endodermis	131 ± 11	6.51 ± 0.20
Pericycle	130 ± 11	6.57 ± 0.20
Protoxylem	127 ± 10	6.90 ± 0.05

Data are means ± 95 per cent confidence limits.
[a] Data were taken with a ruler from Bowling's graph.

lation (Laties and Budd, 1964; Lüttge and Laties, 1967). The fact that steles *in situ* do contain considerable amounts of ions (Yu and Kramer, 1969) is not necessarily in contradiction with a leaky stele, because this could be just the consequence of a steady state in which a constant flow of ions occurs from the cortex and via the stele to the vessels (cf. also Baker, 1973a,b). Of course, it is difficult to answer the question of whether freshly isolated stele or aged stele resembles more closely the nonisolated stele *in situ*. During aging the stele does develop a capacity for ion accumulation (Fig. 9.12). Experiments with leaf slices suggest that the freshly sliced material is more like the tissue *in situ* (Pitman et al., 1974a,b; Ullrich-Eberius et al., 1976a,b). In general, aging after slicing appears to be an adaptive process (''adaptive aging''; van Steveninck, 1976b; see Sect. 9.6).

The importance of the cortex is also underscored by experiments on long-distance transport by decorticated steles. These do not produce root pressure exudate and also do not transport ions between two chambers with experimental solution. Long-distance transport by isolated steles left on the shoot is solely passive and directly proportional to evaporation and transpiration (Lüttge and Laties, 1967). It should be noted, though, that Anderson (1972) has criticized this experiment because steles could have been damaged during decortication, and the possibility of exudation by decorticated stele is not excluded.

Kinetics of long-distance transport by the roots have been compared with system 1 and system 2 kinetics of ion uptake (see Sects. 6.1 and 6.2). If ions leak into the vessels and if according to the Torii-Laties hypothesis loading of the symplast is solely due to system 1, long-distance transport should reflect the kinetic characteristics of system 1. Indeed, the specificities and counterion effects in long-distance transport indicate that system 1 alone is involved. Furthermore, Laties and co-workers observed a hyperbolic concentration isotherm for long-distance transport only in the low concentration range of system 1. In the high concentration range a linear concentration dependence was found suggesting largely passive transport. These findings are in agreement with both the hypothesis of localization of system 1 but not system 2 at the plasmalemma and the hypothesis of passive leakage into the vessels (Laties, 1967, 1969). Conversely, however, Epstein and co-workers also claim that their hypothesis on localization of system 1 and 2 at the plasmalemma is in agreement with the properties of long-distance transport. Results of their exudation experiments in which long-distance transport reflected properties of both system 1 and system 2 appeared to be consistent with a plasmalemma localization of both systems (Läuchli and Epstein, 1971; Läuchli, 1972). This argument is possible only when there is passive leakage into the vessels. Thus it is somewhat surprising that the same authors at the same time appeared to promote the stelar parenchyma pump hypothesis (Läuchli et al., 1971).

More recently electrophysiological work by Bowling and co-workers has made important contributions to the problem (Dunlop and Bowling, 1971a,b,c; Bowling and Ansari, 1972; Bowling, 1973a; Dunlop, 1973, 1974, 1976; Bowling, 1976). This group used electropotential electrodes and ion-selective electrodes to measure gradients across the root and to calculate electrochemical potential gradients. The profiles shown in Figure 12.2 lead to the following conclusions about the distribution and transport of ions across the roots of *Helianthus annuus* and *Zea mays* (where

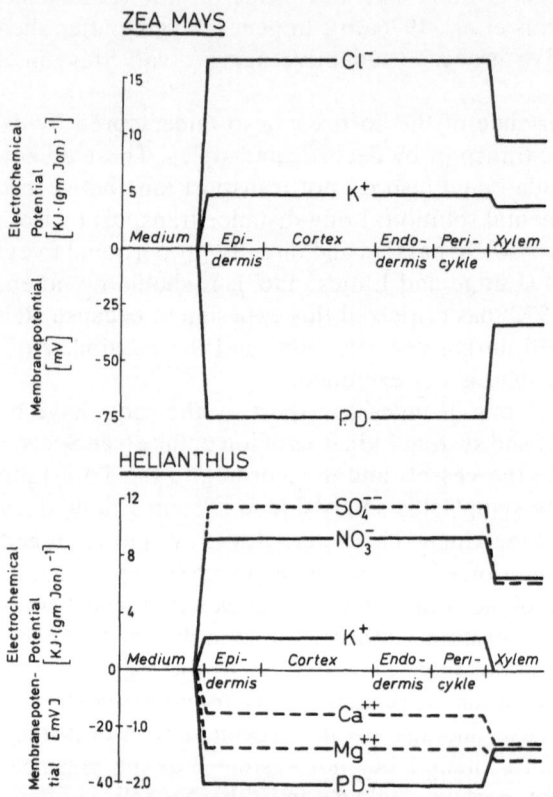

Fig. 12.2. Profiles of membrane potential (*PD*) and electrochemical potential for various cations and anions across *Zea mays* and *Helianthus annuus* roots. (After Fig. 6 in Dunlop and Bowling, 1971c, and Figs. 2 and 3 in Bowling, 1973a.) The *dotted lines* are based on less detailed measurements of many cells across the roots than the *solid lines,* i.e., only on measurements of medium vs. root tissue and root exudate. For maize a solution of 1 mM KCl + 0.1 mM $CaCl_2$, for sunflower a culture solution was used. Changes of external ion concentrations (maize: 0.1 mM–10 mM KCl; sunflower: $^1/_{10}$ culture solution) led to qualitatively similar results. The electrochemical potentials are given in kilo-joules [see Eq. (2.19)].

the Na^+ values for *H. annuus* are taken from Bowling and Ansari, 1972; not shown in Fig. 12.2):

1. Uptake of K^+ and Cl^- in *Zea mays* and of Na^+, K^+, SO_4^{2-} and NO_3^- in *Helianthus annuus* into the vacuoles of root cells is uphill against a steep electrochemical potential gradient; Ca^{2+} and Mg^{2+} are taken up passively.
2. Transport of K^+ and Cl^- in *Zea mays* and of Na^+, SO_4^{2-} and NO_3^- in *H. annuus* across the root is against an overall electrochemical potential gradient between the medium and the solution in the vessels.
3. In none of the cases investigated is the exit of ions into the vessels against an electrochemical potential gradient. (An exception is Na^+ at 1.0 mM and 10 mM but not at 0.1 and 0.25 mM in the medium.) All ions can move into the vessels passively down the electrochemical gradient, and thus according to the data of Bowling's group no active mechanism needs to be assumed in the stele.

This leads back to the question of how polar transport *across* the root, i.e., from the medium to the vessels, is brought about when there is neither an electrochemical potential gradient nor a metabolic gradient (assuming that the P_{O_2}-gradient shown in Table 12.2 is too small). Dunlop and Bowling speculated that the highly different areas of total membrane surfaces of epidermal plus cortical cells might be important. Thus, in the cortex there is much more space for ion pump entities than in the stele, and a simple pump-leak antagonism might lead to centripetal ion movement. The stelar cells then need not be more leaky than the cortical cells. It is enough that for spatial reasons active ion uptake of the living stelar cells is smaller to explain centripetal transport driven by active ion uptake in the epidermis and cortex.

For Na^+, however, at high external concentrations the direction is reversed. Hence the two opposing forces of uptake into the root cells at the membranes of cortical and stelar cells and of the overall electrochemical gradient dependent on external concentration appear to allow a control and regulation of Na^+ transport to the shoot.

Bowling (1973c) also has observed a peculiar gradient of vacuolar pH across the root (Table 12.2) which might help to explain polarity of transport. It is not clear yet what causes this pH gradient and whether it also exists in the root symplast.

The work by Bowling and co-workers seems to supply evidence in favor of passive transport into the root vessels which is difficult to disprove. Their measurements have been criticized, however, for technical reasons. Anderson and Higinbotham (1975) have pointed out that blindly pushing a microelectrode gradually across the root—as done in the experiments of Bowling and co-workers—causes artifacts due to the wound created by the electrode. Furthermore, different results have been obtained in other laboratories. Differentiating between plasmalemma and tonoplast potentials Ginsburg and Ginzburg (1974) found the vacuolar potential in maize roots to be initially more positive than the cytoplasmic potential by 30 to 50 mV and after 24 h in 1 mM KCl + 0.01 mM $CaCl_2$ by 5 to 15 mV. The cytoplasmic phase was equipotential through the root in agreement with the symplast theory, but the potential of the stelar vacuoles was negative relative to the cortical ones. Observations of Mertz and Higinbotham (1976) on barley roots are also in contrast to those shown in Figure 12.2. These authors recorded:

1. a radial PD gradient where the cells of the 3rd cortical cell layer were 10 to 58 mV more negative than the epidermal cell layer;
2. cells within 1 mm of a cut surface (wound!) were depolarized by 90 mV and recovered during aging, becoming even more negative by 25 mV as compared to cells of intact roots;
3. the vacuolar potential is more positive by 9 mV than the cytoplasmic potential, and during ion accumulation in 1 mM KCl + 0.5 mM $CaSO_4$ this may rise to a difference of 35 mV.

Due to the controversy on passive leakage into the vessels, alternatives such as the hypothesis of xylem element differentiation and the concept of the stelar parenchyma pump have concentrated on other possible mechanisms of ion transport from the living root cells into the xylem vessel elements.

12.2.3 The Hypothesis of Xylem Element Differentiation

This hypothesis originally was developed by Hylmö (1953) and was revived more recently by Anderson and Higinbotham and their co-workers (Anderson and House, 1967; Higinbotham, Davis et al., 1973; Davis and Higinbotham, 1976). According to this idea ions eventually appearing in the dead xylem elements are initially accumulated in the vacuoles of developing xylem elements in the tip region of the roots. In this region developing vessel cells still contain normal cytoplasm with organelles, plasmalemma, and tonoplast, and their cytoplasm is part of the root symplast. As the vessel elements mature, the lateral walls become thickened and the cross walls between the vertically adjacent elements differentiate into perforation plates or are removed altogether. Thus at maturity the vessel is a long continuous tube. In corn roots Higinbotham, Davis et al. (1973) found that the cytoplasm lined the early metaxylem walls up to 10 cm from the apex, although the perforation plates were formed in the 5–10 cm zone. Thus vacuolar sap appears to be continuous with the xylem sap above and, in effect, forms a bridge between the dead apoplast above and the symplast below. The salient conclusion of this study is that ion delivery into the vessels may be via the symplast in the 0–10 cm zone, in which absorption and translocation is most active (Burley et al., 1970).

Davis and Higinbotham (1976) have examined in detail electrochemical gradients and arrive at the following scheme for radial transport of K^+ and Cl^- in maize roots: "Both K^+ and Cl^- are actively pumped across the plasmalemma of epidermal and cortical cells. Subsequently, the ions move via the symplast into the vessels. Once in the vessel, cytoplasmic K^+ is actively pumped across the tonoplast into the open-ended vessel vacuole (vessel lumen). Movement of Cl^- across the vessel tonoplast is passive." This means there are two pumps for K^+ on the way across the root as in the two pump hypothesis (Sect. 12.2.5) and one pump for Cl^- as in the assumption of a passive leak to the vessels (Sect. 12.2.2).

It is important for an evaluation of this hypothesis, which assumes an important role of xylem element differentiation, to consider ion transport along the length of roots. The point raised by Anderson and House (1967), that the zones of xylem element differentiation and of maximum ion transport along the root are highly correlated, is not a simple argument. The zones of maximum transport are not identical for various ionic species, and in addition to quantitative differences along the length of a root there may be qualitative differences of transport mechanisms (Eshel and Waisel, 1972, 1973). An anatomical analysis must also consider epidermal and hypodermal suberizations along the length of the roots, which certainly play a role in radial transport (Ferguson and Clarkson, 1976a). It remains essential, though, to relate radial transport to a careful structural analysis of xylem development along the length of the root. Läuchli, Pitman et al. (1978) observed the largest rates of radial Cl^- transport in barley roots in a zone where all xylem elements were fully mature and apoplastic. The amino acid analog p-fluorophenylalanine (FPA), which inhibits ion transfer to the vessels more than ion uptake into the symplast (see Sect. 12.2.5), exerts a similar inhibition of transport to the exudate in all root zones.

If the hypothesis of xylem element differentiation is correct, longitudinal transport in the root xylem should be strictly polar occurring only in the upward direction away from the tip. There are conflicting data on this polarity (Evans and Vaughan, 1966; R.C. Smith, 1970). Lundegårdh's (1950) classical experiment in which a cut cylinder of root exuded at both ends shows, however, that there is no absolute polarity. Furthermore, polarity may be different for different ions. In onion roots only Cl^- appears to be transported in both directions, K^+ transport is more strongly upward than downward toward the tip, Ca^{2+} is transported solely upward, and Na^+ is largely transported downward (Macklon, 1975a,b). The phloem may participate in downward movement, hence the absolute polarity observed for Ca^{2+}, which is not mobile in the phloem. Another argument important in the discussion of the hypothesis of xylem element differentiation is that transport through barley roots is not stopped by removal of that part of the root tip containing metaxylem vessels with living contents (Läuchli, Kramer et al., 1974).

Measurements of longitudinal electrical resistances of root segments have also been used to argue in favor of (Davis, 1968; Katoǔ and Oka-moto, 1970) or against (Ginsburg and Laties, 1973) the hypothesis, but it seems that the anatomical state of the xylem cannot be inferred from electrical resistance measurements (Anderson and Higinbotham, 1976).

12.2.4 The Hypothesis of the Stelar Parenchyma Pump

The hypothesis of the stelar parenchyma pump was initially strongly supported by localizations of ions cytologically using microautoradiography (Weigl and Lüttge, 1962, 1965) and electron probe microanalysis (Läuchli, 1967, 1972; Läuchli et al., 1971). These showed the highest ion concentrations in the vessels and in the xylem parenchyma cells, i.e., the living cells surrounding the dead xylem elements, suggesting an active secretion from the xylem parenchyma cells into the vessels.

The cytological localization techniques, however, have two principal disadvantages and need to be handled with caution: First, the units of reference are always areas and not volumes; calibration to measure absolute quantities is extremely difficult (cf. Lüttge, ed., 1972). Second, the resolution in most investigations so far is not good enough to allow proper distinction between cell compartments such as cytoplasm and vacuole. The latter problem permits two different interpretations of the apparent large accumulation of ions in the stelar parenchyma cells:

1. Ion concentration in the cytoplasm or in the vacuoles, or both, of xylem parenchyma cells is larger than in the more peripheral root cells. This makes an active role of the xylem parenchyma cells very likely. Such an active contribution of xylem parenchyma cells could be either an accumulation within the xylem parenchyma cells followed by a passive release into the vessels or—as envisaged by the stelar pump hypothesis—by an active secretion into the vessels. It should be noted here, that in the latter case an ion accumulation in the xylem parenchyma cells themselves is not a logical prerequisite, because active secretion can, of course, occur from a low concentration uphill into a high concentration.
2. Ion concentration in the whole symplast of the root is high due to loading in the cortex. Xylem parenchyma cells in cytological observations only appear to accumulate ions to higher concentrations than peripheral root cells, because they contain particularly dense cytoplasm and much smaller vacuoles.

This then would support the hypothesis of cortical loading followed by symplastic transport and passive leakage into the vessels (Sect. 12.2.2).

This ambiguity reduces the value of such cytological localization techniques. Furthermore, these techniques show only relative amounts of radioactive label or relative elemental distribution in a tissue. They do not give information on electrochemical potential gradients which are important in ion transport. Nevertheless, the hypothesis of a stelar parenchyma pump has gained support from improved preparation and observation techniques for electron probe analysis (Läuchli, 1975; Yeo et al., 1977).

Important indirect evidence in favor of the hypothesis also comes from detailed investigations of ultrastructure of root steles. It turns out that xylem parenchyma cells very often structurally resemble transfer cells (Sect. 7.4), the occurrence of which is generally correlated with active ion pumping (cf. Läuchli, 1976c; Letvenuk and Peterson, 1976). Läuchli and co-workers have also obtained cytological and electron probe analysis data suggesting that xylem parenchyma transfer cells not only function in active ion secretion into the vessels, but also in selective reabsorption of Na^+ moving upward to the shoot. This can in some cases explain selective transport of K^+ over Na^+ to the shoots, and it may mitigate adverse effects of salinity (Läuchli, 1976c; Kramer et al., 1977; Yeo et al., 1977).

Presumably the best evidence for pumping by the xylem parenchyma cells is provided by observations of differential effects of inhibitors on symplast loading and vessel loading, respectively. This is largely due to the research of Pitman and co-workers; but this work is not simply dedicated to proving operation of a xylem parenchyma pump. Pitman's interpretations focus on a model with two pumps, one for loading the symplast and one for loading the vessels. Therefore, his work will be discussed in a separate section.

12.2.5 The Two-Pump Hypothesis

To measure separately the uptake of ions and loading of the symplast via the root epidermis and cortex, and the efflux of ions into the xylem exudate, respectively, Pitman used a method first introduced by Weigl. Isolated roots are arranged between partitions in plastic boxes, so that ions moving out from the xylem at the cut surface are separated from the uptake medium bathing the external root surface (Weigl, 1969d, 1970; Pitman, 1971, 1972a; Fig. 12.3). The number of partitions of the boxes can be increased, and thus fluxes in different zones along the length of the root can be studied (Läuchli et al., 1977). The set-up then is much akin to that used by Arisz with *Vallisneria* leaves (Fig. 11.4).

Pitman (1977) has recently reviewed the evidence in favor of the two-pump hypothesis with largely active fluxes Φ_{oc} (symplast loading) and Φ_{cx} (vessel loading; see Fig. 12.3b). The concept of the two-pump hy-

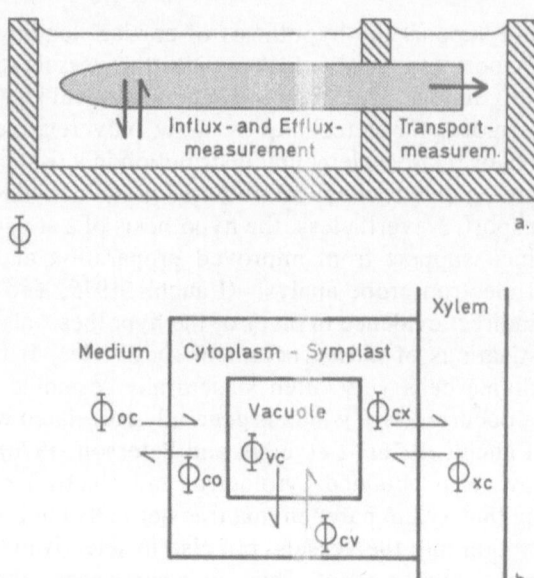

Fig. 12.3. a Separate measurements of ion transport between the medium and the cortex, and the stelar parenchyma and the xylem respectively, of roots; after Weigl (1969d, 1970) and Pitman (1971, 1972a,b); b fluxes that can be investigated with this system; after Pitman (1971, 1972a,b); c symbols see Figure 6.12 with additionally x, xylem.

pothesis does not exclude the possibility that the fluxes shown in Figure 12.3b may have considerable passive components, but active components of Φ_{oc} *and* Φ_{ox} are the decisive driving forces for directed radial transport of ions across the root. Pitman's early evidence for this was as follows (Pitman, 1971, 1972a):

1. One can reduce Φ_{oc} rapidly and considerably by a reduction of the *ion concentration* in the medium; flux into the vessels, Φ_{cx} then still continues unchanged for some while showing that it is independent of Φ_{oc}.
2. Both Φ_{oc} and Φ_{cx} are similarly inhibited by the *uncoupler* Cl-CCP (Pitman, 1977), suggesting that both fluxes are metabolically energy-dependent.

In a large number of subsequent investigations with barley roots, showing selective effects of phytohormones (ABA, cytokinins) and inhibitors of functional protein synthesis on Φ_{oc} and Φ_{cx} the independent nature of the two fluxes was confirmed (summary in Pitman, 1977). Some of the effects are briefly recorded, continuing the above enumeration:

3. The synthetic kinin *benzyladenine* inhibits translocation (Φ_{cx}) of $^{86}Rb^+$ in barley roots but not uptake into the roots (Φ_{oc}) (Pitman et al., 1974c).
4. The phytohormone *abscisic acid* (ABA) can inhibit Φ_{cx} and increase

Φ_{cv} (i.e., accumulation in the root vacuoles) while concomitantly Φ_{oc} remains unaffected (Cram and Pitman, 1972). It needs to be noted that the ABA effects depend on environmental conditions. Translocation (Φ_{cx}) in roots of seedlings grown in 0.1 mM $CaSO_4$ is more strongly inhibited at 28° than at 22°C. In roots of seedlings grown on full culture solution transport at 15° and 22°C is stimulated by ABA; transport at 28°C is highly inhibited after an initial transient stimulation (Pitman et al., 1974c).

5. *Cycloheximide* (CHM), known to inhibit protein synthesis at ribosomes, highly reduces Φ_{cx} while uptake and accumulation in the root cells are not affected (Läuchli, Lüttge et al., 1973; Lüttge, Läuchli et al., 1974; Jackson et al., 1974; Wildes et al., 1976). Simultaneous measurements of protein synthesis, ATP-levels and respiratory O_2 uptake rule out a possible action of CHM via energy metabolism, as suggested for example by Kelday and Bowling (1975). While protein synthesis is inhibited, O_2 uptake is affected very little and ATP levels increase, presumably due to lesser energy consumption in the absence of protein synthesis. Thus CHM action on energy metabolism is very indirect. (See also Cocucci and Marrè, 1973, who have shown this with the yeast *Rhodotorula gracilis*.) Glass (1976b) questions the CHM-work because he observed not only inhibition of translocation but also of uptake of K^+ into barley roots. Since Glass worked with 0.05 mM K^+ solutions and Läuchli, Lüttge et al. (1973) and Lüttge, Läuchli et al. (1974) used 5 mM KCl, the discrepancy may be due to different CHM sensitivity of uptake mechanisms operating at lower and higher external concentrations, respectively.

6. The amino acid analog *p-fluorophenylalanine* (FPA) does not inhibit protein synthesis, but is incorporated into proteins and thus leads to formation of ineffective proteins. FPA also inhibits Φ_{cx} selectively (Wildes et al., 1975; Schaefer et al., 1975; similar experiments with the proline analog azetidine 2-carboxylic acid are described by Pitman, Wildes et al., 1977).

In the case of inhibitors acting on proteins, e.g., CHM and FPA, the time factor is important for determining selective inhibition of Φ_{cx} as compared with Φ_{oc}. CHM, for example, inhibits Φ_{cx} within 45 min, but after more than 120 min an inhibition of Φ_{oc} also becomes apparent (Wildes et al., 1976). It is to be anticipated that a protein with a high turnover is inhibited by these agents earlier than a protein with low turnover. This has been confirmed by experiments on FPA-effects on inducible NO_3^--reductase (high turnover) and noninducible acid phosphatase (low turnover) of barley roots (Schaefer et al., 1975). Separations of cortex and stele and kinetic considerations clearly confirm a primary action on a flux in the interior of the root, Φ_{cx}. Thus Pitman's experiments, in addition to making it very likely indeed that Φ_{oc} and Φ_{cx} are two independent active carrier mechanisms, also show that proteins must be important parts of the carrier entities, the Φ_{oc}-carrier-protein having a much slower turnover than the Φ_{cx}-carrier-protein. (Turnover of Φ_{cx} at 23°C was determined to be 50% in 100 min with a rate constant of $k \simeq 10^{-4}$ s^{-1}; turnover is much increased at 26°C; Schaefer et al., 1975).

In conclusion, we can definitely rule out the hypothesis of the endodermis pump. The other hypotheses can neither be proven to be consistent with all experimental findings nor clearly be ruled out. The two-pump hypothesis combines many features of the hypotheses assuming active symplast loading in the epidermis and cortex and of the hypotheses postulating pumping by stelar parenchyma cells. It also needs symplastic transport allowing co-operation of the two membrane transport mechanisms having key functions of coupling, i.e., active Φ_{oc} and Φ_{cx}. The rate constant for ion equilibration in the cytoplasmic phase and the rate constant for attainment of a constant rate of radial ion transport across the root are identical (Pitman, 1971, 1972a). Thus, the two-pump hypothesis among all the possibilities discussed here explains the largest amount of experimental data and has the smallest number of shortcomings.

12.3 The Model of the Leaf

In the introduction the general features of a model for leaves have already been described (Sect. 12.1). The analogies between a leaf with salt glands or salt hairs and a root have been mentioned. In the following sections we will describe in some detail particularly instructive leaf models with various glands and with salt hairs and then end with discussing stomatal guard cells as very special cells in the leaf epidermis.

Questions regarding transport and distribution of solutes, of course, are also important in considering "average" leaves without glands. There is the problem of partitioning between the tissues of epidermis, palisade parenchyma, and spongy parenchyma in leaves (e.g., Outlaw and Fisher, 1975a,b; Outlaw et al., 1975). Coupling between short-distance, medium-distance and long-distance transport is essential in ionic relations of leaves, where input of ions occurs via the xylem but retranslocation of ions out of the leaves takes place predominantly in the phloem. Similar considerations are pertinent to movements of assimilates from chloroplasts in green leaf cells to other parts of a plant. These are key problems of long-distance translocation in the phloem (Sect. 11.4).

A more trivial and simple example for coupling between different tissues is given by variegated leaves, which have mutated pale or white regions and normal green photosynthesizing regions. It is obvious that the cells of the white regions depend on those of the green regions for supply of energy and photosynthetic products.

12.3.1 The System Leaf-Mesophyll/Stalk Cell/Bladder Cell of Atriplex and Chenopodium

This system is depicted in the schemes of Figure 11.6 and Figure 12.1a, and has been used several times previously. Electrochemical studies demonstrate that Cl^- transport from the leaf apoplast into the large vacuoles of the epidermal bladders is active, against an electrochemical potential gradient. Figure 7.20 shows that although stalk and bladder cells do not perform photosynthesis there is a much higher light-dependent active Cl^- accumulation in the bladder cells than in the green cells of leaf blades. This appears to be impossible without symplastic coupling, which can occur in two ways. Thermodynamically it is sufficient that active chloride transport occurs at only one membrane site in the system (Fig. 12.1a). Thus ions could be accumulated in the symplast by active uptake from the apoplast at the plasmalemma of the green mesophyll cells. Photosynthesis could more or less directly provide the energy for this by one of the mechanisms discussed in Section 8.2. Ions could then be passively released from the symplast into the vacuoles. The large vacuoles of the bladders would lead to a large relative accumulation in the bladders as compared with the mesophyll cells. The possibility that the important energy-dependent step in active Cl^- excretion into the bladder vacuoles resides in the mesophyll cells is unlikely, however, for a number of reasons. The light stimulation of Cl^- uptake is much smaller in the mesophyll cells (about 2-fold) than in the bladder cells (about 5-fold; Fig. 7.20). Furthermore, salt accumulation in the bladder vacuoles is envisaged as an adaptive process to salinity, keeping NaCl levels low in the metabolically active cytoplasm. Hence, active Cl^- accumulation should occur in the bladders and not in the leaf mesophyll cytoplasm. This would imply energetic coupling by transport of energy-rich photosynthetic intermediates (transport metabolites, see Sect. 10.2) from the mesophyll to the bladders. (Reviews Lüttge, 1974, 1975; Hill and Hill, 1976.)

12.3.2 The Salt Glands of Limonium

A case similar to that in Atriplex and Chenopodium can be observed in Limonium vulgare during salt excretion by the leaf glands. Limonium is a littoral halophyte investigated by Arisz et al. (1955), Ziegler and Lüttge (1966, 1967) and in many papers by Hill and Hill (see reviews of Lüttge, 1975; Hill and Hill, 1976). Under certain experimental conditions active excretion can be energetically powered by light, but the cells of the gland complex are photosynthetically inactive. Hill and Hill argue (see review of Lüttge, 1975) that in salt excretion by Limonium glands the active pumping of ions occurs out of the symplast during excretion from the gland cells and not, or not only, into the symplast during salt uptake by the leaf cells. First, the transit half-time of Cl^- from the leaf apoplast into the

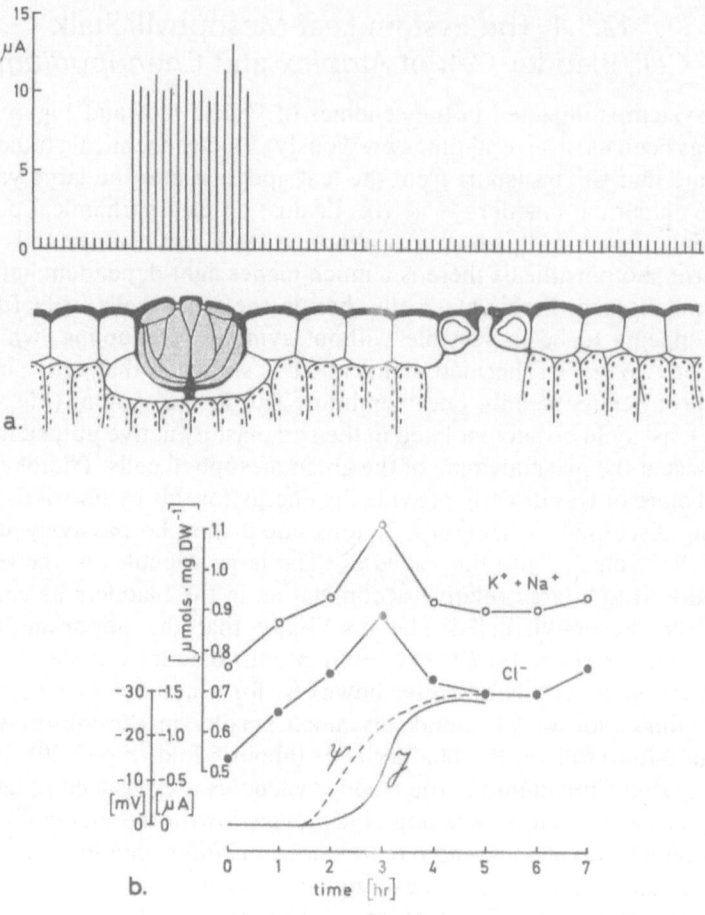

Fig. 12.4. a Leaf epidermis of *Limonium vulgare* with a salt gland. Probing the surface of a *Limonium* leaf with constant voltage pulses gives current pulses of 10–13 μA only above the glands. The ionic conductivity through other parts of the leaf surface (i.e., stomata, normal epidermal cells) is less than a tenth. The cuticle as indicated by *bold black lines* has negligible ionic conductivity (Lüttge, 1975; electrical properties after Hill, 1967). **b** Rise of excretory gland activity and ion content in the chloroplasts of *Limonium* leaves induced to excrete Cl⁻ by treatment with 100 mM NaCl. The excretory gland activity is expressed as potential (Ψ[mV]) and as short circuit current (J[μA]) respectively across as leaf disk. (After Hill, 1970b.) Ion content of the chloroplasts (μmol · mg DW⁻¹) was obtained by analysis of nonaqueously isolated chloroplasts (Larkum and Hill, 1970). (From Lüttge, 1975.)

excreted fluid is correlated with the half-time of filling of the cytoplasm. Second, NaCl excretion is inducible by chloride. After addition of NaCl to disks of leaves, the ion concentration of cytoplasmic constituents such as chloroplasts stops rising when the pumping out of the glands reaches maximum activity (Fig. 12.4). If pumping occurred only into the symplast, one would expect an increase of symplastic ion concentration as pumping activity becomes most effective. A little microscopic observation also suggests that the glands themselves are pumping. The kinetics of fluid excretion from individual glands can be measured when the leaf surface is covered with some oil under which the aqueous excretion droplets can accumulate. It turns out that the kinetics of excretion by individual neighboring glands can be very different.

12.3.3 The *Nepenthes* Pitcher Glands

The strange pitchers of *Nepenthes* which capture insects are modified leaves (Fig. 12.5). The roofed glands in the lower third of the pitcher interior (Fig. 4.10d) may serve several transport processes between the pitcher tissue and the secreted fluid inside the pitcher. There is not only secretion of the digestive enzymes and uptake of low molecular substances from the digested prey, but also water and ion transport. The pitcher fluid usually contains 20–30 mM Cl⁻.

Ion transport across the *Nepenthes* pitcher tissue was extensively investigated because the pitchers with their large volume of secretion are quite suitable for this purpose. As long as the cap of the pitcher is closed,

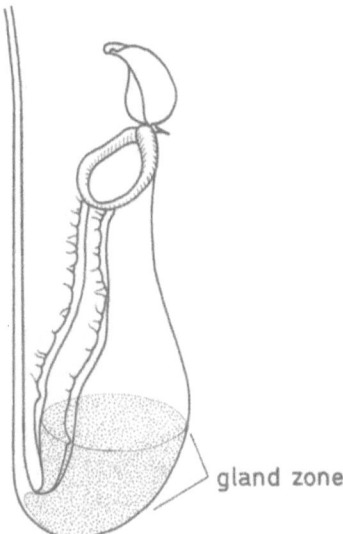

gland zone

Fig. 12.5. *Nepenthes* pitcher with secreted fluid.

the pitcher content is sterile in a bacteriological sense. The cylindrical cells of the gland epithelium contain very dense cytoplasm. A quantitative examination of microautoradiographs of pitcher wall cross sections shows that during tracer Cl^- transport and secretion the symplast of the pitcher tissue is similarly labeled throughout (Lüttge, 1966b).

From inhibitor studies with isolated pitcher tissue disks it is known that the Cl^- excretion by the cells of the interior wall of the pitcher, as well as the Cl^- uptake from a solution which bathes the exterior wall of the pitcher, is dependent upon metabolism (Lüttge, 1966a). Membrane potential and current measurements in short-circuiting experiments suggest that various ion pumps are involved (Nemček et al., 1966). At the surface between the pitcher fluid and cells of the interior pitcher wall Na^+ and Cl^- are pumped actively against the gradient of electrochemical potential from the pitcher fluid into the cells. For K^+, which appears to be near electrochemical equilibrium between the pitcher fluid and the pitcher wall tissue, the situation is somewhat less clear (Nemček et al., 1966).

12.3.4 Nectar Secretion

Nectar glands are very often found on floral parts (sporophylls) and may also occur as extrafloral nectaries on normal leaves. Thus, with a few modifications, the model of Figure 12.1b can be also used for discussion of nectar secretion (Fig. 12.6).

The secretion of assimilates—for instance of sugar in nectar—for two reasons is more difficult to investigate than salt excretion. First, the transported particles are not only moving along complex and tortuous routes, they are at the same time subject to metabolic modifications. Second, the methods developed for salt transport (flux measurements, compartmentation analysis, electropotential measurements) cannot be applied straightforwardly.

The typical nectar is a very specific secretion product. Apart from sugars, mainly glucose, fructose and sucrose, and also oligosaccharides composed of glucose and fructose monomers with traces of other hexoses, there are other substances in the nectar, only in traces. Qualitatively these substances accompanying sugars in the nectar comprise a large variety of chemically different compounds, e.g., mineral ions, amino acids, organic acids, vitamins, etc. (Baker and Baker, 1973a,b; 1975, 1976; review Lüttge, 1977). A comparative anatomical and analytical investigation shows that the perfection with which such accompanying substances are excluded from the nectar depends on the degree of differentiation of the gland tissue. The nectar of highly developed glands contains only relatively small amounts of ninhydrin positive substances (amino acids) whereas this "contamination" in the nectar of anatomically primitive glands is considerably larger (Fig. 12.6; Lüttge, 1961).

Plant Species	Nectary	Ratio: amino acids : sugars
Platycerium div. spec.	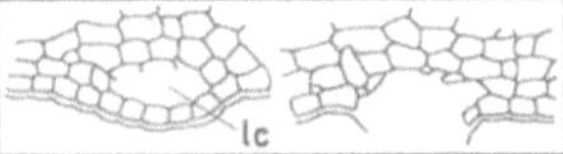 lc	$20\,000 \cdot 10^{-6}$
Sambucus nigra L.	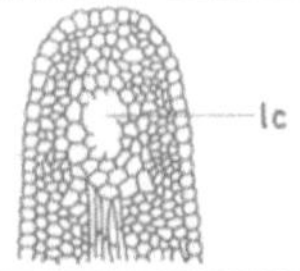 lc	$5\,000 \cdot 10^{-6}$
Sambucus racemosa L.	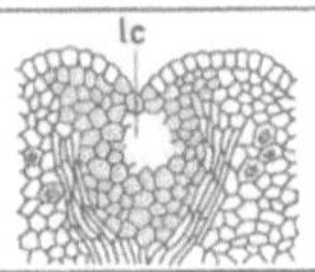 lc	$4\,000 \cdot 10^{-6}$
Pteridium aquilinum Kuhn	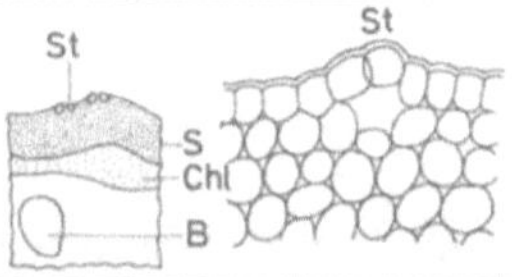 St St S Chl B	$100 \cdot 10^{-6}$
Abutilon striatum Dicks. Robinia pseudo-acacia L. Hoya carnosa R. Br.	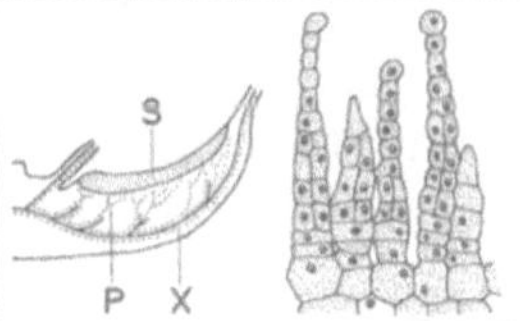 S P X	$5 \cdot 10^{-6}$

Fig. 12.6. Relation of anatomical gland organization to chemical specificity of nectar (e.g., decreasing ratio amino acids : sugars with increasing structural specialization). The degree of structural specialization increases from top to bottom, i.e., from lysigenous secretion (*Platycerium, Sambucus* spp.), via glands where secretion occurs through modified stomata (*Pteridium*), to glands with highly specialized glandular tissues and cells (the example drawn shows *Abutilon* calyx nectaries). Lines *1–4* extrafloral, line *5* floral nectaries. *B*, bundle; *Chl*, chlorenchyma; *lc*, lysigenous cavity; *S*, secretory tissue; *St*, modified stomata; *P*, phloem elements; *X*, xylem elements.

Nectar secretion doubtlessly is under metabolic control. Nectaries are highly active organs metabolically (Fig. 7.19). Phytohormones also interact with the regulation of nectar secretion.

The sugars secreted in the nectar are derived from reserves in the secretory tissue itself only to a small extent. The major part of secreted sugar comes from the assimilation stream in the sieve tubes. As a consequence, we have, again, as in salt excretion, various possible sites for metabolic control of secretion (see Fig. 12.7; reviews: Lüttge and Schnepf, 1976; Schnepf, 1977):

1. The unloading from the sieve tubes. Here the companion cells may play a particular role.

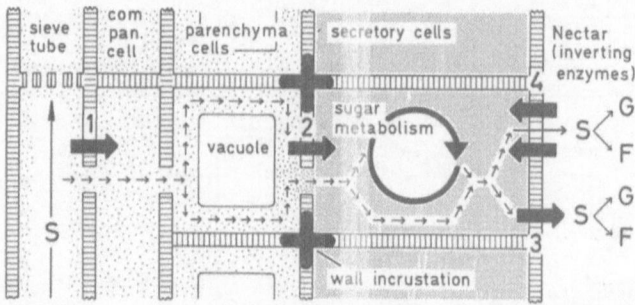

Fig. 12.7. Model of nectar secretion. Possible sites of active sugar membrane transport; *1,* active sieve-tube unloading; *2,* pump concentrating sugars in the secretory cells; *3,* active secretion; *4,* passive leakage (pressure filtration) of nectar fluid accompanied by specific active reabsorption processes. *S,* sucrose; *G,* glucose; *F,* fructose. *Small arrows,* transport in plasmodesmata and symplastic transport. *Thick arrows,* metabolically controlled membrane transport. (Lüttge and Schnepf, 1976.)

2. The active step supplying sugar to the gland tissue may occur farther within the gland, as was suggested by Findlay et al. (1971) and Reed et al. (1971).
3. The site of active transport may be at the plasmalemma of the gland cells, where the secretion product is released to the external gland surface.
4. In addition a possible driving force is the metabolism of sugar in the gland cytoplasm. The fact that all important enzymes of carbohydrate metabolism can be detected in nectar glands shows that metabolism of sugar certainly plays a role. This is corroborated by experiments with labeled glucose.

     Not all sugar molecules which are in fact transported are subject to metabolic modification, however. About 70% of ^{14}C-labeled glucose fed to secreting *Abutilon* nectaries is secreted unchanged as glucose (Ziegler, 1968).

     As shown in Figure 12.7, like ion transport across the root and ion transport by salt glands, symplastic transport plays an essential role in sugar secretion. Evidence has been obtained that the endoplasmic reticulum (ER) might be involved (Gunning and Hughes, 1976; see review of Schnepf, 1977), also, at least in some cases, the eventual elimination of nectar to the external gland surface occurs by granulocrine secretion of vesicles budding off from the symplastic ER system. This, then, would lead to a further possible site of active membrane transport in sugar secretion, as listed next.
5. Active loading into the ER cisternae in the gland symplast may occur (not shown in Fig. 12.7).

     It has also been considered, that nectar might be eliminated through the gland surface by a turgor pressure mechanism forcing outwards a volume of solution similar to a filtration under pressure. The high chemical specificity of nectar, however, still would require membrane-controlled transport to explain the very low amounts of non-sugar compounds in the nectar. In the pressure filtration hypothesis this is thought to be brought about by the following mechanism.

6. Specific and active reabsorption of nonsugar compounds may take place from the secreted nectar, which thus sets up a further possibility of an active step in nectar secretion (Fig. 12.7).

As in the other models (i.e., roots, salt glands) it is difficult to make a definite decision, which of these possible sites really operates as a decisive active transport process in secretion. There is evidence now that different mechanisms of nectar secretion may operate in different glands (Schnepf, 1977). An involvement of membrane transport of sugar across the plasmalemma at the entire surface of all gland cells appears to be likely in the trichome nectaries of *Abutilon*. This conclusion is based on quantitative arguments considering flux rates. Nectary gland cells may or may not have the transfer-cell-like wall protuberances (see Sect. 7.4). *Abutilon* trichome cells do not have protuberances. Figure 12.8 shows sugar flux rates which were calculated on the basis of various assumptions on the surface active in nectar secretion. We can see that extraordinarily large flux rates are obtained if only the cross-sectional area of the trichome stalk cell would contribute to and rate-limit active sugar transport as con-

ABUTILON sucrose secretion	assumption in estima-tion of active secretory membrane surface	rate of mem-brane transport $\mu mol\ m^{-2}\ s^{-1}$
average nectar volume per tri-chome: 30 $\mu m^3\ s^{-1}$ (range 8-80)	distal wall of stalk cell	110
nectar concen-tration 0.4 M (observed up to 0.6 M)	spherical apical cell	65
	total trichome plus apoplastic route to apex	3
	all surfaces of tri-chome cells plus apoplastic route	0.6

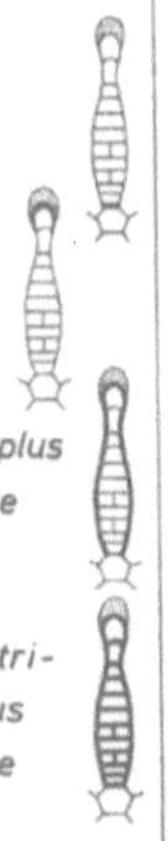

Fig. 12.8. Rates of sugar membrane fluxes in *Abutilon* nectaries on the basis of various assumptions on active secretory membrane surface (*bold lines* in sketches of trichomes). (After data from Findlay and Mercer, 1971; Gunning and Hughes, 1976; from Lüttge, 1978.)

Table 12.3. Membrane flux rates of sugar transport in various plants systems (from Lüttge, 1977).

Object	Sugar transported	Flux rate $\mu mol\ m^{-2}\ s^{-1}$	Concentration from which transport occurs mM	Reference
Abutilon nectary trichomes	Sucrose	between 0.6 and 110[a]	500–1000[b]	Findlay and Mercer, 1971; Gunning and Hughes, 1976
Vicia faba leaves, phloem loading	Sucrose	0.14	?	Gunning et al., 1974
Allium cepa onion epidermis	Glucose	0.03	230	Steinbrecher and Lüttge, 1969
Zea mays scutellum	Sucrose	0.125	>200[c]	Humphreys, 1973
Hydrodictyon africanum	Glucose	0.007 / 0.028 / 0.042	0.1 / 1 / 10	Raven, 1976
Nitella flexilis	Glucose	0.01 / 0.06	1 / 8	Wallen, 1974
	Sucrose	0.005 / 0.03	1 / 8	
Nitella translucens	Glucose	0.025	5	Smith, 1967
Chlorella vulgaris	6-deoxyglucose	0.42	10	Tanner et al., 1974
Neurospora crassa	Glucose	0.5[d]	1[d]	Slayman and Slayman, 1974
Saccharomyces cerevisiae	Glucose	1		Kotyk, 1967 and personal communication

[a] Taken from Figure 12.8.
[b] Assumed on the basis of a 18%–30% sucrose solution supplied to the nectary gland via the transport system.
[c] The rate given represents a V_{max} value of the transport system.
[d] At the extreme and at high sugar concentration an occasionally observed maximum of total glucose influx may be 2 $\mu mol\ m^{-2}\ s^{-1}$ (C. Slayman, personal communication).

sidered by possibility (2) above (2 in Fig. 12.7). Similarly high rates are obtained when exclusively the tip of the trichome secretes actively. If the sugar moves symplastically from the phloem to the trichomes, however, and is secreted into the apoplast of the trichomes via the plasmalemma surface of all trichome cells, then the flux rate becomes comparable to sugar membrane fluxes in other plant systems as listed in Table 12.3. The mechanism of membrane transport of sugar in nectar secretion is not clear. Acid phosphatases may play a role. Their activity in nectar glands is so high that this has even been used as a cytological criterion for nectaries (Frey-Wyssling and Häusermann, 1960). The phosphatases in the nectaries and also in sieve tube companion cells are particularly associated with the plasmalemma (Figier, 1968). It is possible that the secreted sugars move in a phosphorylated form and that phosphatases catalyze vectorial group transfer reactions leading to secretion (Lüttge, 1966c). In *Abutilon* it also seems possible that there is a H^+-sugar co-transport mechanism (Lüttge, 1978; see also Sect. 10.2.3.2).

12.3.5 Stomatal Guard Cells: Special Cells in the Leaf Epidermis

12.3.5.1 The Isolation of Guard Cells From the Leaf Symplast

It has long been known that *selective adjustment of turgor* by stomatal guard cells is the mechanism underlying regulation of the stomatal aperture. This has been demonstrated very elegantly recently by Meidner and Edwards (1975) who applied pressure externally to stomatal guard cells and adjacent subsidiary cells, respectively. Glass capillaries were inserted into both types of cells. If the pressure in the guard cells was increased via the capillaries stomata opened; an increase of pressure in the subsidiary cells led to stomatal closure. A selective regulation of turgor by the guard cells must mean that they function as rather isolated osmotic cells and should not be part of the leaf symplast. It was stressed in Section 11.3. that the symplastic continuity, maintained by plasmodesmata, is important in cooperating systems. Conversely, plasmodesmatal connections might be expected to be absent in systems whose function is based on a relative isolation. The question of whether there are plasmodesmata between the guard cells and the neighboring cells has been carefully investigated by electronmicroscopy. It would be incorrect to state that nowhere are there plasmodesmatal contacts between guard cells and neighboring epidermis cells (cf. Pallas and Mollenhauer, 1972), but it is clear that plasmodesmata are absent in many cases or at least are very rare (Thomson and Journett, 1970; Allaway and Setterfield, 1972). Contrasting with this are cell wall interruptions with very wide cytoplasmic connections between the two co-operating guard cells of grass stomata.

12.3.5.2 The Role of K⁺-Transport for Osmotic Regulation

The outstanding role of K^+ as an osmoticum in stomatal regulation has been discovered by Imamura (1943) and then was much supported by R.A. Fischer and T.C. Hsiao (reviews: Raschke, 1975; Hsiao, 1976). From a large number of investigations especially by using stripped leaf epidermis preparations floating on specifically composed external media and utilizing particular staining techniques and electron probe microanalysis (Fig. 12.9) it became clear that the uptake and release of K^+ decisively contributes to stomatal regulation (Fischer, 1968, 1971; Humble and Hsiao, 1969, 1970; Humble and Raschke, 1971; Sawhney and Zelitch, 1969). As shown in Figure 12.10, the system is very specific for K^+. However, the potassium specificity can be shown only when small amounts of calcium ions are present in the external medium. In the absence of calcium, sodium ions can replace potassium ions during stomatal opening (Pallaghy, 1970). Dayanandan and Kaufmann (1975) have surveyed a large number of plant species and shown that K^+ in all taxonomic groups is involved in stomatal regulation. (Unfortunately the moss sporophytes are missing from the list of taxonomic groups in this publication.) An exemption are the unusual guard cells of the orchid *Paphiopedilum leeanum* which do not contain chlorophyll although they are functional. The epidermis of this orchid is very poor in K^+ but normal K^+ levels are observed in the green mesophyll. Presumably another osmoticum (Na^+?) replaces K^+ in stomatal regulation of this plant (Nelson and Mayo, 1977).

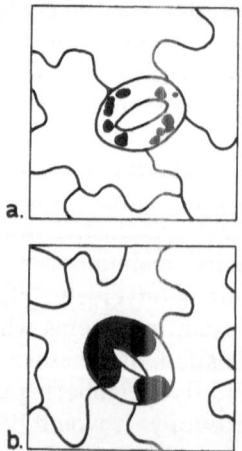

Fig. 12.9. Localization of K^+ in the guard cells of opening and closing stomata of *Vicia faba* by staining with cobalt-sodium-nitrite/ammonium sulfide (black precipitation; Macallum, 1905). Drawn after R.A. Fischer (1971). **a** K-localization shortly before an illumination leading to an opening movement; **b** 3 h after illumination.

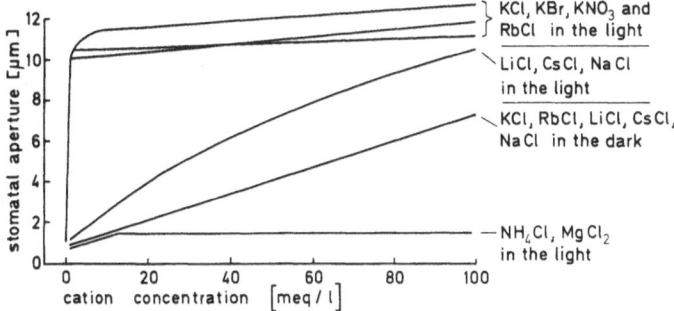

Fig. 12.10. Stomatal opening in isolated strips of leaf epidermis of *Vicia faba* floating on various ionic solutions. Curves (without points) taken from Figures 2, 3, 4 and 5 of Humble and Hsiao (1969). It becomes clear that opening in the light specifically depends on the presence of K^+ in the medium (all solutions contained Ca^{2+}) and is independent of the anion. Only Rb^+ can replace K^+. (K^+ and Rb^+ are often not distinguished by transport mechanisms in plant cells; see Sect. 6.2.3.2, Footnote p. 153.) At high external ion concentrations the specificity is much reduced.

The K^+ pump of the guard cells leads to opening movements within 3 h. The concentration increase is from 50 to 300 mM (Fischer, 1968, 1971), or, in some more recent investigations with rolled epidermis preparations of *Vicia faba*, where stomatal guard cells are the only viable cells, from 110 to 550 mM (volume of one guard cell 4.93 ± 0.09 pl; closed stomata 0.55 p Eq K^+/guard cell; open stomata 2.72 p Eq K^+/guard cell) (Allaway and Hsiao, 1973). With a rate of 10–15 pmol s^{-1} cm^{-2} the pump rate is in the upper range of the normally observed ion transport fluxes in plant cells (Fischer, 1972; see also the list of ion transport rates in Sect. 11.4.2). Based on fresh weight, the rate is 150 μmol h^{-1} g_{FW}^{-1} and thus 12 times the maximum rate of uptake of system 2 of plant tissues (Raschke, 1975).

12.3.5.3 Stoichiometry of the Guard Cell Ion Pump

After establishment of the dominating role of K^+ transport in stomatal regulation questions remaining open are:

1. Where does the K^+ which is taken up or released in guard cells come from, and go to, during opening or closing?
2. How is the positive charge of K^+ balanced electrochemically?
3. Are the movements of K^+ plus counterions sufficient to explain osmotically the turgor and volume changes in stomatal regulation?

The potassium ions which are taken up by the guard cells could originate from the leaf apoplast. Regrettably there are only very few good analyses of the concentration of ions in the free space solution of leaves. As far as we can see so far, the concentration of ions in the free space of leaves can be between 2 and 30 mM (Table 12.1). There is evidence that in the grasses and in *Commelina* the subsidiary cells of the stomatal cell complex can serve as potassium reservoirs and that a transport of K^+ between the guard cells and the subsidiary cells is important in stomatal regulation (Raschke and Fellows, 1971; see below for data on the monocotyledonous plant *Commelina*). In species without subsidiary cells, e.g., *Vicia faba*, K^+ must be supplied in the medium for stomata of isolated epidermis strips to open widely (Hsiao, 1976).

Electron probe analysis, Ag^+ precipitation techniques, and electrochemical considerations led to the discovery that Cl^- is involved in many cases partially fulfilling the requirement of electrical charge balance of K^+ (Humble and Raschke, 1971; Raschke and Fellows, 1971; Pallaghy and Fischer, 1974; Dayanandan and Kaufmann, 1975; Penny et al., 1976). Only few cases, however, have been reported in which Cl^- transport is equivalent to K^+ transport. This is perhaps so in the stomata of *Allium cepa* (Schnabl and Ziegler, 1977). In most of the other cases it turns out that the charge of K^+ influx into guard cells of opening stomata is largely balanced by an H^+ efflux. Just as for the K^+-H^+ exchange mechanism of *Avena* coleoptiles (Sect. 9.2.1, Fig. 9.9) this raises the question of cytoplasmic pH regulation, and just as in *Avena* this appears to be solved by organic acid metabolism leading to an accumulation of malate (references see below). The relative contributions of Cl^- and malate for balancing the charge of K^+ can be manipulated experimentally. If Cl^- is available, the guard cells accumulate largely Cl^- together with K^+. If K^+ is supplied with a nontransportable synthetic anion (e.g., iminodiacetate) as counterion and in the absence of Cl^- a stoichiometric amount of malate is synthesized to balance K^+ (van Kirk and Raschke, 1978).

Penny and Bowling (1974, 1975) have used microelectrodes to analyze K^+- and H^+-activities in the guard cells of opened and closed stomata, respectively, of *Commelina*. They have also calculated the driving forces for K^+ movements. These data are compiled in Figure 12.11, which also gives a good illustration of the K^+ shuttle between guard cells and subsidiary cells as stomata open and close.

Organic acid metabolism and stoichiometric malate accumulation in relation to K^+-H^+ exchange by stomatal guard cells has been suggested and demonstrated by many authors (e.g., Pearson, 1973; Raschke and Humble, 1973; Willmer et al., 1973; Pallaghy and Fischer, 1974; Pearson and Milthorpe, 1974; Willmer and Dittrich, 1974; Pearson, 1975; Penny et al., 1976).

More recently, Raschke and Dittrich (1977; Dittrich and Raschke, 1977a,b) in a remarkable series of three papers report various aspects of K^+, Cl^-/H^+-exchange and malate transport in *Commelina* guard cells. They have tried in a comprehensive discussion to unify a large number of observations and hypotheses developing their view of stomatal function. This includes the fact that stomatal regulation to a high degree is asso-

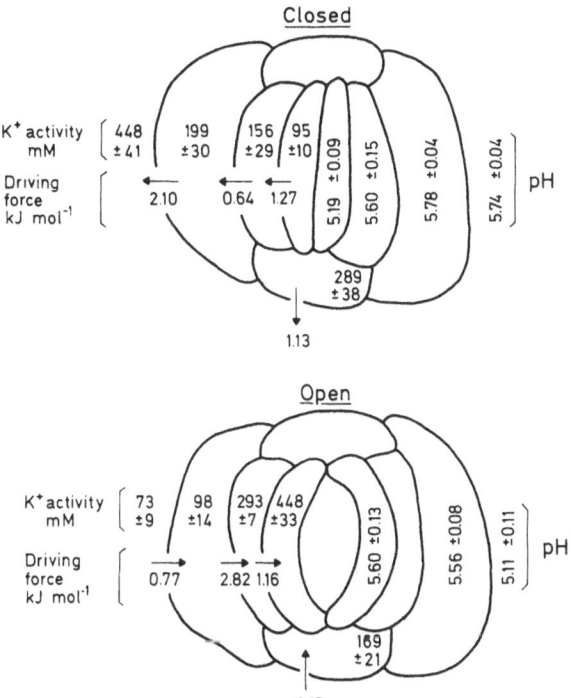

Fig. 12.11. K$^+$ activities, driving forces on K$^+$ required to maintain the observed differences in cellular K$^+$ activities, and pH-gradients in cells of the stomatal complex of *Commelina communis* with closed (*top*) and opened (*bottom*) stomata, respectively. Activities were measured with K$^+$-sensitive microelectrodes calibrated against standard KCl solutions, and are expressed as equivalent KCl concentrations. Driving forces were calculated from the Nernst potentials for K$^+$ and the electrical potential differences between cells measured with microelectrodes. Arrows indicate the direction of active transport. Activities are the means ($\pm$ standard error of mean) of at least five estimates. Values outside the cells of the stomatal complex are of epidermal cells. (Adapted from Penny and Bowling, 1974, 1975; see also Hsiao, 1976.)

ciated with sensing of CO_2 concentration, and high CO_2 concentrations leading to stomatal closure (cf. Raschke, 1975). It also integrates the role of guard cell starch, the breakdown and synthesis of which has often been assumed to be involved in osmotic regulation during guard cell movements, and been rejected as a cause since it is often too much out of phase with the actual movement (reviews Raschke, 1975; Hsiao, 1976). The discussion of Raschke and Dittrich also incorporates metabolic contributions of the green mesophyll cells, especially in view of the nongreen but functional orchid guard cells discovered by Nelson and Mayo (1975, 1977). Figure 12.12 is an attempt to draw a scheme from the discussion of Dittrich and Raschke (1977b). This scheme shows the plasmalemma and

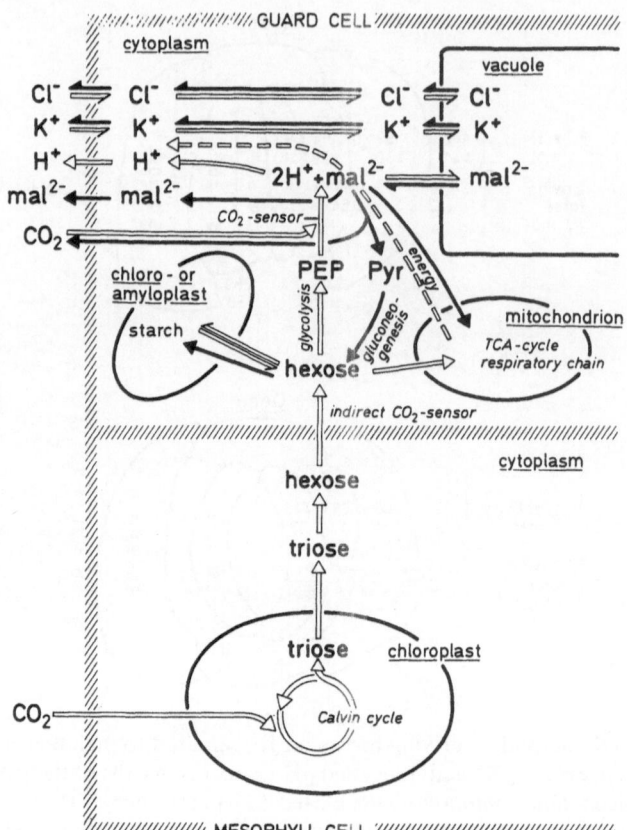

Fig. 12.12. Ion transport and metabolism of green and nongreen (e.g., orchid) sto-matal guard cells and relation to green mesophyll cells during opening (e.g., in the light, *open arrows*) and closing movements (e.g., in the dark, *closed arrows*), respectively. *mal,* malate; *Pyr,* pyruvate; *PEP,* phosphoenolpyruvate. (Drawn after the comprehensive discussion in Dittrich and Raschke, 1977.)

tonoplast fluxes of H^+, K^+, Cl^- and malate^{2-} of opening and closing sto-mata. Carbohydrate metabolism, i.e., CO_2 fixation via the Calvin cycle (largely in the mesophyll), starch synthesis and breakdown, glycolysis and gluconeogenesis, and respiration are important for the provision of energy and of phosphoenol-pyruvate for malate synthesis by CO_2 dark fixation (which is possible both in green and nongreen guard cells). CO_2 dark fixation plays a direct role and CO_2 light fixation in the Calvin cycle plays an indirect role in sensing of CO_2 concentration. With these brief re-marks, the scheme will explain itself. Of course, it is a scheme only and many more subtle points are disregarded. Also, although incorporating a wealth of information, the scheme may not apply universally. Schnabl and Ziegler (1977) have pointed out that *Allium cepa* leaves store neither starch nor soluble polymer carbohydrates to replace it. This may be the reason why in this species most likely K^+ transport is entirely balanced by Cl^- transport. Phosphoenol-pyruvate cannot be formed from starch as a substrate for CO_2 dark fixation, and production of protons and malate ions cannot occur. It can be seen, however, this is a special case of what is depicted in Figure 12.12, and it is inherent in the scheme.

12.3.5.4 Abscisic Acid Regulation of the Guard Cell Ion Pump

In Chapter 9 we discussed the interactions between membrane transport and phytohormones. We should briefly note here that stomata are highly dependent on abscisic acid (ABA). Water stress leads to a considerable accumulation of ABA which causes stomatal closure and hence provides protection (references and further discussion in Sect. 13.3.4). It is possible but not yet conclusive, that ABA affects stomatal movements via action on the K^+-H^+ exchange mechanism (see Hsiao, 1976).

Chapter 13

Transport Regulation in the Plant as a Whole

13.1 Comparison of Isolated Organs With Organs in the Intact Plant

The last two chapters have brought us a little closer to the aim set out in the beginning, i.e., to an understanding of how the integration of individual transport processes allows the intact plant to function as a whole (Fig. 1.1). Work with isolated organs, however, such as the systems of roots and leaves described in Chapter 12 and the elucidation of coupling and interconnections between short-distance, medium-distance and long-distance transports in these systems will never allow us to reach this aim. Isolated organs, although showing many of the same features as in the intact plant, essentially are artificial systems.

Let us first consider *roots*. In intact plants only about 3% of the Ca^{2+}, 16% of the K^+, and 25% of the P taken up by the roots are retained in the roots; the major portions of nutrients taken up are translocated to the shoot (Pitman, 1975). Xylem transfer and exudation in isolated roots largely depend on ion transport across the root and development of root pressure (Sect. 12.2), whereas the major driving force for long-distance transport to the shoots is transpiration in the intact plant (Sect. 11.2.2). There is much early evidence suggesting that transpiration may affect ion transport across the roots (reviews: Sutcliffe, 1962, 1976a,b; Lüttge, 1969; Pitman, 1975) although the interactions are very complex and quantitatively not clear. The mechanism of this interaction was very controversial in the early stages of development of hypotheses of radial ion movement across the root (see Sect. 12.2). Transpiration could directly speed up passive movements within the roots. This appears to be supported by observations of increased effects of transpiration at increased external ion concentration, because passive components of transport become increasingly important as external concentrations rise (see Sect. 6.2.1). Conversely, transpiration could indirectly speed up active transport into the vessels in the roots by increasing the sink. Furthermore, transpiration may affect diffusion in a soil toward the roots and resistances of flow in the root (see reviews cited above). Another point is K^+/Na^+ selectivity of higher plants. This is not due only to the preferential uptake of K^+ over Na^+ at the level of the root. The K^+ and Na^+ input to the shoot and leaves very much depends on modifications along the route during translocation (e.g., Waisel and Kuller, 1972; Yeo et al., 1977; cf. also Pitman, 1975).

Or let us consider *leaves*. It is a tremendous oversimplification that cells of leaves just take up ions from the solution supplied via the transpiration stream as suggested by the models of Figure 12.1. Supply of ions via the xylem and uptake from the leaf apoplast is only one aspect of ionic relations. Retranslocation of ions from leaves in the phloem is just as important. The input and output of ions in leaves of intact higher plants is also very closely related to growth and hence to age as illustrated in Table 13.1 and Figure 13.1. This relation to growth is very complex. It does not only reflect an increased requirement of solutes and water to fill the vacuoles of growing and enlarging cells in the leaves. It is also a result of the change of a developing leaf from a sink for substrates to a source when photosynthetic products begin to exceed the needs of growth and metabolism within an individual leaf. Thus a stage of net import of materials via the phloem is followed by a stage in which export dominates.

The enumeration of points stressing the discrepancies between an isolated organ and an organ of an intact plant could be continued. This would be interesting as it would reveal much additional information on whole plant physiology. The conclusion is quite clear, however, that co-operation and co-ordination between organs introduce a new and very important dimension which we have not discussed thus far. If we comprehend the general importance of supply and demand or source–sink relationships for transport (Sect. 11.4.2.1), then in turn we easily realize the complexities of organ interactions and the indispensable role of transport in the functions of whole plants. We shall describe a few selected whole plant systems, which may illustrate how future research could pro-

Table 13.1. Rates ($\mu mol/day$) of input of potassium ions through the xylem and phloem into barley leaves of different ages. (Greenway and Pitman, 1965; from Sutcliffe, 1976a.)

	Age of leaf		
	Young	Intermediate	Old
Intake *via* xylem	2.0	2.7	1.9
Intake *via* phloem	1.3	0.7	− 1.6
Total intake	3.3	3.4	0.3

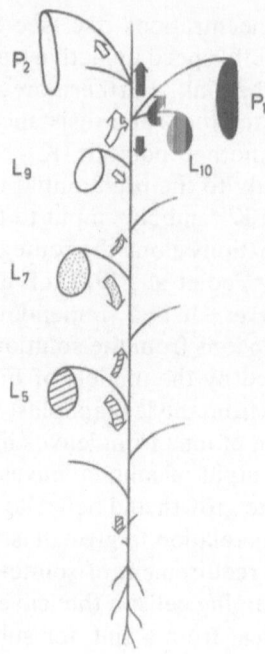

Fig. 13.1. Diagram of a pea (*Pisum sativum*) plant showing distribution of labeled phosphate from individual leaves as a function of their position on the stem. The *size of the arrow* indicates the amount of radioactivity moving in a particular direction from the fed leaves (L_5–L_{10}). Movement from the lower leaves (L_5 and L_7) is predominantly toward the roots, while that from higher leaves (L_9 and L_{10}) is mainly toward the apex. A large fraction of the translocate from leaf 9 moves into the developing pods and most of it goes into the pod at node 11 (P_2). On the other hand, ^{32}P applied to leaf 10 is exported mainly into the pod developing in the axil of this leaf (P_1). (Sutcliffe, 1976a; re-drawn from Linck, 1955.)

ceed in elucidation of the role of transport in the function of whole plants. This work should be aided by the development of general feedback models (e.g., Cram, 1976; Pitman and Cram, 1977). Although we will not go into the mathematical details of such models, in the next section we will list some components of such schemes before describing individual systems.

13.2 Signals in Feedback Systems
of Whole Plants

Summarizing earlier chapters we may conclude that in the functioning and co-operation of different parts of plants various types of messages may be involved:

1. messages inherent in the supply of energy (Chap. 10);

2. messages inherent in the supply of material (e.g., carbon skeletons of substrates);
3. electrical signals (Sect. 11.3.2.2);
4. hormonal signals (Chap. 9).

Three comments on this enumeration seem appropriate. First, transport is involved in dual ways: it responds to the messages and it is responsible for transfer of the messages (e.g., transport of energy-rich metabolites, of substrates, and of hormones). Second, in the transfer or transport of energy it seems that the larger the distance along which energy has to be transferred, the more stable is the form in which energy is transported. Electrochemical gradients are built up at membranes only over very short distances, e.g., at thylakoid membranes, at the plasmalemma, etc. More stable energy-rich transport metabolites move over short and medium dis-

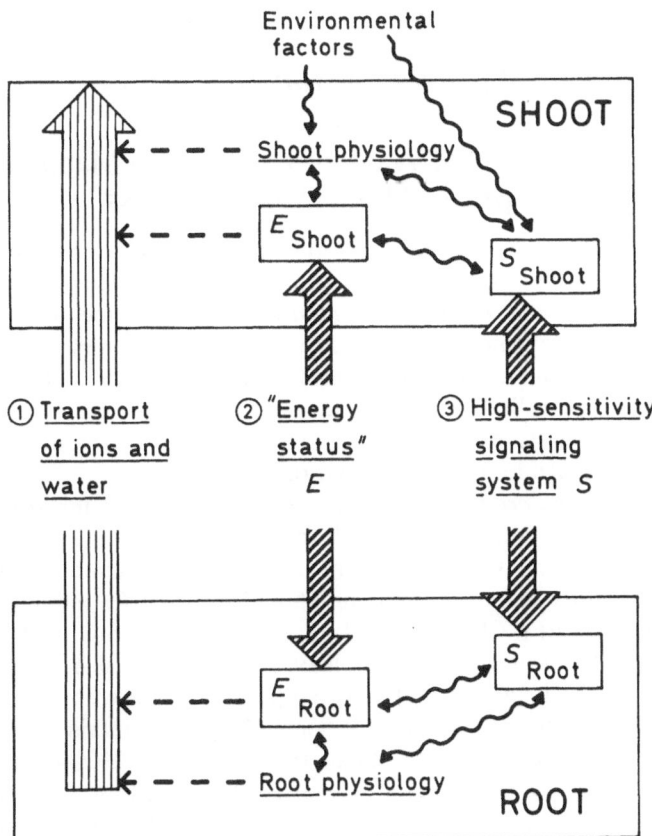

Fig. 13.2. Three components of a feedback system regulating transport of ions and water from root to shoot: *1*, transport itself; *2*, "energy status" (*E*); *3*, high sensitivity signalling system (*S*). *Wavy arrows*, interactions between *E* and *S*, and with general physiological activities; *dashed arrows*, action upon transport. (Lüttge, 1974.)

tances within cells or between individual cells in a parenchyma (Sect. 10.2). In a still more stable form energy is moved in long-distance transport in the phloem, i.e., in the form of substrates like sucrose or other di- and oligosaccharides. In the latter cases (transport metabolites, sucrose) both energy and material are carried by the same molecules. Third, genetic information is not among the messages transferred over larger distances than within a cell. Macro-molecules are not translocated.

Figure 13.2 is an attempt to integrate the above remarks in a scheme showing that we can consider three different components in feedback systems of whole plants:

1. transport itself,
   and two components acting on it, namely:
2. the energy status of the plant; and
3. a highly sensitive signalling system (Lüttge, 1974).

13.3 Particular Systems

13.3.1 Different Ways of Life of Two Mangrove Species Under Salinity Stress

Mangroves must be ecologically adapted to high salinity in the root medium. The different ways of life of the two mangrove species *Rhizophora mucronata* and *Aegialitis annulata* provide a fine illustration for some of the more general points made above (Atkinson et al., 1967). The problem is the same for both species, i.e., to keep the salt load of the leaves at a controlled level. The co-operation between shoot and root allows two entirely different strategies leading to the same end, which would be impossible for the isolated organs alone.

Rhizophora mucronata partially excludes salt from the transpiration stream. The Cl^- concentration of the xylem sap is 17 Eq m^{-3} (i.e., $\mu Eq/ml$) delivering $17 \mu Eq$ to a leaf in one day. By contrast, the xylem sap of *Aegialitis annulata* contains 85–122 Eq m^{-3} Cl^- supplying a leaf with about 100 μEq Cl^- per day. *Aegialitis annulata,* but not *Rhizophora mucronata,* has salt glands on its leaves, and a typical *A. annulata* leaf can eliminate by its glands 100 μEq salt per day. Thus with a considerably different input of salt, the two mangrove species retain about similar amounts. In *R. mucronata* a mean growth rate of 3% increase in dry weight per day keeps the salt at the observed level. In *A. annulata* elimination of NaCl by the salt glands is largely responsible for the balance. As a result, NaCl contents and concentrations in the leaves of *R. mucronata* and *A. annulata* are quite similar; in mature leaves the levels are even lower in *A. annulata* with its much higher xylem sap concentrations, than in *R. mucronata* (Table 13.2; Atkinson et al., 1967; see also review by Lüttge, 1975).

Table 13.2. Na^+ and Cl^- amounts and concentrations in the leaves of a salt-excluding mangrove (*Rhizophora*) and of a salt-excreting mangrove (*Aegialitis*). Increasing sample numbers refer to leaves of increasing age (data from Atkinson et al., 1967; Table from Lüttge, 1975).

Sample	Na^+						Cl^-					
	1	2	3	4	5	6	1	2	3	4	5	6
Rhizophora:												
amounts [µEq leaf^{-1}]	61	290	420	480	520	645	74	520	510	585	580	730
conc. [mEq l^{-1}]	305	313	431	435	461	461	370	562	522	530	515	522
Aegialitis:												
amounts [µEq leaf^{-1}]	420	325	275	280	235	—	415	290	270	260	255	—
conc. [mEq l^{-1}]	518	480	411	388	356	—	512	429	405	361	386	—

13.3.2 Shoot–Root Co-Operation in NO_3^- Uptake and Reduction

Ben-Zioni et al. (1971) developed a model for NO_3^- uptake and reduction in intact plants which relates the activities of roots and leaves connected by xylem and phloem.

To explain this model, we first need to recall that NO_3^- reduction to NH_3, and N-assimilation after uptake of KNO_3, cause problems of pH regulation in the cells due to production of OH^- (Fig. 13.3). This can be overcome by concomitant CO_2 dark fixation and formation of dicarboxylic organic acids, mostly malic acid (Fig. 13.4; reviews: Smith and Raven, 1976; Osmond, 1976). A consequence of this is that 2 K^+ must be taken up for 1 NO_3^- and that 1 H^+ is released, i.e., the external solution acidifies. The second fact that we need is that the capacity of roots to reduce NO_3^- is limited and saturates at relatively low external KNO_3 concentration. Excess NO_3^- taken up by the roots is translocated to the shoot which appears to be the principal site of NO_3^--reduction (Fig. 13.5).

The model of Ben-Zioni et al. (1971) combines these facts in an integrated operation of roots and shoots as shown in Figure 13.6. KNO_3 arrives in the shoots via the transpiration stream in the xylem. NO_3^- is reduced in the leaves, and CO_2 is fixed to give malic acid just as in Figure 13.4. However, K-malate is not stored in the vacuoles but shuttled to the roots via the phloem. In the roots the decarboxylation of malate provides HCO_3^- for exchange with NO_3^-, and K^+ remains in the root cells for transport back to the shoot as a counterion of NO_3^-. This shuttle mechanism between shoot and root allows NO_3^- uptake in excess of K^+ and solves the problem of pH regulation during NO_3^- reduction. (The pyruvate may flow into the Krebs cycle of root cells and thus may be used for energy supply.)

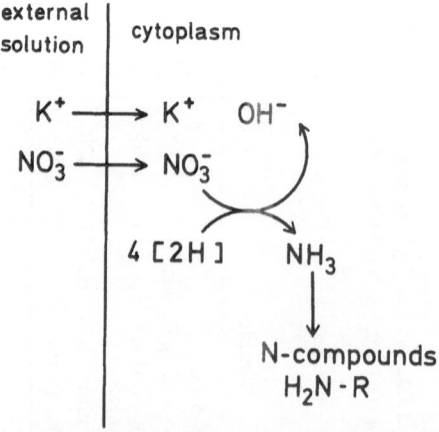

Fig. 13.3. Problem of cell alkalization during NO_3^- uptake and reduction.

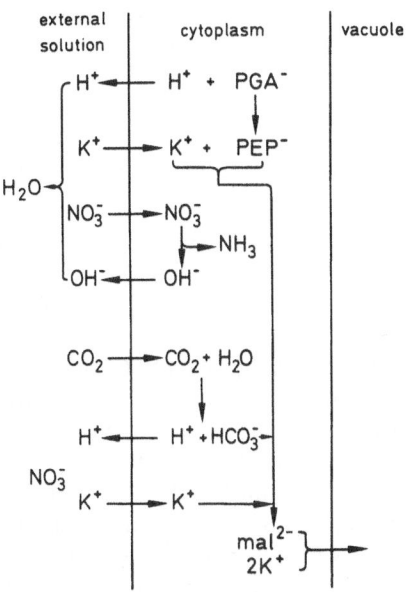

Fig. 13.4. Stoichiometry of NO_3^- uptake and reduction, K^+ uptake, H^+ release, and malate synthesis by CO_2 dark fixation. For abbreviations see usage in Chapter 7. (Osmond, 1976.)

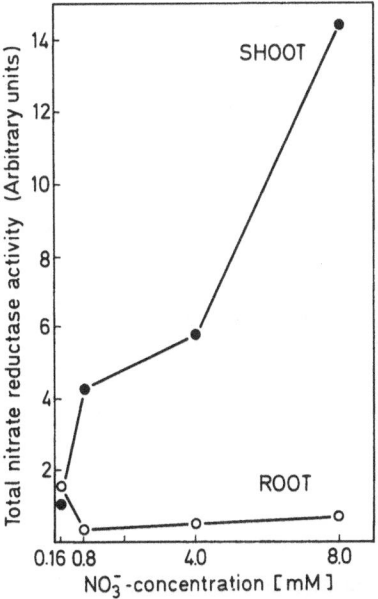

Fig. 13.5. Influence of external NO_3^- concentration on the distribution and total activity of nitrate reductase in *Pisum*. (After Wallace and Pate, 1967; from Osmond, 1976.)

Some reservations against the scheme of Figure 13.6 have been raised by other authors. Neyra and Hageman (1976) suggest that the decarboxylation and oxidation of malate coming from the shoot in the root is not so important as a source of HCO_3^- to exchange with NO_3^- but much more so to obtain reducing potential (NADH + H$^+$) for NO_3^- reduction within the root. Since Neyra and Hageman worked with a C_4-plant, maize, the situation in their experiments may have been somewhat different from that in the C_3-plant, tobacco (*Nicotiana*), used by Ben-Zioni et al. In C_4-plants malate is not only synthesized by dark fixation but large anounts of malate are generated by light-dependent fixation via PEP-carboxylase and transported toward the bundles.

The major objection against the scheme of Ben-Zioni et al. comes from quantitative evaluations of K$^+$ translocation. K$^+$ downward transport in the phloem should be at least as fast as KNO_3 upward transport in the xylem. Normally such rates of K$^+$ transport in the phloem are not observed (Pitman, 1975; Kirkby and Knight, 1977; Pearson and Steer, 1977); but cations other than K$^+$ could also be involved (Frost et al., 1978). It seems that in many cases malic acid synthesis in leaves, required to neutralize the OH$^-$ formed during NO_3^- reduction (Fig. 13.6), is followed by a

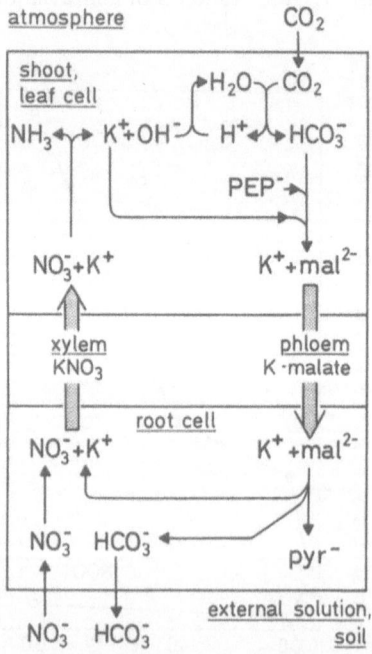

Fig. 13.6. Integrated operation of roots and shoots in whole plants during NO_3^- uptake and reduction by a KNO_3-K-malate shuttle system in the xylem and phloem. (Drawn according to the ideas of Ben-Zioni et al., 1971.) For abbreviations see usage in Chapter 7.

transfer of the organic acid anion together with K^+ to the vacuole. Since the sieve tube elements are also living cells, for reasons of pH control in their cytoplasm they cannot carry large loads of OH^- ions away from the leaves. This must have been important in the evolution of land plants (Raven, 1977a), and the synthesis of organic acids and transfer to the vacuoles appears to have been evolved as the best solution. Interestingly enough, the vacuoles of land plants often contain high levels of organic acids as osmotically active material (see Sect. 6.2.1.3), whereas in the vacuoles of algae mineral ions predominate. In contrast to cells of aerial leaves of higher plants, algal cells can regulate pH by direct exchange of H^+ and OH^- with the external medium (see Sect. 8.2.4.1.2), and internal pH can regulate the uptake and release of Cl^-, HCO_3^-, etc. (Findenegg, 1977a,b).

Notwithstanding these alternatives, the scheme of Ben-Zioni et al. (Fig. 13.6) may prove useful and perhaps can be improved in the future. A somewhat different point suggesting shoot–root co-operation in NO_3^- uptake and reduction comes from the work of Jackson et al. (1974). The induction of an NO_3^- uptake mechanism in the roots and of NO_3^- reductase can be separated under suitable conditions (e.g., also Rao and Rains, 1976; see Sect. 5.3.2). The development of an accelerated NO_3^- uptake by the roots is restricted in isolated roots and needs a factor from the shoot, as yet unidentified. Pearson and Steer (1977) observed close correlations of a daily rhythm of photosynthate translocation from the leaves and NO_3^- uptake by the roots, and they conclude that supply of C-substrate to the roots is essential for NO_3^- uptake.

13.3.3 Source–Sink Studies With Germinating Seedlings

Germinating seedlings not only obtain most of their carbon for initial growth from reserves in the seeds but also most of their mineral nutrients. One can enforce this by growing seedlings on mineral-free culture solutions containing only some Ca^{2+} to maintain membrane integrity of root cells. From the depletion kinetics of dry matter (C) and nutrient elements of the seed reserves one can work out source–sink relations between the depleting reserves and the developing roots and shoots in the intact plantlets.

Such systems have been utilized by Sutcliffe and coworkers; and Figure 13.7 shows their results obtained with oats (Baset and Sutcliffe, 1975; review: Sutcliffe, 1976a,b). The reserves of the endosperm are largely depleted within about seven days of germination. The kinetics of export of four major nutrient elements (K^+, N, Mg^{2+}, P) from the endosperm are largely similar to the export of dry matter. There is a smaller initial rate, a larger maximum rate builds up within 2–3 days, and then the rate decreases as export, and thus depletion of reserves, proceeds. When export

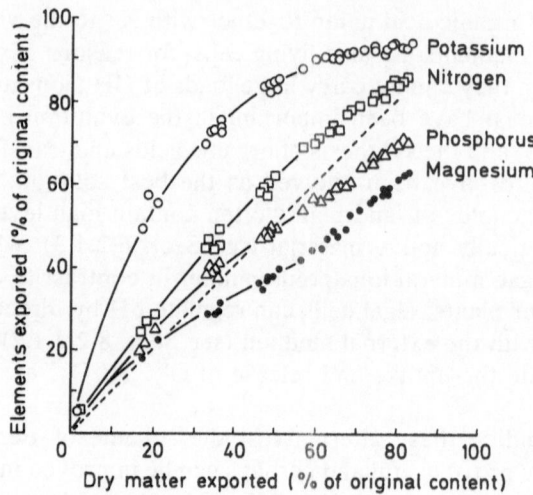

Fig. 13.7. Export of potassium, nitrogen, phosphorus, and magnesium relative to export of dry matter from the endosperm of oat seedlings during germination (Baset and Sutcliffe, 1975, from Sutcliffe, 1976a.)

of K^+, N, Mg^{2+}, and P is plotted against export of dry matter, some interesting differences become obvious (Fig. 13.7) although K^+ initially is exported much more rapidly than dry matter. There is probably an immediate demand by the growing seedling, and K^+ is available immediately in the endosperm. The falling rate can be explained by a diminishing supply becoming the regulating factor. The N curve in Figure 13.7 is qualitatively similar to that of the K^+ curve, but quantitatively the initial rate is lower and falls more slowly with time. In part, this may be due to the fact that the growing seedling initially is a smaller sink. This is difficult to test, and more likely the lower initial rate of N export is due to limited availability of N. Most of the N of oat endosperm is stored in the form of protein in the aleurone layer. N must be set free in a transportable form (amino acids) from this protein. The proteolytic activity of the endosperm required for this process gradually increases during the early stages of germination. Thus, it appears that the growing seedling controls the mobilization and transport of N not only by becoming a sink, but also by regulating the synthesis of proteolytic enzymes in the aleurone layer, presumably via hormones. This is paralleled by an effect of the growing seedlings on amylases catalyzing the mobilization of starch from the endosperm. In this case we have evidence that hormones, i.e., the production of gibberellic acid (Varner, 1964), are involved. The fact that N is initially exported from the endosperm more rapidly than dry matter, may be explained by a more rapid development of proteolytic activity as compared with amylase activity. The P and Mg^{2+} curves in Figure 13.7 are more complex. Export of these elements from the endosperm initially is more rapid than that of dry matter, then it slows down, and later it becomes faster again. Mobilization of P and Mg^{2+} from stored phytin and other sources in the endosperm (e.g., RNA for P), a differing demand of the sink, and other causes may be involved. Export of Mg^{2+} from the endosperm to the developing seedling also seems to be enhanced by light (Sutcliffe, 1976a,b).

In conclusion, the growing seedling can control depletion of the seed reserves in various ways:

1. by the strength of the sinks created in itself during the course of growth, i.e., by utilization of materials for growth and by sequestration in vacuoles;
2. by regulating the strength of the sources, i.e., by affecting the mobilization of reserve materials (e.g., from proteins, starch, phytin) via transport of hormonal signals to the endosperm.

   "Thus both utilization and mobilization are under precise control, and transport can be altered quickly in response to changing needs during early growth of the seedling" (Sutcliffe, 1976a,b).

Similar experiments have been performed studying the depletion of materials from reserves in the cotyledons of germinating peas, *Pisum sativum*. The distribution of mobilized reserves between roots and shoots depends very much on the relative growth rates of the two organs. It is also affected by light. Seedlings grown in the dark transport a higher proportion of the reserves to the shoot than seedlings grown in the light (Fig. 13.8). A competition between roots and shoots for the cotyledonary re-

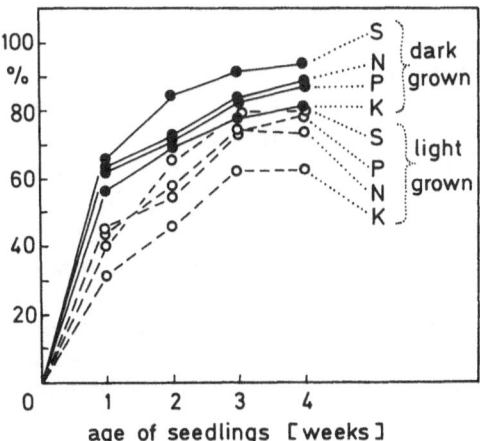

Fig. 13.8. Transport of mineral elements from pea cotyledons into the shoot of dark-grown (*solid lines, closed circles*) and light-grown seedlings (*dotted lines, open circles*) during 4 weeks of growth. Data taken from Guardiola (1973) and Sutcliffe (1976a) are expressed as percent of the total amount transported from the cotyledons recovered in the shoot.

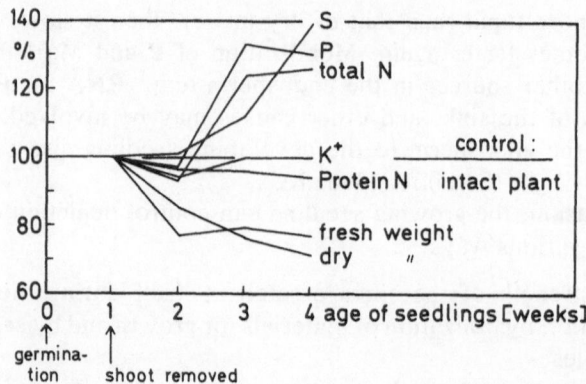

Fig. 13.9. The influence of the shoot of *Pisum sativum* seedlings grown without an external source of nutrients on the growth and content of mineral elements in the roots. The shoots were excised one week after germination. The *curves* indicate the deviation of growth parameters and mineral element contents in the roots of plants with excised shoots from the values observed in the roots of intact control plants which were set equal to 100%. (Data from Guardiola, 1973, and Sutcliffe, 1976a.)

serves is also apparent in the experiment of Figure 13.9, in which the shoot of *P. sativum* plants was removed one week after germination. K^+ and protein-N of the roots are hardly affected by the absence of shoots, fresh weight and dry weight of roots are smaller, i.e., growth is reduced, but total S, P and N in roots of de-shooted plants are considerably larger than in roots of the intact controls. Of course, this may not only be explained by root–shoot competition for the reserves in the cotyledons; secondary redistribution in the intact plant may also be involved (Guardiola, 1973; Sutcliffe, 1976a,b).

The situation can be illustrated by an electrical analog (Fig. 13.10) with a source A (referring to the seed reserves) and two sinks B (shoot) and C (root). Source and sinks are given by capacitors indicating they have a capacity related to their volume and the ability to hold a charge, i.e., a potential. If the amount of an individual solute transported from the source to an individual sink depends on the difference between the source and the sink potentials and on the particular transport resistances (R_1, R_2) in Figure 13.10, the removal of one sink should not affect the transport from the source to the other sink. This is born out by the data on K^+ transport in Figure 13.9. Rates of transport may be altered due to changing potentials of sinks during growth and development, changing potential of the source during depletion, additional sources and sinks coming in, changing resistances during development of vascular bundles, etc. With these and other additions the analog of Figure 13.10 may prove quite useful for future work (Sutcliffe, 1976a,b).

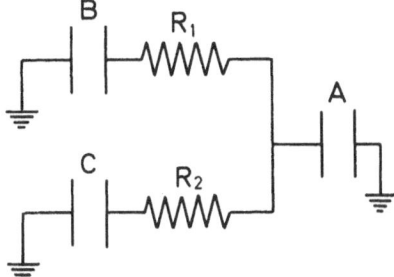

Fig. 13.10. An electrical analog to illustrate movement of solutes from a source (*A*) to shoot (*B*) and root (*C*) sinks in a germinating seed. (Sutcliffe, 1976a.)

13.3.4 Shoot–Root Co-Operation in Ion Input to the Shoot

A model for shoot–root co-operation in ion input to the shoot as drawn in Figure 13.11 was initially proposed by Cram and Pitman (1972; Pitman, 1972b). Figure 13.11 essentially is a more specialized version of Figure 13.2. In this model concentrations of ions and of energy-providing substrates and the levels of hormones are decisive elements in a feedback system between leaves and roots. The long-distance transport pathways of the xylem and phloem connect the leaves and the roots so that ion uptake processes in leaf and in root cells are not independent of each other. Ions flow not only in the xylem in the direction from the root to the leaf, but also in the phloem in the opposite direction. In this way a feedback from the shoot system to the root possibly provides the root with information about the ion content in the shoot system, which may influence ion uptake by the root. Products of photosynthesis, mainly sugars, are transported in the phloem from shoot to root supplying the latter with carbon compounds and energy, which, among other purposes, are required for active ion uptake and transport. Hormones, such as abscisic acid, may be transported also in the phloem from the shoot to the root.

Some of these elements of the figure which appear less self-evident than ion translocation in the xylem and ion retranslocation in the phloem, as well as sugar transport in the phloem, have been confirmed experimentally.

Let us first consider the problem of energy supply from the shoot to the root, i.e., the energy status of the plant as depicted in Figure 13.2. Hatrick and Bowling (1973) have shown that Rb^+ uptake and respiration of barley roots were correlated with translocation of sugar from the shoot to the root. The problem has been assessed by Pitman and co-workers (Pitman, 1972; Pitman et al., 1974a) in a number of experiments with barley seedlings grown under long and under very short daily photoperiods,

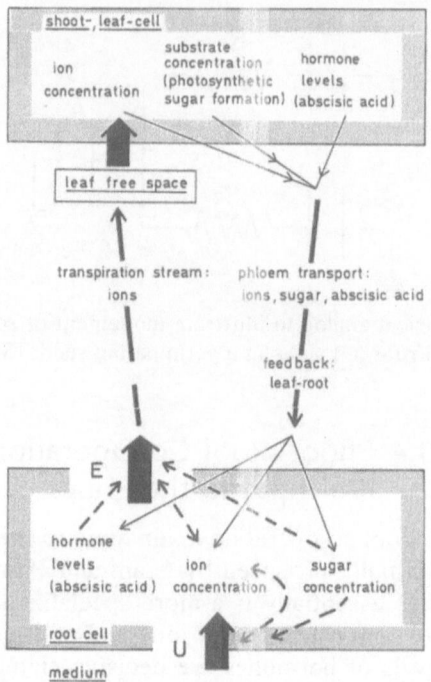

Fig. 13.11. Regulation of ionic and water relations of whole plants by communication and coupling between shoot and roots. *Solid arrows*, short-, medium-, and long-distance transport; *broken arrows*, other interactions. *U*, ion uptake from the medium across the plasmalemma into the root symplast. *E* (export), ion flux out of the root symplast into the xylem vessels. (Drawn after the ideas of Cram and Pitman, 1972; Pitman, 1972b.)

respectively. Some data are summarized in Table 13.3. Clearly growth rate is much reduced when plants grow with a daily photoperiod of 2 h as compared with 16 h. The rate of photosynthesis per gram fresh weight of leaf is somewhat reduced in the 2 h photoperiod but not proportionally to the very much reduced chlorophyll levels. Respiration of leaf slices obtained from plants grown with 16 h and 2 h photoperiod, respectively, is not different. Interestingly, rates of ion uptake by leaf slices of the two sets of plants are also not very different, but the reduced energy supply during a 2 h photoperiod has dramatic effects on active ion uptake and transport by the roots. Thus the changed energy status of the plant due to short periods of photosynthesis in the leaves affects ionic relations of the plants via input from the root, and not at the level of ion uptake by the leaf cells. During a short daily photoperiod of 2 h the sugar transport into the roots limits ion uptake. Only during a long daily photoperiod of 16 h is the sugar concentration in the roots built up to such an extent that the ion uptake is not limited by the provision of energy. Graham and Bowling (1977)

Table 13.3. Various physiological parameters in young barley seedlings grown under a 2 h daily photoperiod as compared with a 16 h daily photoperiod. From data of Pitman, 1972b, and Pitman et al., 1974b.

	Photoperiod ratio $\dfrac{2\ h\ L:22\ h\ D}{16\ h\ L:8\ h\ D}$
Relative growth rate:	0.25
Ionic relations of roots:	
active ion influx to the root	0.38
active ion influx to the vacuoles	0.16
passive ion efflux	0.91
export from the root to the xylem	0.13
Physiological properties of leaves:	
chlorophyll content	0.17
photosynthetic O_2 evolution	0.67
respiratory O_2 uptake	0.99
Ionic relations of leaves:	
K^+ content	0.88
K^+ influx to leaf slices	1.00
Cl^- influx to leaf slices in the light	1.22
Cl^- influx to leaf slices in the dark	0.94

also showed that the membrane potentials of root cells are closely related to processes going on in the whole plant and not just in the root.

Let us then consider the role of hormones. We have already discussed hormonal control of membrane transport processes in Chapter 9. We also have mentioned that abscisic acid (ABA), the aging hormone which leads to leaf fall, may inhibit the potassium pump of stomatal guard cells. In the leaves small changes of ABA levels lead to stomatal guard cell reactions. Only a doubling of the ABA content is required to trigger closure of stomata (Kriedemann et al., 1972). Conversely, under conditions of water stress and wilting there can be an increase up to 40-fold of the ABA level in leaves (Wright and Hiron, 1969; Most, 1971; Mizraki et al., 1970; Milborrow and Noddle, 1970). Thus, in leaves with the ABA system the plant has an instrument for fine regulation of its water relations. ABA formed in the leaves can be transported to the roots. This translocation most likely occurs in the phloem (Hocking et al., 1972; Bellandi and Dörffling, 1974; Hoad, 1975; Zeevaart, 1977). It is possible that by long-distance transport of ABA to the root a corresponding regulation of water relations on the level of the root is obtained. We have to consider, though, that hormonal balance of the root is much more delicate than just an input from the shoot. ABA can also be formed in the root, translocation of ABA or ABA-precursors from the shoot may not be required (Walton et al., 1976). Cytokinins, gibberellic acid, and unknown inhibitors can be exported from the root toward the shoot (Atkin et al., 1973). ABA clearly moves in the stele of roots, but in addition there is also radial transport of ABA in the root (Hartung and Behl, 1975a,b). Nevertheless, the transport of ABA formed under water stress in the shoot via the phloem to the root may be an important signal. Together with the selective effects of ABA on ion export from the root described in Section 12.2.5, this could constitute an essential element in a shoot–root cooperation during water stress.

A partial test of these ideas is provided by the experiment of Figure 13.12 (Pitman, Lüttge et al., 1974a). Leaves of intact barley seedlings were gently wilted so that, after a very short period of recovery from water stress, physiological activities of leaf and root cells such as photosynthesis, respiration and protein synthesis, and ion uptake by leaf cells (not shown in the figure) were indistinguishable from nonwilted control plants. It was assumed, however, that wilting had caused formation of a hormonal factor (ABA) in the leaves which was transported to the roots. It was anticipated that the half-life of such a factor in the roots would be longer than the half-life of discernible wilting effects during recovery. So-called after-effects of wilting clearly seem to be due to delays in resumption of the pre-stress hormonal balance in the plant (e.g., Boussiba and Richmond, 1976). ABA degradation after wilting of pea seedlings occurs more slowly than recovery of turgor pressure (Dörffling et al., 1974). In the experiment of Figure 13.12 roots were cut from prewilted

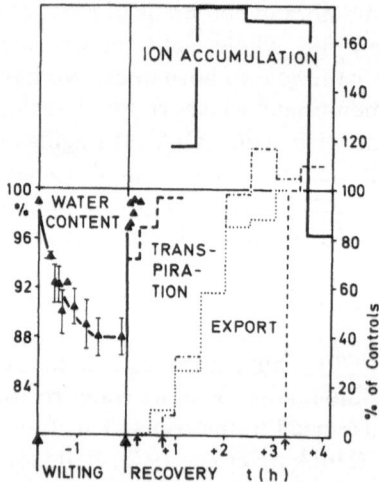

Fig. 13.12. Water content (*triangles*) of barley leaves during mild wilting and during recovery; transpiration and ion accumulation of plants during recovery; export of ions from roots cut from prewilted plants after various times of recovery (i.e., at times indicated by *small arrows* on the abscissa: 15 min ......, 45 min - · - · · · -, 195 min -----). Water content in % = 100 × (water content of leaves: water content of fully turgid leaves). Transpiration, ion accumulation and export are given as % of nonwilted controls. (After Pitman et al., 1974b, from Lüttge, 1974.)

plants at different times after the onset of recovery to perform experiments like those described by Figure 12.3. An appreciable time after obvious complete physiological recovery the ion export out of these roots was still inhibited, while ion accumulation in the root tissue was enhanced. This behavior resembles very much that of ABA-treated roots, in which, under certain conditions, xylem transport but not accumulation was inhibited (Sect. 12.2.5).

13.4 Concluding Remarks

With these examples we have a few instructive models for the transport regulation in the whole plant. Naturally, a lot of data are still required, especially to elucidate quantitative aspects of the feedback systems. Nevertheless, these investigations show how to approach the objective of understanding the interaction and co-operation of individual transport processes as they proceed in the plant as a whole. With this we return to the starting points and close the circle of our presentation (cf. Figs. 1.1 and 13.11).

References

References to articles in the Encyclopedia of Plant Physiology, New Series, A. Pirson and M.H. Zimmermann, editors, Springer-Verlag, have been abbreviated as with journals. For a full citation of the volumes cited see: Zimmermann, M.H., Milburn, J.A., eds., Vol. 1, 1975; Lüttge, U., Pitman, M.G., eds., Vol. 2A, 2B, 1976; Stocking, C.R., Heber, U., eds., Vol. 3, 1976; Heitefuss, R., Williams, P.H., eds., Vol. 4, 1976; Trebst, A., Avron, M., eds., Vol. 5, 1977.

Adamson, A.W.: A textbook of physical chemistry. New York-London: Academic Press (1973).

Albert, R., Kinzel, H.: Unterscheidung von Physiotypen bei Halophyten des Neusiedlerseegebietes (Österreich). Z. Pflanzenphysiol. **70**, 138–157 (1973).

Allaway, W.G., Hsiao, T.C.: Preparation of rolled epidermis of *Vicia faba* L. so that stomata are the only viable cells: Analysis of guard cell potassium by flame photometry. Australian J. Biol. Sci. **26**, 309–318 (1973).

Allaway, W.G., Setterfield, G.: Ultrastructural observations on guard cells of *Vicia faba* and *Allium porrum*. Canad. J. Bot. **50**, 1405–1413 (1972).

Andersen, K.S., Bain, J.M., Bishop, D.G., Smillie, R.M.: Photosystem II activity in agranal bundle sheath chloroplasts from *Zea mays*. Plant Physiol. **49**, 461–466 (1972).

Anderson, W.P.: Ion transport in the cells of higher plant tissues. Ann. Rev. Plant Physiol. **23**, 51–72 (1972).

Anderson, W.P., Ed.: Ion Transport in Plants. London-New York: Academic Press (1973).

Anderson, W.P., Collins, J.C.: The exudation from excised maize roots bathed in sulphate media. J. Exp. Bot. **20**, 72–80 (1969).

Anderson, W.P., Hendrix, D.L., Higinbotham, N.: Higher plant cell membrane resistance by a single intracellular electrode method. Plant Physiol. **53**, 122–124 (1974).

Anderson, W.P., Higinbotham, N.: A cautionary note on plant root electrophysiology. J. Exp. Bot. **26**, 533–536 (1975).

Anderson, W.P., Higinbotham, N.: Electrical resistances of corn root segments. Plant Physiol. **57**, 137 -141 (1976).

Anderson, W.P., House, C.R.: A correlation between structure and function in the root of *Zea mays*. J. Exp. Bot. **18**, 544–555 (1967).

Anderson, W.P., Reilly, E.J.: A study of the exudation of excised maize roots after removal of the epidermis and outer cortex. J. Exp. Bot. **19**, 19–30 (1968).

Andrianov, V.K., Bulychev, A.A., Kurella, G.A., Litvin, F.F.: Effect of light on the resting potential and cation (K^+, Na^+) activity in the vacuole sap of *Nitella* cells. Biofizika **16**, 1031–1036 (1971).

Applewhite, P.B., Satter, R.L., Galston, A.W.: Protein synthesis during endogenous rhythmic leaflet movement in *Albizzia*. J. Gen. Physiol. **62**, 707–713 (1973).

Arisz, W.H.: Transport of chloride in the symplasm of *Vallisneria* leaves. Nature (London) **174**, 223 (1954).

Arisz, W.H.: Significance of the symplasm theory for transport across the root. Protoplasma **46**, 5–62 (1956).

Arisz, W.H.: Symplasmatischer Salztransport in *Vallisneria*-Blättern. Protoplasma **52**, 309–343 (1960).

Arisz, W.H.: Intercellular polar transport and the role of the plasmodesmata in coleoptiles and *Vallisneria* leaves. Acta Botan. Neerl. **18**, 14–38 (1969).

Arisz, W.H., Camphuis, I.J., Heikens, H., van Tooren, A.J.: The secretion of the salt glands of *Limonium latifolium* Ktze. Acta Botan. Neerl. **4**, 322–338 (1955).

Arisz, W.H., Sol, H.H.: Influence of light and sucrose on the uptake and transport of chloride in *Vallisneria* leaves. Acta Botan. Neerl. **5**, 218–247 (1956).

Arisz, W.H., Wiersema, E.P.: Symplasmatic long-distance transport in *Vallisneria* plants investigated by means of autoradiograms. Proc. Koninkl. Ned. Akad. Wetenschap., Ser. C, **69**, 223–241 (1966).

Arnon, D.I.: Growth and function as criteria in determining the essential nature of inorganic nutrients, pp. 313–341. In: Mineral nutrition of plants, Truog, E., (ed.), Madison: Univ. Wisconsin Press (1951).

Atkin, R.K., Barton, G.E., Robinson, D.K.: Effect of root growing temperature on growth substances in xylem exudate of *Zea mays*. J. Exp. Bot. **24**, 475–487 (1973).

Atkinson, M.R., Findlay, G.P., Hope, A.B., Pitman, M.G., Saddler, H.D.W., West, K.R.: Salt regulation in the mangroves *Rhizophora mucronata* Lam. and *Aegialitis annulata* R. Br. Aust. J. Biol. Sci. **20**, 589–599 (1967).

Atkinson, M.R., Polya, G.M.: Effects of L-ethionine on adenosine triphosphate levels, respiration, and salt accumulation in carrot xylem tissue. Aust. J. Biol. Sci. **21**, 409–420 (1968).

Baker, D.A.: The radial transport of ions in maize roots. In: Ion transport in plants, Anderson, W.P., (ed.), pp. 511–517. London-New York: Academic Press (1973a).

Baker, D.A.: The effect of CCCP on ion fluxes in the stele and cortex of maize roots. Planta **112**, 293–299 (1973b).

Baker, D.A., Hall, J.L.: Pinocytosis, ATP-ase and ion uptake by plant cells. New Phytol. **72**, 1281–1291 (1973).

Baker, D.A., Hall, J.L.: Ion transport in plant cells and tissues. Amsterdam: Elsevier (1975).

Baker, H.G., Baker, I.: Amino-acids in nectar and their evolutionary significance. Nature (London) **241**, 543–544 (1973a).

Baker, H.G., Baker, I.: Some anthecological aspects of the evolution of nectar-producing flowers, particularly amino acid production in nectar, pp. 243–264. In: Taxonomy and ecology, V.H., Heywood, (ed.), London: Academic Press (1973b).

Baker, H.G., Baker, I.: Studies of nectar-constitution and pollinator-plant coecolution, pp. 100–139. In: Coevolution of animals and plants, Gilber, L.E., Raven, P.H., (eds.), Austin: Univ. of Texas Press (1975).

Baker, I., Baker, H.G.: Analyses of amino acids in flower nectars of hybrids and their parents, with phylogenetic implications. New Phytol. **76**, 87–98 (1976).

Ballio, A., Graniti, A., Pocchiari, F., Silano, V.: Some effects of "fusicoccin A" on tomato leaf tissues. Life Sci. **7**, 751–760 (1968).

Barber, J., Shieh, Y.J.: Effects of light on net Na^+ and K^+ transport in *Chlorella* and evidence for *in vivo* cyclic photophosphorylation. Plant Sci. Lett. **1**, 405–411 (1973).

Barnes, E.M., Kaback, H.R.: β-galactoside transport in bacterial membrane preparations: energy coupling *via* membrane-bound D-lactic dehydrogenase. Proc. Natl. Acad. Sci. U.S. **66**, 1190–1198 (1970).

Barry, P.H.: Volume flows and pressure changes during an action potential in cells of *Chara australis*. II. Theoretical considerations. J. Membrane Biol. **3**, 335–371 (1970).

Baset, Q.A., Sutcliffe, J.F.: Regulation of the export of potassium, nitrogen, phosphorus, magnesium and dry matter from the endosperm of etiolated oat seedlings (*Avena sativa* cv. Victory). Ann. Bot. (London) N.S. **39**, 31–41 (1975).

Bassham, J.A., Calvin, M.: The path of carbon in photosynthesis. Englewood Cliffs: Prentice-Hall (1957).

Batt, S., Venis, M.A.: Separation and localization of two classes of auxin binding sites in corn coleoptile membranes. Planta **130**, 15–21 (1976).

Batt, S., Wilkins, M.B.: Auxin binding to corn coleoptile membranes: kinetics and specificity. Planta **130**, 7–13 (1976).

Baudhin, P., Beaufay, H., de Duve, C.: Combined biochemical and morphological study of particulate fractions from rat liver. J. Cell. Biol. **26**, 219–243 (1965).

Baule, H., Fricker, C.: Die Düngung von Waldbäumen. München: Bayerischer Landwirtschaftsverlag GmbH (1967).

Baumeister, W.: Die einzelnen Elemente (Aufnehmbarkeit und Aufnahme durch die Pflanzen, Funktionen in der Pflanze, toxische und morphogenetische Wirkungen, Schädigungen durch Mangel oder Übermaß): Hauptnährstoffe, pp. 482–557. In: Handbuch der Pflanzenphysiologie IV, Ruhland, W., (ed.), Berlin-Göttingen-Heidelberg: Springer (1958).

Beck, R.E., Schultz, J.S.: Hindered diffusion in microporous membranes with known pore geometry. Science **170**, 1302–1305 (1970).

Beevers, H., Theimer, R.R., Feierabend, J.: Microbodies (Glyoxysomen, Peroxisomen), pp. 127–146. In: Biochemische Cytologie der Pflanzenzelle, Jacobie, G., (ed.), Stuttgart: Georg Thieme Verlag (1974).

Bellandi, D.M., Dörffling, K.: Transport of abscisic acid-2-^{14}C in intact pea seedlings. Plant Physiol. **32**, 365–368 (1974).

Benner, U., Schnepf, E.: Die Morphologie der Nektarausscheidung bei Bromelia-
ceen: Beteiligung des Golgi-Apparates. Protoplasma **85**, 337–349 (1975).

Benson, A.A.: On the orientation of lipids in chloroplast and cell membranes.
J. Am. Oil Chem. Soc. **43**, 265–270 (1966).

Benson, A.A.: The cell membrane: a lipoprotein monolayer, pp. 190–202. In:
Membrane models and the formation of biological membranes, Bolis, L.,
Pethica, B.A., (eds.), Amsterdam: North Holland Publishing Co. (1968).

Bentrup, F.W.: Einfluss von K^+-Ionen auf die Polaritätsinduktion bei *Fucus ser-
ratus*. Ber. Dtsch. Bot. Ges. **87**, 215–221 (1974a).

Bentrup, F.W.: Lichtabhängige Membranpotentiale bei Pflanzen. Ber. Dtsch.
Bot. Ges. **87**, 515–528 (1974b).

Bentrup, F.W., Pfrüner, H., Wagner, G.: Evidence for differential action of indole acetic acid upon ion fluxes in single cells of *Petroselinum sativum*. Planta
110, 369–372 (1973).

Bentwood, B.J., Cronshaw, J.: Cytochemical localization of adenosine triphosphatase in the phloem of *Pisum sativum* and its relation to the function of
transfer cells. Planta **140**, 111–120 (1978).

Ben-Zioni, A., Vaadia, Y., Lips, W.: Nitrate uptake by roots as regulated by nitrate reduction products of the shoot. Plant Physiol. **24**, 288–290 (1971).

Bernstein, L.: Method for determining solutes in the cell wall of leaves. Plant
Physiol. **47**, 361–365 (1971).

Berry, Y.B.: Western forest trees. New York: Dover Publ. 1964.

Bertalanffy, L., von: Biophysik des Fliessgleichgewichtes. Braunschweig: F.
Vieweg und Sohn (1953).

Betz, A.: Metabolic flux in yeast cells with oscillatory controlled glycolysis.
Plant Physiol. **19**, 1049–1054 (1966).

Betz, A.: Pulsed incorporation of ^{32}P by the soluble oscillating glycolytic system
from *Saccharomyces carlsbergensis*. Eur. J. Biochem. **4**, 354–356 (1968).

Betz, A., Hinrichs, R.: Incorporation of glucose into an insoluble polyglycoside
during oscillatory controlled glycolysis in yeast cells. Eur. J. Biochem. **5**,
154–157 (1968).

Biddulph, S.F.: Visual indications of S^{35} and P^{32} translocation in the phloem. Am.
J. Bot. **43**, 143–148 (1956).

Birt, L.M., Hird, F.J.R.: Kinetic aspects of the uptake of amino acids by carrot
tissue. Biochem. J. **70**, 286–292 (1958).

Bishop, D.G., Andersen, K.S., Smillie, R.M.: Photoreduction and oxidation of cytochrome f in bundle sheath cells of *Zea mays*. Plant Physiol. **49**, 467–470
(1972).

Bishop, N.I.: Mutations of unicellular green algae and their application to studies
on the mechanism of photosynthesis. Record of Chem. Progr. **25**, 181–195
(1964).

Blount, R.W., Levedahl, B.H.: Active Na and Cl transport in the single celled
marine alga *Halicystis ovalis*. Acta Physiol. Scand. **49**, 1–9 (1960).

Blumenthal, R., Changeux, J.-P., LeFever, R.: J. Membrane Biol. **2**, 351 ff.
(1971).

Boerner, F.: Nadelgehölze für Garten und Park. Stuttgart: Stichnote (1969).

Boisard, J., Marmé, D., Briggs, W.R.: *In vivo* properties of a membrane-bound
phytochrome. Plant Physiol. **54**, 272–276 (1974).

Bonnett, H.T., Jr.: The root endodermis: fine structure and function. J. Cell Biol.
37, 199–205 (1968).

374 References

Booij, H.L.: An interpretation of Arisz' experiments on symplasm transport. In: Proc. 1st Eur. Biophys. Congr., Vol. III, Broda, E., Locker, A., Springer-Lederer, H., (eds.), pp. 125–129. Wien: Verlag der Wiener Med. Akad. (1971).

Bornefeld, T., Lee-Kaden, J., Simonis, W.: Lichtabhängige Phosphataufnahme und Metabolismus bei *Anacystis nidulans*. Ber. Dtsch. Bot. Ges. **87**, 493–500 (1974).

Bornefeld, T., Simonis, W.: Effects of light, temperature, pH, and inhibitors on the ATP level of the bluegreen alga, *Anacystis nidulans*. Planta **115**, 309–318 (1974).

Borstlap, A.C.: Antagonisms between amino acids in the growth of *Spirodela polyrhiza* due to competitive amino acid uptake. Acta Bot. Neerl. **23**, 723–738 (1974).

Bostrom, T.E., Walker, N.A.: Intercellular transport in plants I. The flux of chloride and the electric resistance in *Chara*. J. Exp. Bot. **26**, 767–782 (1975).

Bostrom, T.E., Walker, N.A.: Intercellular transport in plants II. Cyclosis and the rate of intercellular transport of chloride in *Chara*. J. Exp. Bot. **27**, 347–357 (1976).

Boussiba, S., Richmond, A.E.: Abscisic acid and the after-effect of stress in tobacco plants. Planta **129**, 217–219 (1976).

Bowling, D.J.F.: The origin of the trans-root potential and the transfer of ions to the xylem of sunflower roots. In: Ion transport in plants, Anderson, W.P., (ed.), pp. 483–491. London-New York: Academic Press (1973a).

Bowling, D.J.F.: Measurement of a gradient of oxygen partial pressure across the intact root. Planta **111**, 323–328 (1973b).

Bowling, D.J.F.: A pH gradient across the root. J. Exp. Bot. **24**, 1041–1045 (1973c).

Bowling, D.J.F.: Ionic gradients in higher plant tissues. Perspectives in Exp. Biol. **2**, 391–399 (1976).

Bowling, D.J.F., Ansari, A.Q.: Control of sodium transport in sunflower roots. J. Exp. Bot. **23**, 241 (1972).

Brachet, J.: Interactions between nucleus and cytoplasm. Enc. Plant Physiol. **3**, 53–84 (1976).

Branton, D., Deamer, D.W.: Membrane structure. Protoplasmatologia. Handbuch der Protoplasmaforschung. vol.II/El. Wien-New York: Springer (1972).

Brauner, L., Brauner, M.: Untersuchungen über den photoelektrischen Effekt in Membranen I. Weitere Beiträge zum Problem der Lichtpermeabilitätsreaktionen. Protoplasma (Wien) **28**, 230–261 (1937).

Brauner, L., Brauner, M.: Untersuchungen über den photoelektrischen Effekt in Membranen II. Rev. Fac. Sci. Univ. Instanbul **3**, 1–66 (1938).

Brauner, L., Bünning, E.: Geoelektrischer Effekt und Elektrotropismus. Ber. Dtsch. Bot. Ges. **48**, 470–476 (1930).

Brauner, L., Diemer, R.: Über den Einfluss von Wuchsstoff auf die Entwicklung bioelektrischer Potentiale in Pflanzengeweben. Planta **77**, 1–31 (1967).

Bräutigam, E., Müller, E.: Transportprozesse in *Vallisneria*-Blättern und die Wirkung von Kinetin und Kolchizin I. Aufnahme von α-Aminoisbuttersäure in Gewebe von *Vallisneria* und die Wirkung von Kinetin. Biochem. Pflanzenphysiol. **167**, 1–15 (1975a).

Bräutigam, E., Müller, E.: II. Symplastischer Transport von α-Aminobuttersäure in *Vallisneria*-Blättern und die Wirkung von Kinetin. Biochem. Pflanzenphysiol. **167**, 17–28 (1975b).

Bräutigam, E., Müller, E.: III. Induktion einer Senke durch Kinetin. Biochem. Pflanzenphysiol. **167**, 29–39 (1975c).

Breidenbach, R.W., Beevers, H.: Association of the glyoxalate cycle enzymes in a novel subcellular particle from castor bean endosperm. Biochem. Biophys. Res. Commun. **27**, 462–469 (1967).

Breidenbach, R.W., Kahn, A., Beevers, H.: Characterization of glyoxysomes from castor bean endosperm. Plant Physiol. **43**, 705–713 (1968).

Briggs, G.E., Haldane, I.B.S.: Biochem. J. **19**, 338 ff. (1925).

Briggs, G.E., Hope, A.B., Pitman, M.G.: Exchangeable ions in beet disks at low temperature. J. Exp. Bot. **9**, 128–141 (1958a).

Briggs, G.E., Hope, A.B., Pitman, M.G.: Measurement of ionic fluxes in red beet tissues using radioisotopes. Radioisotopes Sci. Res. Proc. Int. Conf. Paris **4**, 391–400 (1958b).

Briggs, G.E., Hope, A.B., Robertson, R.N.: Electrolytes and plant cells. Oxford: Blackwell, 1961.

Brinckmann, E., Lüttge, U.: Vorübergehende pH-Änderungen im umgebenden Medium intakter grüner Zellen bei Beleuchtungswechsel. Z. Naturforsch. **27b**, 277–284 (1972).

Brinckmann, E., Lüttge, U.: Lichtabhängige Membranpotentialschwankungen und deren interzelluläre Weiterleitung bei panaschierten Photosynthese-Mutanten von *Oenothera*. Planta **119**, 47–57 (1974).

Brownell, P.F., Crossland, C.J.: The requirement for sodium as a micronutrient by species having the C_4 dicarboxylic photosynthetic pathway. Plant Physiol. **49**, 794–797 (1972).

Brownell, P.F., Crossland, C.J.: Growth responses to sodium by *Bryophyllum tubiflorum* under conditions inducing Crassulacean acid metabolism. Plant Physiol. **54**, 416–417 (1974).

Broyer, T.C., Carlton, A.B., Johnson, C.M., Stout, P.R.: Chlorine–A micronutrient element for higher plants. Plant Physiol. **29**, 526–532 (1954).

Budd, K., Laties, G.G.: Ferricyanide-mediated transport of chloride by anaerobic corn roots. Plant Physiol. **36**, 648–654 (1964).

Bulychev, A.A., Andrianov, V.K., Kurella, G.A., Litvin, F.F.: Transmembrane potential of the cell and chloroplast in higher terrestrial plants. Fiziologia Rastenii **18**, 248–256 (1971).

Bulychev, A.A., Andrianov, V.K., Kurella, G.A., Litvin, F.F.: Microelectrode measurements of the transmembrane potential of chloroplasts and its photoinduced changes. Nature (London) **236**, 175–177 (1972).

Bünning, E.: Die Physiologie des Wachstums und der Bewegungen. Berlin: Springer 1939.

Bünning, E.: The physiological clock. London-New York: Springer (1973).

Bünning, E.: Wilhelm Pfeffer. Stuttgart: Wissenschaftliche Verlagsgesellschaft MBH (1975).

Burley, J.W.A., Nwoke, F.I.O., Leister, G.L., Popham, R.A.: The relationship of xylem maturation to the absorption and translocation of P-32. Am. J. Bot. **57**, 504–511 (1970).

Buser, C., Matile, P.: Malic acid in vacuoles isolated from *Bryophyllum* leaf cells. Z. Pflanzenphysiol. **82,** 462–466 (1977).

Butcher, H.C., Wagner, G.J., Siegelman, H.W.: Localization of acid hydrolases in protoplasts. Examination of the proposed lysosomal function of the mature vacuole. Plant Physiol. **59,** 1098–1103 (1977).

Butz, R.G., Jackson, W.A.: A mechanism for nitrate transport and reduction. Phytochemistry **16,** 409–417 (1977).

Canny, M.J.P.: Translocation: Mechanisms and kinetics. Ann. Rev. Plant Physiol. **22,** 237–260 (1971).

Canny, M.J.P.: Mass transfer. Enc. Plant Physiol. **1,** 139–153 (1975).

Canny, M.J.P.: Protoplasmic streaming. Enc. Plant Physiol. **1,** 289–300 (1975).

Carolin, R.C., Jacobs, S., Vesk, M.: Bot. J. Linn. Soc. **66,** 259 ff. (1973).

Changeux, J.-P., Blumenthal, R., Jasai, M., Podleski, T.: Molecular properties of drug receptors, Porter, R., O'Connor, M., (eds.), pp. 197 ff. London: J. & A. Churchill (1970).

Changeux, J.-P., Thiéry, J.: In: Regulatory functions of biological membranes, Järnefelt, J., (ed)., pp. 115 ff. Amsterdam: Elsevier (1968).

Changeux, J.-P., Thiéry, J., Tung, Y., Kittel, C.: On the cooperativity of biological membranes. Proc. Natl. Acad. Sci. U.S. **57,** 335–341 (1967).

Chapman, E.A., Bain, J.M., Gove, D.W.: Mitochondria and chloroplast peripheral reticulum in the C_4 plants *Amaranthus edulis* and *Atriplex spongiosa*. Aust. J. Plant Physiol. **2,** 207–223 (1975).

Clarkson, D.T., Robards, A.W.: The endodermis, its structural development and physiological role. In: The development and function of roots, Torrey, J.G., Clarkson, D.T., (eds.), London: Academic Press (1975).

Clarkson, D.T., Robards, A.W., Sanderson, J.: The tertiary endodermis in barley roots: fine structure in relation to radial transport of ions and water. Planta **96,** 292–305 (1971).

Cleland, R.E.: Auxin-induced hydrogen ion excretion correlation with growth, and control by external pH and water stress. Planta **127,** 233–242 (1975).

Cleland, R.E.: Kinetics of hormone-induced H^+ excretion. Plant Physiol. **58,** 210–213 (1976).

Cleland, R.E.: Fusicoccin-induced growth and hydrogen ion excretion of *Avena* coleoptiles. Relation to auxin response. Planta **128,** 201–206 (1976).

Cleland, R.E., Prins, H.B.A., Harper, J.R., Higinbotham, N.: Rapid hormone-induced hyperpolarization of the oat coleoptile transmembrane potential. Plant Physiol. **59,** 395–397 (1977).

Cockburn, M., Earnshaw, P., Eddy, A.A.: The stoicheiometry of the absorption of protons with phosphate and L-glutamate by yeasts of the genus *Saccharomyces*. Biochem. J. **146,** 705–712 (1975).

Cocucci, M. C., Marrè, E.: The effects of cycloheximide in respiration, protein synthesis and adenosine nucleotide levels in *Rhodotorula gracilis*. Plant Sci. Lett. **1,** 293–301 (1973).

Cocucci, M., Marrè, E., Ballarin Denti, A., Scacchi, A.: Characteristics of fusicoccin-induced changes of transmembrane potential and ion uptake in maize root segments. Plant Sci. Lett. **6,** 143–156 (1976).

Cole, J.E.: Cone-bearing trees of Yosemite National Park. Yosemite Nat. Hist. Assoc. (1963).

Collins, J.C.: Hormonal control of ion and water transport in the excised maize root, pp. 441–443. In: Membrane transport in plants, Zimmermann, U., Dainty, J., (eds.), Berlin-Heidelberg-New York: Springer (1974).

Colombo, R., De Michelis, M.I., Lado, P.: 3-O-Methyl glucose uptake stimulation by auxin and by fusicoccin in plant materials and its relationships with proton extrusion. Planta **138**, 249–256 (1978).

Conway, E.J.: Evidence for a redox pump in active transport of cations. Int. Rev. Cytol. **4**, 377–396 (1955).

Cooper, M.J., Digby, J., Cooper, P.J.: Effects of plant hormones on the stomata of barley: A study of the interaction between abscisic acid and kinetin. Planta **105**, 43–49 (1972).

Coster, H.G.L., Zimmermann, U.: Dielectric breakdown in the membranes of *Valonia utricularis*. The role of energy dissipation. Biochim. Biophys. Acta **382**, 410–418 (1975).

Costerton, J.W.F., MacRobbie, E.A.C.: Ultrastructure of *Nitella translucens* in relation to ion transport. J. Exp. Bot. **21**, 535–542 (1970).

Crafts, A.S., Broyer, T.C.: Migration of salts and water into xylem of the roots of higher plants. Am. J. Bot. **24**, 415–431 (1938).

Crafts, A.S., Crisp, C.E.: Phloem transport in plants. San Francisco: W.H. Freeman and Co. (1971).

Cram, W.J.: The control of cytoplasmic and vacuolar ion contents in higher plant cells. Dtsch. Akad. Wissensch. Berlin Nr. **4**, 117–126 (1968a).

Cram, W.J.: Compartmentation and exchange of chloride in carrot root tissue. Biochim. Biophys. Acta **163**, 339–353 (1968b).

Cram, W.J.: The effects of ouabain on sodium and potassium fluxes in excised root tissue of carrot. J. Exp. Bot. **19**, 611–616 (1968c).

Cram, W.J.: Short term influx as a measure of influx across the plasmalemma. Plant Physiol. **44**, 1013–1015 (1969a).

Cram, W.J.: Respiration and energy-dependent movements of chloride at plasmalemma and tonoplast of carrot root cells. Biochim. Biophys. Acta **173**, 213–222 (1969b).

Cram, W.J.: Chloride fluxes in cells of the isolated root cortex of *Zea mays*. Aust. J. Biol. Sci. **26**, 757–779 (1973).

Cram, W.J.: Negative feedback regulation of transport in cells. The maintenance of turgor, volume and nutrient supply. Enc. Plant Physiol. **2A**, 284–316 (1976).

Cram, W.J., Laties, G.G.: The use of short-term and quasisteady influx in estimating plasmalemma and tonoplast influx in barley root cells at various external and internal chloride concentrations. Aust. J. Biol. Sci. **24**, 633–646 (1971).

Cram, W.J., Laties, G.G.: The kinetics of bicarbonate and malate exchange in carrot and barley root cells. J. Exp. Bot. **25**, 11–27 (1974).

Cram, W.J., Pitman, M.G.: The action of abscisic acid on ion uptake and water flow in plant roots. Aust. J. Biol. Sci. **25**, 1125–1132 (1972).

Cseh, E., Böszörmenyi, Z.: Investigation of the mechanism of amino acid absorption in higher plants, pp. 277–292. In: Proc. 6th Hungarian Meet. Biochem. (1964).

Cummins, W.R., Kende, H., Raschke, K.: Specificity and reversibility of the rapid stomatal response to abscisic acid. Planta **99**, 347–351 (1971).

Dainty, J.: Water relations of plant cells. Adv. Bot. Res. **I**, 279–326 (1963).

Dainty, J.: Water relations of plant cells. Enc. Plant Physiol. **2A,** 12–35 (1976).

Dainty, J., Ginzburg, B.Z.: The measurement of hydraulic conductivity (osmotic permeability to water) of internodal Characean cells by means of transcellular osmosis. Biochim. Biophys. Acta **79,** 102–111 (1964a).

Dainty, J., Ginzburg, B.Z.: The permeability of the protoplasts of *Chara australis* and *Nitella translucens* to methanol, ethanol and isopropanol. Biochim. Biophys. Acta **79,** 122–128 (1964b).

Dainty, J., Ginzburg, B.Z.: The reflection coefficient of plant cell membranes for certain solutes. Biochim. Biophys. Acta **79,** 129–137 (1964c).

Dainty, J., Hope, A.B.: Ionic relations of cells of *Chara australis* I. Ion exchange in the cell wall. Aust. J. Biol. Sci. **12,** 395–411 (1959).

Danielli, J.F., Davson, H.A.: A contribution to the theory of the permeability of thin films. J. Cell. Comp. Physiol. **5,** 495–508 (1935).

Danielli, J.F., Harvey, E.N.: The tension at the surface of mackerel egg oil, with remarks on the nature of the cell surface. J. Cell. Comp. Physiol. **5,** 483–494 (1935).

Danon, A., Stoeckenius, W.: Photophosphorylation in *Halobacterium halobium.* Proc. Natl. Acad. Sci. U.S. **71,** 1234–1238 (1974).

Davies, D.D.: Control of and by pH. Symp. Soc. Exp. Biol. **27,** 513–529 (1973a).

Davies, D.D.: Metabolic control in higher plants. In: Biosynthesis and its control in plants, Milborrow, B.V., (ed.), pp. 1–20. London-New York: Academic Press (1973b).

Davis, R.F.: Ion transport across excised corn roots. Ph.D. thesis, Washington State Univ., Pullman (1968).

Davis, R.F.: Photoinduced changes in electrical potentials and H^+ activities of the chloroplast, cytoplasm, and vacuole of *Phaeoceros laevis*. In: Membrane transport in plants, Zimmermann, U., Dainty, J., (eds.), pp. 197–201. Berlin-Heidelberg-New York: Springer (1974).

Davis, R.F., Higinbotham, N.: Electrochemical gradients and K^+ and Cl^- fluxes in excised corn roots. Plant Physiol. **57,** 129–136 (1976).

Davson, H.A., Danielli, J.F.: The permeability of natural membranes. Cambridge: Univ. Press (1943).

Davson, H., Danielli, J.F.: The permeability of natural membranes. London: Cambridge Univ. Press 2nd ed. (1952).

Dawson, R.F.: Accumulation of nicotine in reciprocal grafts of tomato and tobacco. Am. J. Bot. **29,** 66–71 (1942).

Day, D.A., Hanson, J.B.: Pyruvate and malate transport and oxidation in corn mitochondria. Plant Physiol. **59,** 630–635 (1977).

Dayanandan, P., Kaufman, P.: Stomatal movement associated with potassium fluxes. Amer. J. Bot. **62,** 221–231 (1975).

Decker, M., Tanner, W.: Respiratory increase and active hexose uptake of *Chlorella vulgaris*. Biochim. Biophys. Acta **266,** 661–669 (1972).

De Duve, C., Baudhuin, P.: Peroxisomes: (microbodies and related particles). Physiol. Rev. **46,** 323 ff. (1966).

Delwiche, C.C.: The cycling of carbon and nitrogen in the biosphere, pp. 29–58. In: Microbiology and soil fertility, Gilmour, C.M., Allen, O.N., (eds.), Corvallis: Oregon State Univ. Press (1965).

Denny, P., Weeks, D.C.: Electrochemical potential gradients of ions in an aquatic angiosperm, *Potamogeton schweinfurthii* (Benn.). New Phytol. **67**, 875–882 (1968).

Denny, P., Weeks, D.C.: Effects of light and bicarbonate on membrane potential in *Potamogeton schweinfurthii* (Benn.). Ann. Bot. (London), N.S. **34**, 483–496 (1970).

Dewar, M.A., Barber, J.: Cation regulation in *Anacystis nidulans*. Planta **113**, 143–155 (1973).

Dewar, M.A., Barber, J.: Chloride transport in *Anacystis nidulans*. Planta **117**, 163–172 (1974).

Dhindsa, R.S., Beasley, C.A., Ting, I.O.: Osmo-regulation in cotton fiber. Accumulation of potassium and malate during growth. Plant Physiol. **56**, 394–398 (1975).

Diamond, J.M., Wright, E.M.: Biological membranes: the physical basis of ion and nonelectrolyte selectivity. Ann. Rev. Physiol. **31**, 581–646 (1969).

Die, J. van, Tammes, P.M.L.: Phloem exudation from monocotyledonous axes. Enc. Plant Physiol. **1**, 196–222 (1975).

Dittrich, P., Raschke, K.: Malate metabolism in isolated epidermis of *Commelina communis* L. in relation to stomatal functioning. Planta **134**, 77–81 (1977a).

Dittrich, P., Raschke, K.: Uptake and metabolism of carbohydrates by epidermal tissue. Planta **134**, 83–90 (1977b).

Dixon, H.H., Joly, J.: On the ascent of sap. Phil. Trans. Roy. Soc. Lond. Ser. B **186**, 563–576 (1895).

Dodd, W.A., Pitman, M.G., West, K.R.: Sodium and potassium transport in the marine alga *Chaetomorpha darwinii*. Aust. J. Biol. Sci. **19**, 341–354 (1966).

Dogar, M.A., Hai, T. van: Multiphasic uptake of ammonium by intact rice roots and its relationship with growth. Z. Pflanzenphysiol. **84**, 25–35 (1977).

Dolzmann, P.: Elektronenmikroskopische Untersuchungen an den Saughaaren von *Tillandsia usneoides* II. Einige Beobachtungen zur Feinstruktur der Plasmodesmen. Planta **64**, 76–80 (1965).

Dörffling, K., Bellandi, D.M., Böttger, M., Lückel, H., Menzer, U.: Abscisic acid: Properties of transport and effect on distribution of potassium and phosphorus, pp. 259–272. In: Trans. 3rd. Symp. Accumulation and Translocation of Nutrients and Regulators in Plant Organisms. Res. Inst. Pomol., Skierniewica, Poland (1973a).

Dörffling, K., Menzer, U., Gerlach-Lüssow, A.: Antagonistische Wirkungen von Abscisinsäure einerseits und Indol-3-essigsäure sowie Gibberellinsäure anderseits auf den Transport von Kalium und Phosphor in *Helianthus*-Epicotylen. Mitt. Staatsinst. Allg. Bot. Hamburg **14**, 19–23 (1973b).

Dörffling, K., Sonks, B., Tietz, D.: Variation and metabolism of abscisic acid in pea seedlings during and after water stress. Planta **121**, 57–66 (1974).

Downton, W.J.S., Berry, J.A., Tregunna, E.B.: C_4-photosynthesis: Non-cyclic electron flow and grana development in bundle sheath chloroplasts. Z. Pflanzenphysiol. **63**, 194–198 (1970).

Dunlop, J.: The transport of potassium to the xylem exudate of ryegrass I. Membrane potentials and vacuolar potassium activities in seminal roots. J. Exp. Bot. **24**, 995–1002 (1973).

Dunlop, J.: The transport of potassium to the xylem exudate of ryegrass II. Exudation. J. Exp. Bot. **25**, 1–10 (1974).

Dunlop, J.: The electrical potential difference across the tonoplast of root cells. J. Exp. Bot. **27**, 908–915 (1976).

Dunlop, J., Bowling, D.J.F.: The movement of ions to the xylem exudate of maize roots I. Profiles of membrane potential and vacuolar potassium activity across the root. J. Exp. Bot. **22**, 434–444 (1971a).

Dunlop, J., Bowling, D.J.F.: The movement of ions to the xylem exudate of maize roots II. A comparison of the electrical potential and electrochemical potentials of ions in the exudate and in the root cells. J. Exp. Bot. **22**, 445–452 (1971b).

Dunlop, J., Bowling, D.J.F.: The movement of ions to the xylem exudate of maize roots III. The location of the electrical and electrochemical potential differences between the exudate and the medium. J. Exp. Bot. **22**, 453–464 (1971c).

Durbin, R.P.: J. Gen. Physiol. **44**, 315–326 (1960).

Dymock, I.J., Hill, B., Bown, A.W.: An investigation into the influence of IAA and malate on *in vivo* and *in vitro* rates of dark carbon dioxide fixation in coleoptile tissue. Can. J. Bot. **55**, 1641–1645 (1977).

Egneus, H., Heber, U., Matthieson, U., Kirk, M.: Reduction of oxygen by the electron transport chain of chloroplasts during assimilation of carbon dioxide. Biochim. Biophys. Acta **408**, 252–268 (1975).

Eisele, R.: Photosynthetische Nitrataufnahme und Nitratreduktion bei *Ankistrodesmus braunii*. Dissertation, Darmstadt, (1976).

Eisele, R., Ullrich, W.R.: Stoichiometry between photosynthetic nitrate reduction and alkalinisation by *Ankistrodesmus braunii* in vivo. Planta **123**, 117–123 (1975).

Eppley, R.W., Bovell, C.R.: Sulfuric acid in *Desmarestia*. Biol. Bull. **115**, 101–106 (1958).

Epstein, E.: Dual pattern of ion absorption by plant cells and by plants. Nature (London) **212**, 1324–1327 (1966).

Epstein, E.: Mineral nutrition of plants: principles and perspectives. New York: Wiley and Sons (1972).

Epstein, E.: Kinetics of ion transport and the carrier concept. Enc. Plant Physiol. **2B**, 70–94 (1976).

Epstein, E.: Genetic potentials for solving problems of soil mineral stress: Adaptation of crops to salinity, pp. 73–82. In: Plant adaptation to mineral stress in problem soils, Ithaca: Cornell Univ. Press (1977).

Epstein, E., Hagen, C.E.: A kinetic study of the absorption of alkali cations by barley roots. Plant Physiol. **27**, 457–474 (1952).

Epstein, E., Jefferies, R.L.: The genetic basis of selective ion transport in plants. Ann. Rev. Plant Physiol. **15**, 169–184 (1964).

Epstein, E., Norlyn, J.D.: Seawater-based crop production: A feasibility study. Science **197**, 249–251 (1977).

Epstein, E., Rains, D.W., Elzam, O.E.: Resolution of dual mechanisms of potassium absorption by barley roots. Proc. Natl. Acad. Sci. U.S. **49**, 684–692 (1963).

Eschrich, W.: Biochemistry and fine structure of phloem in relation to transport. Ann. Rev. Plant Physiol. **21**, 193–214 (1970).

Eschrich, W.: Bidirectional transport. Enc. Plant Physiol. **1**, 245–255 (1975).

Eschrich, W., Evert, R.F., Young, J.H.: Solution flow in tubular semipermeable membranes. Planta **107**, 279–300 (1972).

Eschrich, W., Fritz, E.: Microautoradiography of water soluble organic compounds. In: Microautoradiography and electron probe analysis, Lüttge, U., (ed.), pp. 99–122. Berlin-Heidelberg-New York: Springer (1972).

Eschrich, W., Heyser, W.: Biochemistry of phloem constituents. Enc. Plant Physiol. **1**, 101–138 (1975).

Eschrich, W., Steiner, M.: Autoradiographische Untersuchungen zum Stofftransport bei *Polytrichum commune*. Planta **74**, 330–349 (1967).

Eschrich, W., Steiner, M.: Die Struktur des Leitgewebesystems von *Polytrichum commune*. Planta **82**, 33–49 (1968a).

Eschrich, W., Steiner, M.: Die submikroskopische Struktur der Assimilatleitbahnen von *Polytrichum commune*. Planta **82**, 321–336 (1968b).

Eshel, A., Waisel, Y.: Variations in sodium uptake along primary roots of corn seedlings. Plant Physiol. **49**, 585–589 (1972).

Eshel, A., Waisel, Y.: Variations in uptake of sodium and rubidium along barley roots. Plant Physiol. **28**, 557–560 (1973).

Etherton, B.: Relationship of cell transmembrane electropotential to potassium and sodium accumulation ratios in oat and pea seedlings. Plant Physiol. **38**, 581–585 (1963).

Etherton, B.: Effect of indole-3-acetic acid on membrane potentials of oat coleoptile cells. Plant Physiol. **45**, 527–528 (1970).

Etherton, B., Dedolph, R.R.: Gravity and intercellular differences in membrane potentials of plant cells. Plant Physiol. **49**, 1019–1020 (1972).

Etherton, B., Higinbotham, N.: Transmembrane potential measurements of cells of higher plants as related to salt uptake. Science **131**, 409–410 (1960).

Evans, E.C., III, Vaughan, B.E.: Wounding response in relation to polar transport of radiocalcium in isolated root segments of *Zea mays*. Plant Physiol. **41**, 1145–1151 (1966).

Evans, H.J., Sorger, G.: Role of mineral elements with emphasis on the univalent cations. Ann. Rev. Plant Physiol. **17**, 47–76 (1966).

Evans, M.L.: Rapid responses to plant hormones. Ann. Rev. Plant Physiol. **25**, 195–223 (1974).

Evert, R.F., Eschrich, W., Heyser, W.: Distribution and structure of the plasmodesmata in mesophyll and bundle-sheath cells of *Zea mays* L. Planta **136**, 77–89 (1977).

Falk, H., Lüttge, U., Weigl, J.: Untersuchung zur Physiologie plasmolysierter Zellen II. Ionenaufnahme, O_2-Wechsel, Transport. Z. Pflanzenphysiol. **54**, 446–462 (1966).

Falk, H., Sitte, P.: Untersuchungen am Caspary-Striefen. Proc. Eur. Reg. Conf. on Electron Microscopy, Delft. vol. II pp. 1063–1066 (1960).

Falkner, G., Werdan, K., Horner, F., Heldt, H.W.: Energieabhängige Phosphataufnahme der Blaualge *Anacystis nidulans*. Ber. Dtsch. Bot. Ges. **87**, 263–266 (1974).

Felle, H., Bentrup, F.-W.: Lichtabhängigkeit des Membranpotentials der Rhizoidzelle von *Riccia fluitans:* Versuche zur Signalübertragung. Ber. Dtsch. Bot. Ges. **87**, 223–228 (1974a).

Felle, H., Bentrup, F.W.: Light-dependent changes of the membrane potential and conductance in *Riccia fluitans*. In: Membrane transport in plants, Dainty, J., Zimmermann, U., (eds.), Berlin-Heidelberg-New York: Springer (1974b).

Fensom, D.S.: The bio-electric potentials of plants and their functional significance. Can. J. Bot. **35**, 573–582 (1957).

Fensom, D.S.: Work with isolated phloem strands. Enc. Plant Physiol. **1**, 223–244 (1975).

Fensom, D.S.: Possible mechanisms of phloem transport: Other possible mechanisms. Enc. Plant Physiol. **1**, 354–366 (1975).

Fensom, D.S., Dainty, J.: Electro-osmosis in *Nitella*. Can. J. Bot. **41**, 685–691 (1963).

Fensom, D.S., Meylan, S., Pilet, P.E.: Induced electro-osmosis in root tissues. Can. J. Bot. **43**, 452–467 (1965).

Fensom, D.S., Ursino, D.J., Nelson, C.D.: Determination of relative pore size in living membranes of *Nitella* by the techniques of electroosmosis and radioactive tracers. Can. J. Bot. **45**, 1267–1275 (1967).

Fensom, D.S., Wanless, I.R.: Further studies of electro-osmosis in *Nitella* in relation to pores in membranes. J. Exp. Bot. **18**, 563–577 (1967).

Ferguson, I.B., Clarkson, D.T.: Ion uptake in relation to the development of a root hypodermis. New Phytol. **77**, 11–14 (1976a).

Ferguson, I.B., Clarkson, D.T.: Simultaneous uptake and translocation of magnesium and calcium in barley (*Hordeum vulgare* L.) roots. Planta **128**, 267–269 (1976b).

Figier, J.: Localisation infrastructurale de la phosphomonoestérase acide dans la stipule de *Vicia faba* L. au niveau du nectaire. Rôles possibles de cet enzyme dans les mécanismes de la sécrétion. Planta **83**, 60–79 (1968).

Filippis, de, F.L., Pallaghy, C.K.: Effect of light on the volume and ion relations of chloroplasts in detached leaves of *Elodea densa*. Aust. J. Biol. Sci. **26**, 1251–1265 (1973).

Findenegg, G.R.: Beziehungen zwischen Carboanhydraseaktivität und Aufnahme von Bicarbonat und Chlorid bei der Photosynthese von *Scenedesmus obliquus*. Planta **116**, 123–131 (1974).

Findenegg, G.R.: Interactions of glycolate-, HCO_3^--, Cl^--, and H^+-balance of *Scenedesmus obliquus*. Planta **135**, 33–38 (1977a).

Findenegg, G.R.: Estimation of bicarbonate fluxes in *Scenedesmus obliquus*, pp. 275–281. In: Transmembrane ionic exchanges in plants, Thellier, M., Monnier, A., DeMarty, M., Dainty, J., (eds.), Paris-Rouen: C.N.R.S. and Univ. Rouen (1977b).

Findlay, G.P., Hope, A.B.: Electrical properties of plant cells: methods and findings. Enc. Plant Physiol. **2A**, 53–92 (1976).

Findlay, G.P., Hope, A.B., Walker, N.A.: Quantization of a flux ratio in Charophytes? Biochim. Biophys. Acta **233**, 155–162 (1971).

Findlay, N., Mercer, F.V.: Nectar production in *Abutilon* I. Movement of nectar through the cuticle. Aust. J. Biol. Sci. **24**, 647–656 (1971).

Findlay, N., Reed, M.L., Mercer, F.V.: Nectar production in *Abutilon* III. Sugar secretion. Aust. J. Biol. Sci. **24**, 665–675 (1971).

Fischer, E.: Membranpotential und Aminosäuretransport bei *Lemna gibba* G1. Dipl.-biol.-Thesis, Darmstadt (1978).

Fischer, E., Haschke, H.-P., Hilsdorf, J., Lüttge, U., Weikert, A., Zirke, G.:

Wirkung von Cyanid auf das Membranpotential von Blattzellen von *Mnium cuspidatum*. Ber. Dtsch. Bot. Ges. **88**, 355–360 (1975).

Fischer, E., Lüttge, U., Higinbotham, N.: Effect of cyanide on the plasmalemma potential of *Mnium*. Plant Physiol. **58**, 240–241 (1976).

Fischer, R.A.: Stomatal opening: role of potassium uptake by guard cells. Science **160**, 784–785 (1968).

Fischer, R.A.: Role of potassium in stomatal opening in the leaf of *Vicia faba*. Plant Physiol. **47**, 555–558 (1971).

Fischer, R.A.: Aspects of potassium accumulation by stomata of *Vicia faba*. Aust. J. Biol. Sci. **25**, 1107–1123 (1972).

Fischer, R.A., MacAlister, T.J.: A quantitative investigation of symplasmic transport in *Chara corallina* III. An evaluation of chemical and freeze-substituting techniques in determining the in vivo condition of the plasmodesmata. Can. J. Bot. **53**, 1988–1993 (1975).

Fisher, D.B., Hansen, D., Hodges, T.K.: Correlation between ion fluxes and ion-stimulated adenosine triphosphatase activity of plant roots. Plant Physiol. **46**, 812–814 (1970).

Flowers, T.J., Hall, J.L.: Properties of membranes from the halophyte *Suaeda maritima* II. Distribution and properties of enzymes in isolated membrane fractions. J. Exp. Bot. **27**, 673–689 (1976).

Flowers, T.J., Troke, P.F., Yeo, A.R.: The mechanism of salt tolerance in halophytes. Ann. Rev. Plant Physiol. **28**, 89–121 (1977).

Floyd, R.A., Rains, D.W.: Investigation of respiratory and ion transport properties of ageing bean stem slices. Plant Physiol. **47**, 663–667 (1971).

Fondeville, J.C., Schneider, M.J., Borthwick, H.A., Hendricks, S.B.: Photocontrol of *Mimosa pudica* L. leaf movement. Planta **75**, 228–238 (1967).

Forde, J., Steer, M.W.: Cytoplasmic streaming in *Elodea*. Can. J. Bot. **54**, 2688–2694 (1976).

Fork, D.C., Heber, U.W.: Studies on electron-transport reactions of photosynthesis in plastome mutants of *Oenothera*. Plant Physiol. **43**, 606–612 (1968).

Fox, C.F., Kennedy, E.P.: Specific labelling and partial purification of the M protein, a component of the β-galactosidase transport system of *Escherichia coli*. Proc. Natl. Acad. Sci. U.S. **54**, 891–899 (1965).

Franke, W.W.: On the universality of nuclear pore complex structure. Z. Zellforsch. **105**, 405–429 (1970).

Frey, W.: Neue Vorstellungen über die Verwandtschaftsgruppen und die Stammesgeschichte der Laubmoose. In: Beiträge zur Biologie der niederen Pflanzen, Frey, W., Hurka, H., Oberwinkler, F., (eds.), pp. 117–139. Stuttgart-New York: Fischer (1977).

Frey-Wyssling, A.: Die Stoffausscheidungen der höheren Pflanzen. Berlin: Springer (1935).

Frey-Wyssling, A.: Die pflanzliche Zellwand. Berlin-Göttingen-Heidelberg: Springer (1959).

Frey-Wyssling, A., Häusermann, L.: Deutung der gestaltlosen Nektarien. Ber. Schweiz. Botan. Ges. **70**, 150–162 (1960).

Frey-Wyssling, A., Mühlethaler, K.: Ultrastructural plant cytology. Amsterdam: Elsevier (1965).

Fried, M., Noggle, J.C.: Multiple site uptake of individual ions by roots as affected by hydrogen ion. Plant Physiol. **33**, 139–144 (1958).

Fritz, E., Eschrich, W.: 14C-Mikroautoradiographie wasserlöslicher Substanzen im Phloem. Planta **92**, 267–281 (1970).

Frosch, S., Wagner, E.: Endogenous rhythmicity and energy transduction II. Phytochrome action and the conditioning of rhythmicity of adenylate kinase, NAD- and NADP-linked glyceraldehyde-3-phosphate dehydrogenase in *Chenopodium rubrum* by temperature and light intensity cycles during germination. Can. J. Bot. **51**, 1521–1528 (1973a).

Frosch, S., Wagner, E.: Endodenous rhythmicity and energy transduction III. Time course of phytochrome action in adenylate kinase, NAD- and NADH-linked glyceraldehyde-3-phosphate dehydrogenase in *Chenopodium rubrum*. Can. J. Bot. **51**, 1529–1535 (1973b).

Frosch, S., Wagner, E., Cumming, B.G.: Endogenous rhythmicity and energy transduction I. Rhythmicity in adenylate kinase, NAD- and NADP-linked glyceraldehyde-3-phosphate dehydrogenase in *Chenopodium rubrum*. Can. J. Bot. **51**, 1355–1367 (1973).

Frost, W.B., Blevins, D.G., Barnett, N.M.: Cation pretreatment effects on nitrate uptake, xylem exudate, and malate levels in wheat seedling. Plant Physiol. **61**, 323–326 (1978).

Fujino, M.: Role of adenosinetriphosphate and adenosinetriphosphatase in stomatal movement. Sci. Bull. Fac. Educ. Nagasaki Univ. **18**, 1–47 (1967).

Gabriel, M.L., Fogel, S.: Great experiments in biology. Englewood Cliffs: Prentice-Hall, Inc. (1955).

Gauch, H.G.: Inorganic plant nutrition. Stroudsberg: Dowden-Hutchinson and Ross (1972).

Geiger, D.R.: Phloem loading. Enc. Plant Physiol. **1**, 395–431 (1975).

Gerhart, B.P., Beevers, H.: Developmental studies on glyoxysomes in *Ricinus* endosperm. J. Cell Biol. **44**, 94–102 (1970).

Gerson, D.F., Poole, R.J.: Anion absorption by plants: a unary interpretation of "dual mechanisms". Plant Physiol. **48**, 509–511 (1971).

Gerson, D.F., Poole, R.J.: Chloride accumulation by mung bean root tips: a low-affinity active transport system at the plasmalemma. Plant Physiol. **50**, 603–607 (1972).

Giaquinta, R.: Evidence for phloem loading from the apoplast. Chemical modification of membrane sulfhydryl groups. Plant Physiol. **57**, 872–875 (1976).

Giaquinta, R.: Phloem loading of sucrose: pH dependence and selectivity. Plant Physiol. **59**, 750–755 (1977).

Gibbs, M., Latzko, E., Eds.: Transport in Plants VI. Photosynthesis II, Regulation of photosynthetic carbon metabolism and related processes, vol. 6, Encyclopedia of plant physiology, New Series. New York-Heidelberg-Berlin: Springer (1978).

Gimmler, H., Schäfer, G., Kraminer, H., Heber, U.: Amino acid permeability of the chloroplast envelope as measured by light scattering, volumetry and amino acid uptake. Planta **120**, 47–61 (1974).

Ginsburg, H., Ginzburg, B.Z.: Radial water and solute flows in roots of *Zea mays*. J. Exp. Bot. **21**, 593–604 (1970).

Ginsburg, H., Ginzburg, B.Z.: Radial water and solute flows in roots of *Zea mays*. J. Exp. Bot. **25**, 28–35 (1974).

Ginsburg, H., Laties, G.G.: Longitudinal electrical resistance of maize roots. J. Exp. Bot. **24**, 1035–1040 (1973).

Glass, A.: The regulation of potassium absorption in barley roots. Plant Physiol. **56**, 377–380 (1975).

Glass, A.D.M.: Regulation of potassium absorption in barley roots. Plant Physiol. **58**, 33–37 (1976a).

Glass, A.: Potassium uptake by barley roots: effect of cycloheximide. Z. Pflanzenphysiol. **79**, 446–449 (1976b).

Glidewell, S.M., Raven, J.A.: Measurement of simultaneous oxygen evolution and uptake in *Hydrodictyon africanum*. J. Exp. Bot. **26**, 479–488 (1975).

Glinka, Z.: Effects of abscisic acid and of hydrostatic pressure gradient on water movement through excised sunflower roots. Plant Physiol. **59**, 933–935 (1977).

Glinka, Z., Reinhold, L.: Abscisic acid raises the permeability of plant cells to water. Plant Physiol. **48**, 103–105 (1971).

Glover, G.I., d'Ambrosio, S.M., Jensen, R.A.: Versatile properties of a nonsaturable, homogeneous transport system in *Bacillus subtilis:* Genetic, kinetic, and affinity labeling studies. Proc. Natl. Acad. Sci. U.S. **72**, 814–818 (1975).

Glynn, I.M., Hoffman, J.F., Lew, V.L.: Some "partial reactions" of the sodium pump. Phil. Trans. Roy. Soc. Lond. B. **262**, 91–102 (1971).

Goldin, S.M., Tong, S.W.: Reconstitution of active transport catalyzed by the purified sodium and potassium ion-stimulated adenosine triphosphatase from canine renal medulla. J. Biol. Chem. **249**, 5907–5915 (1974).

Goldman, D.E.: Potential, impedance and rectification in membranes. J. Gen. Physiol. **27**, 37–60 (1943).

Goldsmith, M.H.M.: The polar transport of auxin. Ann. Rev. Plant Physiol. **28**, 439–478 (1977).

Goldsmith, M.H.M., Fernandez, H.R., Goldsmith, T.H.: Electrical properties of parenchymal cell membranes in the oat coleoptile. Planta **102**, 302–323 (1972).

Goldstein, D.A., Solomon, A.K.: Determination of equivalent pore radius for human red cells by osmotic pressure measurement. J. Gen. Physiol. **44**, 1 ff. (1960).

Gonzalez, E., Beevers, H.: Role of the endoplastic reticulum in glyoxysome formation in castor bean endosperm. Plant Physiol. **57**, 406–409 (1976).

Goodwin, P.B.: Physiological and electrophysiological evidence for intercellular communication in plant symplasts, pp. 121–129. In: Intercellular communication in plants: studies on plasmodesmata, Gunning, B.E.S., Robards, A.W., (eds.), Berlin-Heidelberg-New York: Springer (1976).

Göring, H.: Die Bedeutung ungerührter Schichten für die Kinetik des trägervermittelten Membrantransportes in pflanzlichen Geweben. Wiss. Z. der Humboldt-Univ. zu Berlin **25**, 61–65 (1976).

Göring, H., Mardanov, A.A.: Beziehung zwischen dem K^+/Ca^{++} Verhältnis im Gewebe und der Wirkung von Zytokinin in höheren Pflanzen. Biol. Rdsch. **14**, 177–189 (1976).

Gorter, E., Grendel, F.: On bimolecular layers of lipoids on the chromocytes of the blood. J. Exp. Med. **41**, 439–443 (1925).

Gottlieb, M.H., Sollner, K.: Failure of the Nernst-Einstein equation to correlate electrical resistances and rates of ionic self-exchange across certain fixed charge membranes. Biophys. J. **8**, 515–535 (1968).

Gračanin, M.: Zur Rolle osmotischer und nicht osmotischer Kräfte bei Guttation und Exsudation. Flora (Jena) **154**, 21–35 (1964).

Gradmann, D.: Einfluß von Licht, Temperatur und Außenmedium auf das elektrische Verhalten von *Acetabularia crenulata*. Planta **93**, 323–353 (1970).

Gradmann, D.: Analog circuit of the *Acetabularia* membrane. J. Membrane Biol. **25**, 183–208 (1975).

Gradmann, D.: Potassium and turgor pressure in plants. J. Theor. Biol. **65**, 597–599 (1977).

Gradmann, D., Mayer, W.-E.: Membrane potentials and ion permeabilities in flexor cells of the laminar pulvini of *Phaseolus coccineus* L. Planta **137**, 19–24 (1977).

Gradmann, D., Slayman, C.L.: Oscillations of an electrogenic pump in the plasma membrane of *Neurospora*. J. Membrane Biol. **23**, 181–212 (1975).

Gradmann, D., Wagner, G., Gläsel, R.H.: Chloride efflux during light-triggered action potential in *Acetabularia mediterranea*. Biochim. Biophys. Acta **323**, 151–155 (1973).

Graham, R.D., Bowling, D.J.F.: Effect of the shoot on the transmembrane potentials of root cortical cells of sunflower. J. Exp. Bot. **28**, 886–893 (1977).

Grahm, L., Hertz, C.H.: Measurements of the geoelectric effect in coleoptiles by a new technique. Plant Physiol. **15**, 96–114 (1962).

Green, P.B., Stanton, F.W.: Turgor pressure: direct manometric measurement in single cells of *Nitella*. Science **155**, 1675–1676 (1968).

Green, D.E., Yi, S., Brucker, R.F.: Bioenergetics **4**, 527 ff. (1972).

Greenway, H., Pitman, M.G.: Potassium retranslocation in seedlings of *Hordeum vulgare*. Aust. J. Biol. Sci. **18**, 235–247 (1965).

Grombein, S., Rüdiger, W., Pratt, L.H., Marmé, D.: Phytochrome pellatibility in extracts of *Avena* shoots. Plant Sci. Lett. **5**, 275–280 (1975).

Gross, J., Marmé, D.: ATP-dependent Ca^{2+} uptake into plant membrane vesicles. Private communications.

Guardiola, J.L.: Growth and accumulation of mineral elements in the axis of young pea (*Pisum sativum* L.) seedlings. Acta Bot. Neerl. **22**, 55–68 (1973).

Gunning, B.E.S., Hughes, J.E.: Quantitative assessment of symplastic transport of pre-nectar into the trichomes of *Abutilon* nectaries. Aust. J. Plant Physiol. **3**, 619–637 (1976).

Gunning, B.E.S., Pate, J.S.: "Transfer cells". Plant cells with wall ingrowths, specialized in relation to short distance transport of solutes—their occurrence, structure, and development. Protoplasma **68**, 107–133 (1969).

Gunning, B.E.S., Pate, J.S., Minchin, F.R., Marks, I.: Quantitative aspects of transfer cell structure in relation to vein loading in leaves and solute transport in legume nodules. In: Transport at the cellular level. Soc. Exp. Biol. Symp., vol. 28, pp. 87–126. Cambridge: Cambridge University Press (1974).

Gunning, B.E.S., Robards, A.W., Eds.: Plasmodesmata: Current knowledge and outstanding problems, pp. 297–311. In: Intercellular communication in plants: studies on plasmodesmata, Gunning, B.E.S., Robards, A.W., (eds.), Berlin-Heidelberg-New York: Springer (1976).

Gutierrez, M., Gracen, V.E., Edwards, G.E.: Biochemical and cytological relationships in C_4 plants. Planta **119**, 279–300 (1974).

Gutknecht, J.: Permeability of *Valonia* to water and solutes: apparent absence of aqueous membrane pores. Biochim. Biophys. Acta **163**, 20–29 (1968).

Haas, D.L., Carothers, Z.B.: Some ultrastructural observations on endodermal cell development in *Zea mays* roots. Am. J. Bot. **62**, 336–348 (1975).

Haass, D., Tanner, W.: Regulation of hexose transport in *Chlorella vulgaris*. Characteristics of induction and turnover. Plant Physiol. **53**, 14–20 (1974).

Hager, A., Menzel, H., Krauss, A.: Versuche und Hypothese zur Primärwirkung des Auxins beim Streckungswachstum. Planta **100**, 47–75 (1971).

Hales, S.: (1727) Vegetable staticks. Reprint, Hoskin, M.A., (ed.), New York: American Elsevier (1969).

Hall, J.L.: Pinocytotic vesicles and ion transport in plant cells. Nature (London) **226**, 1253–1254 (1970).

Hall, J.L., Baker, D.A.: Cell Membranes, pp 39–77. In: Ion transport in plant cells and tissues, Baker, D.A., Hall, J.L., (eds.), North Holland/American Elsevier Amsterdam-Oxford-New York (1975).

Hall, J.L., Flowers, T.J.: Properties of membranes from the halophyte *Suaeda maritima* I. Cytochemical staining of membranes in relation to the validity of membrane markers. J. Exp. Bot. **27**, 658–671 (1976).

Hall, J.L., Roberts, R.M.: Biochemical characteristics of membrane fractions isolated from maize (*Zea mays* L.) roots. Ann. Bot. **39**, 983–993 (1975).

Hanson, J.B., Bertagnolli, B.L., Shepherd, W.D.: Phosphate-induced stimulation of acceptorless respiration in corn mitochondria. Plant Physiol. **50**, 347–354 (1972).

Hanson, J.B., Koeppe, D.E.: Mitochondria, pp. 79–99. In: Ion transport in plant cells and tissues, Baker, D.A., Hall, J.L., (eds.), Amsterdam-Oxford-New York: North Holland/American Elsevier (1975).

Hanson, J.B., Leonard, R.T., Mollenhauer, H.H.: Increased electron density of tonoplast membranes in washed corn root tissue. Plant Physiol. **52**, 298–300 (1973).

Hansson, G.: Patterns of ionic influences on sugar beet ATPases. Dissertation, University of Stockholm (1975).

Hansson, G., Kylin, A.: ATPase activities in homogenates from sugar-beet roots, relation to Mg^{2+} and ($Na^+ + K^+$)-stimulation. Z. Pflanzenphysiol. **60**, 270–275 (1969).

Harrison-Murray, R.S., Clarkson, D.T.: Relationships between structural development and the absorption of ions by the root system of *Cucurbita pepo*. Planta **114**, 1–16 (1973).

Hartung, W., Behl, R.: Die Wirkung von Licht auf den Transport von 2-(^{14}C) Abscisinsäure in Wurzeln von *Phaseolus coccineus* L. für einen Radialtransport von ABA zwischen Zentralzylinder und Rindenzylinder. Planta **122**, 53–59 (1975a).

Hartung, W., Behl, R.: Die Wirkung von Licht auf den Transport von 2-(^{14}C) Abscisinsäure in Bohnenwurzelsegmenten. Planta **122**, 61–65 (1975b).

Harvey, E.N.: Tension at the cell surface. In: Protoplasmatologia. Bd. II, E5. Wien: Springer (1954).

Haschke, H.-P., Lüttge, U.: β-Indolylessigsäure(-IES)-abhängiger K^+-H^+-Austauschmechanismus und Strekungswachstum bei *Avena*-Koleoptilen. Z. Naturforsch. **28C**, 555–558 (1973).

Haschke, H.-P., Lüttge, U.: Interactions between IAA, potassium, and malate accumulation, and growth in *Avena* coleoptile segments. Z. Pflanzenphysiol. **76**, 450–455 (1975a).

Haschke, H.-P., Lüttge, U.: Stoichiometric correlation of malate accumulation with auxin-dependent K^+-H^+ exchange and growth in *Avena* coleoptile segments. Plant Physiol. **56**, 696–698 (1975b).

Haschke, H.-P., Lüttge, U.: Action of auxin on CO_2 dark fixation in *Avena* coleoptile segments as related to elongation growth. Plant Sci. Lett. **8**, 53–58 (1977a).

Haschke, H.-P., Lüttge, U.: Auxin action on K^+-H^+-exchange and growth, $^{14}CO_2$ fixation and malate accumulation in *Avena* coleoptile segments. In: Regulation of cell membrane activities in plants, Marré. E., Ciferri, O., (eds.), pp. 243–248. Amsterdam: Elsevier/North-Holland Biomedical Press (1977b).

Hastings, D.F., Gutknecht, J.: Turgor pressure regulation: Modulation of active potassium transport by hydrostatic pressure gradients, pp. 79–83. In: Membrane transport in plants, Zimmermann, U., Dainty, J., (eds.), Berlin-Heidelberg-New York: Springer (1974).

Hatch, M.D., Kagawa, T., Craig, S.: Subdivision of C_4-pathway species based on differing C_4 acid decarboxylating systems and ultrastructural features. Aust. J. Plant Physiol. **2**, 111–128 (1975).

Hatch, M.D., Osmond, C.B.: Compartmentation and transport in C_4 photosynthesis. Enc. Plant Physiol. **3**, 144–184 (1976).

Hatch, M.D., Osmond, C.B., Slatyer, R.O., Eds.: Photosynthesis and photorespiration. New York-London-Sydney-Toronto: Wiley-Interscience (1971).

Hatch, M.D., Slack, C.R.: Photosynthetic CO_2-fixation pathways. Ann. Rev. Plant Physiol. **21**, 141–162 (1970).

Hatrick, A.A., Bowling, D.J.F.: A study of the relationship between root and shoot metabolism. J. Exp. Bot. **24**, 607–613 (1973).

Haupt, W.: Die Orientierung der Phytochrom-Moleküle in der *Mougeotia* zelle: Ein neues Modell zur Deutung der experimentellen Befunde. Z. Pflanzenphysiol. **58**, 331–346 (1968).

Haupt, W.: Über den Dichroismus von Phytochrom$_{660}$ und Phytochrom$_{730}$ bei *Mougeotia*. Z. Pflanzenphysiol. **62**, 287–298 (1970a).

Haupt, W.: Localization of phytochrom in the cell. Physiol. Vég. **8**, 551–563 (1970b).

Haupt, W.: Bewegungsphysiologie der Pflanzen. Stuttgart: G. Thieme (1977).

Haupt, W., Mörtel, G., Winkelnkemper, I.: Demonstration of different dichroic orientation of phytochrome P_R and P_{FR}. Planta **88**, 183–186 (1969).

Haupt, W., Trump, K.: Lichtorientierte Chloroplastenbewegung bei *Mougeotia*. Die Größe des Phytochromgradienten steuert die Bewegungsgeschwindigkeit. Pflanzenphysiol. Biochem. **168**, 131–140 (1975).

Heber, U.: Energy transfer within leaf cells, pp. 1335–1348. In: Proc. 3rd Int. Congr. Photosynth. Res. Avron, M., (ed.), Amsterdam: Elsevier (1975).

Heber, U., Kirk, M.R.: Aufnahme und Umsatz exportierter Photosyntheseprodukte durch die Chloroplasten während der Assimilation von CO_2. Biochem. Pflanzenphysiol. **168**, 211–233 (1975).

Heber, U., Krause, G.: Transfer of carbon, phosphate energy, and reducing equivalents across the chloroplast envelope. In: Photosynthesis and photorespiration, Hatch, M.D., Osmond, C.B., Slatyer, R.O., (eds.), New York-London-Sydney-Toronto: Wiley and Sons (1971).

Heber, U., Santarius, K.A.: Compartmentation and reduction of pyridine nucleotides in relation to photosynthesis. Biochim. Biophys. Acta **109**, 390–408 (1965).

Heber, U., Santarius, K.A.: Direct and indirect transfer of ATP and ADP across the chloroplast envelope. Z. Naturforsch. **25b**, 718–728 (1970).

Heimer, Y.M.: Nitrite-induced development of the nitrate uptake system in plant cells. Plant Sci. Lett. **4**, 137–139 (1975).

Heimer, Y.M., Filner, P.: Regulation of the nitrate assimilation pathway in cultured tobacco cells III. The nitrate uptake system. Biochim. Biophys. Acta **230**, 362–372 (1971).

Heimer, Y.M., Wray, J.L., Filner, P.: The effect of tungstate on nitrate assimilation in higher plant tissues. Plant Physiol. **44**, 1197–1199 (1969).

Heitefuss, R., Williams, P.H., Eds.: Transport in Plants IV. Physiological Plant Pathology, vol. 4, Encyclopedia of plant physiology, New Series. New York-Heidelberg-Berlin: Springer (1976).

Helder, R.J.: Transport across the root tissue and transfer to the shoot. 10th Int. Bot. Congr. Edinburgh (1964).

Helder, R.J.: Translocation in *Vallisneria spiralis*. Handbuch Pflanzenphysiol. **13**, 20–43 (1967).

Heldt, H.W.: Metabolite carriers of chloroplasts. Enc. Plant Physiol. **3**, 137–143 (1976a).

Heldt, H.W.: Transport of metabolites between cytoplasm and the mitochondrial matrix. Enc. Plant Physiol. **3**, 235–254 (1976b).

Heldt, H.W., Chon, C.J., Maronde, D., Herold, A., Stankovic, Z.S., Walker, D.A., Kraminer, A., Kirk, M.R., Heber, U.: Role of orthophosphate and other factors in the regulation of starch formation in leaves and isolated chloroplasts. Plant Physiol. **59**, 1146–1155 (1977).

Helgerson, S.L., Cramer, W.A., Morré, D.J.: Evidence for an increase in microviscosity of plasma membranes from soybean hypocotyls induced by the plant hormone, indole-3-acetic acid. Plant Physiol. **58**, 548–551 (1976).

Heller, R.: Electric potentials and ion transport in free cells of *Acer pseudoplatanus* L. Proc. 4th Winter School Biophys. Membrane Transport Poland. Part III, pp. 5–21, Wrocław, (1977).

Heller, R., Grignon, C., Rona, J.-P.: Importance of the cell wall in the thermodynamic equilibrium of ions in free cells of *Acer pseudoplatanus* L., pp. 239–243. In: Membrane transport in plants, Zimmermann, U., Dainty, J., (eds.), Berlin-Heidelberg-New York: Springer (1974).

Hendriks, T.: Iodination of maize coleoptiles: a possible method for identifying plant plasma membranes. Plant Sci. Lett. **7**, 347–357 (1976).

Hendriks, T.: Multiple location of K-ATPase in maize coleoptiles. Plant Sci. Lett. **9**, 351–363 (1977).

Hendrix, D.L., Higinbotham, N.: Effects of filipin and cholesterol on K^+ movement in etiolated stem cells of *Pisum sativum* L. Plant Physiol. **52**, 93–97 (1973).

Hertel, R., Thomson, K.-St., Russo, V.E.: In vitro auxin binding to particulate cell fractions from corn coleoptiles. Planta **107**, 325–340 (1972).

Hess, B.: Trends in Biochemical Sciences **2**, 193 ff. (1977).

Hewitt, E.J.: Mineral nutrition of plants in culture media. In: Plant physiology, vol. III, Stewart, F.C., (ed.), pp. 97–133. New York-London: Academic Press (1963).

Hiatt, A.J.: Reactions *in vitro* of enzymes involved in CO_2 fixation accompanying salt uptake by barley roots. Z. Pflanzenphysiol. **56**, 233–245 (1967).

Higinbotham, N.: Movement of ions and electrogenesis in higher plant cells. Am. Zool. **10**, 393–403 (1970).

Higinbotham, N.: The mineral absorption process in plants. Bot. Rev. **39**, 15–69 (1973a).

Higinbotham, N.: Electropotential of plant cells. Ann. Rev. Plant Physiol. **24**, 25–46 (1973b).

Higinbotham, N., Anderson, W.P.: Electrogenic pumps in higher plant cells. Can. J. Bot. **52**, 1011–1021 (1974).

Higinbotham, N., Davis, R.F., Mertz, S.M., Shumway, L.K.: Some evidence that radial transport in maize roots is into living vessels. In: Ion transport in plants, Anderson, W.P., (ed.), pp. 493–506. London-New York: Academic Press (1973).

Higinbotham, N., Etherton, B., Foster, R.J.: Effect of external K, NH_4, Na, Ca, Mg, and H ions on cell transmembrane electropotential of *Avena* coleoptile. Plant Physiol. **39**, 196–203 (1964).

Higinbotham, N., Etherton, B., Foster, R.J.: Mineral ion contents and cell transmembrane electropotentials of pea and oat seedling tissue. Plant Physiol. **42**, 37–46 (1967).

Higinbotham, N., Graves, J.S., Davis, R.F.: Evidence for an electrogenic ion transport pump in cells of higher plants. J. Membrane Biol. **3**, 210–222 (1970).

Higinbotham, N., Hope, A.B., Findlay, G.P.: Electrical resistance of cell membranes of *Avena* coleoptiles. Science **143**, 1448–1449 (1964).

Higinbotham, N., Latimer, H., Eppley, R.: Stimulation of rubidium absorption by auxins. Science **118**, 243–245 (1953).

Higinbotham, N., Pierce, W.S.: Uptake with respect to cation-anion balance in pea epicotyl segments, pp. 406–411. In: Membrane transport in plants, Zimmermann, U., Dainty, J., (eds.), Berlin-Heidelberg-New York: Springer (1974).

Higinbotham, N., Pratt, M.J., Foster, R.J.: Effects of calcium, indole-acetic acid, and distance from stem apex on potassium and rubidium absorption by excised segments of etiolated pea epicotyl. Plant Physiol. **37**, 203–214 (1962).

Hilden, S., Hokin, L.E.: Active potassium transport coupled to active sodium transport in vesicles reconstituted from purified sodium and potassium ion-activated adenosine triphosphatase from the rectal gland of *Squalus acanthias*. J. Biol. Chem. **25**, 6296–6303 (1974).

Hill, A.E.: Ion and water transport in *Limonium*. I. Active transport by the leaf gland cells. Biochim. Biophys. Acta **135**, 454–460 (1967).

Hill, A.E.: Ion and water transport in *Limonium*. III. Time constants of the transport system. Biochim. Biophys. Acta **196**, 66–72 (1970a).

Hill, A.E.: Ion and water transport in *Limonium*. IV. Delay effects in the transport process. Biochim. Biophys. Acta **196**, 73–79 (1970b).

Hill, A.E., Hill, B.S.: The *Limonium* salt gland: a biophysical and structural study. Int. Rev. Cytol. **35**, 299–319 (1973).

Hill, A.E., Hill, B.S.: Mineral ions. Enc. Plant Physiol. **2B**, 225–243 (1976).

Hill, B.S., Hill, A.E.: ATP-driven chloride pumping and ATPase activity in the *Limonium* salt gland. J. Membrane Biol. **12**, 145–158 (1973).

Hillman, W.S., Koukari, W.L.: Phytochrome effects in the nyctinastic leaf movements of *Albizzia julibrissin* and some other legumes. Plant Physiol. **42**, 1413–1418 (1967).

Hind, G., Nakatani, H.Y., Izawa, S.: Light-dependent redistribution of ions in suspensions of chloroplast thylakoid membranes. Proc. Natl. Acad. Sci. U.S. **71**, 1484–1488 (1974).

Hoad, G.V.: Effect of osmotic stress on abscisic acid levels in xylem sap of sunflower (*Helianthus annuus*). Planta **124**, 25–29 (1975).

Hoagland, D.R., Arnon, D.I.: The water-culture method for growing plants without soil. Calif. Agric. Exp. Sta. Circ. **347** (1950).

Hocking, T.J., Hillman, J.R., Wilkins, M.B.: Nomenclature for isolated chloroplasts. Nature (London) **235**, 124–125 (1972).

Hodges, T.K.: Ion absorption by plant roots. Adv. Agron. **25**, 163–207 (1973).

Hodges, T.K.: ATPases associated with membranes of plant cells. Enc. Plant Physiol. **2A**, 260–283 (1976).

Hodgkin, A.L.: The ionic basis of electrical activity in nerve and muscle. Biol. Rev. **26**, 339 ff. (1951).

Hodgkin, A.L., Katz, B.: The effect of sodium ions on the electrical activity of the giant axon of the squid. J. Physiol. **108**, 37–77 (1949).

Hodgkin, A.L., Keynes, R.D.: The potassium permeability of a giant nerve fibre. J. Physiol. (London) **128**, 61–68 (1955).

Hoelzl-Wallach, D.F., Fischer, H., Eds.: The dynamic structure of cell membranes 22. Coll. Ges. Biol. Chem. Mosbach. Berlin-Heidelberg-New York: Springer (1971).

Hoelzl-Wallach, D.F., Knüfermann, H.G.: Plasmamembranen: Chemie, Biologie und Pathologie. Berlin-Heidelberg-New York: Springer (1973).

Höfler, K.: Permeabilitätsstudien an Parenchymzellen der Blattrippe von *Blechnum spicant*. S.B. Österr. Akad. Wiss., math.-nat. Kl., I. Abt. **167**, 237–295 (1958).

Höfler, K.: Permeabilität und Plasmabau. Ber. Dtsch. Bot. Ges. **72**, 236–245 (1959).

Höfler, K.: Permeability of protoplasm. Protoplasma **52**, 145–156 (1960).

Höfler, K.: Grundplasma und Plasmalemma. Ihre Rolle beim Permeationsvorgang. Ber. Dtsch. Bot. Ges. **74**, 233–242 (1961).

Hofmann, K.P., Zundel, G.: Stepwise protonation of PO_4^{3-}, ADP and ATP salts, IR investigations. Z. Naturforsch. **29c**, 19–28 (1974a).

Hofmann, K.P., Zundel, G.: Effect of the protons arising in aqueous hydrolysing ATP solutions, IR investigations. Z. Naturforsch. **29c**, 29–35 (1974b).

Hokin, M.R., Hokin, L.E.: The mechanism of phosphate exchange in phosphatidic acid in response to acetylcholine. J. Biol. Chem. **234**, 1387–1390 (1959).

Hokin, M.R., Hokin, L.E.: Studies on the enzymic mechanism of the sodium pump. In: Membrane transport and metabolism. London-New York: Academic Press (1961).

Hokin, M.R., Hokin, L.E.: Phosphatidic acid metabolism and active transport of sodium. Fed. Proc. **22**, 8–18 (1963).

Holm-Hansen, O.: ATP levels in algal cells as influenced by environmental conditions. Plant Cell Physiol. **11**, 689–700 (1970).

Holm-Hansen, O., Gerloff, G.C., Skoog, F.: Cobalt as an essential element for blue-green algae. Plant Physiol. **7**, 665–675 (1954).

Holmern, K., Vange, M.S., Nissen, P.: Multiphasic uptake of sulfate by barley roots II. Effects of washing, divalent cations, inhibitors, and temperature. Plant Physiol. **31**, 302–310 (1974).

Honert, van den: Over Eigenschappen van Plantenwortels, welke een Rol spelen bij de Opname van Voedingszouten. Natuurk. Tijdschr. Nederl. Ind. **97**, 150–162 (1937).

Hope, A.B.: Ion transport and membranes. London: Butterworth (1971).

Hope, A.B., Walker, N.A.: Ionic relations of cells of *Chara australis* IV. Membrane potential differences and resistances. Aust. J. Biol. Sci. **14**, 26–44 (1961).

Hope, A.B., Findlay, G.P.: The action potential in *Chara*. Plant Cell Physiol. (Tokyo) **5**, 377–379 (1964).

Hope, A.B., Lüttge, U., Ball, E.: Photosynthesis and apparent proton fluxes in *Elodea canadensis*. Z. Pflanzenphysiol. **68**, 73–81 (1972).

Hope, A.B., Lüttge, U., Ball, E.: Chloride uptake in strains of *Scenedesmus obliquus*. Z. Pflanzenphysiol. **72**, 1–10 (1974).

Hope, A.B., Walker, N.A.: The physiology of giant algal cells. Cambridge: Cambridge Univ. Press (1975).

Horton, R.F., Moran, L.: Abscisic acid inhibition of potassium influx into stomatal guard cells. Z. Pflanzenphysiol. **66**, 193–196 (1972).

House, C.R.: Water transport in cells and tissues. London: Edward Arnold (1974).

Hsiao, T.C.: Stomatal ion transport. Enc. Plant Physiol. **2B**, 195–221 (1976).

Humble, G.D., Hsiao, T.C.: Specific requirement of potassium for light-activated opening of stomata in epidermal strips. Plant Physiol. **44**, 230–234 (1969).

Humble, G.D., Hsiao, T.C.: Light-dependent influx and efflux of potassium of guard cells during stomatal opening and closing. Plant Physiol. **46**, 483–487 (1970).

Humble, G.D., Raschke, K.: Stomatal opening quantitatively related to potassium transport. Evidence from electron probe analysis. Plant Physiol. **48**, 447–453 (1971).

Humphreys, T.E.: Sucrose transport at the tonoplast. Phytochemistry **12**, 1211–1219 (1973).

Hüsken, D., Steudle, E., Zimmermann, U.: Pressure probe technique for measuring water relations of cells of higher plants. Plant Physiol. **61**, 158–163 (1978).

Hutchings, V.M.: Sucrose and proton cotransport in *Ricinus* cotyledon. Ph.D. thesis. Cambridge (1976).

Hutchings, V.M.: Sucrose and proton cotransport in *Ricinus* cotyledons. I. H^+ influx associated with sucrose uptake. Planta **138**, 229–235 (1978a).

Hutchings, V.M.: Sucrose and proton cotransport in *Ricinus* cotyledons. II. H^+ efflux and associated K^+ uptake. Planta **138**, 237–241 (1978b).

Hutner, S.H., Provasoli, L., Stockstad, E.L.R., Hoffman, C.E., Belt, M., Franklin, A.L., Jukes, J.H.: Assay of antipernicious anemia factor with *Euglena*. Proc. Soc. Exp. Biol. Med. **70**, 117–120 (1949).

Hybl, A., Dorset, D.: Biophys. Soc. Abstract. **49a** (1970).

Hylmö, B.: Transpiration and ion absorption. Plant Physiol. **6**, 333–405 (1953).

Ighe, U., Pettersson, S.: Metabolism-linked binding of rubidium in the free space of wheat roots and its relation to active uptake. Plant Physiol. **30**, 24–29 (1974).

Imamura, S.: Untersuchungen über den Mechanismus der Turgorschwankung der Spaltöffnungsschließzellen. Jap. J. Bot. **12**, 251–346 (1943).

Ingelsten, B., Hylmö, B.: Apparent free space and surface film determined by a centrifugation method. Plant Physiol. **14**, 157–170 (1961).

Iren, F. van, Spiegel, A. van der: Subcellular localization of inorganic ions in plant cells by in vivo precipitation. Science **187**, 1210–1211 (1975).

Jackman, M.E., Steveninck, R.F.M. van: Changes in the endoplasmic reticulum of beetroot slices during ageing. Aust. J. Biol. Sci. **20**, 1063–1068 (1967).

Jackson, W.A., Flesher, D., Hageman, R.H.: Nitrate uptake by dark-grown corn seedlings. Some characteristics of apparent induction. Plant Physiol. **51**, 120–127 (1973).

Jackson, W.A., Johnson, R.E., Volk, R.J.: Nitrite uptake by nitrogen-depleted wheat seedlings. Plant Physiol. **32**, 37–42 (1974a).

Jackson, W.A., Johnson, R.E., Volk, R.J.: Nitrite uptake patterns in wheat seedlings as influenced by nitrate and ammonium. Plant Physiol. **32**, 108–114 (1974b).

Jacob, F., Neuman, St., Strobel, U.: Studies on the mobility of exogen-applied substances in plants. Transactions 3rd Symp. on Accumulation and Translocation of Nutrients and Regulators in Plant Organisms, pp. 315–330. Warszawa, Jablonna, Skierniewice, Brzezna, Krabow (1973).

Jacobs, M., Ray, P.M.: Rapid auxin-induced decrease in free space pH and its relationship to auxin-induced growth in maize and pea. Plant Physiol. **58**, 203–209 (1976).

Jacobs, M.H.: Diffusion processes. New York: Springer (1967).

Jacobson, S.L.: A method for extraction of extracellular fluid: Use in development of a physiological saline for Venus's flytrap. Can. J. Bot. **49**, 121–127 (1971).

Jacoby, B.: The effect of dichlorophenyldimethylurea on light-stimulated ^{22}Na, ^{42}K, and ^{86}Rb absorption in different tissues of *Phaseolus vulgaris* leaves. Plant Physiol. **35**, 1–4 (1975).

Jacoby, B., Abas, S., Steinitz, B.: Rubidium and potassium absorption by bean-leaf slices compared to sodium absorption. Physiol. Plant. **28**, 209–214 (1973).

Jacoby, B., Laties, G.G.: Bicarbonate fixation and malate compartmentation in relation to salt-induced stoichiometric synthesis of organic acid. Plant Physiol. **47**, 525–531 (1971).

Jacoby, B., Nissen, P.: Potassium and rubidium interaction in their absorption by bean leaf slices. Plant Physiol. **40**, 42–44 (1977).

Jaffe, L.F., Nuccitelli, R.: Electrical controls of development. Ann. Rev. Biophys. Bioeng. **6**, 445–476 (1977).

Jaffe, L.F., Robinson, K.R., Nuccitelli, R.: Transcellular currents and ion fluxes through developing Fucoid eggs. In: Membrane transport in plants, Zimmermann, U., Dainty, J., (eds.), pp. 226–233 (1974).

Jaffe, M.J.: Phytochrome-mediated bioelectric potentials in mung bean seedlings. Science **162**, 1061–1067 (1968).

Jaffe, M.J.: Evidence for the regulation of phytochrome-mediated processes in bean roots by the neurohumor, acetylcholine. Plant Physiol. **46**, 768–777 (1970).

Jaffe, M.J., Galston, A.W.: Phytochrome control of rapid nyctinastic movements and membrane permeability in *Albizzia julibrissin*. Planta **77**, 135–141 (1967).

Jaffe, M.J., Thoma, L.: Rapid phytochrome-mediated changes in the uptake by bean roots of sodium acetate (1-^{14}C) and their modification by cholinergic drugs. Planta **113**, 283–291 (1973).

Jagendorf, A.T., Uribe, E.: ATP formation caused by acid-base transition of spinach chloroplasts. Proc. Natl. Acad. Sci. U.S. **55**, 170–177 (1966).

James, R.B., Pierce, W.S., Higinbotham, N.: The effect of indoleacetic acid on cell electropotential and potassium flux in etiolated oat coleoptile tissue, pp. 521–527. In: Transmembrane ionic exchanges in plants, Thellier, M., Monnier, A., DeMarty, M., Dainty, J., (eds.), Paris-Rouen: C.N.R.S. and Univ. Rouen (1977).

Jarvis, P., House, C.R.: Evidence for symplasmic ion transport in maize roots. J. Exp. Bot. **21**, 83–90 (1970).

Jennings, D.H.: The absorption of solutes by plant cells. Edinburgh-London: Oliver & Boyd (1963).

Jeschke, W.D.: Die cyclische und die nichtcyclische Photophosphorylierung als Energiequellen der lichtabhängigen Chloridionenaufnahme bei *Elodea*. Planta **73**, 161–174 (1967).

Jeschke, W.D.: Der Influx von Kaliumionen bei Blättern von *Elodea densa*, Abhängigkeit vom Licht, von der Kaliumkonzentration und von der Temperatur. Planta **91**, 111–128 (1970a).

Jeschke, W.D.: Über die Verwendung von ^{86}Rb als Indikator für Kalium, Untersuchungen am lichtgeförderten ^{42}K/K- und ^{86}Rb/Rb-Influx bei *Elodea densa*. Z. Naturforsch. **25b**, 624–630 (1970b).

Jeschke, W.D.: Lichtabhängige Veränderungen des Membranpotentials bei Blattzellen von *Elodea densa*. Z. Pflanzenphysiol. **62**, 158–172 (1970c).

Jeschke, W.D.: Energetic linkages of individual ion fluxes in leaf cells of *Elodea densa*. In: 1st Eur. Biophys. Congr. Broda, E., Locker, A., Springer-Lederer, H. (eds.). Wien: Verlag Wiener Med. Akad. (1971).

Jeschke, W.D.: Wirkung von K$^+$ auf die Fluxe und den Transport von Na$^+$ in Gerstenwurzeln, K$^+$-stimulierter Na$^+$-Efflux in der Wurzelrinde. Planta **106**, 73–90 (1972a).

Jeschke, W.D.: The effect of DNP and CCCP on photosynthesis and light-dependent Cl$^-$ influx in *Elodea densa*. Z. Pflanzenphysiol. **66**, 409–419 (1972b).

Jeschke, W.D.: Über den licht-geförderten Influx von Ionen in Blättern von *Elodea densa*. Vergleich der Influxe von K$^+$- und Cl$^-$-Ionen. Planta **103**, 164–180 (1972c).

Jeschke, W.D.: The effect of the inhibitor of photophosphorylation Dio-9 and the uncoupler atebrin on the light-dependent Cl$^-$ influx of *Elodea densa*: direct inhibition of membrane transport? Z. Pflanzenphysiol. **66**, 379–408 (1972d).

Jeschke, W.D.: Ionic relations of leaf cells. Enc. Plant Physiol. **2B**, 160–194 (1976).

Jeschke, W.D., Simonis, W.: Effect of CO$_2$ on photophosphorylation *in vivo* as revealed by the light-dependent Cl$^-$ uptake in *Elodea densa*. Z. Naturforsch. **22b**, 873–876 (1967).

Jeschke, W.D., Simonis, W.: Über die Wirkung von CO$_2$ auf die lichtabhängige Cl$^-$-Aufnahme bei *Elodea densa*: Regulation zwischen nichtcyclischer und cyclischer Photophosphorylierung. Planta **88**, 157–171 (1969).

Jeschke, W.D., Stelter, W.: K$^+$-dependent net Na$^+$ efflux in roots of barley plants. Planta **114**, 251–258 (1973).

Johansen, C., Loneragan, J.F.: Effects of anions and cations on potassium absorption by plants of high potassium chloride content. Aust. J. Plant Physiol. **2**, 75–83 (1975).

Johansen, C., Lüttge, U.: Respiration and photosynthesis as alternative energy sources for chloride uptake by *Tradescantia albiflora* leaf cells. Z. Pflanzenphysiol. **71**, 189–199 (1974).

Johansen, C., Lüttge, U.: A comparison of potassium and chloride uptake by *Tradescantia albiflora* leaf cells at different KCl concentrations. Aust. J. Plant Physiol. **2**, 471–479 (1975).

Johnson, C.M., Stout, P.R., Broyer, T.G., Carlton, A.B.: Comparative chlorine requirements of different plant species. Plant and Soil **8**, 337–353 (1957).

Jones, M.G.K.: The origin and development of plasmodesmata, pp. 81–105. In: Intercellular communication in plants: studies on plasmodesmata, Gunning, B.E.S., Robards, A.W., (eds.), Berlin-Heidelberg-New York: Springer (1976).

Jones, M.G.K., Novacky, A., Dropkin, V.H.: Transmembrane potentials of parenchyma cells and nematode-induced transfer cells. Protoplasma **85**, 15–37 (1975).

Jones, R.G.W., Lunt, O.R.: The function of calcium in plants. Bot. Rev. **33**, 407–426 (1967).

Jones, R.J., Mansfield, T.A.: Suppression of stomatal opening in leaves treated with abscisic acid. J. Exp. Bot. **21**, 714–719 (1970).

Jones, R.J., Mansfield, T.A.: Effects of abscisic acid and its esters on stomatal aperture and the transpiration ratio. Plant Physiol. **26**, 321–327 (1972).

Jose, A.M.: Phytochrome modulation of ATPase activity in a membrane fraction from *Phaseolus*. Planta **137**, 203–206 (1977).

Joseph, R.A., Hai, T. van: Uptake of phosphate by intact soybean roots: Mediated by a single multiphasic mechanism. Z. Pflanzenphysiol. **78**, 222–227 (1976).

Junge, W.: Membrane potentials in photosynthesis. Ann. Rev. Plant Physiol. **28**, 503–536 (1977).

Junghans, H., Jaffe, M.J.: Rapid respiratory changes due to red light or acetylcholine during the early events of phytochrome-mediated photomorphogenesis. Plant Physiol. **49**, 1–7 (1972).

Juniper, B.E., Barlow, P.W.: The distribution of plasmodesmata in the root tip of maize. Planta **89**, 352–360 (1960).

Kaback, H.R.: The transport of sugars across isolated bacterial membranes. In: Current topics in membranes and transport, Bronner, F., Kleinzeller, A., (eds.), vol. I. New York-London: Academic Press (1970a).

Kaback, H.R.: Transport. Ann. Rev. Biochem. **39**, 561–598 (1970b).

Kaback, H.R., Milner, L.S.: Relationship of a membrane-bound D-(-)-lactic dehydrogenase to amino acid transport in isolated bacterial membrane preparations. Proc. Natl. Acad. Sci. U.S. **66**, 1008–1015 (1970).

Kamiya, N.: Protoplasmic Streaming. Protoplasmalogia Bd. VIII-3a. Wien: Springer (1959).

Kandler, O.: Über die Beziehung zwischen Phosphathaushalt und Photosynthese II. Gesteigerter Glucoseeinbau im Licht als Indikator einer lichtabhängigen Phosphorylierung. Z. Naturforsch. **9b**, 625–644 (1954).

Kandler, O.: III. Hemmungsanalyse der lichtabhängigen Phosphorylierung. Z. Naturforsch. **10b**, 38–46 (1955).

Karlsson, J., Kylin, A.: Properties of Mg^{2+}-stimulated and $(Na^+ + K^+)$-activated adenosine-5′-triphosphatase from sugar beet cotyledons. Plant Physiol. **32**, 136–142 (1974).

Kasamo, K., Yamaki, T.: The stimulative effects of auxins in vitro on Mg^{++}-activated ATPase activity in crude enzyme extract from mung bean hypocotyls. Sci. Pap. Coll. Gen. Educ. Univ. Tokyo **23**, 131–138 (1973).

Kasemir, H., Mohr, H.: Involvement of acetylcholine in phytochrome-mediated processes. Plant Physiol. **49**, 453–454 (1972).

Kashket, E.R., Wilson, T.H.: Proton-coupled accumulation of galactoside in *Streptococcus lactis* 7962. Proc. Natl. Acad. Sci. U.S. **70**, 2866–2869 (1973).

Katchalsky, A., Curran, P.F.: Nonequilibrium thermodynamics in biophysics. Cambridge: Harvard University Press (1965).

Katou, K., Okamoto, H.: Distribution of electric potential and ion transport in the hypocotyl of *Vigna sesquipedalis* I. Distribution of overall ion concentration and the role of hydrogen ion in generation of potential difference. Plant Cell Physiol. **11**, 385–402 (1970).

Kausch, W.: Saugkraft und Wassernachleitung im Boden als physiologische Faktoren. Unter besonderer Berücksichtigung des Tensiometers. Planta **45**, 217–263 (1955).

Kedem, O.: In: Membrane transport and metabolism: pp. 87 ff. Kleinzeller, A., Kotyk, A., (eds.), New York: Academic Press (1961).

Keegstra, K., Talmadge, K.W., Bauer, W.D., Albersheim, P.: The structure of plant cell walls III. A model of the walls of suspension-cultured sycamore cells based on the interconnections of the macromolecular components. Plant Physiol. **51**, 188–196 (1973).

Kelday, L.S., Bowling, D.J.F.: The effect of cycloheximide on uptake and transport of ions by sunflower roots. Ann. Bot. **39**, 1023–1027 (1975).

Keller, H.: Untersuchung über die Funktion des Zinks in den roten Blutkörperchen. Ber. Ges. Physiol. **215**, 43–44 (1960).

Kelly, G.J., Gibbs, M.: A mechanism for the indirect transfer of photosynthetically reduced nicotinamide adenine dinucleotide phosphate from chloroplasts to the cytoplasm. Plant Physiol. **52**, 674–676 (1973).

Kimpel, J.A., Hanson, J.B.: Activation of·endogenous respiration and anion transport in corn mitochondria by acidification of the medium. Plant Physiol. **60**, 933–934 (1977).

Kirk, C.A. van, Raschke, K.: Presence of chloride reduces malate production in epidermis during stomatal opening. Plant Physiol. **61**, 361–364 (1978).

Kirkby, E.A., Knight, A.H.: Influence of level of nitrate nutrition on ion uptake and assimilation, organic acid accumulation, and cation-anion balance in whole tomato plants. Plant Physiol. **60**, 349–353 (1977).

Kishimoto, U., Tazawa, M.: Ionic composition of the cytoplasm of *Nitella flexilis*. Plant Cell Physiol. **6**, 507–518 (1965a).

Kishimoto, U., Tazawa, M.: Ionic composition and the electric response of *Lamprothamnium succinctum*. Plant Cell Physiol. **6**, 529–536 (1965b).

Kitasato, H.: The influence of H^+ on the membrane potential and ion fluxes in *Nitella*. J. Gen. Physiol. **52**, 60–87 (1968).

Klingenberg, M.: The ADP, ATP carrier in mitochondrial membranes, pp. 383–438. In: The enzymes of biological membranes, vol. III, Martonosi, A. (ed.). New York: Plenum Press (1976).

Kluge, M., Ziegler, H.: Der ATP-Gehalt der Siebröhrensäfte von Laubbäumen. Planta **61**, 167–177 (1964).

Kollmann, R.: Sieve element structure in relation to function, pp. 225–242. In: Phloem transport, Aronoff, S., Dainty, J., Gorham, P.R., Srivastava, L.M., Swanson, C.A., (eds.), New York: Plenum Press (1975).

Komnick, H.: Elektronenmikroskopische Lokalisation von Na^+ und Cl^- in Zellen und Geweben. Protoplasma **55**, 414–418 (1962).

Komor, E.: Proton-coupled hexose transport in *Chlorella vulgaris*. FEBS Lett. **38**, 16–18 (1973).

Komor, E.: Sucrose uptake by cotyledons of *Ricinus communis* L.: Characteristics, mechanism, and regulation. Planta, **137**, 119–131 (1977).

Komor, E., Haass, D., Tanner, W.: Unusual features of the active hexose uptake system of *Chlorella vulgaris*. Biochim. Biophys. Acta **266**, 649–660 (1972).

Komor, E., Rotter, M., Tanner, W.: A proton cotransport system in a higher plant: Sucrose transport in *Ricinus communis*. Plant Sci. Lett. **9**, 153–162 (1977).

Komor, E., Tanner, W.: Characterisation of the active hexose transport system of *Chlorella vulgaris*. Biochim. Biophys. Acta **241**, 170–179 (1971).

Komor, E., Loos, E., Tanner, W.: A confirmation of the proposed model for the hexose uptake system of *Chlorella vulgaris*. Anaerobic studies in the light and in the dark. J. Membrane Biol. **12**, 89–99 (1973).

Komor, E., Tanner, W.: The nature of the energy metabolite responsible for sugar accumulation in *Chlorella vulgaris*. Z. Pflanzenphysiol. **71**, 115–128 (1974a).

Komor, E., Tanner, W.J.: The hexose-proton cotransport system of *Chlorella*. pH-dependent change in K_m values and translocation constants of the uptake system. J. Gen. Physiol. **64**, 568–581 (1974b).

Komor, E., Tanner, W.J.: Simulation of a high- and low-affinity sugar-uptake system in *Chlorella* by a pH-dependent change in the K_m of the uptake system. Planta **123**, 195–198 (1975).

Korn, E.D.: Structure of biological membranes. Science **153**, 1491–1498 (1966).

Koshland, D.E., Jr.: The molecular basis for enzyme regulation. In: The enzymes Boyer, P.D., (ed.), 3rd ed., vol. I, pp. 341–396. New York: Academic Press (1970).

Kotyk, A.: Properties of the sugar carrier in Baker's yeast. II. Specificity of transport. Folia Microbiol. **12**, 121–131 (1967).

Kotyk, A., Janáček, K.: Membrane transport, Prague: Academia (1977).

Kramer, D., Läuchli, A., Yeo, A.R., Gullasch, J.: Transfer cells in roots of *Phaseolus coccineus*: Ultrastructure and possible function in exclusion of sodium from the shoot. Ann. Bot. **41**, 1031–1040 (1977).

Kramer, P.J.: Plant and soil water relationships: A modern synthesis. New York: McGraw-Hill (1969).

Krause, G.H.: Indirekter ATP-Transport zwischen Chloroplasten und Zytoplasma während der Photosynthese. Z. Pflanzenphysiol. **65**, 13–23 (1971).

Krichbaum, R., Lüttge, U., Weigl, J.: Mikroautoradiographische Untersuchung der Auswaschung des „anscheinend freien Raumes" von Maiswurzeln. Ber. Dtsch. Bot. Ges. **80**, 167–176 (1967).

Kriedemann, P.E., Loveys, B.R., Fuller, G.L., Leopold, A.C.: Abscisic acid and stomatal regulation. Plant Physiol. **49**, 842–847 (1972).

Kröger, H.: Hormones, ion balances and gene activity in Dipteran chromosomes. Mem. Soc. Endocrin. **15**, 55–66 (1967).

Kylin, A., Hansson, G.: Transport of sodium and potassium, and properties of (sodium + potassium)-activated adenosine triphosphatases: possible connection with salt tolerance in plants. In: Proc. 8th Colloq. Intern. Potash Inst., pp. 64–68. Bern: Int. Potash Inst. (1971).

Lado, P., Michelis, de, M.I., Cerana, R., Marrè, E.: Fusicoccin-induced, K^+-stimulated proton secretion and acid-induced growth of apical root segments. Plant Sci. Lett. **6**, 5–20 (1976a).

Lado, P., Rasi-Caldogno, F., Colombo, R., Michelis, de, M.I., Marrè, E.: Effects of monovalent cations on IAA- and FC-stimulated proton-cation exchange in pea stem segments. Plant Sci. Lett. **7**, 199–209 (1976b).

Lado, P., Rasi-Caldogno, F., Colombo, R.: Fusicoccin-activated proton extrusion coupled with K^+ uptake, and its role in the regulation of growth, germination, opening of stomata and mineral nutrition. Accademia nazionale dei lincei. Ser. VIII, vol. LVII, fasc. **6**, 690–700, (1974).

Laetsch, W.M.: Chloroplast structural relationships in leaves of C_4 plants. In: Photosynthesis and photorespiration, Hatch, M.D., Osmond, C.B., Slatyer, R.O., (eds.), pp. 323–349. New York-London-Sydney-Toronto: Wiley-Interscience (1971).

Laetsch, W.M.: The C_4 syndrome: A structural analysis. Ann. Rev. Plant Physiol. **25**, 27–52 (1974).

Lange, O.L., Kappen, L., Schulze, E.-D.: Water and plant life. Ecological studies 19. Berlin-Heidelberg-New York: Springer (1976).

Langmuir, I.: The shapes of molecules forming the surfaces of liquids. Proc. Natl. Acad. Sci. U.S. **3**, 251–257 (1917a).

Langmuir, I.: The constitution and fundamental properties of solids and liquids. II. Liquids. J. Am. Chem. Soc. **39**, 1848–1906 (1917b).

Langmuir, I.: Surface Chemistry. Chem. Rev. **13**, 147–191 (1933).

Lannoye, J., Tarr, S.E., Dainty, J.: The effects of pH on the ionic and electrical properties of the internodal cells of *Chara australis*. J. Exp. Bot. **21**, 543–551 (1970).

Lanyi, J.K.: Salt-dependent properties of proteins from extremely halophilic bacteria. Bacteriol. Rev. **38**, 272–290 (1974).

Larcher, W.: Ökologie der Pflanzen. Stuttgart: Ulmer-Verlag (1973).

Larcher, W.: Physiological plant ecology. Berlin-Heidelberg-New York: Springer (1975).

Larkum, A.W.D.: Ionic relations of chloroplasts *in vivo*. Nature (London) **218**, 447–449 (1968).

Larkum, A.W.D., Hill, A.E.: Ion and water transport in *Limonium* V. The ionic status of chloroplasts in the leaf of *Limonium vulgare* in relation to the activity of salt glands. Biochim. Biophys. Acta **203**, 133–138 (1970).

Latimer, W.M., Hildebrand, J.H.: Reference book of inorganic chemistry. 3rd ed. New York: Macmillan (1951).

Laties, G.G.: Metabolic and physiological development in plant tissues. Aust. J. Sci. **30**, 193–203 (1967).

Laties, G.G.: Dual mechanisms of salt uptake in relation to compartmentation and long-distance transport. Ann. Rev. Plant Physiol. **20**, 89–116 (1969).

Laties, G.G.: Solute transport in relation to metabolism and membrane permeability in plant tissues, pp. 98–151. In: Historical and current aspects of plant physiology: A symposium honoring F.C. Steward. Davies, P.J., (ed.), Ithaca: New York State College Agriculture and Life Sciences (1975).

Laties, G.G., Budd, K.: The development of differential permeability in isolated steles of corn roots. Proc. Natl. Acad. Sci. U.S. **52**, 462–469 (1964).

Läuchli, A.: Untersuchungen über Verteilung und Transport von Ionen in Pflanzengeweben mit der Röntgen-Mikrosonde I. Versuche an vegetativen Organen von *Zea mays*. Planta **75**, 185–206 (1967).

Läuchli, A.: Translocation of inorganic solutes. Ann. Rev. Plant Physiol. **23**, 197–218 (1972).

Läuchli, A.: X-ray microanalysis in botany. J. Microscop. **22**, 433–440 (1975).

Läuchli, A.: Apoplasmic transport in tissues. Enc. Plant Physiol. **2B**, 3–34 (1976a).

Läuchli, A.: Genotypic variation in transport. Enc. Plant Physiol. **2B**, 372–393 (1976b).

Läuchli, A.: Symplasmic transport and ion release to the xylem, pp. 101–112. In: Transport and transfer processes in plants. Wardlaw, I.F., Passioura, J.B., (eds.), New York-Sydney-San Francisco-London: Academic Press (1976c).

Läuchli, A., Epstein, E.: Lateral transport of ions into the xylem of corn roots I. Kinetics and energetics. Plant Physiol. **48**, 111–117 (1971).

Läuchli, A., Kramer, D., Pitman, M.G., Lüttge, U.: Ultrastructure of xylem parenchyma cells of barley roots in relation to ion transport to the xylem. Planta **119**, 85–99 (1974).

Läuchli, A., Kramer, D., Sluiter, E., Gullasch, J.: Function of xylem parenchyma cells in ion transport through barley roots: Localization of ions and ATPases, pp. 469–476. In: Transmembrane ionic exchanges in plants. Thellier, M., Monnier. A., DeMarty, M., Dainty J., (eds.), Paris-Rouen: C.N.R.S. and Univ. Rouen (1977).

Läuchli, A., Lüttge, U., Pitman, M.G.: Ion uptake and transport through barley seedlings: differential effect of cycloheximide. Z. Naturforsch. **28c**, 431–434 (1973).

Läuchli, A., Pitman, M.G., Lüttge, U., Kramer, D., Ball, E.: Are developing xylem vessels the site of ion exudation from root to shoot? Plant, Cell and Environment **1** (1978).

Läuchli, A., Spurr, A.R., Epstein, E.: Lateral transport of ions into the xylem of corn roots II. Evaluation of a stelar pump. Plant Physiol. **48**, 118–124 (1971).

Lehninger, A.L.: The mitochondrion. New York-Amsterdam: W.A. Benjamin (1964).

Leigh, R.A., Wyn Jones, R.G.: The effect of increased internal ion concentration upon the ion uptake isotherms of excised maize root segments. J. Exp. Bot. **24**, 787–795 (1973).

Leigh, R.A., Wyn Jones, R.G.: Correlations between ion-stimulated adenosine triphosphatase activities and ion influxes in maize roots. J. Exp. Bot. **26**, 508–520 (1975).

Lenard, J., Singer, S.J.: Protein conformation in cell membrane preparations as studied by optical rotatory dispersion and circular dichroism. Proc. Natl. Acad. Sci. U.S. **56**, 1828–1835 (1966).

Leonard, R.T., Hodges, T.K.: Characterization of plasma membrane-associated adenosine triphosphatase activity of oat roots. Plant Physiol. **52**, 6–12 (1973).

Leonard, R.T., Hotchkiss, C.W.: Cation-stimulated adenosine triphosphatase activity and cation transport in corn roots. Plant Physiol. **58**, 331–335 (1976).

Leonard, R.T., Hotchkiss, C.W.: Plasma membrane-associated adenosine triphosphatase activity of isolated cortex and stele from corn roots. Plant Physiol. **61**, 175–179 (1978).

Leonard, R.T., VanDerWoude, W.J.: Isolation of plasma membranes from corn roots by density gradient centrifugation. Plant Physiol. **57**, 105–114 (1976).

Letvenuk, L.J., Peterson, R.L.: Occurrence of transfer cells in vascular parenchyma of *Hieraceum florentium* roots. Can. J. Bot. **54**, 1458–1471 (1976).

Levine, R.P.: The analysis of photosynthesis using mutant strains of algae and higher plants. Ann. Rev. Plant Physiol. **20**, 523–540 (1969).

Levitzki, A., Koshland, D.E., Jr.: Negative cooperativity in regulatory enzymes. Proc. Natl. Acad. Sci. U.S. **62**, 1121–1128 (1969).

Lezzi, M.: Induktion eines Ecdyson-aktivierbaren Puff in isolierten Zellkernen von *Chironomus* durch KCl. Exp. Cell Res. **43**, 571–577 (1966).

Lin, W., Hanson, J.B.: Increase in electrogenic membrane potential with washing of corn root tissue. Plant Physiol. **54**, 799–801 (1974).

Lin, W., Wagner, G.J., Siegelman, H.W., Hind, G.: Membrane-bound ATPase of intact vacuoles and tonoplasts isolated from mature plant tissue. Biochim. Biophys. Acta **465**, 110–117 (1977).

Linask, J., Laties, G.G.: Multiphasic absorption of glucose and 3-O-methyl glucose by aged potato slices. Plant Physiol. **51**, 289–294 (1973).

Linck, A.J.: Studies on the distribution of phosphorus-32 in *Pisum sativum* in relation to fruit development. Ph.D. Dissertation. Ohio State University, Columbus, Ohio (1955).

Linden, van der, A.C., Thijsse, G.J.E.: The mechanisms of microbial oxidation of petroleum hydrocarbons. Adv. Enzymol. **27**, 469 ff. (1965).

Lineweaver, H., Burk, D.: The determination of enzyme dissociation constants. J. Am. Chem. Soc. **56**, 658–666 (1934).

Ling, G.N.: A physical theory of the living state: The association-induction hypothesis. New York: Blaisdell Publ. Co. (1962).

Ling, G.N., Cope, F.W.: Potassium ion: Is the bulk of intracellular K^+ adsorbed? Science **163**, 1335–1336 (1969).

Loneragan, J.F., Snowball, K.: Calcium requirements of plants. Aust. J. Agr. Res. **20**, 465–467 (1969).

Lookeren Campagne, R.N. van: Light-dependent chloride absorption in *Vallisneria* leaves. Acta Botan. Neerl. **6**, 543–582 (1957).

Löppert, H., Kronberger, W., Kandeler, R.: Phytochrome-mediated changes in the membrane potential of subepidermal cells of *Lemna paucicostata* 6746. Planta **138**, 133–136 (1978).

Lorimer, G.H., Andrews, T.J.: Plant photorespiration: An inevitable consequence of the existence of atmospheric oxygen. Nature (London) **243**, 359–360 (1973).

Lowe, A.G.: Enzyme mechanism for the active transport of sodium and potassium ions in animal cells. Nature (London) **219**, 934–936 (1968).

Lucy, J.A., Glauert, A.M.: Structure and assembly of macromolecular lipid complexes composed of globular micelles. J. Mol. Biol. **8**, 727–748 (1964).

Lundegårdh, H.: The translocation of salts and water through wheat roots. Plant Physiol. **2**, 103–151 (1950).

Lundegardh, H.: Mechanisms of absorption, transport, accumulation, and secretion of ions. Ann. Rev. Plant Physiol. **6**, 1–24 (1955).

Lundegårdh, H.: Investigations on the mechanism of absorption and accumulation of salts I. Initial absorption and continued accumulation of potassium chloride by wheat roots. Plant Physiol. **11**, 332–346 (1958a).

Lundegårdh, H.: II. Absorption of phosphate by potato tissue. Plant Physiol. **11**, 564–571 (1958b).

Lundegårdh, H.: III. Quantitative relations between salt uptake and respiration. Plant Physiol. **11**, 525–598 (1958c).

Lundegårdh, H., Burström, H.: Untersuchungen über die Salzaufnahme der Pflanzen III. Quantitative Beziehungen zwischen Atmung und Anionenaufnahme. Biochem. Z. **261**, 235–251 (1933).

Lundegårdh, H., Burström, H.: Untersuchungen über die Atmungsvorgänge in Pflanzenwurzeln. Biochem. Z. **277**, 223–249 (1935).

Lunt, O.R.: Sodium, In: Diagnostic criteria for plants and soils. Chapman, H.D., (ed.) pp. 409–432. Berkeley: Div. Agr. Sci., Univ. California (1966).

Lüttge, U.: Über die Zusammensetzung des Nektars und den Mechanismus seiner Sekretion I. Planta **56**, 189–212 (1961).

Lüttge, U.: Untersuchungen zur Physiologie der Carnivoren-Drüsen IV. Die Kinetik der Chloridsekretion durch das Drüsengewebe von *Nepenthes*. Planta **68**, 44–56 (1966a).

Lüttge, U.: Untersuchungen zur Physiologie der Carnivoren-Drüsen. V. Mikroautoradiographische Untersuchung der Chloridsekretion durch das Drüsengewebe von *Nepenthes*. Planta **68**, 269–285 (1966b).

Lüttge, U.: Funktion und Struktur pflanzlicher Drüsen. Naturwissenschaften **53**, 96–103 (1966c).

Lüttge, U.: Die Kinetik von Parenchymtransporten. In: Symposium Stofftransport. Vorträge Gesamtgebiet der Botanik. Dtsch. Bot. Ges. N.F., Ziegler, H., (ed.), **2**, 66–78, Stuttgart: Fischer (1968).

Lüttge, U.: Aktiver Transport (Kurzstreckentransport bei Pflanzen). Protoplasmatologia **VIII**, 7b, 1–146 (1969).

Lüttge, U.: Structure and function of plant glands. Ann. Rev. Plant Physiol. **22**, 23–44 (1971a).

Lüttge, U.: Localized ion transport in complex systems of higher plants as related to respiration and photosynthesis. In: Proc. 1st Eur. Biophys. Congr., vol. III, Broda, E., Locker, A., Springer-Lederer, H., (eds.), pp. 119–123. Wien: Verlag der Wiener Med. Akad. (1971b).

Lüttge, U. (ed.): Microautoradiography and electron probe analysis. Berlin-Heidelberg-New York: Springer (1972).

Lüttge, U.: Stofftransport der Pflanzen. Berlin-Heidelberg-New York: Springer (1973a).

Lüttge, U.: Proton and chloride uptake in relation to the development of photosynthetic capacity in greening etiolated barley leaves. In: Ion transport in plants, Anderson, W.P., (ed.), pp. 205–221. London-New York: Academic Press (1973b).

Lüttge, U.: Co-operation of organs in intact higher plants: a review. In: Membrane transport in plants, Zimmermann, U., Dainty, J., (eds.), pp. 353–362. Berlin-Heidelberg-New York: Springer (1974).

Lüttge, U.: Salt glands. In: Ion transport in plant cells and tissues, Baker, D.A.,

Hall, J.L., (eds.), pp. 335–376. Amsterdam-Oxford-New York: North-Holland Elsevier (1975).

Lüttge, U.: Nectar composition and membrane transport of sugars and amino acids: A review of the present state of nectar research. Apidologie **8**, 305–319 (1977).

Lüttge, U., Ball, E.: Light-independent uncoupler-sensitive ion uptake by green and by pale cells of variegated leaves of higher plants in relation to protein content and chloroplast integrity. Z. Naturforsch. **26b**, 158–161 (1971).

Lüttge, U., Ball, E.: Ion uptake by slices from greening etiolated barley and maize leaves. Plant Sci. Lett. **1**, 275–280 (1973).

Lüttge, U., Ball, E.: ATP levels and energy requirement of ion transport in cells of slices of greening barley leaves. Z. Pflanzenphysiol. **80**, 50–59 (1976).

Lüttge, U., Ball, E.: Concentration and pH dependence of malate efflux and influx in leaf slices of CAM plants. Z. Pflanzenphysiol. **83**, 43–54 (1977).

Lüttge, U., Ball, E., Willert, K. von: Gas exchange and ATP levels of green cells of leaves of higher plants as affected by FCCP and DCMU in *in vitro* experiments. Z. Pflanzenphysiol. **65**, 326–335 (1971a).

Lüttge, U., Ball, E., Willert, K. von: A comparative study of the coupling of ion uptake to light reactions in leaves of higher plant species having the C_3- and C_4-pathway of photosynthesis. Z. Pflanzenphysiol. **65**, 336–350 (1971b).

Lüttge, U., Bauer, K.: Die Kinetik der Ionenaufnahme durch junge und alte Sprosse von *Mnium cuspidatum*. Planta **78**, 310–320 (1968).

Lüttge, U., Cram, W.J., Laties, G.G.: The relationship of salt stimulated respiration to localised ion transport in carrot tissue. Z. Pflanzenphysiol. **64**, 418–426 (1971).

Lüttge, U., Higinbotham, N., Pallaghy, C.K.: Electrochemical evidence of specific action of indole acetic acid on membranes in *Mnium* leaves. Z. Naturforsch. **27b**, 1239–1242 (1972).

Lüttge, U., Kramer, D., Ball, E.: Photosynthesis and apparent proton fluxes in intact cells of greening etiolated barley and maize leaves. Z. Pflanzenphysiol. **71**, 6–21 (1974).

Lüttge, U., Krapf, G.: Die Ultrastruktur der Blattzellen junger und alter *Mnium*-Sprosse und ihr Zusammenhang mit der Ionenaufnahme. Planta **81**, 132–139 (1968).

Lüttge, U., Laties, G.G.: Dual mechanisms of ion absorption in relation to long distance transport in plants. Plant Physiol. **41**, 1531–1539 (1966).

Lüttge, U., Laties, G.G.: Absorption and long distance transport by isolated stele of maize roots in relation to the dual mechanisms of ion absorption. Planta **74**, 173–187 (1967).

Lüttge, U., Läuchli, A., Ball, E., Pitman, M.G.: Cycloheximide: A specific inhibitor of protein synthesis and intercellular ion transport in plant roots. Experientia **30**, 470–471 (1974).

Lüttge, U., Pallaghy, C.K.: Light-triggered transient changes of membrane potentials in green cells in relation to photosynthetic electron transport. Z. Pflanzenphysiol. **61**, 58–67 (1969).

Lüttge, U., Pallaghy, C.K.: Unerwartete Kinetik des Efflux und der Aufnahme von Ionen bei verschiedenen Pflanzengeweben. Z. Pflanzenphysiol. **67**, 359–366 (1972).

Lüttge, U., Pallaghy, C.K., Osmond, C.B.: Coupling of ion transport in green cells of *Atriplex spongiosa* leaves to energy sources in the light and in the dark. J. Membrane Biol. **2,** 17–30 (1970).

Lüttge, U., Pitman, M.G.: Transport and energy. Enc. Plant Physiol. **2A,** 251–259 (1976).

Lüttge, U., Pitman, M.G., Eds.: Transport in plants II. Part A, Cells. vol. 2, Encyclopedia of plant physiology, New Series. New York-Heidelberg-Berlin: Springer (1976).

Lüttge, U., Pitman, M.G., Eds.: Transport in plants II. Part B, Tissues and organs. vol. 2, Encyclopedia of plant physiology, New Series. New York-Heidelberg-Berlin: Springer (1976).

Lüttge, U., Schnepf, E.: Organic substances. Enc. Plant Physiol. **2B,** 244–277 (1976).

Lüttge, U., Schöch, E.V., Ball, E.: Can externally applied ATP supply energy to active ion uptake mechanisms of intact plant cells? Aust. J. Plant Physiol. **1,** 211–220 (1974).

Macallum, A.B.: On the distribution of potassium in animal and vegetable cells. J. Physiol. **32,** 95–118 (1905).

MacDonald, I.R.: Effect of vacuum infiltration on photosynthetic gas exchange in leaf tissue. Plant Physiol. **56,** 109–112 (1975).

MacDonald, I.R., Macklon, A.E.S.: Light-enhanced chloride uptake by wheat laminae. A comparison of chopped and vacuum-infiltrated tissue. Plant Physiol. **56,** 105–108 (1975).

MacDonald, I.R., Macklon, A.E.S., MacLeod, R.W.G.: Energy supply and light-enhanced chloride uptake in wheat laminae. Plant Physiol. **56,** 699–702 (1975).

Macklon, A.E.S.: Cortical cell fluxes and transport to the stele in excised root segments of *Allium cepa* L. I. Potassium, sodium and chloride. Planta **122,** 109–130 (1975a).

Macklon, A.E.S.: Cortical cell fluxes and transport to the stele in excised root segments of *Allium cepa* L. II. Calcium. Planta **122,** 131–141 (1975b).

Macklon, A.E.S., Higinbotham, N.: Potassium and nitrate uptake and cell trans-membrane electro-potential in excised pea epicotyls. Plant Physiol. **43,** 888–892 (1968).

Macklon, A.E.S., Sim, A.: Cortical cell fluxes and transport to the stele in excised root segments of *Allium cepa* L. III. Magnesium. Planta **128,** 5–9 (1976).

Macnicol, P.K.: Rapid metabolic changes in the wound response of leaf discs following excision. Plant Physiol. **57,** 80–84 (1976).

MacRobbie, E.A.C.: Factors affecting the fluxes of potassium and chloride ions in *Nitella translucens*. J. Gen. Physiol. **47,** 859–877 (1964).

MacRobbie, E.A.C.: The nature of the coupling between light energy and active ion transport in *Nitella translucens*. Biochim. Biophys. Acta **94,** 64–73 (1965).

MacRobbie, E.A.C.: Metabolic effects on ion fluxes in *Nitella translucens*. I. Active influxes. Aust. J. Biol. Sci. **19,** 363–370 (1966).

MacRobbie, E.A.C.: Ion fluxes to the vacuole of *Nitella translucens*. J. Exp. Bot. **20,** 236–256 (1969).

MacRobbie, E.A.C.: The active transport of ions in plant cells. Quart. Rev. Biophys. **3,** 251–294 (1970a).

MacRobbie, E.A.C.: Quantized fluxes of chloride to the vacuole of *Nitella translucens*. J. Exp. Bot. **21,** 335–344 (1970b).

MacRobbie, E.A.C.: Phloem translocation. Facts and mechanisms: a comparative survey. Biol. Rev. **46**, 429–481 (1971a).

MacRobbie, E.A.C.: Fluxes and compartmentation in plant cells. Ann. Rev. Plant Physiol. **22**, 75–96 (1971b).

MacRobbie, E.A.C.: Vacuolar fluxes of chloride and bromide in *Nitella translucens*. J. Exp. Bot. **22**, 487–502 (1971c).

MacRobbie, E.A.C.: Intracellular kinetics of tracer chloride and bromide in *Nitella translucens*. J. Exp. Bot. **26**, 489–507 (1975).

MacRobbie, E.A.C., Dainty, J.: Ion transport in *Nitellopsis obtusa*. J. Gen. Physiol. **42**, 335–353 (1958).

Magnuson, J.A., Magnuson, N.S., Hendrix, D.L., Higinbotham, N.: Nuclear magnetic resonance studies of sodium and potassium in etiolated pea stem. Biophys. J. **3**, 763–771 (1973).

Malek, F., Baker, D.A.: Proton co-transport of sugars in phloem loading. Planta **135**, 297–299 (1977).

Malone, C.P., Burke, J.J., Hanson, J.B.: Histochemical evidence for the occurrence of oligomycin-sensitive ATPase in corn roots. Plant Physiol. **60**, 916–922 (1977).

Malone, C., Koeppe, D.E., Miller, R.J.: Corn mitochondrial swelling and contraction—an alternate interpretation. Plant Physiol. **53**, 918–927 (1974).

Marchant, H.J.: Plasmodesmata in algae and fungi, pp. 59–78. In: Intercellular communication in plants: studies on plasmodesmata. Gunning, B.E.S., Robards, A.W., (eds.), Berlin-Heidelberg-New York: Springer (1976).

Marinos, N.G.: Studies on submicroscopic aspects of mineral deficiencies. I. Calcium deficiency in the shoot apex of barley. Am. J. Bot. **49**, 834–841 (1962).

Marmé, D.: Phytochrome: membranes as possible sites of primary action. Ann. Rev. Plant Physiol. **28**, 173–198 (1977).

Marmé, D., Boisard, J., Briggs, W.R.: Binding properties *in vitro* of phytochrome to a membrane fraction. Proc. Natl. Acad. Sci. U.S. **70**, 3861–3865 (1973).

Marmé, D., Mackenzie, J.M., Jr., Boisard, J., Briggs, W.R.: The isolation and partial characterization of a membrane fraction containing phytochrome. Plant Physiol. **54**, 263–271 (1974).

Marmé, D., Schäfer, E.: On the localization and orientation of phytochrome molecules in corn coleoptiles (*Zea mays* L.). Z. Pflanzenphysiol. **67**, 192–194 (1972).

Marrè, E.: Effects of fusicoccin and hormones on plant cell membrane activities. Observations and hypotheses. In: Regulation of cell membrane activities in plants, Marrè, E., Ciferri, O., (eds.), pp. 185–202. Amsterdam: Elsevier/North-Holland Biomedical Press (1977).

Marrè, E., Ciferri, O., Eds.: Regulation of Cell Membrane Activities in Plants, (Proc. Int. Workshop, Pallanza, Italy, 26–29 Aug. 1976). North/Holland Publ. Co., Amsterdam-Oxford-New York (1977).

Marrè, E., Colombo, R., Lado, P., Rasi-Caldogno, F.: Correlation between proton extrusion and stimulation of cell enlargement. Effects of fusicoccin and of cytokinins on leaf fragments and isolated cotyledons. Plant Sci. Lett. **2**, 139–150 (1974).

Marrè, E., Lado, P., Ferroni, A., Ballarin Denti, A.: Transmembrane potential increase induced by auxin, benzyladenine and fusicoccin. Correlation with proton extrusion and cell enlargement. Plant Sci. Lett. **2**, 257–265 (1974a).

Marrè, E., Lado, P., Rasi-Caldogno, F., Colombo, R., Michelis, de, M.I.: Evidence for the coupling of proton extrusion to K^+ uptake in pea internode segments treated with fusicoccin or auxin. Plant Sci. Lett. **3**, 365–379 (1974b).

Marrè, E., Lado, P., Rasi-Caldogno, F., Colombo, R., Cocucci, M., Michelis, de, M.I.: Regulation of proton extrusion by plant hormones and cell elongation. Physiol. Vég. **13**, 797–811 (1975).

Marschner, H., Günther, I.: Ionenaufnahme und Zellstruktur bei Gerstenwurzeln in Abhängigkeit von der Calcium-Versorgung. Ztschr. Pflanzenern., Düngung, Bodenkunde. **107**, 118–136 (1964).

Marschner, H., Schimansky, C.: Suitability of using rubidium-86 as a tracer for potassium in studying potassium uptake by barley plants. Z. für Pflanzenernährung u. Bodenk. **128**, 129–143 (1971).

Matile, P.: Enzyme der Vakuolen aus Wurzelzellen von Maiskeimlingen. Ein Beitrag zur funktionellen Bedeutung der Vakuole bei der intrazellulären Verdauung. Z. Naturforsch. **21**, 871–878 (1966).

Matile, P.: Lysosomes of root tip cells in corn seedlings. Planta **79**, 181–196 (1968).

Matile, P., Moor, H.: Vacuolation: origin and development of the lysosomal apparatus in root-tip cells. Planta **80**, 159–175 (1968).

Matile, P., Wiemken, A.: Interactions between cytoplasm and vacuole. Enc. Plant Physiol. **3**, 255–287 (1976).

Mayer, W.-E.: Kalium- und Chloridverteilung im Laminargelenk von *Phaseolus coccineus* L. während der circadianen Blattbewegung im tagesperiodischen Licht-Dunkelwechsel. Z. Pflanzenphysiol. **83**, 127–135 (1977).

Mayne, B.C., Dee, A.M., Edwards, G.E.: Photosynthesis in mesophyll protoplasts and bundle sheath cells of various types of C-4 plants III. Fluorescence emission spectra, delayed light emission, and P700 content. Z. Pflanzenphysiol. **74**, 275–291 (1974).

McCarty, R.E.: Ion transport and energy conservation in chloroplasts. Enc. Plant Physiol. **3**, 347–376 (1976).

McFarlane, J.C., Berry, W.L.: Cation penetration through isolated leaf cuticles. Plant Physiol. **53**, 723–727 (1974).

McGivan, J.D., Klingenberg, M.: Correlation between H^+ and anion movement in mitochondria and the key role of the phosphate carrier. Eur. J. Biochem. **20**, 392–399 (1971).

McLaren, J.S., Barber, D.J.: Evidence for carrier-mediated transport of L-leucine into isolated pea (*Pisum sativum* L.) chloroplasts. Planta **136**, 147–151 (1977).

Meidner, H., Edwards, M.: Direct measurements of turgor pressure potentials of guard cells. I. J. Exp. Bot. **26**, 319–330 (1975).

Mertz, S.M., Jr.: Electrical potentials and kinetics of potassium, sodium, and chloride uptake in barley roots. Doctoral thesis, Washington State University, Pullman (1973).

Mertz, S.M. Jr., Higinbotham, N.: The cellular electropotential isotherm as related to the kinetic K^+ absorption isotherm in low-salt barley roots. pp. 343–346. In: Membrane transport in plants, Zimmermann, U., Dainty, J., (eds.), Berlin-Heidelberg-New York: Springer (1974).

Mertz, S.M., Jr., Higinbotham, N.: Transmembrane electropotential in barley roots as related to cell type, cell location, and cutting and aging effects. Plant Physiol. **57**, 123–128 (1976).

Michaelis, L., Menten, M.L.: Die Kinetik der Invertasewirkung. Biochem. Z. **49**, 333–369 (1913).

Michel, J.P., Thibault, P.: Étude cinétique de la synthèse d'ATP in vivo en lumière mono-et bichromatique chez *Zea mays:* Effet antagoniste rouge-rouge lointain. Biochem. Biophys. Acta **305**, 390–396 (1973).

Migliaccio, F., Weigl, J.: Akkumulation, Transport und Efflux von Cl^-, SO_4^{--} und K^+ in abgeschnittenen Maiswurzeln. Z. Pflanzenphysiol. **69**, 318–328 (1973).

Milborrow, B.V., Noddle, R.C.: Conversion of 5-(1,2-Epoxy-2,6,6-trimethylcyclohexyl)-3-methylpenta-cis-2-trans-4-dienoic-acid into abscisic acid in plants. Biochem. J. **119**, 727–734 (1970).

Miller, E.C.: Plant physiology. New York: McGraw-Hill (1938).

Mitchell, P.: Coupling of phosphorylation to electron and hydrogen transfer by a chemiosmotic type of mechanism. Nature (London) **191**, 144–148 (1961).

Mitchell, P.: Molecules, group and electron translocation through natural membranes. Biochem. Soc. Symp. (Great Britain) **22**, 142–169 (1962).

Mizraki, Y., Blumenfeld, A., Richmond, A.E.: Abscisic acid and transpiration in leaves in relation to osmotic root stress. Plant Physiol. **46**, 169–171 (1970).

Mohr, H.: Lehrbuch der Pflanzenphysiologie. Berlin-Heidelberg-New York: Springer (1969).

Mollenhauer, H.H., Morré, D.J.: Golgi apparatus and plant secretion. Ann. Rev. Plant Physiol. **17**, 27–46 (1966).

Moore, W.J.: Physical Chemistry. London: Longmans (1957).

Morath, M., Hertel, R.: Lateral electrical potential following asymmetric auxin application to maize coleoptiles. Planta **140**, 31–35 (1978).

Morré, D.J., Bracker, C.E.: Ultrastructural alteration of plant plasma membranes induced by auxin and calcium ions. Plant Physiol. **58**, 544–547 (1976).

Morré, D.J., Mollenhauer, H.H.: Interactions among cytoplasm, endomembranes, and the cell surface. Enc. Plant Physiol. **3**, 288–346 (1976).

Most, B.H.: Abscisic acid in immature apical tissue of sugar cane and in leaves of plants subjected to drought. Planta **101**, 67–75 (1971).

Muir, R.M., Fujita, T., Hansch, C.: Structure-activity relationship in the auxin activity of mono-substituted phenylacetic acids. Plant Physiol. **42**, 1519–1526 (1967).

Münch, E.: Die Stoffbewegungen in der Pflanze. Jena: Gustav Fischer (1930).

Murakami, S., Packer, L.: Light-induced changes in the conformation and configuration of the thylakoid membrane of *Ulva* and *Porphyra* chloroplasts in vivo. Plant Physiol. **45**, 289–299 (1970).

Nagahashi, G., Thomson, W.W., Leonard, R.T.: The casparian strip as a barrier to the movement of lanthanum in corn roots. Science **183**, 670–671 (1974).

Nastuk, W.L., Ed.: Physical techniques in biological research. vol. VI, Electrophysiological methods, Part A. (1963).

Néel, J.: Les membranes artificielles. La Recherche **5**, 33–43 (1974).

Nelles, A.: Einfluss von Indolylessigsäure (IES) auf das Membranpotential in Maiskoleoptilen (*Zea mays* L.). Biochem. Pflanzenphysiol. **167**, 182–184 (1975a).

Nelles, A.: Das Membranpotential von Zellen der Maiskoleoptile unter dem Einfluss von Kalium- und Kalzium-Ionen I. Empirische Gleichungen. Biochem. Pflanzenphysiol. **167**, 541–552 (1975b).

Nelles, A.: Das Membranpotential von Zellen der Maiskoleoptile unter dem

Einfluss von Kalium- und Kalzium-Ionen II. Wirkungen von Gibberellinsäure und Indolessigsäure. Biochem. Pflanzenphysiol. **169**, 385–391 (1976a).

Nelles, A.: Das Membranpotential von Zellen der Maiskoleoptile unter dem Einfluss von Kalium- und Kalzium-Ionen III. Experimentelle Prüfung der Wirkung der Kalziumionen. Biochem. Pflanzenphysiol. **169**, 501–506 (1976b).

Nelles, A.: Das Membranpotential von Zellen der Maiskoleoptile unter dem Einfluss von Kalium- und Kalzium-Ionen IV. Experimentelle Prüfung der Wirkung von Kaliumionen. Biochem. Pflanzenphysiol. **170**, 542–544 (1976c).

Nelles, A., Müller, E.: Die Wirkung von Kaliumionen auf das Zellpotential von Maiskoleoptilen und ihre Beeinflussung durch Indolylessigsäure (IES). Biochem. Pflanzenphysiol. **167**, 185–187 (1975a).

Nelles, A., Müller, E.: Ioneninduzierte Depolarisation der Zellen von Maiskoleoptilen unter dem Einfluss von Kalzium and β-Indolylessigsäure (IES). Biochem. Pflanzenphysiol. **167**, 253–260 (1975b).

Nelson, S.D., Mayo, J.M.: The occurrence of functional non-chlorophyllous guard cells in *Paphiopedilum* spp. Can. J. Bot. **53**, 1–7 (1975).

Nelson, S.D., Mayo, J.M.: Low K in *Paphiopedilum leeanum* leaf epidermis: implication for stomatal functioning. Can. J. Bot. **55**, 489–495 (1977).

Nemček, O., Sigler, K., Kleinzeller, A.: Ion transport in the pitcher of *Nepenthes henryana*. Biochim. Biophys. Acta **126**, 73–80 (1966).

Netter, H.: Theoretische Biochemie. Berlin-Göttingen-Heidelberg: Springer (1959).

Netter, H.: Mögliche Mechanismen und Modelle für aktive Transportvorgänge, pp. 15–44. In: Biochemie des Aktiven Transports, Berlin-Göttingen-Heidelberg: Springer (1961).

Neumann, D., Janossy, A.G.S.: Effect of gibberellic acid on the ion ratios in a dwarf maize mutant (*Zea mays* L. d_1): An electron microprobe study. Planta **134**, 151–153 (1977).

Neumann, J., Levine, R.P.: Reversible pH changes in cells of *Chlamydomonas reinhardii* resulting from CO_2 fixation in the light and its evolution in the dark. Plant Physiol. **47**, 700–704 (1971).

Newman, I.A., Sullivan, J.K.: Auxin transport in oats: a model for the electric changes. In: Transport and transfer processes in plants, Wardlaw, I.F., Passioura, J.B., (eds.), New York-San Francisco-London: Academic Press (1976).

Neyra, C.A., Hageman, R.H.: Nitrate uptake and induction of nitrate reductase in excised corn roots. Plant Physiol. **56**, 692–695 (1975).

Neyra, C.A., Hageman, R.H.: Relationships between carbon dioxide, malate, and nitrate accumulation and reduction in corn (*Zea mays* L.) seedlings. Plant Physiol. **58**, 726–730 (1976).

Nikaido, H.: Phospholipid as a possible component of carrier system in β-galactoside permease of *Escherichia coli*. Biochem. Biophys. Res. Commun. **9**, 486–492 (1962).

Nissen, P.: Choline sulfate permease: transfer of information from bacteria to higher plants? Biochem. Biophys. Res. Commun. **32**, 696–703 (1968).

Nissen, P.: Choline sulfate permease: transfer of information from bacteria to higher plants? II. Induction processes. In: Informative molecules in biological systems, Ledoux, L.G.H., (ed.), pp. 201–212. Amsterdam: North-Holland Publ. Co. (1971a).

Nissen, P.: Uptake of sulfate by roots and leaf slices of barley: mediated by single, multiphasic mechanisms. Plant Physiol. **24**, 315–324 (1971b).

Nissen, P.: Kinetics of ion uptake in higher plants. Plant Physiol. **28**, 113–120 (1973a).

Nissen, P.: Multiphasic uptake in plants II. Mineral cations, chloride, and boric acid. Plant Physiol. **29**, 298–354 (1973b).

Nissen, P.: Multiphasic uptake in plants I. Phosphate and sulfate. Plant Physiol. **28**, 304–316 (1973c).

Nissen, P.: Bacteria-mediated uptake of choline sulfate by plants. Sci. Rept. Agr. Univ. Norway **52**, 1–53 (1973d).

Nissen, P.: Uptake mechanisms: inorganic and organic. Ann. Rev. Plant Physiol. **25**, 53–79 (1974).

Nissl, D., Zenk, M.H.: Evidence against induction of protein synthesis during auxin-induced initial elongation of *Avena* coleoptiles. Planta **89**, 323–341 (1969).

Nobel, P.S.: Light-induced changes in the ionic content of chloroplasts in *Pisum sativum*. Biochim. Biophys. Acta **172**, 134–143 (1969).

Nobel, P.S.: Introduction to biophysical plant physiology. San Francisco: W.H. Freeman and Co. (1974).

Nobel, P.S.: Chloroplasts, pp. 101–124. In: Ion transport in plant cells and tissues, Baker, D.A., Hall, J.L., (eds.), Amsterdam-Oxford-New York: North Holland/Elsevier (1975).

Nobel, P.S., Wang, C.T.: Biochim. Biophys. Acta **211**, 79 ff. (1970).

Noeske, O., Läuchli, A., Lange, O.L., Vieweg, G.H., Ziegler, H.: Konzentration und Lokalisierung von Schwermetallen in Flechten der Erzschlackenhalden des Harzes. Vorträge Gesamtgebiet der Botanik. Dtsch. Bot. Ges., N.F. **4**, 67–79 Stuttgart: G. Fischer (1970).

Northcote, D.H.: Chemistry of the plant cell wall. Ann. Rev. Plant Physiol. **23**, 113–132 (1972).

Novacky, A., Fischer, E., Ullrich-Eberius, C.I., Lüttge, U., Ullrich, W.R.: Membrane potential changes during transport of glycine as a neutral amino acid and nitrate in *Lemna gibba* G1. FEBS Lett. **88**, 264–267 (1978a).

Novacky, A., Ullrich-Eberius, C.I., Lüttge, U.: Membrane potential changes during transport of hexoses in *Lemna gibba* G1. Planta **138**, 263–270 (1978b).

Nuccitelli, R., Jaffe, L.F.: Current pulses involving chloride and potassium efflux relieve excess pressure in *Pelvetia* embryos. Planta **131**, 315–320 (1976).

O'Brien, T.P., Carr, D.J.: A suberized layer in the cell walls of the bundle sheath of grasses. Aust. J. Biol. Sci. **23**, 275–287 (1970).

Oda, K., Linstead, P.J.: Changes in cell length during action potentials in *Chara*. J. Exp. Bot. **26**, 228–239 (1975).

Oesterhelt, D., Gottschlich, R., Hartmann, R., Michel, H., Wagner, G.: Light energy conversion in halobacteria. Symp. Soc. General Microbiology **27**, 333–349 (1977).

Oesterhelt, D., Stoeckenius, W.: Functions of a new photoreceptor membrane. Proc. Natl. Acad. Sci. U.S. **70**, 2853–2857 (1973).

O'Kelly, J.C.: Inorganic nutrients. In: Algal physiology and biochemistry, Stewart, W.D.P., (ed.), pp. 610–635. Oxford: Blackwell Scientific Publications (1974).

Olesen, P.: Plasmodesmata between mesophyll and bundle sheath cells in relation to the exchange of C_4-acids. Planta **123**, 199–202 (1975).

Oparin, A.I.: Origin and evolution of metabolism. In: Evolutionary biochemistry, Oparin, A.I. (ed.) Oxford-London-New York-Paris: Pergamon Press (1963a).

Oparin, A.I.: Das Leben. Seine Natur, Herkunft und Entwicklung. Stuttgart: Fischer (1963b).

Osmond, C.B.: Ion absorption in *Atriplex* leaf tissue I. Absorption by mesophyll cells. Aust. J. Biol. Sci. **21**, 1119–1130 (1968).

Osmond, C.B.: Metabolite transport in C_4 photosynthesis. Aust. J. Biol. Sci. **24**, 159–163 (1971).

Osmond, C.B.: Ion absorption and carbon metabolism in cells of higher plants. Enc. Plant Physiol. **2A**, 347–372 (1976).

Osmond, C.B., Laties, G.G.: Compartmentation of malate in relation to ion absorption in beet. Plant Physiol. **44**, 7–14 (1969).

Osmond, C.B., Laties, G.G.: Effect of poly-L-lysine on potassium fluxes in red beet tissue. J. Membrane Biol. **2**, 85–94 (1970).

Osmond, C.B., Lüttge, U., West, K.R., Pallaghy, C.K., Schachar-Hill, B.: Ion absorption in *Atriplex* leaf tissue. II. Secretion of ions to epidermal bladders. Aust. J. Biol. Sci. **22**, 797–814 (1969).

Osmond, C.B., Smith, F.A.: Symplastic transport of metabolites during C_4-photosynthesis, pp. 229–240. In: Intercellular communication in plants: studies on plasmodesmata, Gunning, B.E.S., Robards, A.W., (eds.), Berlin-Heidelberg-New York: Springer (1976).

Outlaw, W.H., Fisher, D.B.: Compartmentation in *Vicia faba* leaves I. Kinetics of ^{14}C in the tissues following pulse labelling. Plant Physiol. **55**, 699–703 (1975a).

Outlaw, W.H., Fisher, D.B.: Compartmentation in *Vicia faba* leaves III. Photosynthesis in the spongy and palisade parenchyma. Aust. J. Plant Physiol. **2**, 435–439 (1975b).

Outlaw, W.H., Fisher, D.B., Christy, A.L.: Compartmentation in *Vicia faba* leaves II. Kinetics of ^{14}C-sucrose redistribution among individual tissues following pulse labelling. Plant Physiol. **55**, 704–711 (1975).

Overton, E.: Über die allgemeinen osmotischen Eigenschaften der Zelle, ihre vermutlichen Ursachen und ihre Bedeutung für die Physiologie. Vierteljahrschr. Naturforsch. Ges. Zurich **44**, 88–135 (1899).

Packer, L., Murakami, S., Mehard, C.W.: Ion transport in chloroplasts and plant mitochondria. Ann. Rev. Plant Physiol. **21**, 271–304 (1970).

Pallaghy, C.K.: The effect of Ca^{++} on the ion specificity of stomatal opening in epidermal strips of *Vicia faba*. Z. Pflanzenphysiol. **62**, 58–62 (1970).

Pallaghy, C.K., Fischer, R.A.: Metabolic aspects of stomatal opening and ion accumulation by guard cells in *Vicia faba*. Z. Pflanzenphysiol. **71**, 332–344 (1974).

Pallaghy, C.K., Lüttge, U.: Light-induced H^+-ion fluxes and bioelectric phenomena in mesophyll cells of *Atriplex spongiosa*. Z. Pflanzenphysiol. **62**, 417–425 (1970).

Pallas, J.E., Mollenhauer, H.H.: Physiological implications of *Vicia faba* and *Nicotiana tabacum* guard-cell ultrastructure. Am. J. Bot. **59**, 504–514 (1972).

Palmer, L.G., Gulati, J.: Potassium accumulation in muscle: A test of the binding hypothesis. Science **194**, 521–523 (1976).

Pardee, A.B.: Crystallization of a sulfate-binding protein (permease) from *Salmonella typhimurium*. Science **156**, 1627–1628 (1967).

Pardee, A.B.: Membrane transport proteins. Science **162**, 632–637 (1968).

Pardee, A.B., Prestidge, L.S.: Cell-free activity of a sulfate binding site involved in active transport. Proc. Natl. Acad. Sci. U.S. **55**, 189–191 (1966).

Parker, B.C.: Translocation in *Macrocystis* III. Composition of sieve-tube exudate and identification of the major C^{14}-labeled products. J. Physiol. (London) **2**, 38–41 (1966).

Passow, H.: Passive Permeabilität von Zellmembranen. Zur Frage der Penetration durch Poren. Verh. Ges. Dtsch. Naturforsch. u. Ärzte. Berlin-Göttingen-Heidelberg: Springer (1963).

Paszewski, A., Zawadzki, T., Dziubińska, H.: Higher plant biopotentials and the integration of biological sciences. Folia Soc. Scientiarum Lubliensis **19**, 95–116 (1977).

Pate, J.S.: Exchange of solutes between phloem and xylem and circulation in the whole plant. Enc. Plant Physiol. **1**, 451–473 (1975).

Pate, J.S.: Transport in symbiotic systems fixing nitrogen. Enc. Plant Physiol. **2B**, 278–306 (1976).

Pate, J.S., Gunning, B.E.S.: Transfer cells. Ann. Rev. Plant Physiol. **23**, 173–196 (1972).

Pauling, L.: Nature of chemical bond. Ithaca, New York: Cornell University Press (1960).

Pearson, C.J.: Daily changes in stomatal aperture and in carbohydrates and malate within epidermis and mesophyll of leaves of *Commelina cyanea* and *Vicia faba*. Aust. J. Biol. Sci. **26**, 1035–1044 (1973).

Pearson, C.J.: Fluxes of potassium and changes in malate within epidermis of *Commelina cyanea* and their relationships with stomatal aperture. Aust. J. Plant Physiol. **2**, 85–89 (1975).

Pearson, C.J., Milthorpe, F.L.: Structure, carbon dioxide fixation and metabolism of stomata. Aust. J. Plant Physiol. **1**, 221–236 (1974).

Pearson, C.J., Steer, B.T.: Daily changes in nitrate uptake and metabolism in *Capsicum annuum*. Planta **137**, 107–112 (1977).

Penel, C., Greppin, H.: Photocontrôle immédiat de l'activité peroxydasique et corrélation entre les feuilles de l'epinard. Saussurea **6**, 287–291 (1975).

Penny, M.G., Bowling, D.J.F.: A study of potassium gradients in the epidermis of intact leaves of *Commelina communis* in relation to stomatal opening. Planta **119**, 17–25 (1974).

Penny, M.G., Bowling, D.J.F.: Direct determination of pH in the stomatal complex of *Commelina*. Planta **122**, 209–212 (1975).

Penny, P., Dunlop, J., Perley, J.E., Penny, D.: pH and auxin-induced growth: a causal relationship? Plant Sci. Lett. **4**, 35–40 (1975).

Penny, M.G., Kelday, L.S., Bowling, D.J.F.: Active chloride transport in the leaf epidermis of *Commelina communis* in relation to stomatal activity. Planta **130**, 291–294 (1976).

Penth, B., Weigl, J.: Unterschiedliche Wirkung von Licht auf die Aufnahme von Chlorid und Sulfat in *Limnophilia*blätter. Abhängigkeit der Lichtwirkung von der Konzentration der Anionen. Z. Naturforsch. **24b**, 342–348 (1969).

Penth, B., Weigl, J.: Anionen-Influx, ATP-Spiegel und CO_2-Fixierung in *Limnophila gratioloides* und *Chara foetida*. Planta **96**, 212–223 (1971).

Perrin, A.: Contribution à l'étude de l'organisation et du fonctionnement des hydathodes: Recherches anatomiques, ultrastructurales et physiologiques. Thesis, Lyon (1972).

Pfeffer, W.: Osmotische Untersuchung. Studien zur Zellmechanik. Leipzig (1877).

Pfeffer, W.: The physiology of plants, vol. III, 2nd ed., transl. by Ewart, A.J. London: Oxford (1906).

Pickard, B.G.: Action potentials in higher plants. Bot. Rev. **39**, 172–201 (1973).

Pierce, W.S., Higinbotham, N.: Compartments and fluxes of K^+, Na^+ and Cl^- in *Avena* coleoptile cells. Plant Physiol. **46**, 666–673 (1970).

Pike, C.S.: Lack of influence of phytochrome on membrane permeability to tritiated water. Plant Physiol. **57**, 185–187 (1976).

Pike, C.S., Richardson, A.E.: Phytochrome-controlled hydrogen ion excretion by *Avena* coleoptiles. Plant Physiol. **59**, 615–617 (1977).

Pitman, M.G.: The determination of the salt relations of the cytoplasmic phase in cells of beetroot tissue. Aust. J. Biol. Sci. **16**, 647–668 (1963).

Pitman, M.G.: Ion exchange and diffusion in roots of *Hordeum vulgare*. Aust. J. Biol. Sci. **18**, 541–546 (1965).

Pitman, M.G.: Conflicting measurements of sodium and potassium uptake by barley roots. Nature (London) **216**, 1343–1344 (1967).

Pitman, M.G.: Simulation of Cl^- uptake by low-salt barley roots as a test of models of salt uptake. Plant Physiol. **44**, 1417–1427 (1969).

Pitman, M.G.: Active H^+ efflux from cells of low-salt barley roots during salt accumulation. Plant Physiol. **45**, 787–790 (1970).

Pitman, M.G.: Uptake and transport of ions in barley seedlings I. Estimation of chloride fluxes in cells of excised roots. Aust. J. Biol. Sci. **24**, 407–421 (1971).

Pitman, M.G.: Uptake and transport of ions in barley seedlings II. Evidence for two active stages in transport to the shoot. Aust. J. Biol. Sci. **25**, 243–257 (1972a).

Pitman, M.G.: Uptake and transport of ions in barley seedlings III. Correlation of potassium transport to the shoot with plant growth. Aust. J. Biol. Sci. **25**, 905–919 (1972b).

Pitman, M.G.: Whole plants. In: Ion transport in plant cells and tissues, Baker, D.A., Hall, J.L., (eds.), pp. 267–308. Amsterdam-Oxford-New York: North-Holland Publishing Company/American Elsevier (1975).

Pitman, M.G.: Ion transport in the xylem. Ann. Rev. Plant Physiol. **28**, 71–88 (1977).

Pitman, M.G., Courtice, A.C., Lee, B.: Comparison of potassium and sodium uptake by barley roots at high and low salt status. Aust. J. Biol. Sci. **21**, 871–881 (1968).

Pitman, M.G., Cram, W.J.: Regulation of ion content in whole plants, pp. 391–424. In: Integration of activity in the higher plant, Jennings, D.H., (ed.), Soc. Exp. Biol. Cambridge: University Press (1977).

Pitman, M.G., Lüttge, U., Kramer, D., Ball, E.: Free space characteristics of barley leaf slices. Aust. J. Plant Physiol. **1**, 65–75 (1974a).

Pitman, M.G., Lüttge, U., Läuchli, A., Ball, E.: Ion uptake to slices of barley leaves, and regulation of K content in cells of the leaves. Z. Pflanzenphysiol. **72**, 75–88 (1974b).

Pitman, M.G., Lüttge, U., Läuchli, A., Ball, E.: Action of abscisic acid on ion transport as affected by root temperature and nutrient status. J. Exp. Bot. **25**, 147–155 (1974c).

Pitman, M.G., Mertz, S.M. Jr., Graves, J.S., Pierce, W.S., Higinbotham, N.:

Electrical potential differences in cells of barley roots and their relation to ion uptake. Plant Physiol. **47**, 76–80 (1970).

Pitman, M.G., Mowat, J., Nair, H.: Interaction of processes for accumulation of salt and sugar in barley plants. Aust. J. Biol. Sci. **24**, 619–631 (1971).

Pitman, M.G., Saddler, H.D.W.: Active sodium and potassium transport in cells of barley roots. Proc. Natl. Acad. Sci. U.S. **57**, 44–49 (1967).

Pitman, M.G., Schaefer, N., Wildes, R.A.: Stimulation of H^+ efflux and cation uptake by fusicoccin in barley roots. Plant Sci. Lett. **4**, 323–329 (1975a).

Pitman, M.G., Schaefer, N., Wildes, R.A.: Relation between permeability to potassium and sodium ions and fusicoccin-stimulated hydrogen-ion efflux in barley roots. Planta **126**, 61–73 (1975b).

Pitman, M.G., Wildes, R.A., Schaefer, N., Wellfare, D.: Effect of azetidine 2-carboxylic acid on ion uptake and ion release to the xylem of excised barley roots. Plant Physiol. **60**, 240–246 (1977).

Polya, G.M., Atkinson, M.R.: Evidence for a direct involvement of electron transport in the high affinity ion accumulation system of aged beet parenchyma. Aust. J. Biol. Sci. **22**, 573–584 (1969).

Polya, G.M., Osmond, C.B.: Photophosphorylation by mesophyll and bundle sheath chloroplasts by C_4 plants. Plant Physiol. **49**, 267–269 (1972).

Poole, R.J.: Effect of sodium on potassium fluxes at the cell membrane and vacuole membrane of red beet. Plant Physiol. **47**, 731–734 (1971a).

Poole, R.J.: Development and characteristics of sodium selective transport in red beet. Plant Physiol. **47**, 735–739 (1971b).

Pope, D.G.: Does indoleacetic acid promote growth via cell wall acidification? Planta **140**, 137–142 (1978).

Prins, H.B.A.: The action spectrum of photosynthesis and the rubidium chloride uptake by leaves of *Vallisneria spiralis*. Kon. Ned. Akad. Wetensch. Amsterdam Ser. C, **76**, 495–499 (1973).

Prosser, C.L.: Excitable membranes. In: Comparative animal physiology, Prosser, C.L., (ed.), Philadelphia-London-Toronto: W.B. Saunders Co. (1973).

Quail, P.H.: Particle-bound phytochrome: spectral properties of bound and unbound fractions. Planta **118**, 345–355 (1974a).

Quail, P.H.: *In-vitro* binding of phytochrome to a particulate fraction: a function of light dose and steady-state P_{fr} level. Planta **118**, 357–360 (1974b).

Quail, P.H.: Particle-bound phytochrome: association with a ribonucleoprotein fraction from *Cucurbita pepo* L. Planta **123**, 223–234 (1975a).

Quail, P.H.: Particle-bound phytochrome: the nature of the interaction between pigment and particulate fractions. Planta, **123**, 235–246 (1975b).

Quail, P.H., Marmé, D., Schäfer, E.: Particle-bound phytochrome from maize and pumpkin. Nature (New Biol) **245**, 189–191 (1973).

Quail, P.H., Schäfer, E.: Particle-bound phytochrome: a function of light dose and steady-state level of the far-red-absorbing form. J. Membrane Biol. **15**, 393–404 (1974).

Rachmilevitz, T., Fahn, A.: The floral nectary of *Tropaeolum majus* L.: the nature of the secretory cells and the manner of nectar secretion. Ann. Bot. **39**, 721–728 (1975).

Racusen, R.H.: Phytochrome control of electrical potentials and intercellular coupling in oat-coleoptile tissue. Planta **132**, 25–29 (1976).

Racusen, R.H., Etherton, B.: Role of membrane-bound, fixed charge changes in phytochrome-mediated mung bean root adherence phenomena. Plant Physiol. **55**, 491–495 (1975).

Racusen, R.H., Galston, A.W.: Electrical evidence for rhythmic changes in the cotransport of sucrose and hydrogen ions in *Samanea* pulvini. Planta **135**, 57–62 (1977).

Racusen, R.H., Kinnersley, A.M., Galston, A.W.: Osmotically induced changes in electrical properties of plant protoplast membranes. Science **198**, 405–407 (1977).

Racusen, R.. Miller. K.: Phytochrome-induced adhesion of mung bean root tips to platinum electrodes in a direct current field. Plant Physiol. **49**, 654–655 (1972).

Racusen, R.H.. Satter, R.L.: Rhythmic and phytochrome-regulated changes in transmembrane potential in *Samanea* pulvini. Nature (London) **255**, 408–410 (1975).

Radmer, R.J.. Kok. B.: Photoreduction of O_2 primes and replaces CO_2 assimilation. Plant Physiol. **58**, 336–340 (1976).

Rains, D.W., Epstein, E.: Sodium absorption by barley roots: role of the dual mechanisms of alkali cation transport. Plant Physiol. **42**, 314–318 (1967).

Rains, D.W.. Floyd, R.A.: Influence of calcium on sodium and potassium absorption by fresh and aged bean stem slices. Plant Physiol. **46**, 93–98 (1970).

Ramani, S., Kannan, S.: Action of Mn^{++} on the absorption of Na^+, K^+, and Rb^+ by excised rice roots. Z. Pflanzenphysiol. **77**, 107–112 (1976).

Rao, K.P.. Rains, D.W.: Nitrate absorption in barley II. Influence of nitrate reductase activity. Plant Physiol. **57**, 59–62 (1976).

Raschke, K.: Stomatal Action. Ann. Rev. Plant Physiol. **26**, 309–340 (1975).

Raschke, K.. Dittrich, P.: [^{14}C] Carbon-dioxide fixation by isolated leaf epidermis with stomata closed or open. Planta **134**, 69–75 (1977).

Raschke, K., Fellows, M.P.: Stomatal movement in *Zea mays*: Shuttle of potassium and chloride between guard cells and subsidiary cells. Planta **101**, 296–316 (1971).

Raschke, K., Humble, G.D.: No uptake of anions required by opening stomata of *Vicia faba*: Guard cells release hydrogen ions. Planta **115**, 47–57 (1973).

Raven, J.A.: Ion transport in *Hydrodictyon africanum*. J. Gen. Physiol. **50**, 1607–1625 (1967a).

Raven, J.A.: Light stimulation of active transport in *Hydrodictyon africanum*. J. Gen. Physiol. **50**, 1627–1640 (1967b).

Raven, J.A.: The linkage of light-stimulated Cl^- influx to K^+ and Na^+ influxes in *Hydrodictyon africanum*. J. Exp. Bot. **19**, 233–253 (1968a).

Raven, J.A.: The action of phlorizin on photosynthesis and light-stimulated transport in *Hydrodictyon africanum*. J. Exp. Bot. **19**, 712–723 (1968b).

Raven, J.A.: Action spectra for photosynthesis and light-stimulated ion transport processes in *Hydrodictyon africanum*. New Phytol. **68**, 45–62 (1969a).

Raven, J.A.: Cyclic and non-cyclic photophosphorylation as energy sources for active K influx in *Hydrodictyon africanum*. J. Exp. Bot. **22**, 420–433 (1971b).

Raven, J.A.: Active influx of hexose in *Hydrodictyon africanum*. New Phytol. **76**, 189–194 (1976).

Raven, J.A.: ATP synthesis coupled to nitrate photoreduction in the alga *Hydrodictyon africanum*. J. Exp. Bot. **28**, 314–319 (1977a).

Raven, J.A.: H^+ and Ca^{2+} in phloem and symplast: Relation of relative immobility of the ions to the cytoplasmic nature of the transport paths. New Phytol. **79,** 465–480 (1977b).

Raven, J.A., Glidewell, S.M.: Sources of ATP for active phosphate transport in *Hydrodictyon africanum:* evidence for pseudocyclic photophosphorylation in vivo. New Phytol. **75,** 197–204 (1975a).

Raven, J.A., Glidewell, S.M.: Effects of CCCP on photosynthesis and an active and passive chloride transport at the plasmalemma of *Hydrodictyon africanum.* New Phytol. **75,** 205–213 (1975b).

Raven, J.A., MacRobbie, E.A.C., Neumann, J.: The effect of Dio-9 on photosynthesis and ion transport in *Nitella, Tolypella,* and *Hydrodictyon.* J. Exp. Bot. **20,** 221–235 (1969).

Raven, J.A., Smith, F.A.: The regulation of intracellular pH as a fundamental biological process. In: Ion transport in plants, Anderson, W.P., (ed.), pp. 271–278. London: Academic Press (1973).

Raven, J.A., Smith, F.A.: Significance of hydrogen ion transport in plants. Can. J. Bot. **52,** 1035–1048 (1974).

Ray, P.M.: Auxin-binding sites of maize coleoptiles are localized on membranes of the endoplasmic reticulum. Plant Physiol. **59,** 594–599 (1977).

Rayle, D.L.: Auxin-induced hydrogen-ion secretion in *Avena* coleoptiles and its implications. Planta **114,** 63–73 (1973).

Reed, M.L., Findlay, N., Mercer, F.V.: Nectar production in *Abutilon* IV. Water and solute relations. Aust. J. Biol. Sci. **24,** 677–688 (1971).

Robards, A.W.: Electron microscopy and plant ultrastructure. London: McGraw-Hill (1970).

Robards, A.W.: Plasmodesmata in higher plants, pp. 15–53. In: Intercellular communication in plants: studies on plasmodesmata, Gunning, B.E.S., Robards, A.W., (eds.), Berlin-Heidelberg-New York: Springer (1976).

Robards, A.W., Clarkson, D.T.: The role of plasmodesmata in the transport of water and nutrients across roots, pp. 181–199. In: Intercellular communication in plants: studies on plasmodesmata, Gunning, B.E.S., Robards, A.W., (eds.), Berlin-Heidelberg-New York: Springer (1976).

Robards, A.W., Robb, M.E.: The entry of ions and molecules into roots: an investigation using electron-opaque tracers. Planta **120,** 1–12 (1974).

Robertson, J.D.: Unit membranes. A review with recent new studies of experimental alterations and a new subunit structure in synaptic membranes. In: Cellular membranes in development. Locke, M., (ed.), New York-London: Academic Press (1964).

Robertson, R.N.: Ion transport and respiration. Biol. Rev. **35,** 231–264 (1960).

Robertson, R.N.: Protons, electrons, phosphorylation and active transport. Cambridge: Cambridge University Press (1968).

Robinson, J.B., Laties, G.G.: Plasmalemma and tonoplast influxes of potassium in barley roots, and interpretation of the absorption isotherm at high concentrations. Aust. J. Plant Physiol. **2,** 177–184 (1975).

Rogers, H.J., Perkins, H.R.: Cell walls and membranes. London: E. and F. Spon (1968).

Rona, J.P., Cornel, L.D., Heller, R.: Determination and interpretation of the electrical profile of free cells of *Acer pseudoplatanus,* pp. 349–356. In: Transmembrane ionic exchanges in plants, Thellier, M., Monnier, A., DeMarty, M., Dainty, J., (eds.), Paris-Rouen: C.N.R.S. and Univ. Rouen (1977).

Rubinstein, B.: Osmotic shock inhibits auxin-stimulated acidification and growth. Plant Physiol. **59**, 369–371 (1977).

Rubinstein, B., Mahar, P., Tattar, T.A.: Effects of osmotic shock on some membrane-regulated events of oat coleoptile cells. Plant Physiol. **59**, 365–368 (1977).

Ruhland, W.: Studien über die Aufnahme von Kolloiden durch die pflanzliche Plasmahaut. Jb. Wiss. Bot. **51**, 376–431 (1912).

Ruhland, W.: Untersuchungen über die Hautdrüsen der Plumbaginaceen. Ein Beitrag zur Biologie der Halophyten. Jahrb. Wiss. Bot. **55**, 409–498 (1915).

Ruhland, W., Hoffmann, C.: Die Permeabilität von *Beggiatoa mirabilis*. Planta **1**, 1–83 (1925).

Rungie, J.M., Wiskich, J.T.: Salt-stimulated adenosine triphosphatase from smooth microsomes of turnip. Plant Physiol. **51**, 1064–1068 (1973).

Saddler, H.D.W.: The membrane potential of *Acetabularia mediterranea*. J. Gen. Physiol. **55**, 802–821 (1970a).

Saddler, H.D.W.: The ionic relations of *Acetabularia mediterranea*. J. Exp. Bot. **21**, 345–359 (1970b).

Saito, K., Senda, M.: The effect of external pH on the membrane potential of *Nitella* and its linkage to metabolism. Plant Cell Physiol. **14**, 1045–1052 (1973).

Saltman, P., Forte, J.G., Forte, G.M.: Permeability studies on chloroplasts from *Nitella*. Exp. Cell Res. **29**, 504–514 (1963).

Satter, R.L., Applewhite, P.B., Galston, A.W.: Phytochrome-controlled nyctinasty in *Albizzia julibrissin* V. Evidence against acetylcholine participation. Plant Physiol. **50**, 523–525 (1972).

Satter, R.L., Applewhite, P.B., Galston, A.W.: Rhythmic potassium flux in *Albizzia*. Effect of aminophylline, cations, and inhibitors of respiration. Plant Physiol. **54**, 280–285 (1974).

Satter, R.L., Applewhite, P.B., Kreis, D.J., Jr., Galston, A.W.: Rhythmic leaflet movement in *Albizzia julibrissin*. Effect of electrolytes and temperature alteration. Plant Physiol. **52**, 202–207 (1973).

Satter, R.L., Galston, A.W.: Leaf movements: rosetta stone of plant behavior? Bioscience **23**, 407–416 (1973).

Satter, R.L., Geballe, G.T., Applewhite, P.B., Galston, A.W.: Potassium flux and leaf movement in *Samanea saman* I. Rhythmic movement. J. Gen. Physiol. **64**, 413–430 (1974a).

Satter, R.L., Geballe, G.T., Galston, A.W.: Potassium flux and leaf movement in *Samanea saman*. II. Phytochrome controlled movement. J. Gen. Physiol. **64**, 431–442 (1974b).

Satter, R.L., Marinoff, P., Galston, A.W.: Phytochrome controlled nyctinasty in *Albizzia julibrissin*. II. Potassium fluxes as a basis for leaflet movement. Am. J. Bot. **57**, 916–926 (1970).

Satter, R.L., Sabnis, D.D., Galston, A.W.: Phytochrome controlled nyctinasty in *Albizzia julibrissin*. I. Anatomy and fine structure of the pulvinule. Am. J. Bot. **57**, 374–381 (1970).

Satter, R.L., Schrempf, M., Chaudri, J., Galston, A.W.: Phytochrome and circadian clocks in *Samanea*. Rhythmic redistribution of potassium and chloride within the pulvinus during long dark periods. Plant Physiol. **59**, 231–235 (1977).

Sauter, J.J., Iten, W., Zimmermann, M.H.: Studies on the release of sugar into the vessels of sugar maple (*Acer saccharum*). Can. J. Bot. **51**, 1–8 (1973).

Sawhney, B.L., Zelitch, I.: Direct determination of potassium ion accumulation in guard cells in relation to stomatal opening in light. Plant Physiol. **44**, 1350–1354 (1969).

Schaefer, N., Wildes, R.A., Pitman, M.G.: Inhibition by p-fluorophenylalanine of protein synthesis and of ion transport across the roots in barley seedlings. Aust. J. Plant Physiol. **2**, 61–74 (1975).

Schäfer, G., Heber, U., Heldt, H.W.: Glucose transport into spinach chloroplasts. Plant Physiol. **60**, 286–289 (1977).

Schilde, C.: Schnelle photoelektrische Effekte der Alge *Acetabularia*. Z. Naturforsch. **23b**, 1369–1376 (1968).

Schloemer, R.H., Garrett, R.H.: Nitrate transport system in *Neurospora crassa*. J. Bacteriol. **118**, 259–269 (1974).

Schlögl, R.: Stofftransport durch Membranen. Darmstadt: Steinkopff (1964).

Schmitz, K., Lüning. K., Willenbrink, J.: CO_2-Fixierung und Stofftransport in benthischen marinen Algen II. Zum Ferntransport ^{14}C-markierter Assimilate bei *Laminaria hyperborea* und *Laminaria saccharina*. Z. Pflanzenphysiol. **67**, 418–429 (1972).

Schnabl, H., Ziegler, H.: The mechanism of stomatal movement in *Allium cepa* L. Planta **136**, 37–43 (1977).

Schnarrenberger, C., Burkhard. C.: In-vitro interaction between chloroplasts and peroxisomes as controlled by inorganic phosphate. Planta **134**, 109–114 (1977).

Schnepf, E.: Zur Feinstruktur von *Geosiphon pyriforme*. Arch. Mikrobiol. **49**, 112–131 (1964a).

Schnepf, E.: Über Zellwandstrukturen bei den Köpfchendrüsen der Schuppenblätter von *Lathraea clandestina* L. Planta **60**, 473–483 (1964b).

Schnepf, E.: Zur Cytologie und Physiologie pflanzlicher Drüsen 4. Teil. Licht. und elektronenmikroskopische Untersuchungen an Septalnektarien. Protoplasma **58**, 137–171 (1964c).

Schnepf, E.: Organellen-Reduplikation und Zellkompartmentierung. In: Probleme der biologischen Reduplikation. Sitte, P., (ed.), 3. Wiss. Konf. Ges. Dtsch. Naturforsch. u. Ärzte. Berlin-Heidelberg-New York: Springer (1966).

Schnepf, E.: Transport by compartments. In: Int. Symp. Stofftransport und Stoffverteilung in Zellen höherer Pflanzen. Jhrg. 1968. Mothes, K., Müller. E., Nelles, A., Neumann, D., (eds.), Nr. 4a. Abh. Dtsch. Akad. Wiss. Berlin, pp. 39–49. Berlin: Akademie Verlag (1968).

Schnepf, E.: Sekretion und Exkretion bei Pflanzen. Protoplasmatologia. vol. VIII/8. Wien-New York: Springer (1969).

Schnepf, E.: Bau und Feinbau der Nektarien und der Mechanismus der Nektarsekretion. Apidologie **8**, 295–304 (1977).

Schöch, E.V., Lüttge, U.: Zur Entstehung einer lichtabhängigen Komponente der Kationenaufnahme bei Blattgewebestreifen mit zunehmendem Zeitabstand von der Präparation. Biochem. Physiol. Pflanzen **165**, 345–350 (1974).

Scholander, P.F., Hammel, H.T., Hemmingsen, E.A., Bradstreet, E.D.: Hydrostatic pressure and osmotic potential in leaves of mangroves and some other plants. Proc. Natl. Acad. Sci. U.S. **52**, 119–125 (1964).

Schönherr, J.: The nature of the pH effect on water permeability of plant cuticles. Ber. Dtsch. Bot. Ges. **87**, 389–402 (1974).

Schönherr, J.: Water permeability of isolated cuticular membranes: the effect of pH and cations on diffusion, hydrodynamic permeability and size of polar pores in the cutin matrix. Planta **128**, 113–126 (1976a).

Schönherr, J.: Water permeability of isolated cuticular membranes: the effect of cuticular waxes on diffusion of water. Planta **131**, 159–164 (1976b).

Schopfer, P.: Phytochrome control of enzymes. Ann. Rev. Plant Physiol. **28**, 223–252 (1977).

Schötz, F., Diers, L.: Vergrösserung der Kontaktfläche zwischen Chloroplasten und ihrer cytoplasmatischen Umgebung durch tubuläre Ausstülpungen der Plastidhülle. Planta **124**, 277–285 (1975).

Schötz, F., Diers, L., Bathelt,: Zur Feinstruktur der Raphidenzellen I. Die Entwicklung der Vakuolen und der Raphiden. Z. Pflanzenphysiol. **63**, 91–113 (1970).

Schrempf, M., Satter, R.L., Galston, A.W.: Potassium-linked chloride fluxes during rhythmic leaf movement of *Albizzia julibrissin*. Plant Physiol. **58**, 190–192 (1976).

Schröter, K., Läuchli, A., Sievers, A.: Mikroanalytische Identifikation von Bariumsulfat-Kristallen in den Statolithen der Rhizoids von *Chara fragilis* Desv. Planta **122**, 213–225 (1975).

Scott, B.I.H.: Electric fields in plants. Ann. Rev. Plant Physiol. **18**, 409–418 (1967).

Setty, S., Jaffe, M.J.: Phytochrome-controlled rapid contraction and recovery of contractile vacuoles in the motor cells of *Mimosa pudica* as an intracellular correlate of nyctinasty. Planta **108**, 121–131 (1972).

Severne, B.C., Brooks, R.R.: A nickel-accumulating plant from Western Australia. Planta **103**, 91–94 (1972).

Shachar-Hill, B., Hill, A.E.: Ion and water transport in *Limonium* VI. The induction of chloride pumping. Biochim. Biophys. Acta **211**, 313–317 (1970).

Shamoo, A.E., MacLennan, D.H.: A Ca^{++}-dependent and -selective ionophore as part of the Ca^{++}- + Mg^{++}-dependent adenosinetriphosphatase of sarcoplasmic reticulum. Proc. Natl. Acad. Sci. U.S. **71**, 3522–3526 (1974).

Short, S.A., Kaback, H.R., Kohn, L.D.: D-lactate dehydrogenase binding in *Escherichia coli* dld-membrane vesicles reconstituted for active transport. Proc. Natl. Acad. Sci. U.S. **71**, 1461–1465 (1974).

Sibaoka, T.: Physiology of rapid movements in plants. Ann. Rev. Plant Physiol. **20**, 165–184 (1969).

Sievers, A.: Funktion des Golgi-Apparates in pflanzlichen und tierischen Zellen. In: Sekretion und Exkretion, Wohlfarth-Bottermann, (ed.), Proc. 2, Wiss. Konf. Ges. Dtsch. Naturf. Arzte, pp. 89–111. Berlin-Heidelberg-New York: Springer (1965).

Simonis, W., Bornefeld, T., Lee, J., Majumdar, K.: Phosphate uptake and photophosphorylation in the blue-green alga *Anacystis nidulans*. In: Membrane transport in plants, Zimmerman, U., Dainty, J., (eds.), pp. 220–225. Berlin-Heidelberg-New York: Springer (1974).

Simonis, W., Urbach, W.: Über eine Wirkung von Natrium-Ionen auf die Phosphataufnahme und die lichtabhängige Phosphorylierung von *Ankistrodesmus braunii*. Arch. Mikrobiol. **46**, 265–286 (1963).

Simonis, W., Urbach, W.: Photophosphorylation *in vivo*. Ann. Rev. Plant Physiol. **24**, 89–114 (1973).

Singer, S.J., Nicolson, G.L.: The fluid mosaic model of the structure of cell membranes. Science **175**, 720–731 (1972).

Sitte, P.: Zum Feinbau der Suberinschichten im Flaschenkork. Protoplasma **54**, 555–559 (1962).

Sitte, P.: Bau und Feinbau der Pflanzenzelle. Jena: Gustav Fischer (1965).

Sitte, P.: Allgemeine Mikromorphologie der Zelle. In: Die Zelle, Struktur und Funktion, Metzner, H., (ed.), Stuttgart: Wissenschaftliche Verlagsgesellschaft (1966).

Sitte, P.: Die Bedeutung der molekularen Lamellen-Bauweise von Kork-Zellwänden. Biochem. Physiol. Pflanzen **168**, 287–297 (1975).

Sitte, P., Rennier, R.: Untersuchungen an cuticularen Zellwandschichten. Planta **60**, 19–40 (1963).

Sjöstrand, F., Elfvin, L.-G.: The granular structure of mitochondrial membranes and of cytomembranes as demonstrated in frozen-dried tissue. J. Ultrastruct. Res. **10**, 263–292 (1964).

Slatyer, R.O.: Plant-water relationships. London-New York: Academic Press (1967).

Slayman, C.L.: Electrical properties of *Neurospora crassa*. Effects of external cations on the intracellular potential. J. Gen. Physiol. **49**, 69–92 (1965).

Slayman, C.L.: Movement of ions and electrogenesis in microorganisms. Am. Zoologist **10**, 377–392 (1970).

Slayman, C.L., Gradmann, D.: Electrogenic proton transport in the plasma membrane of *Neurospora*. Biophysical J. **15**, 968–971 (1975).

Slayman, C.L., Long, W.S., Lu, C.Y.-H.: The relationship between ATP and an electrogenic pump in the plasma membrane of *Neurospora crassa*. J. Membrane Biol. **14**, 305–338 (1973).

Slayman, C.L., Lu, C.Y.-H., Shane, L.: Correlated changes in membrane potential and ATP concentrations in *Neurospora*. Nature (London) **226**, 274–276 (1970).

Slayman, C.L., Slayman, C.W.: Depolarization of the plasma membrane of *Neurospora* during active transport of glucose: evidence for a proton-dependent cotransport system. Proc. Natl. Acad. Sci. U.S. **71**, 1935–1939 (1974).

Slayman, C.W.: The genetic control of membrane transport. In: Current topics in membranes and transport, vol. IV, Bronner, F., Kleinzeller, A., (eds.), pp. 1–174. New York-London: Academic Press (1973).

Smillie, R.M., Andersen, K.S., Tobin, N.F., Entsch, B., Bishop, D.G.: Nicotinamide adenine dinucleotide phosphate photoreduction from water by agranal chloroplasts isolated from bundle sheath cells of maize. Plant Physiol. **49**, 471–475 (1972).

Smith, F.A.: Links between glucose uptake and metabolism in *Nitella translucens*. J. Exp. Bot. **18**, 348–358 (1967).

Smith, F.A.: The mechanism of chloride transport in *Characean* cells. New Phytol. **69**, 903–917 (1970).

Smith, F.A., Fox, A.L.: The free space of *Citrus* leaf slices. Aust. J. Plant Physiol. **2**, 441–446 (1975).

Smith, F.A., Raven, J.A.: Transport and regulation of cell pH. Enc. Plant Physiol. **2A**, 317–346 (1976).

Smith, F.A., West, K.R.: A comparison of the effects of metabolic inhibitors on chloride uptake and photosynthesis in *Chara australis*. Aust. J. Biol. Sci. **22**, 351–363 (1969).

Smith, R.C.: Time course of exudation from excised corn root segments of different stages of development. Plant Physiol. **45**, 571–575 (1970).

Smith, R.C., Epstein, E.: Ion absorption by shoot tissue: technique and first findings with excised leaf tissue of corn. Plant Physiol. **39**, 338–341 (1964).

Solomon, A.K.: Measurement of the equivalent pore radius in cell membranes. In: Membrane, transport, and metabolism, Kleinzeller, A., Kotyk, A., (eds.), London-New York: Academic Press (1961).

Spanner, D.C., Jones, R.L.: The sieve tube wall and its relation to translocation. Planta **92**, 64–72 (1970).

Spanner, D.C.: Electroosmotic flow. Enc. Plant Physiol. **1**, 301–327 (1975).

Spanswick, R.M.: Electrical coupling between cells of higher plants: a direct demonstration of intercellular communication. Planta **102**, 215–227 (1972a).

Spanswick, R.M.: Evidence for an electrogenic ion pump in *Nitella translucens* I. The effects of pH, K^+, Na^+, light and temperature on the membrane potential and resistance. Biochim. Biophys. Acta **288**, 73–89 (1972b).

Spanswick, R.M.: Electrogenesis in photosynthetic tissues. In: Ion transport in plants, Anderson, W.P., (ed.), pp. 113–128. London-New York: Academic Press (1973).

Spanswick, R.M.: Evidence for an electrogenic pump in *Nitella translucens* II. Biochim. Biophys. Acta **332**, 387–398 (1974a).

Spanswick, R.M.: Hydrogen ion transport in giant algal cells. Can. J. Bot. **52**, 1029–1034 (1974b).

Spanswick, R.M.: Symplasmic transport in tissues. Enc. Plant Physiol. **2B**, 35–56 (1976).

Spanswick, R.M., Costerton, J.W.F.: Plasmodesmata in *Nitella translucens*: structure and electrical resistance. J. Cell Sci. **2**, 451–464 (1967).

Spanswick, R.M., Miller, A.G.: Measurements of the cytoplasmic pH in *Nitella translucens*. Comparison of values obtained by microelectrodes and weak acid methods. Plant Physiol. **59**, 664–666 (1977).

Sprague, H.B., Ed.: Hunger signs in crops. 3rd ed. New York: David McKay Co. (1964).

Stadelmann, E.J.: Evaluation of turgidity, plasmolysis and deplasmolysis of plant cells. In: Methods in cell physiology, vol. II, Prescott, D.M., (ed.), pp. 143–216. New York: Academic Press (1966).

Steer, B.T., Beevers, H.: Compartmentation of organic acids in corn roots III. Utilization of exogenously supplied acids. Plant Physiol. **42**, 1197–1201 (1967).

Stein, W.D.: Facilitated diffusion. Recent Prog. in Surface Sci. **1**, 300–337 (1964).

Stein, W.D., Danielli, J.F.: Disc. Faraday Soc. **21**, 238 ff. (1956).

Steinbrecher, W., Lüttge, U.: Sugar and ion transport in isolated onion epidermis. Australian J. Biol. Sci. **22**, 1137–1143 (1969).

Stelzer, R., Läuchli, A., Kramer, D.: Interzelluläre Transportwege des Chlorids in Wurzeln intakter Gerstepflanzen. Cytobiologie **10**, 449–457 (1975).

Stern, K.: Beiträge zur Kenntnis der Nepenthaceen. Flora (Jena) **109**, 213–282 (1917).

Steudle, E., Zimmermann, U.: Hydraulische Leitfähigkeit von *Valonia utricularis*. Z. Naturforsch. **26b**, 1302–1311 (1971).

Steudle, E., Zimmermann, U.: Determination of the hydraulic conductivity and of reflection coefficients in *Nitella flexilis* by means of direct cell-turgor pressure measurements. Biochim. Biophys. Acta **332**, 399–412 (1974).

Steveninck, R.F.M., van: Abscisic acid stimulation of ion transport and alteration in K^+/Na^+ selectivity. Z. Pflanzenphysiol. **67**, 282–286 (1972).

Steveninck, R.F.M., van: The "washing" or "ageing" phenomenon in plant tissues. Ann. Rev. Plant Physiol. **26**, 237–258 (1975).

Steveninck, R.F.M., van: Effect of hormones and related substances on ion transport. Enc. Plant Physiol. **2B**, 307–342 (1976a).

Steveninck, R.F.M., van: Cellular differentiation, aging and ion transport. Enc. Plant Physiol. **2B**, 343–371 (1976b).

Steveninck, R.F.M., van: Cytochemical evidence for ion transport through plasmodesmata, pp. 131–145. In: Intercellular communication in plants: studies on plasmodesmata, Gunning, B.E.S., Robards, A.W., (eds.), Berlin-Heidelberg-New York: Springer (1976c).

Steveninck, R.F.M., van, Chenowith, A.R.F., Steveninck, M.E., van.: Ultrastructural localization of ions, pp. 25–37. In: Ion transport in plants, Anderson, W.P., (ed.), London-New York: Academic Press (1973).

Stiles, W.: Essential micro-(trace) elements, pp. 558–614. In: Handbuch der Pflanzenphysiologie IV. Ruhland, W., (ed.), Berlin-Göttingen-Heidelberg: Springer (1958).

Stocking, C.R., Heber, U., Eds.: Transport in plants III. Intracellular interactions and transport processes. vol. 3, Encyclopedia of plant physiology, New Series. New York-Heidelberg-Berlin: Springer (1976).

Stocking, C.R., Larson, S.: A chloroplast cytoplasmic shuttle and the reduction of extraplastic NAD. Biochem. Biophys. Res. Commun. **37**, 278–282 (1969).

Stocking, C.R., Ongun, A.: The intracellular distribution of some metallic elements in leaves. Am. J. Bot. **49**, 284–289 (1962).

Stoeckenius, W.: An electron microscope study of myelin figures. J. Biophys. Biochem. Cytol. **5**, 491–500 (1959).

Stoeckenius, W.: Osmium tetroxide fixation of lipids. Proc. Eur. Reg. Conf. Electron Micro. Delft **2**, 716–720 (1960).

Stoeckenius, W., Schulman, J.H., Prince, L.M.: The structure of myelin figures and microemulsions as observed with the electron microscope. Kolloid Z. **169**, 170–180 (1960).

Storey, R., Wyn Jones, R.G.: Quarternary ammonium compounds in plants in relation to salt resistance. Phytochemistry **16**, 1–7 (1977).

Stout, R.G., Cleland, R.E.: Effects of fusicoccin on the activity of a key pH-stat enzyme, PEP-carboxylase. Planta **139**, 43–45 (1978).

Stout, R.G., Johnson, K.D., Rayle, D.L.: Rapid auxin- and fusicoccin-enhanced Rb^+ uptake and malate synthesis in *Avena* coleoptile sections. Planta **139**, 35–41 (1978).

Street, H.E.: The physiology of root growth. Ann. Rev. Plant Physiol. **17**, 315–344 (1966).

Strotmann, H., Heldt, H.W.: Phosphate containing metabolites participating in photosynthetic reactions in *Chlorella pyrenoidosa*. In: Progress in photosynthetic research, Metzner, H., (ed.), vol. III, pp. 1131–1140, Tübingen: International Biological Union (1969).

Strotmann, H., Murakami, S.: Energy transfer between cell compartments. Enc. Plant Physiol. **3**, 398–418 (1976).

Sutcliffe, J.F.: Mineral absorption in plants. Oxford-London-New York-Paris: Pergamon Press (1962).

Sutcliffe, J.F.: Regulation in the whole plant. Enc. Plant Physiol. **2B**, 394–417 (1976a).

Sutcliffe. J.F.: Regulation of ion transport in the whole plant. Perspectives in Experimental Biology, Sunderland, N., (ed.), vol. II, Botany p. 542. I. Oxford: Pergamon Press (1976b).

Sutcliffe. J.F., Hackett, D.P.: Efficiency of ion transport in biological systems. Nature (London) **180**, 95–96 (1957).

Sze. H., Hodges, T.K.: Characterization of passive ion transport in plasma membrane vesicles of oat roots. Plant Physiol. **58**, 304–308 (1976).

Sze. H., Hodges, T.K.: Selectivity of alkali cation influx across the plasma membrane of oat roots. Cation specificity of the plasma membrane ATPase. Plant Physiol. **59**, 641–646 (1977).

Takeda, J., Senda. M.: Effect of light on sodium and chloride influxes in partly illuminated *Nitella flexilis* cells. Plant Cell Physiol. **15**, 957–964 (1974).

Tal, M., Imber, D.: Abnormal stomatal behavior and hormonal imbalance in *flacca*, a wilty mutant of tomato II. Auxin and abscisic acid-like activity. Plant Physiol. **46**, 373–376 (1970).

Tal, M., Imber. D., Itai, C.: Abnormal stomatal behaviour and hormonal imbalance in *flacca*, a wilty mutant of tomato. Plant Physiol. **46**, 367–372 (1970).

Talmadge, K.W., Keegstra, K., Bauer. W.D., Albersheim, P.: The structure of plant cell walls I. The macromolecular components of the walls of suspension-cultured sycamore cells with a detailed analysis of the pectic polysaccharides. Plant Physiol. **51**, 158–173 (1973).

Tanada, T.: A rapid photo-reversible response of barley root tips in the presence of 3-indole acetic acid. Proc. Natl. Acad. Sci. U.S. **59**, 376–380 (1968).

Tangl, E.: Über offene Kommunikationen zwischen den Zellen des Endosperms einiger Samen. Jahrb. Wiss. Botan. **12**, 170–190 (1879).

Tanner, W.: Light-driven active uptake of 3-O-methylglucose via an inducible hexose uptake system of *Chlorella*. Biochem. Biophys. Res. Commun. **36**, 278–283 (1969).

Tanner, W., Grünes, R., Kandler, O.: Spezifität und Turnover des induzierbaren Hexose-Aufnahmesystems von *Chlorella*. Z. Pflanzenphysiol. **62**, 376–386 (1970).

Tanner, W., Haass, D., Decker, M., Loos, E., Komor, B., Komor, E.: Active hexose transport in *Chlorella vulgaris*. In: Membrane transport in plants, Zimmermann, U., Dainty, J., (eds.), pp. 202–208. Berlin-Heidelberg-New York: Springer (1974).

Tanner, W., Kandler, O.: Die Abhängigkeit der Adaptation der Glucose-Aufnahme von der oxydativen und der photosynthetischen Phosphorylierung bei *Chlorella vulgaris*. Z. Pflanzenphysiol. **58**, 24–32 (1967).

Tanner, W., Löffler, M., Kandler, O.: Cyclic photophosphorylation in vivo and its relation to photosynthetic CO_2-fixation. Plant Physiol. **44**, 422–428 (1969).

Tanton, T.W., Crowdy, S.H.: Water pathways in higher plants II. Water pathways in roots. J. Exp. Bot. **23**, 600–618 (1972).

Taylor, D.L.: J. Phycol. **3**, 234 ff. (1967).

Taylor, D.L.: J. Mar. Biol. Assoc., U.K. **48**, 1 ff. (1968).

Taylor, D.L.: Int. Rev. Cyt. **27**, 29 ff. (1970).

Taylor, D.L.: Comp. Biochem. Physiol. **38**A, 233 ff. (1971).

Taylor, D.L.: Adv. Mar. Biol. **II**, 9 ff. (1973a).

Taylor, D.L.: Algal symbionts of invertebrates. Ann. Rev. Microbiol. **27**, 171 ff. (1973b).

Taylor, D.L., Lee, C.C.: Arch. Microbiol. **75**, 269 ff. (1971).

Taylor, F.J.R., Blackbourn, D.J., Blackbourn, J.: Ultrastructure of the chloroplasts and associated structures within the marine ciliate *Mesodinium rubrum* (Lohmann). Nature (London) **224**, 819–821 (1969).

Taylor, F.J.R., Blackbourn, D.J., Blackbourn, J.: J. Fish. Res. Bd. Canada **28**, 391 ff. (1971).

Tazawa, M., Kishimoto, U., Kikuyama, M.: Potassium, sodium and chloride in the protoplasm of Characeae. Plant Cell Physiol. **15**, 103–110 (1974).

Teichler-Zallen, D., Hoch, G.: Cyclic electron transport in algae. Arch. Biochem. Biophys. **120**, 227–230 (1967).

Teipel, J., Koshland, D.E., Jr.: The significance of intermediary plateau regions in enzyme saturation curves. Biochemistry **8**, 4656–4663 (1969).

Teorell, T.: Membrane electrophoresis in relation to bio-electrical polarization effects. Arch. Sci. Physiol. **3**, 205–219 (1949).

Tezuka, T., Yamamoto, Y.: Control of ion absorption by phytochrome. Planta **122**, 239–244 (1975).

Thibault, P., Michel, J.P.: Possible role of phytochrome in phosphorylations and photosynthetic oxygen evolution in corn leaves (*Zea mays*). 2nd Int. Cong. Photosynth. Stresa (1971).

Thimann, K.V.: Plant growth substances; past, present, and future. Ann. Rev. Plant Physiol. **14**, 1–18 (1963).

Thomas, D.A.: The regulation of stomatal aperture in tobacco leaf epidermal strips II. The effect of ouabain. Aust. J. Biol. Sci. **23**, 981–989 (1970).

Thomson, W.W., Berry, W.L., Liu, L.L.: Localization and secretion of salt by the salt glands of *Tamarix aphylla*. Proc. Natl. Acad. Sci. U.S. **63**, 310–317 (1969).

Thomson, W.W., Journett, R., de.: Studies on the ultrastructure of guard cells of *Opuntia*. Am. J. Bot. **57**, 309–316 (1970).

Thomson, W.W., Liu, L.L.: Ultrastructural features of the salt gland of *Tamarix aphylla* L. Planta **73**, 201–220 (1967).

Throm, G.: Die lichtabhängige Änderung des Membranpotentials bei *Griffithsia setacea*. Z. Pflanzenphysiol. **63**, 162–180 (1970).

Throm, G.: Einfluss von Hemmstoffen und des Redoxpotentials auf die lichtabhängige Änderung des Membranpotentials bei *Griffithsia setacea*. Z. Pflanzenphysiol. **64**, 281–296 (1971a).

Throm, G.: Aktionsspektrum der Photosysteme II und I für die lichtabhängige Änderung des Membranpotentials bei *Griffithsia setacea*. Z. Pflanzenphysiol. **65**, 389–403 (1971b).

Throm, G.: Untersuchungen zur Beziehung zwischen der lichtabhängigen und der redoxabhängigen Änderung des Membranpotentials bei *Griffithsia setacea*. Planta **112**, 273–284 (1973).

Tolbert, N.E.: Microbodies-peroxisomes and glyoxysomes. Ann. Rev. Plant Physiol. **22**, 45–74 (1971).

Torii, K., Laties, G.G.: Dual mechanisms of ion uptake in relation to vacuolation in corn roots. Plant Physiol. **41**, 863–870 (1966a).

Torii, K., Laties, G.G.: Organic acid synthesis in response to excess cation absorption in vacuolate and non-vacuolate sections of corn and barley roots. Plant Cell Physiol. **7**, 395–403 (1966b).

Toriyama, H., Jaffe, M.J.: Migration of calcium and its role in the regulation of seismonasty in the motor cells of *Mimosa pudica* L. Plant Physiol. **49**, 72–81 (1972).

Trebst, A., Avron, M., Eds.: Photosynthesis I. Photosynthetic electron transport and photophosphorylation. vol. V, Encyclopedia of plant physiology, New Series. New York-Heidelberg-Berlin: Springer (1977).

Trip, P., Gorham, P.R.: Autoradiographic study of the pathway of translocation. Can. J. Bot. **45**, 1567–1573 (1967).

Trip, P., Gorham, P.R.: Bidirectional translocation of sugars in sieve tubes of squash. Plant Physiol. **43**, 877–882 (1968).

Troshin, A.S.: Sorption properties of protoplasm and their role in cell permeability, pp. 45–53. In: Membrane transport and metabolism, Kleinzeller, A., Kotyk, A., (eds.), London: Academic Press (1961).

Tukey, H.B. Jr.: The leaching of substances from plants. Ann. Rev. Plant Physiol. **21**, 305–324 (1970).

Tyree, M.T.: The symplast concept. A general theory of symplastic transport according to the thermodynamics of irreversible processes. J. Theoret. Biol. **26**, 181–214 (1970).

Tyree, M.T., Fischer, R.A., Dainty, J.: A quantitative investigation of symplasmic transport in *Chara corallina* II. The symplasmic transport of chloride. Can. J. Bot. **52**, 1325–1334 (1974).

Tyree, M.T., Tammes, P.M.L.: Translocation of uranin in the symplasm of staminal hairs of *Tradescantia*. Can. J. Bot. **53**, 2038–2046 (1975).

Ullrich, W.R.: Nitratabhängige nichtcyclische Photophosphorylierung bei *Ankistrodesmus branunii* in Abwesenheit von CO_2 und O_2. Planta **100**, 18–30 (1971).

Ullrich, W.R.: Der Einfluss von CO_2 und pH auf die ^{32}P-Markierung von Polyphosphaten und organischen Phosphaten bei *Ankistrodesmus braunii* im Licht. Planta **102**, 37–54 (1972).

Ullrich, W.R., Eisele, R.: Relation between nitrate uptake and nitrate reduction in *Ankistrodesmus braunii*, pp. 307–313. In: Transmembrane ionic exchanges in plants, Thellier, M., Monnier, A., DeMarty, M., Dainty, J., (eds.), Paris-Rouen: C.N.R.S. and Univ. Rouen (1977).

Ullrich-Eberius, C.I.: Beziehungen der Aufnahme von Nitrat, Nitrit und Phosphat zur photosynthetischen Reduktion von Nitrat, Nitrit und zum ATP-Spiegel bei *Ankistrodesmus braunii*. Planta **115**, 25–36 (1973).

Ullrich-Eberius, C.I., Lüttge, U., Neher, L.: CO_2-uptake by barley leaf slices as measured by photosynthetic O_2-evolution. Z. Pflanzenphysiol. **79**, 336–346 (1976a).

Ullrich-Eberius, C.I., Lüttge, U., Neher, L.: Energy relations of phosphate uptake distribution in barley leaf slices as affected by cutting and adaptive ageing. Z. Pflanzenphysiol. **79**, 347–359 (1976b).

Ullrich-Eberius, C.I., Novacky, A., Lüttge, U.: Active hexose uptake in *Lemna gibba* G1. Planta **139**, 149–153 (1978).

Ullrich-Eberius, C.I., Simonis, W.: Der Einfluß von Natrium-und Kaliumionen auf die Phosphataufnahme bei *Ankistrodesmus braunii*. Planta **93**, 214–226 (1970).

Umrath, K.: Der Erregungsvorgang, pp. 24–110. In: Handbuch der Pflanzenphysiologie XVII/1, Ruhland, W., (ed.), Berlin-Göttingen-Heidelberg: Springer (1959).

Urbach, W., Kaiser, W.: Changes of ATP levels in green algae and intact chloroplasts by different photosynthetic reactions. 2nd Int. Cong. Photosynth. Stresa (1971).

Ursprung, A., Blum, G.: Zur Kenntnis der Saugkraft IV. Die Absorptionszone der Wurzel. Der Endodermissprung. Ber. Dtsch. Bot. Ges. **39**, 70–79 (1921).

Ussing, H.H.: The distinction by means of tracers between active transport and diffusion. Acta Physiol. Scand. **19**, 43–56 (1949).

Vanderhoef, L.N., Findley, J.S., Burke, J.J., Blizzard, W.E.: Auxin has no effect on modification of external pH by soybean hypocotyl cells. Plant Physiol. **59**, 1000–1003 (1977a).

Vanderhoef, L.N., Lu, T.Y.S., Williams, C.A.: Comparison of auxin-induced and acid-induced elongation in soybean hypocotyl. Plant Physiol. **59**, 1004–1007 (1977b).

Vanderkooi, G., Green, D.E.: Biological membrane structure I. The protein crystal model for membranes. Proc. Natl. Acad. Sci. U.S. **66**, 615–621 (1970).

Vange, M.S., Holmern, K., Nissen, P.: Multiphasic uptake of sulfate by barley roots I. Effect of analogues, phosphate, and pH. Plant Physiol. **31**, 292–301 (1974).

Varner, J.E.: Gibberellic-acid controlled synthesis of α-amylase in barley endosperm. Plant Physiol. **39**, 413–415 (1964).

Vennesland, B., Guerrero, M.G.: Reduction of nitrate and nitrite. Enc. Plant Physiol. **6** (1978).

Viets, F.G., Jr.: Calcium and other polyvalent cations as accelerators of ion accumulation by excised barley roots. Plant Physiol. **19**, 466–480 (1944).

Villegas, R., Barnola, F.V.: J. Gen. Physiol. **44**, 963–977 (1961).

Vorobiev, L.N.: Electrophysiological peculiarities of plant cells. Abhandl. Dtsch. Akad. Wiss. Berlin **4**, 197–201 (1968).

Vredenberg, W.J.: Energy control of ion fluxes in *Nitella* as measured by changes in potential, resistance and current-voltage characteristics of the plasmalemma. In: Ion transport in plants, Anderson, W.P., (ed.), pp. 153–170. London-New York: Academic Press (1973).

Vredenberg, W.J.: The kinetics of light-induced changes in the electrical potential measured across the thylakoid membranes of intact chloroplasts. Proc. 3rd Int. Cong. Photosynth., Avron, M., (ed.), Amsterdam: Elsevier Scientific Publ. Co., No. 152, 929–939 (1974).

Vredenberg, W.J.: Electrical interactions and gradients between chloroplast compartments and cytoplasm. The intact chloroplast. Barber, J., (ed.) Amsterdam-New York: Elsevier/North-Holland Biomedical Press (1976).

Vredenberg, W.J., Homann, P.H., Tonk, W.J.M.: Light-induced potential changes across the chloroplast enclosing membranes as expressions of primary events at the thylakoid membrane. Biochim. Biophys. Acta **314**, 261–265 (1973).

Vredenberg, W.J., Tonk, W.J.M.: Photosynthetic energy control of an electrogenic ion pump at the plasmalemma of *Nitella translucens*. Biochim. Biophys. Acta **298**, 354–368 (1973).

Vredenberg, W.J., Tonk, W.J.M.: On the steady-state electrical potential difference across the thylakoid membranes of chloroplasts in illuminated plant cells. Biochim. Biophys. Acta **387**, 580–587 (1975).

Wagner, G.: Fluxes and compartmentation of potassium and chloride in the green alga *Mougeotia*. Planta **118**, 145–157 (1974).

Wagner, G., Hope, A.B.: Proton transport in *Halobacterium halobium*. Aust. J. Plant Physiol. **3**, 665–676 (1976).

Wagner, G.J., Siegelman, H.W.: Large-scale isolation of intact vacuoles and isolation of chloroplasts of mature plant tissues. Science **190**, 1298–1299 (1975).

Waisel, Y.: Biology of halophytes. New York and London: Academic Press (1972).

Waisel, Y., Kuller, Z.: Control of selectivity and ion fluxes in bean hypocotyls. Experientia **28**, 1377–1378 (1972).

Walker, D.A.: Plastids and intracellular transport. Enc. Plant Physiol. **3**, 85–136 (1976).

Walker, N.A., Hope, A.B.: Membrane fluxes in electric conductance in characean cells. Aust. J. Biol. Sci. **22**, 1179–1195 (1969).

Walker, N.A., Pitman, M.G.: Measurement of fluxes across membranes. Enc. Plant Physiol. **2A**, 93–128 (1976).

Walker, N.A., Smith, F.A.: Intracellular pH in *Chara corallina* measured by DMO distribution. Plant Sci. Lett. **4**, 125–132 (1975).

Wallace, W., Pate, J.S.: Nitrate assimilation in higher plants with special reference to the cocklebur (*Xanthium pennsylvanium* Wallr.). Ann. Bot. **31**, 213–228 (1967).

Wallach, D.F.H., Zahler, P.H.: Protein conformations in cellular membranes. Proc. Natl. Acad. Sci. U.S. **56**, 1552–1559 (1966).

Wallen, D.G.: Glucose, fructose, and sucrose influx into *Nitella flexilis*. Can. J. Bot. **52**, 1–4 (1974).

Walton, D.C., Harrison, M.A., Cote, P.: The effects of water stress on abscisic-acid levels and metabolism in roots of *Phaseolus vulgaris* L., and other plants. Planta **131**, 141–144 (1976).

Wanless, I.R., Bryniak, W., Fensom, D.S.: The effect of some growth regulating compounds upon electro-osmotic measurements, transcellular water flow, and Na^+, K^+ and Cl^- influx in *Nitella flexilis*. Can. J. Bot. **51**, 1055–1070 (1973).

Warburg, O., Lüttgens, W.: Photochemical reduction of quinone in green cells and granules. Biochimia **11**, 303–322 (1946).

Wartiovaara, V., Collander, T.: Permeabilitätstheorien. Protoplasmatologia Bd. II, C8d. Wien: Springer (1960).

Webb, J.A., Gorham, P.R.: Radial movement of C^{14}-translocates from squash phloem. Can. J. Bot. **43**, 97–103 (1965).

Weigl, J.: Zur Hemmung der aktiven Ionenaufnahme durch Arsenat. Z. Naturforschung **19b**, 646–648 (1964).

Weigl, J.: Austausch-Mechanismus des Ionentransports in Pflanzen am Beispiel des Phosphat-und Chloridtransports bei Maiswurzeln. Planta **79**, 197–207 (1968).

Weigl, J.: Einbau von Auxin in gequollene Lecithin-Lamellen. Z. Naturforsch. **24b**, 365–366 (1969a).

Weigl, J.: Specificität der Wechselwirkung zwischen Wuchsstoffen und Lecithin. Z. Naturforsch. **24b**, 367–368 (1969b).

Weigl, J.: Wechselwirkung pflanzlicher Wachstumshormone mit Membranen. Z. Naturforsch. **24b**, 1046–1052 (1969c).

Weigl, J.: Efflux und Transport von Cl^- und Rb^+ in Maiswurzeln. Wirkung von Außenkonzentration, Ca^{++}, EDTA und IES. Planta **84**, 311–323 (1969d).

Weigl, J.: Wirkung von CCCP und UO_2^{2+} auf die Ionenfluxe in Wurzeln. Planta **91**, 270–273 (1970).

Weigl, J.: Cl^- Fluxe, Akkumulation und Transport in abgeschnittenen Mais-Wurzeln. Planta **98**, 315–322 (1971).

Weigl, J., Lüttge, U.: Mikroautoradiographische Untersuchungen über die Aufnahme von $^{35}SO_4^{--}$ durch Wurzeln von *Zea mays* L. Die Funktion der primären Endodermis. Planta **59**, 15–28 (1962).

Weigl, J., Lüttge, U.: Die Ionenaufnahme durch die Luftwurzeln von *Epidendrum*. Protoplasma **60**, 1–6 (1965).

Weisenseel, M.H., Jaffe, L.F.: Planta **133**, 1–7 (1976).

Weisenseel, M.H., Ruppert, H.K.: Phytochrome and calcium ions are involved in light-induced membrane depolarization in *Nitella*. Planta **137**, 225–229 (1977).

Weisenseel, M., Smeibidl, E.: Phytochrome controls the water permeability in *Mougeotia*. Z. Pflanzenphysiol. **70**, 420–431 (1973).

Weiss, M.G.: Inheritance and physiology of efficiency in iron utilization in soybeans. Genetics **28**, 253–268 (1943).

Wells, A.F.: Structural inorganic chemistry. New York: Oxford University Press (1945).

West, I.C.: Lactose transport coupled to proton movements in *Escherichia coli*. Biochem. Biophys. Res. Commun. **41**, 655–661 (1970).

West, I.C., Mitchell, P.: Proton-coupled β-galactoside translocation in non-metabolizing *Escherichia coli*. J. Bioenerg. **3**, 445–462 (1972).

West, I.C., Mitchell, P.: Stoicheiometry of lactose-H^+ symport across the plasma membrane of *Escherichia coli*. Biochem. J. **132**, 587–592 (1973).

White, J.M., Pike, C.S.: Rapid phytochrome mediated changes in adenosine 5′-triphosphatase content of etiolated bean buds. Plant Physiol. **53**, 76–79 (1974).

White, P.R.: "Root pressure"—an unappreciated force in sap movement. Am. J. Bot. **25**, 223–227 (1938).

Wiessner, W.: Inorganic micronutrients, pp. 267–286. In: Physiology and biochemistry of the algae, Lewin, R.A., (ed.), New York-London: Academic Press (1962).

Wildes, R.A., Pitman, M.G., Schaefer, N.: Comparison of isomers of fluorophenylalanine as inhibitors of ion transport across barley roots. Aust. J. Plant Physiol. **2**, 659–661 (1975).

Wildes, R.A., Pitman, M.G., Schaefer, N.: Inhibition of ion uptake to barley roots by cycloheximide. Planta **128**, 35–40 (1976).

Wilkins, J.A., Thompson, J.E.: Subcellular distribution of basal and cation-sensitive ATPases in roots of *Phaseolus vulgaris*. Plant Physiol. **29**, 181–185 (1973).

Willenbrink, J.: Riesentange der Meeresküsten. Biologie in unserer Zeit **6**, 41–48 (1976).

Willenbrink, J., Rangoni-Kübbeler, M., Tersky, B.: Frond development and CO_2-fixation in *Laminaria hyperborea*. Planta **125**, 161–170 (1975).

Williams, E.J., Fensom, D.S.: Axial and transnodal movement of ^{14}C, ^{22}Na, and ^{36}Cl in *Nitella translucens*. J. Exp. Bot. **26**, 783–807 (1975).

Williams, S.E., Spanswick, R.M.: Propagation of the neuroid action potential of the carnivorous plant *Drosera*. J. Comp. Physiol. **108**, 211–223 (1976).

Williamson, F.Q., Morré, D.J., Jaffe, M.J.: Association of phytochrome with rough-surfaced endoplasmic reticulum fractions from soybean hypocotyls. Plant Physiol. **56**, 738–743 (1975).

Willison, J.H.M.: Plasmodesmata: A freeze-fracture view. Can. J. Bot. **54**, 2842–2847 (1976).

Willmer, C.M., Dittrich, P.: Carbon dioxide fixation by epidermal and mesophyll tissues of *Tulipa commelina*. Planta **117**, 123–132 (1974).

Wilmer, C.M., Pallas, J.E., Jr., Black, C.C., Jr.: Carbon dioxide metabolism in leaf epidermal tissue. Plant Physiol. **52**, 448–452 (1973).

Wilson, R.H., Graesser, R.J.: Ion transport in plant mitochondria. Enc. Plant Physiol. **3**, 377–397 (1976).

Winter-Sluiter, E., Läuchli, A., Kramer, D.: Cytochemical localization of K^+-stimulated adenosine triphosphatase activity in xylem parenchyma cells of barley roots. Plant Physiol. **60**, 923–927 (1977).

Woo, K.C., Anderson, J.M., Boardman, N.K., Downton, W.J.S., Osmond, C.B., Thorne, S.W.: Deficient photosystem II in agranal bundle sheath chloroplasts of C_4-plants. Proc. Natl. Acad. Sci. U.S. **67**, 18–25 (1970).

Wright, S.T.C., Hiron, R.W.P.: (+)-Abscisic acid, the growth inhibitor induced in detached wheat leaves by a period of wilting. Nature (London) **224**, 719–720 (1969).

Wyn Jones, R.G., Storey, R., Leigh, R.A., Ahmad, N., Pollard, A.: A hypothesis on cytoplasmic osmoregulation. In: Regulation of cell membrane activities in plants, Marrè, E., Ciferri, O., (eds.), pp. 121–136. Amsterdam: Elsevier/North-Holland Biomedical Press (1977).

Yapa, P.A.J., Spanner, D.C.: Localisation of adenosine triphosphatase activity in mature sieve elements of *Tetragonia*. Planta **117**, 321–328 (1974).

Yeo, A.R., Kramer, D., Läuchli, A., Gullasch, J.: Ion distribution in salt-stressed mature *Zea mays* roots in relation to ultrastructure and retention of sodium. J. Exp. Bot. **28**, 17–29 (1977).

Yocum, C.F.: Photophosphorylation associated with photosystem II–II. Effects of electron donors, catalyst oxidation, and electron transport inhibitors on photosystem II cyclic photophosphorylation. Plant Physiol. **60**, 592–596 (1977a).

Yocum, C.F.: Photophosphorylation associated with photosystem II–III. Characterisation of uncoupling, energy transfer inhibition, and proton uptake reactions associated with photosystem II cyclic photophosphorylation. Plant Physiol. **60**, 597–601 (1977b).

Yocum, C.F., Guikem, J.A.: Photophosphorylation associated with photosystem II–I. Photosystem II cyclic photophosphorylation catalyzed by p-phenylenediamine. Plant Physiol. **59**, 33–37 (1977).

Young, J.H., Evert, R.F., Eschrich, W.: On the volume-flow mechanism of phloem transport. Planta **113**, 355–366 (1973).

Yu, G.H., Kramer, P.J.: Radial transport of ions in roots. Plant Physiol. **44**, 1095–1100 (1969).

Yu, R.: Characterization of the phytochrome-containing particles obtained by glutaraldehyde pre-fixation of maize coleoptiles. J. Exp. Bot. **26**, 808–822 (1975a).

Yu, R.: Distribution of phytochrome in subcellular fractions from maize coleoptiles following glutaraldehyde treatment. Aust. J. Plant Physiol. **2**, 273–279 (1975b).

Zahler, P.: Modern problems of blood preservation. Stuttgart: G. Fischer (1969).

Zebe, E., Delbrück, A., Bücher, Th.: Über den Glycerin-1-P Cyclus im Flugmuskel von *Locusta migratoria*. Biochem. Z. **331**, 254–272 (1959).

Zeevaart, J.A.D.: Sites of abscisic acid synthesis and metabolism in *Ricinus communis* L. Plant Physiol. **59**, 788–791 (1977).

Ziegler, H.: Untersuchungen über die Leitung und Sekretion der Assimilate. Planta **47**, 447–500 (1956).

Ziegler, H.: Der Ferntransport organischer Stoffe in den Pflanzen. Naturwissenschaften **50**, 177–186 (1963).

Ziegler, H.: Der Stofftransport in der Pflanze. In: Symposium Stofftransport. Ziegler, H., (ed.), Vorträge Gesamtgebiet der Botanik. Dtsch. Bot. Ges. N.F. **2**, 5–14. Stuttgart: G. Fischer (1968).

Zeigler, H., Ed.: Symposium Stofftransport. Vorträge Gesamtgebiet der Botanik. Dtsch. Bot. Ges. N.F. **2**. Stuttgart: G. Fischer (1968).

Ziegler, H.: Nature of transported substances. Enc. Plant Physiol. **1**, 59–100 (1975).

Ziegler, H., Lüttge, U.: Die Salzdrüsen von *Limonium vulgare* I. Die Feinstruktur. Planta **70**, 193–206 (1966).

Ziegler, H., Lüttge, U.: Die Salzdrüsen von *Limonium vulgare* II. Mitteilung, Die Lokalisierung des Chlorids. Planta **74**, 1–17 (1967).

Ziegler, H., Vieweg, G.H.: Der experimentelle Nachweis einer Massenströmung im Phloem von *Heracleum mantegazzianum* Somm. et Lev. Planta **56**, 402–408 (1961).

Ziegler, H., Weigl, J., Lüttge, U.: Mikroautoradiographischer Nachweis der Wanderung von $^{35}SO_4^{--}$ durch die Tertiärendodermis der *Iris*-Wurzel. Protoplasma **56**, 362–370 (1963).

Zimmermann, M. H., Milburn, J.A., Eds.: Transport in Plants I. Phloem Transport. vol. I, Encyclopedia of plant physiology, New Series. New York-Heidelberg-Berlin: Springer (1975).

Zimmermann, M.H., Ziegler, H.: List of sugars and sugar alcohols in sieve-tube exudates. Appendix III: Enc. Plant Physiol. **1**, 480–503 (1975).

Zimmermann, U.: Cell turgor regulation and pressure mediated transport processes. In: Integration of activity in the higher plant, Jennings, D., (ed.), pp. 155–165, Proc. 31st Symp. Soc. Exp. Biol. Cambridge: Univ. Press (1977).

Zimmermann, U., Pilwat, G., Beckers, F., Riemann, F.: Effects of external electrical fields on cell membranes. Bioelectrochem. Bioenerg. **3**, 58–83 (1976).

Zimmermann, U., Steudle, E.: Bestimmung von Reflexionskoeffizienten an der Membran der Alge *Valonia utricularis*. Z. Naturforsch. **25b**, 500–504 (1970).

Zimmermann, U., Steudle, E.: The pressure-dependence of the hydraulic conductivity, the membrane resistance and membrane potential during turgor pressure regulation in *Valonia utricularis*. J. Membrane Biol. **16**, 331–352 (1974).

Zimmermann, U., Steudle, E.: The hydraulic conductivity and volumetric elastic modulus of cells and isolated cell walls of *Nitella* and *Chara* spp.: pressure and volume effects. Aust. J. Plant Physiol. **2**, 1–12 (1975).

Zimmermann, U., Steudle, E.: Physical aspects of water relations of plant cells. Adv. Bot. Res. **6**, 45–117 (1978).

Index

Encyclopedia of Plant Physiology
New Series
Series Editors: **A. Pirson,** University of Göttingen
M.H. Zimmermann, Harvard University

The *Encyclopedia of Plant Physiology, New Series*, is a multi-volume reference work encompassing the entire field of plant physiology. These volumes provide an excellent and comprehensive survey of general biological principles and the state of current research in plant physiology. They will be of great value to research scientists, instructors, and graduate students in this and related fields.

Volume 1: **Transport in Plants I: Phloem Transport**
Editors: M.H. Zimmermann, J.A. Milburn
1975/xix, 535 pp./93 fig./cloth ISBN 0-387-07314-0

Volume 2: **Transport in Plants II**
Editors: U. Luttge, M.G. Pitman
Part A: Cells
1976/xvi, 400 pp./97 fig., 64 tables/cloth ISBN 0-387-07452-X
Part B: Tissues and Organs
1976/xii, 456 pp,/129 fig., 45 tables/cloth
ISBN 0-387-07453-8

Volume 3: **Transport in Plants III: Intracellular Interactions and Transport Processes**
Editors: C.R. Stocking, U. Heber
1976/xxii, 517 pp./123 fig./cloth ISBN 0-387-07818-5

Volume 4: **Physiological Plant Pathology**
Editors: R. Heitefuss, P.H. Williams
1976/xx, 890 pp./92 fig./cloth ISBN 0-387-07557-7

Volume 5: **Photosynthesis I: Photosynthetic Electron Transport and Photophosphorylation**
Editors: A. Trebst, M. Avron
1977/xxiv, 730 pp./128 fig./cloth ISBN 0-387-07962-9

Encyclopedia of Plant Physiology con't.

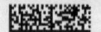